2013 International Conference on Optoelectronics and Microelectronics

(ICOM 2013)

AA001106

Harbin, China
7 – 9 September 2013

IEEE Catalog Number:	CFP1357T-POD
ISBN:	978-1-4799-1215-5

Copyright © 2013 by the Institute of Electrical and Electronic Engineers, Inc
All Rights Reserved

Copyright and Reprint Permissions: Abstracting is permitted with credit to the source. Libraries are permitted to photocopy beyond the limit of U.S. copyright law for private use of patrons those articles in this volume that carry a code at the bottom of the first page, provided the per-copy fee indicated in the code is paid through Copyright Clearance Center, 222 Rosewood Drive, Danvers, MA 01923.

For other copying, reprint or republication permission, write to IEEE Copyrights Manager, IEEE Service Center, 445 Hoes Lane, Piscataway, NJ 08854. All rights reserved.

***This publication is a representation of what appears in the IEEE Digital Libraries. Some format issues inherent in the e-media version may also appear in this print version.**

IEEE Catalog Number: CFP1357T-POD
ISBN 13:   978-1-4799-1215-5

Additional Copies of This Publication Are Available From:

Curran Associates, Inc
57 Morehouse Lane
Red Hook, NY 12571 USA
Phone: (845) 758-0400
Fax: (845) 758-2633
E-mail: curran@proceedings.com
Web: www.proceedings.com

Table of Contents

Performance of All Fiber Optical Current Transducer in Optical Transmission Based on Optisystem 1
Shen Tao, Tang Miao, Wei Xin-lao, Yang Qian-ru, Yang Wen-long, Xiong Yan-ling

A local analysis method of marine gravity aided navigation area 5
Fanming Liu, Jianqi Yao

Study on Polarization Properties of a Novel High Birefringence Photonic Crystal Fiber 10
Xuyou Li, Yingying Yu, Bo Sun

The MEMS gyro stabilized platform design based on Kalman Filter 14
Guangchun Li, Yunfeng He, Yanhui Wei, Shenbo Zhu, Yanzhe Cao

A Novel redundant inertial measurement unit and calibration algorithm 18
Kunpeng He, Jitao Han, Yuping Shao

A MEMS integrated navigation system 24
Jianhui Zeng, Xingzhi Zhang, Kunpeng He, Huiyu Liu

The LPFG temperature characteristic research based on electric heating method 29
Tao Geng, Dongdong Zi, Wenlei Yang, Chengguo Tong

Influence of Si SBD P^+ Ring Junction Depth on ESD Robustness 33
Jinyu Dong, Jinghua Yin, Shiyin Guan, Yue Li, Shuting Gao

A novel pre-amplifying and latching comparator 38
Rongke Ye, Rongbin Hu

A kind of 3-bit flash ADC core 42
Rongbin Hu, Rongke Ye

The Design and Implementation of Oscillograph
Based on STM32 46
Mingxin Song, Yue Li, Yang Yang

A fabrication method of the micro resonant beam compliant to unfitted thickness 50
Jiandong Jin, Mingwei Wang, Yuling Li, Hong Qi

Optical fiber spectroscopy in near infrared spectral range for rapidly discriminating formulated apple vinegar drinks 53
Bingfang Zhang, Libo Yuan, Bingxiu Zhang

A study on MEMS acoustic vibration sensor technology 56
Yonghe Qin, Yingjie Qiao, Wenjiang Zou, Xingyue Xu

Quantitative Analysis and Identification of liver B-scan Ultrasonic Image Based on BP Neural Network 62
Fuzhen Zhu, Bing Zhu, Peihua Li, Zhifang Wang, Liqiu Wei

Modified BP Neural Network Model is Used for Odd-even Discrimination of Integer Number 67
LIAN Tongli, XIE Minxiang, XU Jiren, CHEN Ling, GAO Huaihui

The effect of incident angle of pumping light to Cholesteric Liquid Crystal 71
Xiangbao Yin, Yongjun Liu, Lingli Zhang

Generation of a High-Quality Hollow Laser Beam by a Liquid-Core Optical Fiber 75
Xiaobo HU, Wei GAO, Peijing SUN, Shengnan LIU, Xuelian YU, Shaozhi PU

Microfluidic Device with Compound Structure 79
He Zhang, Xiaowei Liu, Li Tian, Xiaowei Han

The Research and Application of Angular Correlation for Helim Optically-pumped Magnetometer 83
Zong Fabao, Zou Pengyi, Chen En, Wang Jingran, Zhang Jin

A double-ring Mach-Zehnder interferometer for highly sensitive temperature sensing 87
Xiaoqi Liu, Yundong Zhang, Xuenan Zhang, Ping Yuan

Carbon Nanotube-Based Printed Antenna for Conformal Applications 91
Yu-Ming Wu, Xin Lv, Beng Kang Tay, Hong Wang

Modeling and analysis of analog single event transients in an amplifier circuit 94
Yongsheng Wang, Wenjuan Wang, Yunfei Du, bei Cao

Design of the Photo-ionization Detector of 98
Xinping Dong, Peng Zhang, Zhenqi Zhao, Hui Wang, Shouchen Chai

Numerical Analysis of Pulse Laser Deformation on GaAs 101
Haijiao Zhou, Wenjun Sun, Zhong Meng, Zhongyang Liu

Add-drop ring resonator coupled Mach-Zehnder interferometer for highly sensitive sensing 105
Xiaoqi Liu, Yundong Zhang, Xuenan Zhang, Ping Yuan

The application of auto correlation detection technology in magnetic targets searching 109

Li Yuxiang, Sun Weimin, Wang Shuai

The Intensity Image Mosaic of LADER Based on SIFT 112

Dejian Meng, Jianfeng Sun, Jian Gao

Research on the Key Technology of TOC Detection based on Ultraviolet Optical Absorbable 117

Shimin Fu, Haitao Chen, Lijie Chen, Jiannan Yu, Likai Sun

A low-noise MEMS acoustic vector sensor 121

Jinping Li, Lijie Chen, Zhanjiang Gong, Shi Xin, Meng Hong

The experimental research on gas magneto-optic properties 125

Yuanyuan Wang, Lufei Hong, Guoli Song

Research on Capacitance Attitude Self-Correction Vector Hydrophone 128

Peng Zhang, Xin Shi, Haitao Chen

Study of Influence of Pre-pulse Power on Xe Capillary Discharge EUV Source 131

Qiang Xu, Yongpeng Zhao, Yao Xie, Qi Li, Qi Wang

Research of high accuracy straight line parameter estimation algorithm for ultralow-pixel CCD 134

Jie Tang, MaoJun Fan, Jun Hu

CA optimization based on simulation annealing in BIST 138

Bei Cao, Yongsheng Wang, Yanwei Dou, Dan Bu, Bin Zhou

Fabrication and Optical Properties of Linear-core-array Multicore Fiber.. 142

Qiang Dai, Hong Bo Bai, Xiao Liang Zhu, Li Jia Ma, Tao Zhang

Fabricate phase-shifted fiber Bragg grating based on a piezoelectric ceramics 146

Lin Jiping, Gui Yonglei, Cao Zhigang

On the Detection of Sea Wave Using a Slit Streak Tube Imaging Lidar 151

Jian.Gao, J.F.Sun, Qi.Wang

Fitting Operate Mode for Incoherent Mie Doppler Wind Lidar 155

Jun Du, Xiang Yang Cheng, Yan Chen Qu, Wei Jiang Zhao, De Ming Ren, Zhen Lei Chen, Li Jie Geng

High-efficiency Focusing Grating Coupler for 160

Biao Yang, Zhi-Yong Li, Xi Xiao, Jin-Zhong Yu, Yu-De Yu

The Study and Realization of Real-Time Supervision and Control of Medical Image System 163
Chunqiu You

A Micro-machined Optical Fiber Acoustic Sensor..167
Shi-Ning Wang, Mei-Yu Zhang, Yong-Hai Cao, Li-Jie Chen

A Vehicle Laser Doppler Velocimeter Configured With Three Transmitting Beams 170
Meng Shanshan

Investigation of focal ratio degradation caused by stress in Large-Core Astronomical Fibers 174
Yunxiang Yan, Ruichen WANG, Weimin Sun, Yongjun Liu

Broadband optical beam power splitter for wavelength dependent light circuits on silicon substrates 177
Zhiyong Li, Jiejiang Xing, Biao Yang, Yude Yu

The influence on FRD with different encircled energy 180
Ruichen Wang, Yunxiang Yan, Yongjun Liu, Hongquan Zhang, Weimin Sun

Supermodes Coupling Characterizaion of Optical Fiber Brush 183
Xiaoliang Zhu, Qi Yan, Haijiao Yu, Weimin Sun

Tapering technique for an embedded microstructure fiber device 186
Weimin Sun, Qi Yan, Haijiao Yu, Xiaoliang Zhu

Effect of oil molecular contamination on the space infrared optical system...189
Chunlian Lu, He Lv, Shijie Wang, Cuiling Wang, Weimin Sun

Effect of selective saturated absorption on Atomic Line Filter 193
Shuangqiang Liu, Weimin Sun

Photorefractive effect in relaxor ferroelectric $0.88Pb(Zn_{1/3}Nb_{2/3})O_3$-$0.12PbTiO_3$ single crystal 197
Hong Jia, Yang Li, Jun Li, Hao Tian, Liang Sun

Dependence of Cs Atomic Clock CPT Signal on Magnetic Intensity 200
Hongsong Mei, Qiang Huang, Junhai Zhang, Weimin Sun, Zongjun Huang

Bistability of laser-induced thermal radiation in rare earth doped solids..203
Jing Dai, Hong Li, Zhenguo Zhang, Xinlu Zhang, Li Li

Enhanced laser cooling of Tm-doped solids by upconversion pumping 207
Jing Dai, Li Li, Xinlu Zhang

The techniques of fixed pattern noise reduction for high speed digital CMOS image sensor................211
  Na Zhang, Haiyong Zheng

Polarization Characteristics and High Birefringence for Chiral Photonic Crystal Fiber with Squeezed Triangular Lattice.215
  She Li, JunQing Li, DunLiang Ren, Lin Zhang, Hui Liu, HongXin Shi

A real non-contact remote trapping single fiber tweezers 219
  Yu Zhang, Zhihai Liu, Jun Yang, Libo Yuan

Pencil-lead-break transient detecting by phase- optical time domain reflectometer based on coherent detecting...............224
  Yuelan Lu

Anti-Relaxation Coating of the Vapor Cell 227
  Xie Xin, Liu Yunhui, Bai Yuanyuan, Xu Yunfei

Research on the measurement range of particle size with laser backscattering based on PT algorithm 231
  Jian Xing, Yuandong Sui, Weimin Sun

Birefringence properties analysis of a novel three quasi-rectangular cores fiber................234
  Fengjun Tian, Libo Yuan

Coaxial Step Index Large Mode Area Fiber with low propagation loss................238
  Souaci Farida, Li-Bo Yuan, Ya-Xian Fan

The calculation of doped vanadium dioxide thin films 242
  Xue-song Tian, Qi Wang, Jian-feng Sun, Zhi-gang Fan

Linear polarization conversion in planar chiral metamaterial 245
  Yiqun Xu, Xingchen Liu, Zheng Zhu, Zhengping Wang, Jinhui Shi

Beam cleanup of 20kW peak power laser pulses by SBS in 105 μ m large core diameter fiber with high beam quality ($M_2 \approx$ 1.5) 248
  C.Y. Zhu, J.H. Zhang, Y.B. Yuan, D.X. Ba, J. Yan, Q.L. Gao, Z.W. Lu

A practical way of selective mode group excitation in large core graded-index multimode fibers................251
  C.Y. Zhu, Y.B. Yuan, J. Yan, J.H. Zhang, D.X. Ba, Z.W. Lu

An improved location algorithm in Wireless Sensor Network................255
  Jinlong Liu, Zhilu Wu, Zhendong Yin

2013 ICOM Organization

Organized by:

Harbin Engineering University

Sponsors:

Harbin Engineering University
IEEE EDS Harbin Chapter

Patrons:

The 49th Research Institute of China Electronic Technology Group Corporation
Committee on Optoelectronic Technology，Chinese Optical Society
Heilongjiang Optical Society
Heilongjiang Physical Society
Heilongjiang Instrument and Measurement Society
Key Laboratory of In-Fiber Integrated Optics, Ministry of Education
Heilongjiang Fiber Optic Sensors Science and Technology Key Laboratory
National Key Laboratory of Science and Technology on Tunable laser
Sensor technology branch of the Chinese Institute of Electronics

Organizing Committee:

2013 International Conference on Optoelectronics and Microelectronics (2013 ICOM)

Chairman :

Royal Academician   Professor K.T.V.Grattan, City University, UK

Co-Chairman：

Libo Yuan, Harbin Engineering University, China
Qi Wang, Harbin Institute of Technology, China
Yalin Wu, The 49th Research Institute of China Electronics Technology Group Corporation, China
Anna Mignani, Istituto di Fisica Applicata Nello Carrara, Italy
Maojun Fan, The 3rd Research Institute of China Electronics Technology Group Corporation, China
Gerald Farrell, Dublin Institute of Technology, Ireland
Elfed Lewis, University of Limerick, Ireland
Xiudong Sun, Harbin Institute of Technology, China
Xiyuan Peng, Harbin Institute of Technology, China
Weimin Sun, Harbin Engineering University, China

Secretary:

Jianzhong Zhang, Harbin Engineering University, China
Li Li, Harbin Engineering University, China
Jianfeng Sun, Harbin Institute of Technology, China

Performance of All Fiber Optical Current Transducer in Optical Transmission Based on Optisystem

Shen Tao12*, Tang Miao1 , Wei Xin-lao1, Yang Qian-ru2, Yang Wen-long2, Xiong Yan-ling2

1 College of Electrical and Electronic Engineering, Harbin University of Science and Technology, Harbin, China
2 College of Applied Sciences, Harbin University of Science and Technology, Harbin, China
taoshenchina@163.com

Abstract—The all fiber optical current transducer (AFOCT) modulation technique was designed in this paper, and the capabilities of fiber optic sensor was simulated by the software of Optisystem. The direct detection AFOCT transmission over the source function, the optical system, processing of optical signal was addressed. And the performance of different signal from of the direct detection structure was simulated. The pulse signals were processed by using the Optisystem, and the feasibility of the proposal has been validated.

Keywords- all fiber optical current transducer; output signal ; Optisysterm

I. INTRODUCTION

Current transducer (CT) always play a very important role in the electric power industry[1-3]. Upon the invention of laser and the development of fiber communication, more and more attentions are attracted by the optical current transducer (OCT) which has been the focus of the current transducer field nowadays[4]. OCT produce digital and analog signals. In recent years, they have become more available, playing an increasingly important role in electric power grid upgrades. There are basically two linear effects by which the magnetic field can be measured by optical sensors: magneto-optic effect (or Faraday effect) and magnetic force (or Lorentz force). Nowadays there is a great diversity of OCT. Taking into account the sensing mechanism employed and the materials used, OCT can be organized in four main groups [5,6]: all-fiber optic transducer (AFOCT), bulk optic sensor, magnetic force sensors and hybrid sensors.

II. BASIC PRINCIPLE OF AFOCT

The fiber itself acts as a transducer mechanism of AFOCT. The Faraday effect is used to induce a rotation in the angle of polarization of the light propagating in the fiber, which is proportional to the magnetic field. Usually, the fiber is coiled around the electrical conductor, making it immune to external currents and magnetic fields. AFOCT use very simple configurations because the fiber can be simply coiled around the electric conductor to be measured. Also, the sensor sensibility can be changed by simply changing the number of turns of the optical fiber around the conductor [7].

The Sagnac fiber current transducer employing the Faraday effect in a coil of optical fiber are very attractive for metering, control, and protection in high-voltage substations. Advantages include the inherent electric separation of the sensor electronics

at ground potential from the sensing fiber coil at high voltage as well as the small size and weight[8]. They have high accuracy, a superior transient response, a wide bandwidth, and DC measurement capability making them ideal for use low power flow measurements in such applications as wind farms and solar plants. OCT apply Faraday effects, which also have some drawbacks, including mechanical and optical instability, due to vibration and temperature[9-10]. In Figure 1, a typical current sensor with flint fiber is represented.

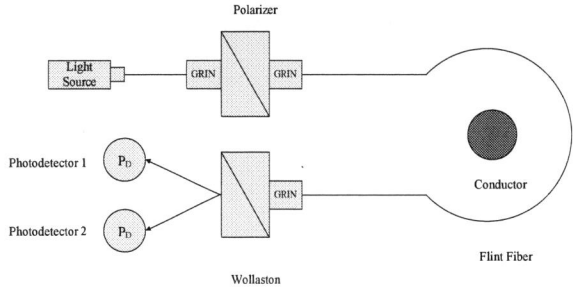

Figure 1. The Structure of AFOCT

Two circular polarizations with the same sense of rotation are counter propagating in the sensing coil to the Sagnac configuration of the sensor. The magnetic field of the current induces a nonreciprocal phase shift between the two waves. The circular polarizations are generated from linear polarizations prior to entering the coil by means of quarter-wave retarders. Upon leaving the coil, the circular waves are converted back to linear ones. Polarization-maintaining elliptical-core fibers serve to transmit the forward- and backward-propagating linear waves. The polarization directions of the linear waves coincide and are for example parallel to the major core axis as indicated.

III. AFOCT SIMULATION MODEL

OptiSystem is an innovative optical communication system simulation software, there are various types of function. Eg: design, test and optimize. OptiSystem had a very great environmental simulation capabilities, extensive device and classification system. Its performance can be achieved by additional extensions device of user devices, and become a widely used simulation tools of fiber optic systems and optical fiber sensing system.

978-1-4799-1215-5/13 $31.00 © 2013 IEEE

A. Theory Model of AFOCT

AFOCT of Sagnac principle diagram as shown in figure 2, a light wave from the light source through the coupler, by linear polarized analyzer was formed, it is divided into two linear polarized light wave after the second coupler. The line polarized light through the lambda 1/4 wave plate making transform into circularly polarized light, polarized light into the optical fiber head of sensing. Faraday effect of magnetic fields generated due to transfer current to produce non reciprocity phase shift, make the two beams of direct and reverse phase change different circularly polarized light, a bunch of positive change, a bunch of negative change. Two circularly polarized light beam again after 1/4 wave plate, and the optic power are obtain by the photoelectric detector through coupler output.

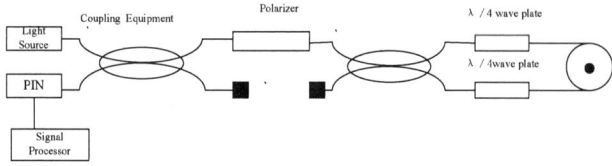

Figure 2. Structure of the AFOCT

Basic theory of the simulation diagram as shown in figure 3, the light wave from the light source through the polarizer and then through the coupler for 1:1 coupling, it divided into two linear polarized beams, two beams of light are identical lambda 1/4 wave plate and the sensing optical fiber. The modulator send out modulation signal modulation, and Comparison with the modulated light wave and not modulated light wave, the output signal processing and analysis.

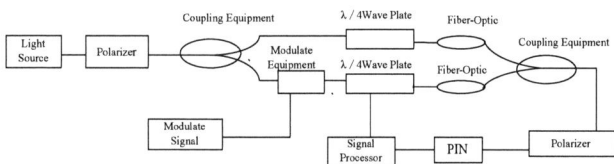

Figure 3. Theory diagram of AFOCT

B. Simulation Model of AFOCT

OptiSystem simulation structure based on the above principle is shown in figure 4.

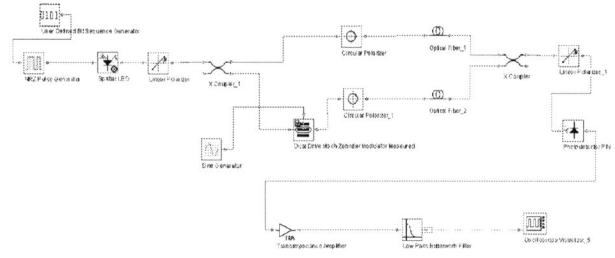

Figure 4. OptiSystem simulation structure of AFOCT

IV. RESULTS OF SIMULATION

A. Analysis of Main Components in The Structure

Light source model as shown in figure 5 of the OptiSystem software, The center wavelength is 1550 nm, spectral width is 40nm, the spectrogram of light source as shown in figure 6.

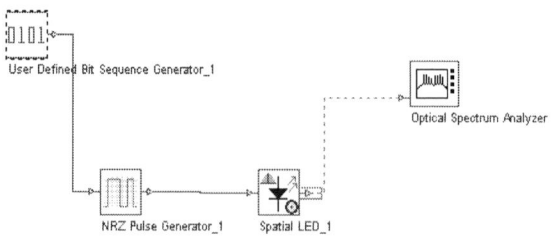

Figure 5. Light source model

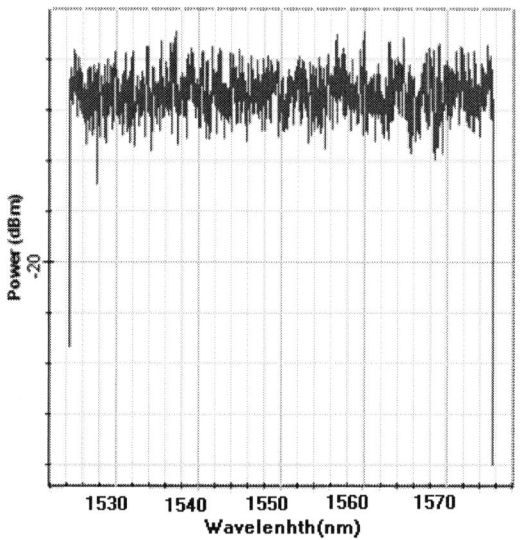

Figure 6. Spectrogram of light source

Polarization source simulation output using Storck said as shown in Figure 7.

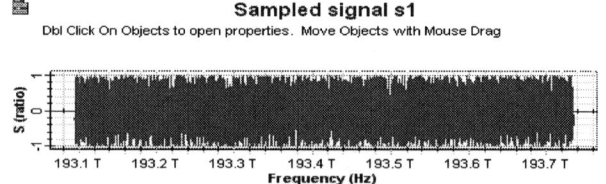

Figure 7. Storck of polarization source simulation

This work was financially supported by the Science and Technology Research Project of Heilongjiang Province Education Bureau (No.12531134).

The spectrogram after polarizer as shown in figure 8. And the simulation effect is ideal, source after the analyzer, almost all transformed into horizontal polarization.

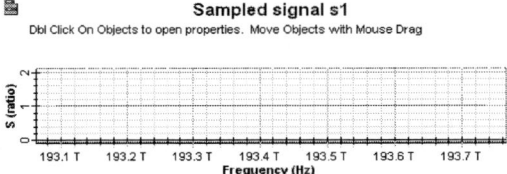

Figure 8.　Storck of polarization source simulation after polarizer

Figure 9 and figure 10 show the circularly polarized light conversion model and fiber loop output light condition respectively, so accuracy of circular and optical visible models fiber.

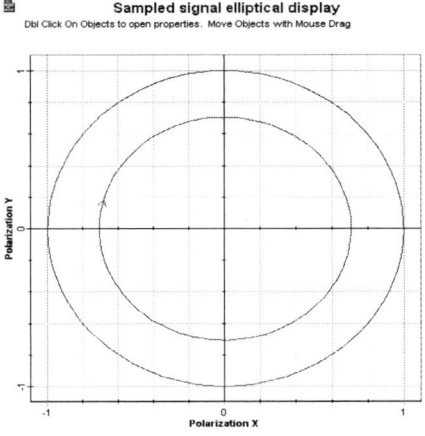

Figure 9.　Polarization ellipsoids planar representation after 1/ 4 wave plate

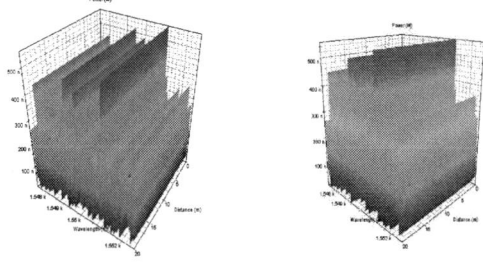

Figure 10.　Transmission spectra of optical fiber ring along X and Y axis

B.　Simulation and Analysis of the Overall System Function

Overall system simulation diagram as shown in figure 4, and set the important parameters. The system adopts sine and square wave conduct signal demodulation, AFOCT signal modulation signal waveform is shown in figure 11.

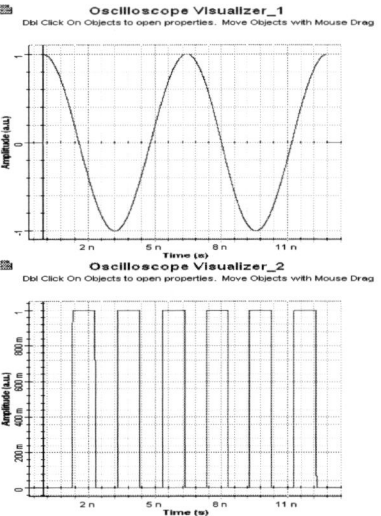

Figure 11. AFOCT signal modulation signal waveform

Optical power time domain analysis and the detection of light of without modulation signal for figure 12, left for the optical time domain analysis, the right for the photoelectric detector output.

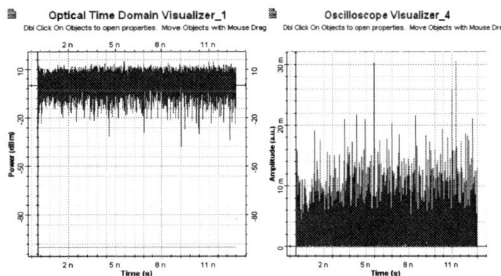

Figure 12. AFOCT output waveform without signal modulation signal

AFOCT simulation system output waveform in sine signal modulation for Figure 13 and filtered for Figure 14.

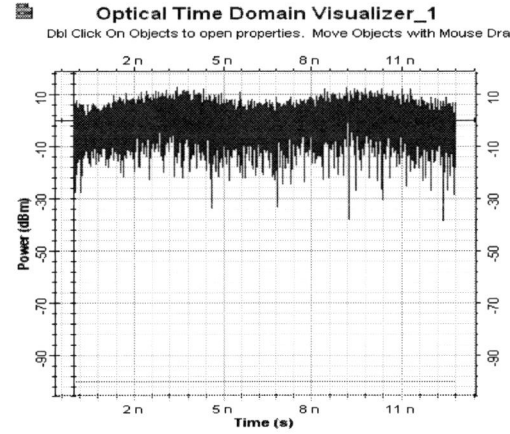

Figure 13. Output waveform of the AFOCT simulation system in sine signal modulation

978-1-4799-1215-5/13 $31.00 © 2013 IEEE

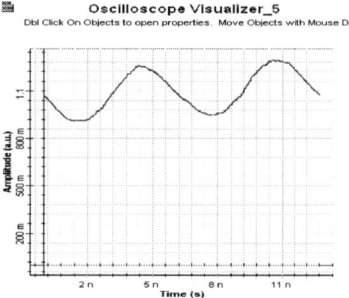

Figure 14. Output waveform of the AFOCT simulation system in sine signal modulation and filtered

The same method and AFOCT simulation system output waveform in square wave signal modulation for Figure 15.

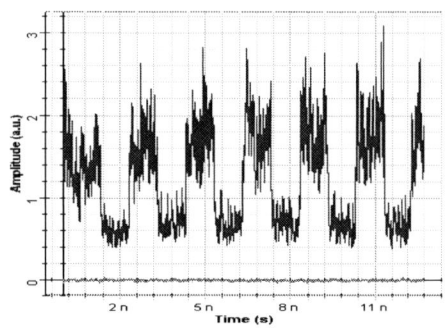

Figure 15. Output waveform of the AFOCT simulation system in square wave signal modulation

V. CONCLUSIONS

The output on AFOCT system is given by using the software of OptiSystem. The results show that the feasible design system, external signal on the output influence. The output effect is very good when the filter processing the output, through which the system performance is improved. The results and discussions we have presented could be referenced by some applications of AFOCT in the future.

ACKNOWLEDGMENT

This work was financially supported by the Science and Technology Research Project of Heilongjiang Province Education Bureau (No.12531134).

REFERENCES

[1]   C. Zhou, D. L. Wang, W. J. Zhang, L. Wu and Y. Yao, "Fiber Bragg grating high-current sensor based on magnetic coupling", Proc. of SPIE, Vol. 8034, pp. 1–8, 2011.

[2]   K. Bohnert, P. Gabus, J. Nehring, H. Brändle and M. G. Brunzel, "Fiber-optic current sensor for electrowinning of metals", Journal of Lightwave Technology, vol.25, pp.3602-3609, 2007.

[3]   B. Lee, "Review of the Present Status of Optical fiber sensors", Optical Fiber Technology, vol. 9, pp.57-59, 2003.

[4]   B. J. Li and L. J. Li, "An Overview of the Optical Current Sensor," IEEE, vol. 161, pp. 201–205, 2012 [2012 International Conference on Computer Science and Electronics Engineering].

[5]   Y. N. Ning and Z. P. Wang, "Recent progress in optical current sensing techniques," Rev. Sci. Instrum, vol. 66, pp. 3097–3111, 1995.

[6]   D. A. Jackson, "An optical system with potential for remote health monitoring of subsea machinery," Meas. Sci. Technol, vol. 20, pp. 1–8, 2009.

[7]   R. M. Silva, H. Martins, I. Nascimento, J. M. Baptista, A. L. Ribeiro, J. L. Santos, P. Jorge and O.,Frazão, "Optical Current Sensors for High Power Systems: A Review," Appl. Sci., vol. 2, pp. 602–628, 2012.

[8]   H. d. Pei, J. Zu and H. Chen, "A Novel Demodulation Method for the Sagnac Interferometric Fiber Current Sensor", IEEE, vol. 4, pp. 208–211, 2007.

[9]   N. A. Jaeger and F. Rahmatian, "Integrateed optics Pockels cell highvoltage sensor," IEEE Trans. Power Del., vol. 10, pp. 127–134, 1995.

[10] C. Li, X. Cui, I. Yamaguchi, Y. Masayuki, and Y. Toshihiko, "Optical voltage sensor using a pulse-controlled electrooptic quarter waveplate," IEEE Trans. Instr. Meas., vol. 54, pp. 273–277, 2005.

A local analysis method of marine gravity aided navigation area

Fanming Liu[1], Jianqi Yao[1]

1 College of Automation, Harbin Engineer University, Harbin, China
yaojianqiyinshan@163.com

Abstract—Because gravity matching accuracy is associated with local navigation area characteristics and carrier's heading, so we firstly calculate the local characteristic of gravity field and select navigation areas, then extract the skeleton of them with fast Euclidean distance field algorithm and skeleton simplified algorithm. So the carrier can select heading according to the statistical characteristics of local areas formed by the skeleton points and their Euclidean distance. The simulation result verifies the validity of the analysis method.

Keywords-marine gravity aided navigation; local navigation area; simplified skeleton extraction; heading; gravity characteristics

I. INTRODUCTION

Long time autonomous underwater navigation mainly relays on inertial navigation system(INS). But INS error accumulates over time, it need regular calibration or re-adjusted to maintain its effectiveness. Gravity aided navigation can revise the INS position error of underwater vehicle based on gravity measurement information and has the advantage of concealment, autonomy and anti-jamming capacity[1].

Because the accuracy of gravity aided navigation is affected by the statistical characteristics of local navigation area[2], so in this paper, we extract the simplified skeleton of navigation area to generate local navigation areas, and then calculate their statistical characteristics to realize the local analysis of them.

II. THE STATISTICAL CHARACTERISTICS OF GRAVITY FIELD AND SELECTION OF NAVIGATION AREA

Gravity field is stored in form of discrete grid, which is similar to DEM data, so the characteristics of gravity field can be described as follows: gravity roughness, standard deviation, abundance, slope, slope direction[3,4].

A. Analysis of Gravity Field Statistical Characteristics

1) Gravity roughness: roughness σ_z reflects the local fluctuation of gravity field, and its definition is shown in (1).

$$\sigma_z = \left(Q_x + Q_y\right)/2 \qquad (1)$$

$$Q_x^2 = \frac{1}{M(N-1)}\sum_{i=1}^{M}\sum_{j=1}^{N-1}\left(g_{i,j} - g_{i,j+1}\right)^2$$

$$Q_y^2 = \frac{1}{(M-1)N}\sum_{i=1}^{M-1}\sum_{j=1}^{N}\left(g_{i,j} - g_{i+1,j}\right)^2 \qquad (2)$$

2) Gravity standard deviation: standard deviation σ_T describes the overall fluctuation of gravity field.

$$\sigma_T^2 = \frac{1}{MN-1}\sum_{i=1}^{M}\sum_{j=1}^{N}\left(g_{i,j} - \bar{g}\right)^2, \bar{g} = \frac{1}{MN}\sum_{i=1}^{M}\sum_{j=1}^{N}g_{i,j} \qquad (3)$$

The ratio of roughness and standard deviation is abundance; it describes the abundance of the gravity field fluctuation.

3) Gravity slope and slope direction: the definition of slope is as follows:

$$f_x = [g_{(i+1,j+1)} - g_{(i-1,j+1)} + g_{(i+1,j)} - g_{(i-1,j)} + g_{(i+1,j-1)} - g_{(i-1,j-1)}]/6$$
$$f_y = [g_{(i-1,j-1)} - g_{(i-1,j+1)} + g_{(i,j-1)} - g_{(i,j+1)} + g_{(i+1,j-1)} - g_{(i+1,j+1)}]/6 \qquad (4)$$

f_x, f_y is gravity change rate in direction X and Y. The definition of slope direction is shown in (5).

$$A = 270^\circ + \arctan\left(f_y/f_x\right) - 90^\circ\, f_x/|f_x| \qquad (5)$$

The slope direction indicates the direction which is perpendicular to gravity contour. Similar to σ_T, we can define its standard deviation A_{std} which expresses the dispersion of A and the gravity change of local navigation area.

B. Navigation Area Selection

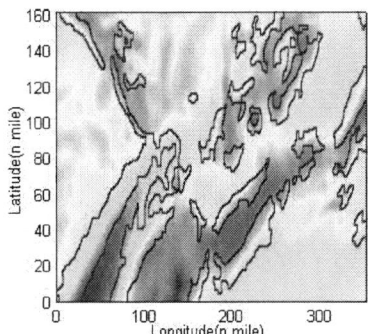

Figure 1.   The selected gravity aided navigation area

The navigation area with large value of σ_z and σ_T has high accuracy of gravity matching for its ability of anti-noise measurements, so by setting a threshold T, the local gravity characteristics, which is calculated by a moving window, can be used to select gravity aided navigation area[3,5] as shown in Fig. 1 in which the gravity navigation areas are surrounded by black lines, and T is defined in (6).

$$T = [\sigma_T/\sigma_N > 5] \wedge [\sigma_Z/\sigma_T > 0.1] \wedge [\sigma_Z/\sigma_N > 1.5] \quad (6)$$

III. LOCALIZATION ANALYSIS OF NAVIGATION AREAS

A. The Impact of Heading to Accuracy of Gravity Aided Navigation

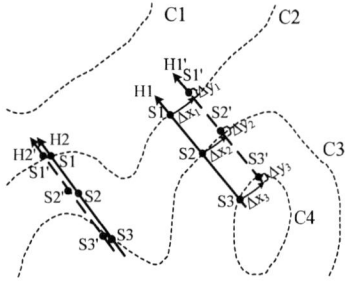

Figure 2. Gravity aided navigation with different heading

As shown in Fig. 2, H1-H2 are true tracks whose headings are obliquely across and perpendicular to gravity contours, H1'-2' are matching tracks of H1-2. S1-S3 and S1'-S3' are their sampling points. With TEROCM algorithm[6], and because the difference between gravity contour directions of S1-3 on H1 is small, so the Δx_{1-3} of H1' relative to H1 may be large for the gravity measurement noise when H1' is moving along the gravity contour by TERCOM; the difference between gravity contour directions of H2 is larger, so Δx_{1-3} of H2' relative to H2 may be smaller than H1, and the matching accuracy of H2 maybe better than H1. In summary, only gravity aided navigation area can't indicate the impact of gravity characteristics to the precision of gravity matching; it needs to analyze the local characteristics of gravity field[7,8].

B. Skeleton Extraction of Gravity Aided Navigation Area

The analysis of local navigation area can be done by a sliding window as shown in Fig. 3.

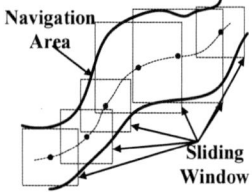

Figure 3. Local analysis of gravity aided navigation area

The center of sliding window is on the center axis of navigation area, which is namely skeleton line[9], the size of the window is corresponding to the minimum Euclidean distance between its center and border. Based on the above, we use the Euclidean distance field to generate the skeleton of navigation area.

If the size of gravity field is $m \times n$, and its boundary is represented by array of $b[\cdot,\cdot]$, then the Euclidean distance field is the minimum Euclidean distance between every grid point (x, y) and the area's boundary, which is represented by $\sqrt{EDT(x,y)}$, $EDT(x,y)$ is defined as (7):

$$EDT(x,y) = \min((x-i)^2 + (y-j)^2 : \\ i \in [0,m] \wedge j \in [0,n] \wedge b[i,j]) \quad (7)$$

It's very time-consuming to directly calculate (7), so we use the exact and linear time Euclidean distance field calculation algorithm[10], where $EDT(x,y)$ is expressed as (8):

$$EDT(x,y) = \min((x-i)^2 + G(i,y)^2 : i \in [0,m)) \\ G(i,y) = \min(|y-j| : j \in [0,n) \wedge b[i,j]) \quad (8)$$

The algorithm firstly calculates G for grids of every column, and G is defined as (9):

$$G(i,y) = GT(i,y) \min GB(i,y) \\ GT(i,y) = \min(y-j : j \in [0,y) \wedge b[i,j]) \\ GB(i,y) = \min(j-y : j \in [y,n) \wedge b[i,j]) \quad (9)$$

GT is calculated from top to bottom, and if $b[i,y] = 1$, then $GT(i,y) = 0$, else $GT(i,y) = GT(i,y-1)+1$. Based on GT, GB can be calculated from bottom to top according to the basis $GB(i,y) = GB(i,y+1)+1$. When y is fixed, then F_i and distance field is defined as (10):

$$F_i = (x-i)^2 + g(i)^2 \quad g(i) = G(i,y) \\ EDT(x,y) = \min(F_i : i \in [0,m)) \quad (10)$$

So $EDT(x,y)$ is equivalent to the piecewise lower envelope of $\{F_i \mid i \in [0,m)\}$ as shown in (11) and Fig. 4.

$$H(x,u) = \min(h \mid F_h \leq F_i, h, i \in [0,u)) \quad (11)$$

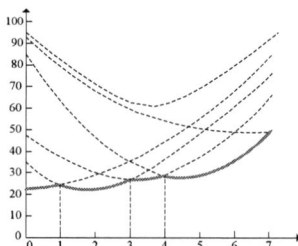

Figure 4. The Lower EnveLope of the Parabola Collection

Then $EDT(x,y) = F_{H(x,m)}$, and define $S(u)$ and $T(u)$:

$$S(u) = \{H(x,u) \mid x \in [0,m)\}$$
$$T(h,u) = \{x \mid H(x,u) = h, x \in [0,m), h \in [0,u)\} \quad (12)$$

$S(u)$ is the index of F_i in $H(x,u)$, $T(h,u)$ is the set of $x \in [0,m)$ corresponding to $S(u)$. $S(u)$, $T(h,u)$ can be calculated as follows: if current $S(u)$ is $\{s[1] \cdots s[q]\}$, when F_{u+1} is added, there will be 3 conditions:

1) Condition 1: if F_{u+1} is above the current lower envelop $F_{H(x,u)}$, then $S(u+1) = S(u)$ and $T(u+1, u+1)$ is empty.

2) Condition *2:* if F_{u+1} is below $F_{H(x,u)}$, then $S(u+1) = \{u+1\}$ and $T(u+1, u+1) = [1, m)$.

3) Condition 3: if F_{u+1} intersect with $F_{H(x,u)}$, then search $S(u)$ from right to left to find that $F_{u+1}(t[l]) \geq F_{s[l]}(t[l])$ and $t[l]$ is the beginning of $s[l]$ on axis x. F_{u+1} and $s[l]$ intersect at x^*, and let *Sep* is to search x^* as defined in (13), then $S(u+1) = \{s[1] \cdots s[l+1]\}$, $s[l+1] = u+1$, $T(u+1, u+1) = [Sep(s[l], u+1), m)$.

$$Sep(s[l], u+1) = ((u+1)^2 - s[l]^2 + g(u+1)^2$$
$$- g(s[l])^2) div(2(u - s[l] + 1)) \quad (13)$$

According to the above steps, we will get $S(m)$ and $T(m)$, since then $EDT(x,y)$ can be obtained by (10) and (11). After we get $EDT(x,y)$, we choose the points, whose distance value is larger than three-fourths of its neighbors' distance as the skeleton points[11], then use morphological thinning algorithm[9] and broken branch connection algorithm[12] to make the skeleton continuous and unit width. With the algorithm above, we get the distance field and skeleton of the navigation area as shown in Fig. 5(a) and 5(b).

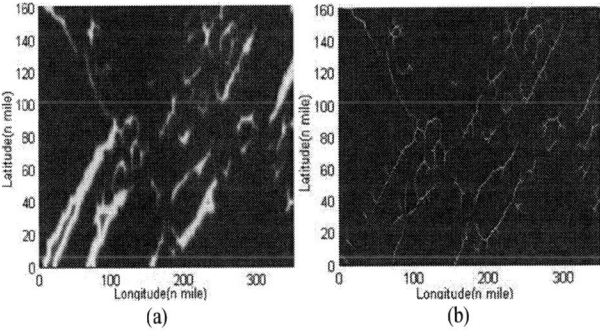

(a) (b)

Figure 5. (a) The Euclidean distance field. (b) Skeleton of gravity aided navigation area

C. Skeleton Simplification and Local Analysis Of Gravity Aided Navigation Area

The skeleton points in Fig. 5(b) are too dense to generate local areas, so we need to simplify them, after simplification,

the local areas formed by skeleton points can cover most of the gravity aided navigation area, and the overlap between them mustn't be too much. Make definition: a local area is a square region which is centered at a skeleton point and its side length is twice the skeleton point's Euclidean distance, then the simplification is defined as follows:

1) Generation of Simplified Skeleton Points: Set overlap ratio γ, when the overlap between local areas formed by the target skeleton point is less than γ, then takes this point as a valid skeleton point. Filter all skeleton points to generate the simplified skeleton according to this criteria.

2) Collect The Omitted Areas: There may be some region omitted by the first step, so calculate the area of them, and when the area is beyond threshold K, generate local area using the skeleton point with the largest distance in that region to collect the omitted gravity aided navigation area.

3) Large Area Dividing: Some local areas are too large, so divide them into small areas according to certain ranks interval.

Based on the above steps, we can analyze the statistical characteristics of local areas by calculating their σ_Z, σ_T, A_{std} and $\overline{A}$. By a large number of matching experiments we find that some areas' σ_Z and σ_T are large, but A_{std} are very small (less than $10°$), and the accuracy of gravity matching is low for the small changes in contour direction, so we should delete them. Finally, every local area can be expressed as a set which is composed of a skeleton point centered at the area, the area's slide length, σ_Z, σ_T, A_{std} and $\overline{A}$, so we can study their statistical properties with these parameters. The final skeleton points and the local areas formed by them are shown in Fig. 6.

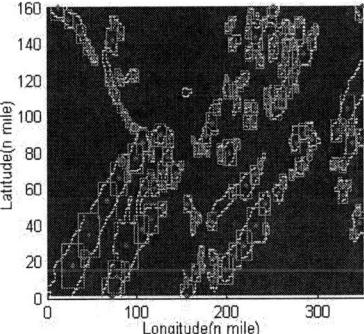

Figure 6. The final skeleton points and the local areas formed by them

IV. SIMULATION ANALYSIS

Select 10 local areas in Fig. 6 to carry out the gravity matching experiment with TERCOM algorithm. In the matching experiments, the carrier uses DR algorithm for dead reckoning, and the sailing speed is 10n mile/h. The sampling period and accuracy of gravimeter is 3 min and 1 mGal. Each matching of TERCOM takes 21 sampling points. The carrier heading is selected in the following intervals defined in (14).

$$H1 = (\bar{A} - 30^\circ, \bar{A} + 30^\circ)$$
$$H2 = (\bar{A} - 60^\circ, \bar{A} - 30^\circ) \cup (\bar{A} + 30^\circ, \bar{A} + 60^\circ) \quad (14)$$
$$H3 = (\bar{A} - 90^\circ, \bar{A} - 60^\circ) \cup (\bar{A} + 60^\circ, \bar{A} + 90^\circ)$$

So H1 is the interval of heading perpendicular to gravity contour of local navigation area, and H2, H3 is obliquely across and parallel to gravity contour respectively. In each local area, randomly select five directions as carrier headings in H1, H2 and H3 respectively, and perform matching experiment for 100 times in each direction. It will be a mismatching if a matching experiment's error exceeds 1 n mile. The matching probability (MP) is the ratio of effective matching number and total matching number, and the matching accuracy (MA) is the average positioning precision of matching experiment. The characteristic parameters of local navigation area and the results of matching experiment are shown in TABLE I.

TABLE I.   CHARACTERISTIC PARAMETERS OF LOCAL NAVIGATION AREAS AND THE RESULTS OF MATCHING EXPERIMENT

Area Number	σ_Z/σ_N	σ_T/σ_N	$A_{std}(^\circ)$	H1		H3		H2	
				MP (%)	MA (n mile)	MP (%)	MA (n mile)	MP (%)	MA (n mile)
1	2.12	12.3	15.10	72.5	0.96	85.2	0.62	88.5	0.51
2	2.42	15.8	17.24	78.3	0.85	83.4	0.72	88.0	0.55
3	2.20	12.24	27.40	86.1	0.56	95.6	0.29	96	0.35
4	2.61	11.65	30.32	90.0	0.51	96.4	0.19	100	0.21
5	8.16	58.2	13.33	83.10	0.72	88.70	0.54	94.12	0.39
6	8.45	56.5	15.45	87.25	0.66	94.2	0.32	97.62	0.27
7	7.21	65.65	35.5	93.28	0.44	95.65	0.36	96.87	0.40
8	6.05	64.22	36.41	92.21	0.54	92.10	0.58	95.50	0.42
9	5.87	51.21	87.56	93.46	0.42	94.80	0.48	93.5	0.39
10	6.23	23.56	115.8	96.8	0.30	98.56	0.15	100	0.26

A_{std} of area 1-2 is small, MP and MA of H2 are better than H1 and H3. The gravity change of area 3, 4 is no larger than area 1 and 2, but A_{std} is larger, and the difference of MP and MA between H2 and H1, H3 is smaller. In area 5 and 6, the change of gravity increases significantly, but A_{std} is no larger than area 1 and 2, then the difference of MP and MA between H2 and H1, H3 still exists. The gravity change and A_{std} of area 7 and 8 both increase, the gap of MP and MA between H2 and H1, H3 decreases. A_{std} of area 9-10 is very large, the contour direction of these area changes dramatically, MP and MA of H1-3 are almost the same. The experimental results show that MP and MA of H3 are better than H1, this is because the directions of gravity contour always changes along the directions of H3, so in fact, H3 is obliquely crossing gravity contour. In summary, MA and MP of H2 are better, and in areas where A_{std} is small, MA and MP of H1 and H3 are poor, otherwise, MA and MP of H1 and H3 become better.

CONCLUSION

In this paper, the statistical characteristics are used to perform local analysis of navigation area which is formed by the simplified skeleton points and their distance value of the Euclidean distance field. The result of simulation experiment verifies that the precision of gravity matching is related to carrier heading and statistical characteristics of local navigation area, and provides the basis for the carrier heading selection in gravity aided navigation area.

REFERENCES

[1] Behzad Kamgar-Parsi, Behrooz Kamgar-Parsi, Vehicle localization on gravity maps. The SPIE Conference on Unmanned Ground Vehicle Technology, SPIE vol.3693. Orlando, Florida, USA, pp.182-191, 1999.

[2] YUAN Shu-ming, SUN Feng, LIU Guang-jun. Application of Gravity Map Matching Technology in Underwater Navigation. Journal of Chinese Inertial Technology, vol.12, no.2, pp.13-17, 2004.

[3] CHENG Li, ZHANG Ya-jie, CAI Ti-jing. Selection Criterion for Matching Area in Gravity Aided Navigation. Journal of Chinese Inertial Technology, vol.15, no.5, pp.559-563, 2007.

[4] TANG Guo-an, LI Fa-yuan, LIU Xue-jun. Digital Elevation Model Tutorial. BeiJing: Science Press, 2005.

[5] LI De-hua, YANG Can, HU Chang-chi. On The Selection Criterion for A Terrain Matching Field. Journal of Huazhong University of Science and Technology, vol.24, no.2, pp.7-8, 1996.

[6] YAN Li, CUI Chen-feng, WU Hua-ling. A Gravity Matching Algorithm Based on TERCOM. Geomatics and Information Science of Wuhan University, vol.34, no.3, pp.261-264, 2009.

[7] CHEN Li-hua, WANG Shi-cheng, SUN Yuan, ZHENG Yu-hang, LIU Zhi-guo. Matching of multi-dimensional feature elements in areas with smooth magnetic fields. Journal of Chinese Inertial Technology, vol.19, no.6, pp.720-724, 2011.

[8] GUO Qing, WEI Rui-xuan, ZHOU Wei, XU Jie, GUO Chuang. Multidimensional Geomagnetic Matching Fusion Algorithm Based on Projection Pursuit. Acta Armamentarii, vol.31, no.2, pp.235-238, 2010.

[9] Blum H. Biological Shape And Visual Science: Part I. Journal of Theoretical Biology, vol.38, pp.205-287, 1973.

[10] A. Meijster, J.B.T.M. Roerdink, W.H. Hesselink. A General Algorithm For Computing Distance Transforms In Linear Time. J. Goutsias, L. Vincent, D.S. Bloomberg, Mathematical Morphology and Its Applications to Image and Signal Processing, Boston, Kluwer, pp.331-340, 2000.

[11] LIU Jun-tao, LIU Wen-yu, WU Cai-hua, YUAN Liang. A New Method of Extracting Objects' Curve-skeleton. Acta Automatica Sinica, vol.34, no. 6, pp.617-622, 2008.

[12] Ge Y, Fitzpatrick JM. On the generation of skeletons from discrete euclidean distancemaps. IEEE Trans. on Pattern Analysis and Machine Intelligence, vol.18, no.11, pp.1055-1066, 1996.

Study on Polarization Properties of a Novel High Birefringence Photonic Crystal Fiber

Xuyou Li[1]*, Yingying Yu[1], Bo Sun[1]

1 Automation College, Harbin Engineering University, Harbin, China

lixuyou8663@163.com

Abstract—A novel kind of photonic crystal fiber (PCF) is proposed and investigated numerically by employing finite element method (FEM). This kind of PCF is composed of silica core and cladding with elliptical air holes of lozenge arrangement. The mode field distribution, effective refractive index, beating length and birefringence is studied. For comparative analysis, the properties of optimized hexagonal photonic crystal fiber with the same air holes size, shape, distance and layer number are also investigated. Simulation results demonstrate that the mode field of fundamental modes is restricted effectively in the core of the proposed PCF. High birefringence of the proposed PCF can be achieved easily as large as the order of 10^{-2} at the wavelength of 1.55μm, which is higher than that of hexagonal PCF. A smaller beating length of the proposed PCF is also obtained at 1.55μm. It is useful for a range of fiber optics sensing, communication in polarization-maintaining applications.

Keywords-component; photonic crystal fiber; finite element method; mode field; biefringence

I. INTRODUCTION

Photonic crystal fiber, i.e., a silica-based fiber with air holes arranged periodically in a regular pattern running along its length, which has attracted considerable attention for its unusual characteristics since its first fabrication in 1996[1]. Compared with the conventional fiber, PCF has flexible and reliable structure, and appropriately designed structure can exhibit excellent properties of endlessly single mode[2], nonlinearity[3], dispersion[4], and birefringence[5]. Already these properties have had a number of applications for fiber optics sensing, nonlinear fiber optics and fiber optics communication fields[6], especially, high birefringence photonic crystal fibers are either preferred or required. Some recent has been aimed at realizing a high birefringence[5,7]. High birefringence PCF has potential applications, like polarization-maintaining and single-polarization single-mode[8], etc.

In this paper, we proposed a new type of lozenge cladding photonic crystal fiber (L-PCF), which is composed of elliptical holes based on rectangle lattice. Meanwhile, properties of a hexagonal cladding (H-PCF) with the same structure parameters are investigated. Mode field distribution of the proposed L-PCF, effective refractive indices, beating length and birefringence of the two kinds of PCF are numerically analyzed by use of FEM. It is found that the mode field is effectively limited in the core of the proposed L-PCF, and it can achieve a higher birefringence than H-PCF with the same

air holes size, shape, pitch and layer number. Moreover, it can also achieve a smaller beating length at 1.55μm. Through theoretical simulation for L-PCF and the H-PCF, the results show the L-PCF has bigger birefringence and better polarization properties. It is predicted that the proposed L-PCF has potential applications in fiber polarization-maintaining optics sensing or communication.

II. NUMERICAL METHOD AND STRUCTURE MODELING

A. Finite Element Method

Here, FEM is employed to analyze the properties of PCF, which is designed specially for the analysis and solving of the arbitrary structures and has been assumed to be more flexible and reliable than other known approaches. The formulation for the FEM based on the vectorial wave equation is given by [8]

$$\nabla \times [\mu_r^{-1} \nabla \times \vec{E}] - k_0^2 \varepsilon_r \vec{E} = 0 \qquad (1)$$

where k_0 is the wave number in the free space, ε_r, μ_r is the relative dielectric permittivity and magnetic permeability tensors, respectively, and $\vec{E}$ is the vectorial electric field.

The FEM method allows an arbitrary cross section domain to be divided into a patchwork of triangular elements with properly presented refractive indices. Equation(1) yields the eigenvalue equation:

$$[K]\{\vec{E}\}^T = k_0^2 n_{eff}^2 [M]\{\vec{E}\} \qquad (2)$$

where $[K]$, $[M]$ are the finite element matrices, $\{\vec{E}\}$ is the discretized electric field vector distribution of the mode, and n_{eff} is effective refractive index. By solving Equation (2), the effective refractive can be obtained, the polarization properties will be analyzed further according to the resolution.

B. Structure Modeling

The cross section of L-PCF we designed is shown in Figure 1(a). For the proposed PCF, the core of the fiber is formed by the absence of one elliptical air hole, elliptical air holes are arranged as a lozenge cladding structure and the layer number is assumed four. Considering the air holes distance of the y axis direction for the L-PCF, the structure of conventional hexagonal PCF is changed, so that it can make sure that the air holes distance of y axis direction is the same as the one of the

L-PCF. In addition, the layer number of H-PCF is also assumed four, and the cross section of H-PCF is shown in Figure 1(b).

For the two kinds of PCF, they are embedded in silica background material, to simplify the analysis, the refractive indices of the background material and the surrounding air is assumed to be 1.45 and 1, respectively, the elliptical air holes diameter of the x and y axis direction $b_1 = 1.9\mu m$, $b_2 = 1.0\mu m$, elliptical air holes pitch of x and y axis direction $\Lambda_x = 2.0\mu m$, $\Lambda_y = (\Lambda_x \tan(\alpha/2))\mu m$, the interior angle of the L-PCF along x axis $\alpha = 60^\circ$.

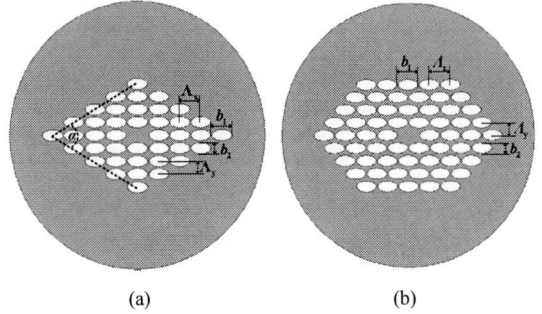

(a)   (b)

Figure 1. Cross section of (a) L-PCF; (b) H-PCF

III. RESULTS AND DISCUSSION

A. Mode Field Distribution

In order to illuminate the effect of confinement on light of the fiber core for a wide wavelength, several intensity and vector of mode field distributions of the two orthogonal fundamental modes for the proposed L-PCF with different wavelength λ are shown in Figure 2. From the figure, we can find that mode field of them are confined effectively in the core region of the proposed L-PCF when the wavelength is equal to 0.30, 1.55, and 2.00µm, respectively. Comparatively, the proposed L-PCF obtains the best effect of confinement when the wavelength is equal to 0.30µm, when the wavelength is 1.55µm, the effect of confinement takes second place, and the effect of confinement is inferior to the former when the wavelength is equal to 2.00µm. In other words, the shorter the wavelength is, the better effect of confinement is. Meanwhile, the mode field distributions of y polarization modes are expanded wider and wider from 0.30µm to 2.00µm, which is a symbol of a higher birefringence toward the high wavelength.

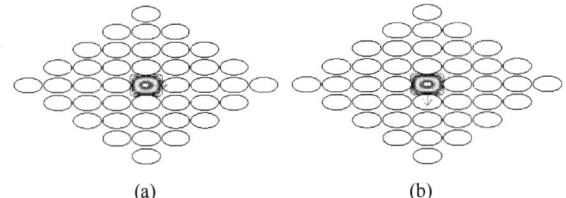

(a)   (b)

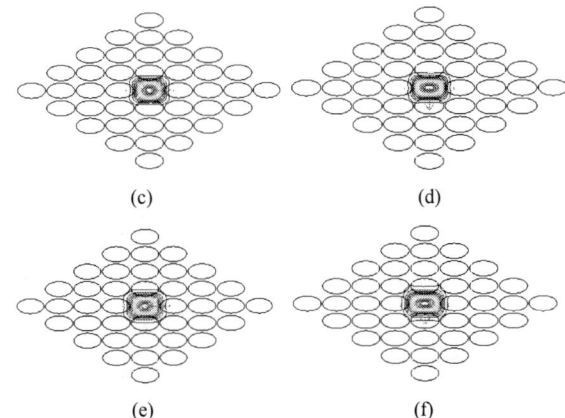

(c)   (d)

(e)   (f)

Figure 2. Contour plot of intensity and vector of mode field distribution for the L-PCF with different wavelength: (a)(c)(e) x polarization mode, $\lambda = 0.30, 1.55, 2.00\mu m$, respectively, (b)(d)(f) y polarization mode, $\lambda = 0.30, 1.55, 2.00\mu m$, respectively

B. Effective Refractive Index

By using the FEM, the effective refractive index can be obtained. Figure 3 shows the effective refractive indices curves of two orthogonal polarized fundamental modes as a function of wavelength from 0.30 to 2.00µm for these two kinds of PCF, by red and black line, respectively. As shown as the figure, we can find that the curves of these two kinds of PCF decrease monotonously with the increase of the wavelength, meanwhile, the effective refractive indices of the L-PCF is lower than that of H-PCF and the decrease of y polarization for the L-PCF is bigger than that of the H-PCF with respect to x polarization for a specific wavelength. That is explained by the different air holes arrangements in the cladding of the PCF.

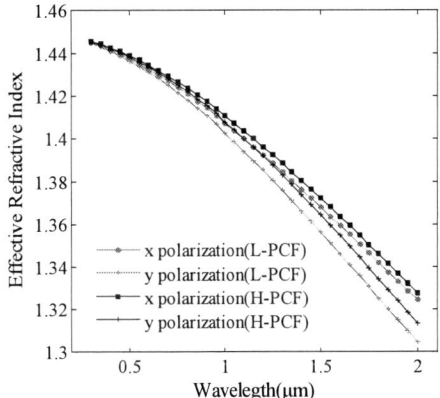

Figure 3. Effective refractive index of x and y polarization as a function of wavelength for the proposed L-PCF and H-PCF

C. Beating Length

The beating length is defined as the period of interference effects in the medium of birefringence and the beating length L is calculated by the following formula

978-1-4799-1215-5/13 $31.00 © 2013 IEEE

$$L = \frac{\lambda}{\left| n_{eff}^x - n_{eff}^y \right|} \quad (3)$$

where n_{eff}^x, n_{eff}^y is the effective refractive index of the x and y polarization modes, respectively. Equation (3) shows that the beating length is inversely proportional to the difference of the effective refractive indices between the two orthogonal polarization modes. As depicted in Figure 4, the curves of beating length of L-PCF and H-PCF as a function of wavelength are shown. From the exhibited figure, these two curves show a decreasing trend toward a higher wavelength. Moreover, the beating lengths of the L-PCF are smaller than that of the H-PCF from 0.30µm to 2.00µm, and the L-PCF can have beating lengths of sub-millimeter and the value of that is about 0.12mm at the wavelength of 1.55µm, which is smaller than the value of the beating length for the H-PCF, ~0.19mm. A shorter beating length can reduce the sensitivity of the PCF to mode coupling effects.

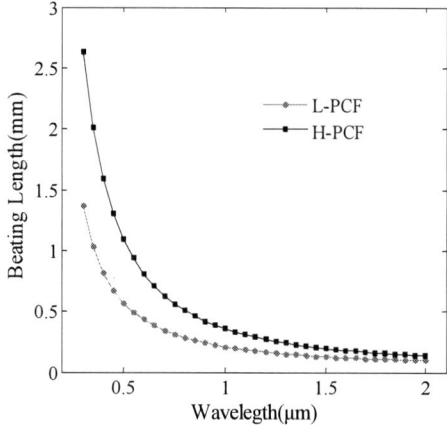

Figure 4. Beating length as a function of wavelength for the proposed L-PCF and H-PCF

D. Birefringence

Generally, in order to describe the birefringence of PCF, the modal birefringence of the PCF is considered. The modal birefringence B is defined as the difference of the absolute value of the effective refractive index of the x polarization mode and y polarization mode, which is calculated by the following equation

$$B = \left| n_{eff}^x - n_{eff}^y \right| \quad (4)$$

From the equation (4), obviously, the value of the birefringence is inversely proportional to the one of the beating length, which is demonstrated in Figure 5. We can find that two birefringence curves of the L-PCF and the H-PCF show an increase trend for the long wavelength side, and the modal birefringence of the L-PCF is higher than that of the H-PCF for the wavelength range 1.70µm. At the wavelength of 1.55µm, the value of the birefringence for the proposed L-PCF is about 0.013, while the value of birefringence for the H-PCF is about 0.0083, which reveals that the structure of elliptical air holes according to lozenge arrangements can easily achieve a higher

birefringence, and better polarization properties will be achieved. This photonic crystal fiber is of the potential application for the fiber optics sensing and communication in polarization-maintaining.

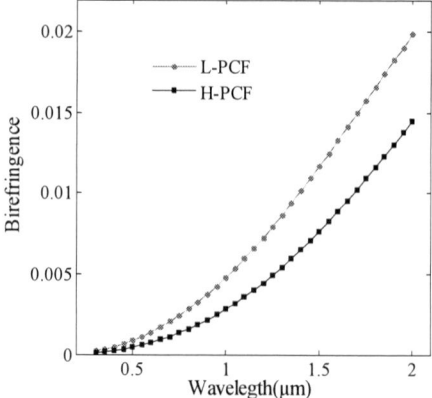

Figure 5. Birefringence as a function of wavelength for the proposed L-PCF and H-PCF

IV. CONCLUSIONS

In this study, we have proposed and investigated a high birefringence PCF comprising elliptical air holes cladding structure based on lozenge arrangement. The comparative analyses of the L-PCF and the H-PCF exhibited that the mode field distribution of the L-PCF is restricted effectively in the core of the fiber for a wider range of wavelength. The L-PCF is of smaller beating length and higher birefringence than that of the H-PCF with the same air holes size, shape, distance and air holes layer number. At the wavelength of 1.55µm, the beating length of the proposed L-PCF can be as low as ~0.12mm, and the birefringence of that can be as high as ~0.013. According to the results, it is believed that the proposed L-PCF will potentially obtain better polarization properties through optimizing the structure parameters.

REFERENCES

[1] J. C. Knight, T. A. Birks, P. St. J. Russell, and D. M. Atkin, "All-silica single-mode optical fiber with photonic crystal cladding," Optics Letters, vol. 21, pp. 1547–1549, October, 1996.

[2] T. A. Birks, J. C. Knight, and P. St. J. Russell, "Endlessly single-mode photonic crystal fiber," Optics Letters, vol. 22, pp. 961–963, July, 1997.

[3] F. Poletti, X. Feng, G. M. Ponzo, M. N. Petrovich, W. H. Loh, and D. J. Richardson, "All-solid highly nonlinear singlemode fibers with a tailored dispersion profile," Optics Express, vol. 19, pp. 66–80, January 2011.

[4] K. Saitoh, M. Koshiba, T. Hasegawa, and E. Sasaoka, "Chromatic dispersion control in photonic crystal fibers: application to ultra-flattened dispersion," Optics Express, vol. 11, pp. 843–852, April 2003.

[5] J. Ju, W. Jin, and M. S. Demokan, "Properties of a highly birefringent photonic crystal fieber," IEEE Phtonics Technology Letters, vol. 15, pp. 1375–1377, October 2003.

[6] Philip St. J. Russell, "Photonic-crystal fibers," Journal of Lightwave Technology, vol. 24, pp. 4729–4749, December 2006.

[7] R. Hao, Z. Q. Li, G. F. Sun, L. Y. Niu, and Y. C. Sun, "Photonic crystal fiber with high birefringence and low confinement loss," Optik-International Journal for Light and Electron Optics, 2013.

[8] K. Saitoh, and M. Koshiba, "Single-polarization single-mode photonic crystal fibers," IEEE Phtonics Technology Letters, vol. 15, pp. 1384–1386, October 2003.

The MEMS gyro stabilized platform design based on Kalman Filter

Guangchun Li, Yunfeng He, Yanhui Wei, Shenbo Zhu, Yanzhe Cao

College Of Automation,Harbin Engineering University,Harbin, China

hyf421@163.com

Abstract—**According to the characteristics of MEMS gyro, this paper designed a MEMS gyro stabilized platform based on Kalman Filter. Firstly, a hardware stabilized platform system is designed based on MEMS gyro. And then the system controller is designed by the motor frequency response transfer function. Finally, Kalman filter designed to estimate the constant deviation of accelerometer and gyro in pitch axis and electronic compass sensor in azimuth axis. The experimental results showed that Kalman filter can suppress the pitch and azimuth gyro drift effectively, and the system kept in good dynamic performances. The adjustment time of pitch axis is 0.143s without overshoot, and the adjustment time of azimuth axis is 0.184s with overshoot of 11.12%, which satisfied the requirements of the stabilized platform control system completely.**

Keywords- MEMS; stabilized platform;servo control;Kalman filter

I. INTRODUCTION

Stabilized platform is a device that enables stable object isolation from outside interference and stabilized in inertial space. Stabilized platform widely used in aerospace, aviation and some other occasions, it is an important part of tracking system and many investigators [1].

Traditionally, stabilized platform designed by mechanical or optical gyroscope gyro, which has high precision and simple control methods characteristics. However, its high cost limits its widespread promotion and applications. In recent years, with the rapid development of MEMS technology, MEMS showing small size, low cost, high reliability characteristics, and has been widely used in automotive, north-finder, micro-satellites[3-5]. Therefore, the application of MEMS gyro stabilized platform has unique advantages and broad application prospects.

Kalman filter is based on the known stochastic signal of mathematical models, and it is adaptive to time-varying non-stationary time series digital filtering method. Its essence is based on the observations data, and through the recursion theory to achieve the estimated of future state, and makes the estimate values close to the true value as much as possible. At present, Kalman filter has been successfully applied in inertial navigation, data fusion, and even in the military applications of radar systems and etc.[6].

However, the paper first discussed a MEMS gyro stabilized platform system. Secondly, identified the transfer function of the motor through frequency response method, on this basis,

designed the system controller. Thirdly, designed angle fusion algorithm on pitch axis and azimuth axis, deduced Kalman filter state equation and measurement equation. Finally, the Kalman filter fusion experiments and step response experiments are conducted with the stabilized platform. The experimental results showed that the MEMS gyro stabilized platform suppressed the MEMS gyro drift successfully by Kalman angle fusion with accelerometer, geomagnetic sensor and gyro. Additionally, the system maintained a good dynamic performance. Subsequently, the proposed system provide a reference to these persons who are engaged in researching stabilized platform.

II. MEMS STABILIZED PLATFORM SYSTEM

A. Stabilized platform design

The designed system adopts gyro angular velocity signal as the rate feedback signal, and the angular velocity integral and accelerometer or geomagnetic measurement data fusion as the position feedback signal. The platform can make the stable object stabilized in inertial space by isolating the vehicle pitch and azimuth axis angular velocity interferences. Designed the processor with digital signal processor TMS320F2812 to complete the sampling data reading and control algorithm complement, and then selected MIMU ADIS16350 as inertial measurement unit, the tri-axial geomagnetic sensor HMC5883 achieve azimuth axis compensation component. Fig. 1 shows the hardware structure of the platform system.

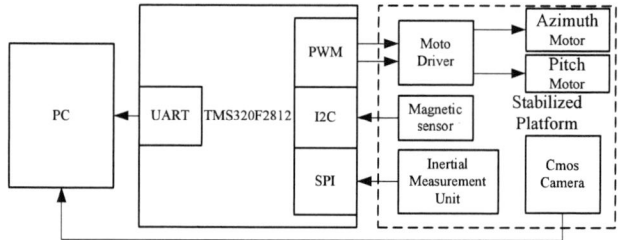

Figure 1. Hardware structure of the platform system

Stabilized platform structure includes the inner pitch frame and the outside azimuth frame, and the DC torque motor is installed in pitch axis and azimuth axis, respectively. To avoid the interference of motor magnetic, the HMC5883 installed in the location far away from the motor's position. An inertial measurement unit ADIS16350 and photovoltaic device needs to be isolated are installed in pitch frame.

---

1. National Natural Science Fund of China (51205074); 2. Higher Specialized Research Fund for the Doctoral Program (20112304120007)

B. System model and controller design

In this paper, we proposed frequency response method get an accurate mathematical model. Fig. 2 is the result of pitch and azimuth motor amplitude-frequency characteristic.

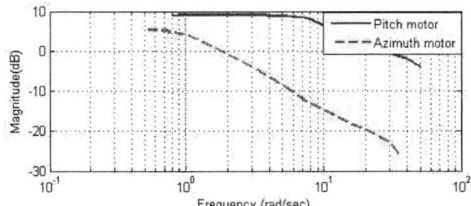

Figure 2. Two-axis motor frequency domain test curve

In Fig. 2, the transfer function can be obtained from the inflection point of the curve. The motor transfer function are shown in following expressions:

$$G_f(s) = \frac{2.8736}{0.0000792s^2 + 0.02349s + 1} \quad (1)$$

$$G_a(s) = \frac{1.6253}{0.000983s^2 + 0.1971s + 1} \quad (2)$$

Where, $G_f(s)$ denotes pitch axis motor transfer function, and $G_a(s)$ denotes azimuth axis motor transfer function.

However, the PWM power amplification coefficient 2.4 is known by the selected motor. To satisfy the accuracy requirements of stabilized platform, the designed PI calibration network are proposed and shown as follows:

$$G_{fc}(s) = \frac{0.0568s + 4.546}{s} \quad (3)$$

$$G_{ac}(s) = \frac{0.196s + 12.678}{s} \quad (4)$$

Where, $G_{fc}(s)$ denotes the pitch axis PI calibration network and $G_{ac}(s)$ denotes azimuth axis PI calibration network. After adding calibration network, the frequency characteristics of the pitch axis and the azimuth axis is shown in Fig. 3.

The Fig. 3 demonstrates that the pitch axis's shear frequency is 11.9rad/sec, phase margin is 69°, and 50dB gain at 1 rad/sec. The azimuth axis's shear frequency is 11.1rad/sec, phase margin is 54°, and 55dB gain at 1 rad/sec. Therefore, both the pitch axis and azimuth axis have a good steady state and dynamic characteristics after adding the calibration network.

III. ANGLE FUSION ALGORITHM

Because of the measurement accuracy of MEMS gyro, its pitch angle and azimuth angle after integral calculation are accurately in a short time, but the two angles would produce drift along with time. Therefore, the platform system only uses MEMS gyros could not work in high-precision properties, and the measurement data of accelerometer and magnetic sensor

contains large fluctuation noises. Therefore, the dynamic characteristics of the two sensors are not in good conditions.

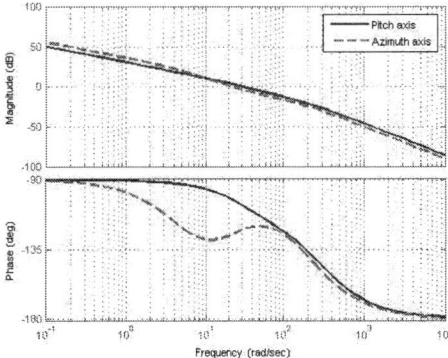

Figure 3. Frequency domain characteristic of pitch axis and azimuth axis

Therefore, we adopt the data fusion algorithm to acquire the azimuth angle and the pitch angle information without accumulate deviations. However, the reference [7] presents a method to calculate the angle between accelerometer input axis and the horizontal angle, and the reference [8] describes a method to calculate the azimuth angle through the geomagnetic sensor data, which provide theoretical foundations to this proposed algorithm.

Finally, we designed the Kalman filter on the pitch axis and azimuth axis by integrating the advantages of each sensor in the designed stabilized platform system. And the pitch axis Kalman filter design processes are expressed in following:

Firstly, selecting the pitch angle as a state vector and using accelerometer to estimate gyro constant deviation b . The discrete system state and measurement equations with the deviation as a state vector are as follows:

$$\begin{cases} x_k = \mathbf{A}x_{k-1} + \mathbf{B}u_{k-1} + \mathbf{W}_{gk} \\ y_k = \mathbf{C}x_k + \mathbf{W}_{ak} \end{cases} \quad (5)$$

Where, x_k denotes state vector at k time, y_k denotes accelerometer angular measured values at k time, $\mathbf{A}$ denotes state transition matrix, $\mathbf{B}$ denotes control matrix, $\mathbf{C}$ denotes observation matrix, u_{k-1} denotes gyro outputs contain fixed angular deviation at k-1 time, $\mathbf{W}_{gk}$ denotes system noise, $\mathbf{W}_{ak}$ denotes accelerometer measurement noise at k time.

More specifically, the expanded form of (5) can be shown as following expressions:

$$\begin{cases} \begin{bmatrix} \varphi_k \\ b_k \end{bmatrix} = \begin{bmatrix} 1 & -T_s \\ 0 & 1 \end{bmatrix} x_{k-1} + \begin{bmatrix} T_s \\ 0 \end{bmatrix} u_{k-1} + \begin{bmatrix} \omega_{gk}T_s \\ 0 \end{bmatrix} \\ y_k = \begin{bmatrix} 1 & 0 \end{bmatrix} x_k + \mathbf{W}_{ak} \end{cases} \quad (6)$$

Where, φ_k is the position angle, b_k is the estimate deviation, T_s is data sampling period, ω_{gk} is the gyro measurement noise at k time.

However, the gyro measurement noise ω_{gk} and the system noise W_{ak} are independent Gauss white noises. More importantly, the Kalman filtering processes are as follows:

Step prediction of the state:

$$x_{k,k-1} = \mathbf{A}x_{k-1,k-1} + \mathbf{B}u_k \qquad (7)$$

Estimate of state is:

$$x_{k,k} = x_{k,k-1} + \mathbf{K}_k(\mathbf{Z}_k - \mathbf{H}x_{k,k-1}) \qquad (8)$$

Step prediction error matrix is:

$$\mathbf{P}_{k,k-1} = \mathbf{A}\mathbf{P}_{k-1,k-1}\mathbf{A}^T + \mathbf{Q} \qquad (9)$$

Kalman gain is:

$$\mathbf{K}_k = \frac{\mathbf{P}_{k,k-1}\mathbf{H}^T}{\mathbf{H}\mathbf{P}_{k,k-1}\mathbf{H}^T + \mathbf{R}} \qquad (10)$$

The prediction error matrix is:

$$\mathbf{P}_{k,k} = (\mathbf{I} - \mathbf{K}_k\mathbf{H})\mathbf{P}_{k,k-1} \qquad (11)$$

Where, $\mathbf{Q} = \begin{pmatrix} q_{acce} & 0 \\ 0 & q_{gyro} \end{pmatrix}$ denotes system process noise covariance matrix, q_{acce} and q_{gyro} are accelerometers and gyros covariance, $\mathbf{R}$ denotes covariance matrix with the measurement noise.

By consecutive updating from (7) to (11), the optimal estimation of the system could be obtained, and the optimal angle results by updating the Kalman gain and covariance system. Similarly, azimuth axis Kalman filter design is basically the same with pitch axis, above φ_{acce} and geomagnetic sensor measurement noise matrix $\mathbf{W}_a$ need to be replaced by geomagnetic sensor data computation angle φ_{mag} and geomagnetic sensor measurement noise matrix $\mathbf{W}_{mag}$, $\mathbf{Q}$ and $\mathbf{R}$ also need to be changed correspondingly.

IV. EXPERIMENTS RESULTS AND ANALYSIS

A. Angular fusion experiments

Based on the Kalman filter model, experiments in pitch axis and azimuth axis carried out as follows:

1) Data collection: collect the IMU and geomagnetic data at the sampling period of 0.01s in static conditions, and save the collected data in computer;

2) Data processing: adopting the MATLAB software read the stored data, and preparation for the pitch and azimuth axis Kalman filter procedures.

The results of Kalman filter are shown in Fig. 4 and Fig. 5, respectively. The Fig. 4 demonstrates the pitch axis gyro and accelerometer fusion, and Fig. 5 demonstrates the fusion result of azimuth axis gyro and geomagnetic sensor. More specifically, in Fig. 4, Fig. 5, we can conclude that MEMS gyro noise is small, and the integration angle value in 300s time has

drift to 26° due to MEMS gyro low precision and bias instability, which could not meet the stability requirements in a stabilized platform system.

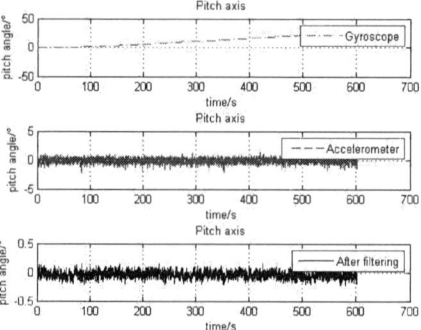

Figure 4.   The results of Kalman filter curve in pitch

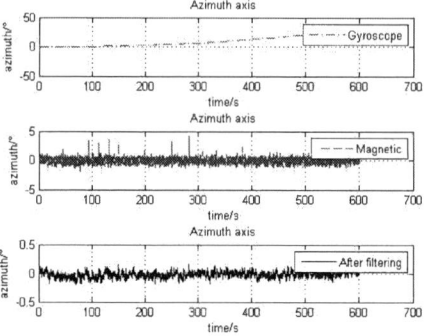

Figure 5.   The results of Kalman filter curve in azimuth

The Table I describes the gyro, accelerometer and geomagnetic sensor data statistic characteristics. The mean value of the accelerometer is $-0.0058°$, the geomagnetism is $-0.0062°$. However, the expected values of these two sensors are constant, and the output noise signals amplitude of the two sensors is relatively large.

TABLE I.   EACH SENSOR STATISTICS

Parameters	Mean(/°)	Variance
Pitch gyro	Divergence	Divergence
Accelerometer	-0.0058	0.0573
Pitch Kalman	-0.0049	0.0028
Azimuth gyro	Divergence	Divergence
Geomagnetic	-0.0062	0.0775
Azimuth Kalman	-0.0051	0.0037

In Fig. 4 of the pitch axis fusion curves, we can know that the output pitch angle after Kalman filter is fluctuate in the range of $\pm 0.2°$ in static conditions. What's more, the azimuth axis fusion curves show that the output azimuth angle after Kalman filter are fluctuate in the range of ± 0.2° in static conditions. More importantly, the variance of pitch angle and azimuth angle after the fusion is smaller than accelerometer and geomagnetic, which indicate that the larger noise signal has been filtered out by Kalman filter. Therefore, the MEMS stabilized platform system could meet the precision requirements in low and mediate occasions by combining the advantages of gyro, accelerometer and geomagnetic.

B. The step response experiments

In order to verify the dynamic properties of the stabilized platform system after the Kalman filter, a step response experiments are underway. Step response test procedure is as follows:

1) Parameters setting: Debug the control parameters of the pitch axis and the azimuth axis based on the proposed calibration network, and make the stabilized platform system achieved optimal control results;

2) Step response curves: Kalman filter added to the control algorithm, the step response signal of 10° was added into the pitch axis and the azimuth axis, and then sent the response position angle to the computer.

Fig. 6 and Table II show the step response curves and curve indicators of the step response curves in platform system. Fig. 6 shows the pitch axis curve without any overshoot, and the pitch adjustment time is longer to 0.2s after Kalman filter. However, there is no acceleration and deceleration process on the platform in the experiments, during the rotation accelerometer output contains a centrifugal acceleration in pitch axis, so the dynamic performance of the system is affected. Moreover, the distance between accelerometer and the pitch axis is relatively short, the pitch axis dynamics are still within an acceptable range.

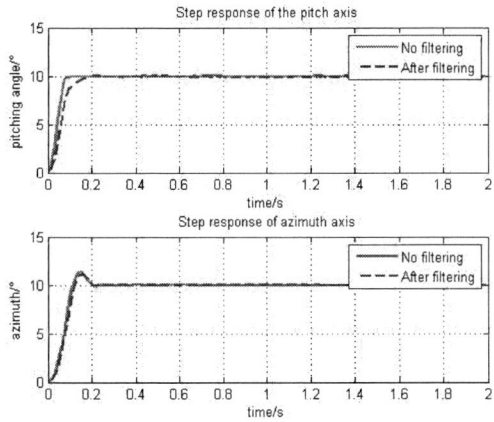

Figure 6.   The step response curves of the platform

Similarly, in Fig. 6 and Table 2, the azimuth axis overshoot becomes smaller after Kalman filter, adjustment time increased slightly, but 0.16s adjustment time is still a good dynamic system performance. Therefore, the design can suppress azimuth stabilization loop disturbance and achieve better isolation.

TABLE II.   TWO AXES STEP RESPONSE INDICATORS

Parameters	Overshoot (/°)	Regulation time (/s)	Rise time (/s)
Pitch no filter	No	0.0757	0.0692
Pitch with filter	No	0.1431	0.1146
Azimuth no filter	11.365%	0.1807	0.1013
Azimuth with filter	11.128%	0.1839	0.1107

V.   CONCLUSIONS

The introduction of Kalman filter in MEMS gyro stabilized platform system, not merely solves the problem of gyro integral angle drift, but still keeps the platform in good dynamic properties. Besides, the dynamic properties can be enhanced by the appropriate filter parameters selection and harmful items elimination. All in all, it is a meaningful attempt in MEMS gyro stabilized platform design and has some instruction meaning to engineering applications.

ACKNOWLEDGMENT

The research work was a team effort,and also the authors would like to thank Guan Lianwu for language modification.

REFERENCES

[1] Hong Huajie, Yun Pingping , Zhao Chuangshe,"The Application Research on Fuzzy PI Control Arithmetic of Photoelectric Stabilized Platform" in Mechatronics and Automation (ICMA),2011 International Conference on , Wuhan,China, pp.1-5,2009.

[2] Somà A., "MEMS Design for reliability: mechanical failure modes and testing" Perspective Technologies and Methods in MEMS Design(MEMSTECH), 2011 Proceedings of VIIth International Conference on , Polyana, Ukraine, pp.91-100, May,2011.

[3] Perlmutter, M. Robin, L.,"Michael Perlmutter, High-performance, low cost inertial MEMS: A market in motion!" Position Location and Navigation Symposium(PLANS), Myrtle Beach, SC, pp.225–229,2012.

[4] Dongqing Gu El-Sheimy,N.,"Heading accuracy improvement of MEMS IMU/DGPS integrated navigation system for land vehicle" Position, Location and Navigation Symposium. Monterey, CA, pp.1292–1296.2008.

[5] Yuan Long Wei . Min Cheol Lee, "Mobile robot autonomous navigation using MEMS gyro north finding method in Global Urban System" in Mechatronics and Automation (ICMA), 2011 International Conference on, Beijing , China, pp.91–96,2011.

[6] Fiedziuszko, S.J.,"Applications of MEMS in communication satellites"Microwaves, Radar and Wireless Communications. 2000. MIKON-2000.13th International Conference on , Wroclaw, Vol.3, pp.201–211,2000.

[7] Jiang Yanwei, Fang Jiancheng , Huang Xuegong,"Design of GMI micro-magnetic sensor and its application for geomagnetic navigation" in Systems and Control in Aerospace and Astronautics, 2008. ISSCAA 2008. 2nd International Symposium on,Shenzhen, pp.1-6,2008.

[8] Roan, P., Deshpande, N.,Wang, Y,"Manipulator state estimation with low cost accelerometers and gyros", Intelligent Robots and Systems (IROS), 2012 IEEE/RSJ International Conference on, pp. 4822–4827,2012.

[9] Efe, M. Atherton, D.P,"Maneuvering target tracking with an adaptive Kalman filter"in Decision and Control, 1998. Proceedings of the 37th IEEE Conference on, Tampa, FL,vol.1,pp.737–742,1998.

[10] Favre, J. Jolles, B.M., Siegrist,"Quaternion-based fusion of gyros and accelerometers to improve 3D angle measurement" Electronics Letters (Volume:42 , Issue: 11),pp.612–614,2006.

A Novel redundant inertial measurement unit and calibration algorithm

Kunpeng He*, Jitao Han, Yuping Shao

College Of Automation, Harbin Engineering University, Harbin, China
*hekunpeng@hrbeu.edu.cn, hjt553095158@163.com

Abstract—A novel redundant inertial measurement unit (RIMU) based on non-redundant inertial measurement unit (NRIMU) as its basic unit is proposed. Low-cost, high-precision and high-reliability are features of the novel RIMU which is different from traditional RIMU based on single inertial sensors as its basic unit. Moreover, a novel stratified calibration algorithm is proposed for the RIMU, which calibrated transformation matrices after calibrating NRIMU. The simulation results show that the stratified calibration algorithm can be implemented on the RIMU, and the presented algorithm features high-precision and simple operation in the RIMU, and the proposed technique can be applied in engineering.

Keywords- RIMU; NRIMU; stratified calibration

I. INTRODUCTION

The RIMU as a kind of novel inertial measurement unit (IMU) that has higher precision than IMU with three-axis orthogonal inertial sensors, and can improve the performance of single IMU greatly. Generally, the RIMU is a sensor assembly system constructed by more than three inertial sensors such as accelerometers and gyros. Redundant sensor configurations contain orthogonal sensors and heterotropic sensors[1]. From the view of the redundancy system reliability, more inertial sensors (accelerometers or gyros) have higher reliability in navigation system. Fault-tolerant capability of heterotropic sensors is stronger than that of orthogonal sensors, when the number of inertial sensors is the same[2]. At present the configuration of RIMU is based on single inertial sensor[3, 4], a kind of RIMU based on NRIMU as its basic unit is proposed in paper.

The NRIMU in the proposed design is made up of MEMS (Micro-Electro-Mechanical System) sensors. More importantly, calibration is necessary before application because of the bad stability of MEMS sensors[5]. Therefore, the stratified calibration method is proposed, and the result of simulation verification is presented.

II. DESIGN OF THE NOVEL RIMU

In Fig. 1, the novel RIMU consist of a regular tetrahedron which three NRIMU is installed on two flanks and one underside, respectively. Each of NRIMU is made up of three orthogonal quartz tuning fork gyroscopes and three orthogonal vibrating inertial accelerometers. Obviously, NRIMU belong to Micro Inertial Measurement Unit (MIMU).

Three NRIMU included in the proposed RIMU is called IMU_1, IMU_2 and IMU_3, respectively. IMU_1 is placed in the middle of underside BCD, the x-axis of it is perpendicular to straight line CB and far from point D, the y-axis of it is parallel to straight line CB, and the z-axis of it is perpendicular to underside BCD and outward. IMU_2 is placed in the middle of plane ABC, the x-axis is parallel to straight line BC, the y-axis is perpendicular to straight line BC and point to point D, and the z-axis is perpendicular to plane ABC and outward. IMU_3 is placed in the middle of plane ABD, the x-axis is parallel to straight line DB, the y-axis is perpendicular to straight line BD and point to point A, the z-axis is perpendicular to plane ABD and outward.

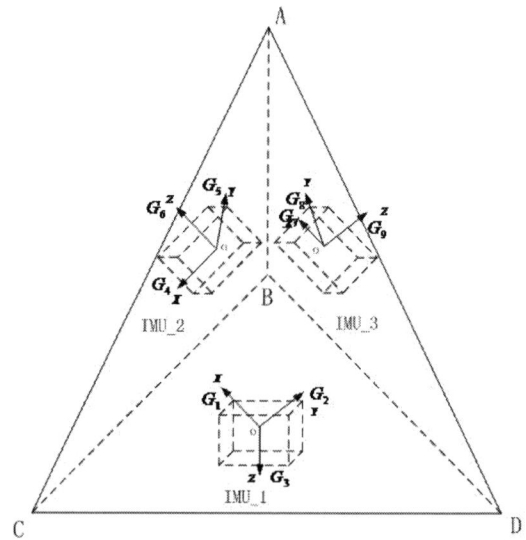

Figure 1. Integrated Design of the Novel RIMU

Integration architecture of orthogonal and heterotropic configuration is the design included nine gyros and nine accelerometers, and improved the reliability and fault tolerance capability of the RIMU design. At the same time, MEMS sensors are the major components of inertial sensors, and ensured the advantages of low-cost and small-sized than traditional inertial sensors. Finally, the complete MIMU instead of single inertial sensor is innovatively adopted in the basic unit of RIMU, and be of great importance to engineering application with the novel RIMU.

---

1. National Natural Science Fund of China (51309059).

III. STRATIFIED CALIBRATION ALGORITHM

Stratified calibration based on regular tetrahedron structure design is presented. Firstly, the process of calibration to error factors of every NRIMU is shown, and then the calibration of transformation matrices is proceed to ensure the accuracy of the calibration results.

A. NRIMU Calibration

Different turn angles provided by high-precision three-axis turntable provide the input of three-axis angular velocity and acceleration for NRIMU, and then accurate calibration in the laboratory is proceed[6]. Specifically, the gyro calibration of IMU_1 is taken as an example.

Error model of gyro is designed in following expressions:

$$\begin{bmatrix} \omega_{x1} \\ \omega_{y1} \\ \omega_{z1} \end{bmatrix} = \begin{bmatrix} K_{xx} & K_{xy} & K_{xz} \\ K_{yx} & K_{yy} & K_{yz} \\ K_{zx} & K_{zy} & K_{zz} \end{bmatrix} \begin{bmatrix} \bar{\omega}_{x1} \\ \bar{\omega}_{y1} \\ \bar{\omega}_{z1} \end{bmatrix} + \begin{bmatrix} b_{gx} \\ b_{gy} \\ b_{gz} \end{bmatrix} \quad (1)$$

$$\mathbf{\Omega}_{IMU\_1} = \mathbf{K}_1 \bar{\mathbf{\Omega}}_{IMU\_1} + \mathbf{B}_1 \quad (2)$$

Where $\mathbf{\Omega}_{IMU\_1}$ denotes the measurement value of the gyro output, $\bar{\mathbf{\Omega}}_{IMU\_1}$ the input of gyro provided by turntable, $\mathbf{K}_1$ the scale factor matrix of the gyro output, $\mathbf{B}_1$ the gyro bias matrix. Moreover, $K_{ii}(i=x,y,z)$ indicates the scale factor of i-axis gyro, $K_{ij}(i,j=x,y,z \mathbf{\text{且}} j \neq i)$ the misalignment error between i-axis and j-axis gyro.

Through extending dimensions of (1), the equation is represented as follows:

$$\underbrace{\begin{bmatrix} \omega_{x1} \\ \omega_{y1} \\ \omega_{z1} \end{bmatrix}}_{\omega} = \underbrace{\begin{bmatrix} K_{xx} & K_{xy} & K_{xz} & b_{gx} \\ K_{yx} & K_{yy} & K_{yz} & b_{gy} \\ K_{zx} & K_{zy} & K_{zz} & b_{gz} \end{bmatrix}}_{K} \underbrace{\begin{bmatrix} \bar{\omega}_{x1} \\ \bar{\omega}_{y1} \\ \bar{\omega}_{z1} \\ 1 \end{bmatrix}}_{\bar{\omega}} \quad (3)$$

Each of IMU_1 gyro axis respectively point to up or down, then the turntable round azimuth axis (i.e., round up axis or down axis) rotates integer circles with constant angular velocity by using high-precision turntable, and six groups of data is collected. Meanwhile, record turntable rotation angular velocity as $\bar{\omega}_i (i=1,2\ldots 6)$ and the output value of every gyro axis as $\omega_i (i=1,2\ldots 6)$. Therefore, (3) can be abbreviated as (4).

$$\mathbf{\Omega} = \mathbf{K}\bar{\mathbf{\Omega}} \quad (4)$$

Where: $\mathbf{\Omega} = \begin{bmatrix} \omega_1 & \omega_2 & \cdots & \omega_6 \end{bmatrix}, \bar{\mathbf{\Omega}} = \begin{bmatrix} \bar{\omega}_1 & \bar{\omega}_2 & \cdots & \bar{\omega}_6 \end{bmatrix}$

The matrix $\mathbf{K}$ is estimated by the least square technique. Lastly, $\mathbf{K}_1$ and $\mathbf{B}_1$ used for (2) is obtained by splitting matrix $\mathbf{K}$.

$$\mathbf{K} = \mathbf{\Omega} \cdot \bar{\mathbf{\Omega}}^T \cdot \left(\bar{\mathbf{\Omega}}\bar{\mathbf{\Omega}}^T \right)^{-1} \quad (5)$$

Error model of accelerometer is similar to gyro, so the calibration algorithm is the same as gyro.

B. Transformation Matrices Calibration

Three NRIMUs are placed in three surfaces of regular tetrahedron. The direction cosine matrix presented between NRIMU frame and RIMU frame is called transformation matrix. The process of computation transformation matrix of NRIMU is called transformation matrices calibration. More importantly, transformation matrices calibration of three NRIMUs is necessary because of error must be exist with NRIMU installed in regular tetrahedron.

Each gyro and accelerometer of NRIMU that misalignment error of has been calibrated in last section. The output of NRIMU in RIMU frame is obtained from the output of NRIMU in NRIMU frame according to the method of Euler Angle. Similarly, the transformation matrices calibration of IMU_1 is taken as an example. Equation (6) or (7) is the transformation matrix of IMU_1.

$$\begin{bmatrix} \omega_x^1 \\ \omega_y^1 \\ \omega_z^1 \end{bmatrix} = \begin{bmatrix} C_{11} & C_{12} & C_{13} \\ C_{21} & C_{22} & C_{23} \\ C_{31} & C_{32} & C_{33} \end{bmatrix} \begin{bmatrix} \omega_{x1} \\ \omega_{y1} \\ \omega_{z1} \end{bmatrix} \quad (6)$$

$$\mathbf{\Omega}_{RIMU}^1 = \mathbf{C}\mathbf{\Omega}_{IMU\_1} \quad (7)$$

Where $\mathbf{\Omega}_{IMU\_1}$ denotes the angular velocity output of IMU_1 gyro calibrated, $\mathbf{\Omega}_{RIMU}^1$ the angular velocity output of IMU_1 gyro in RIMU frame, $\mathbf{C}$ the 3×3 transformation matrix.

Extend the (7), we can obtain the expression as follow:

$$\mathbf{\Omega}_{RIMU}^1 = \mathbf{C}\mathbf{\Omega}_{IMU\_1} = \mathbf{M}_\Omega \mathbf{X}_C \quad (8)$$

Where:

$$\mathbf{M}_\Omega = \begin{bmatrix} \omega_{x1} & \omega_{y1} & \omega_{z1} & 0 & 0 & 0 & 0 & 0 & 0 \\ 0 & 0 & 0 & \omega_{x1} & \omega_{y1} & \omega_{z1} & 0 & 0 & 0 \\ 0 & 0 & 0 & 0 & 0 & 0 & \omega_{x1} & \omega_{y1} & \omega_{z1} \end{bmatrix}$$

$$\mathbf{X}_C = \begin{bmatrix} C_{11} & C_{12} & C_{13} & C_{21} & C_{22} & C_{23} & C_{31} & C_{32} & C_{33} \end{bmatrix}^T$$

Therefore, computation transformation matrix is to obtain the nine unknown variables in $\mathbf{X}_C$.

For calibration, RIMU x-axis point to up, and round azimuth axis (i.e., round x-axis) rotates integer circles with constant angular velocity, and record the rotation angular velocity of IMU_1 gyros calibrated, so the coefficient matrix $\mathbf{M}_\Omega^x$ is obtained as in (9). Meanwhile, record the three axis output of IMU_1 round x-axis of RIMU in RIMU frame, the output matrix $\mathbf{\Omega}_{RIMU}^{1\_x}$ as in (10).

$$\mathbf{M}_\Omega^x = \begin{bmatrix} \omega_{x1}^x & 0 & 0 & 0 & 0 & 0 & 0 & 0 & 0 \\ 0 & 0 & 0 & \omega_{x1}^x & 0 & 0 & 0 & 0 & 0 \\ 0 & 0 & 0 & 0 & 0 & 0 & \omega_{x1}^x & 0 & 0 \end{bmatrix} \quad (9)$$

$$\mathbf{\Omega}_{RIMU}^{1\_x} = \begin{bmatrix} \omega_x^{1\_x} & \omega_y^{1\_x} & \omega_z^{1\_x} \end{bmatrix}^T \quad (10)$$

Similarly, y-axis and z-axis of RIMU point to up, respectively. The coefficient matrix M_Ω^y, M_Ω^z is obtained as in (11) and (13), as the same as the output matrix $\mathbf{\Omega}_{RIMU}^{1\_y}, \mathbf{\Omega}_{RIMU}^{1\_z}$ as in (12) and (14).

$$\mathbf{M}_\Omega^y = \begin{bmatrix} 0 & \omega_{y1}^y & 0 & 0 & 0 & 0 & 0 & 0 & 0 \\ 0 & 0 & 0 & 0 & \omega_{y1}^y & 0 & 0 & 0 & 0 \\ 0 & 0 & 0 & 0 & 0 & 0 & 0 & \omega_{y1}^y & 0 \end{bmatrix} \quad (11)$$

$$\mathbf{\Omega}_{RIMU}^{1\_y} = \begin{bmatrix} \omega_x^{1\_y} & \omega_y^{1\_y} & \omega_z^{1\_y} \end{bmatrix}^T \quad (12)$$

$$\mathbf{M}_\Omega^z = \begin{bmatrix} 0 & 0 & \omega_{z1}^z & 0 & 0 & 0 & 0 & 0 & 0 \\ 0 & 0 & 0 & 0 & 0 & \omega_{z1}^z & 0 & 0 & 0 \\ 0 & 0 & 0 & 0 & 0 & 0 & 0 & 0 & \omega_{z1}^z \end{bmatrix} \quad (13)$$

$$\mathbf{\Omega}_{RIMU}^{1\_z} = \begin{bmatrix} \omega_x^{1\_z} & \omega_y^{1\_z} & \omega_z^{1\_z} \end{bmatrix}^T \quad (14)$$

Construction $\mathbf{M}_\Omega, \mathbf{\Omega}_{RIMU}^1$ of (8) based on (9) ~ (14), as in (15) and (16).

$$\mathbf{M}_\Omega = \begin{bmatrix} \mathbf{M}_\Omega^x & \mathbf{M}_\Omega^y & \mathbf{M}_\Omega^z \end{bmatrix}^T \quad (15)$$

$$\mathbf{\Omega}_{RIMU}^1 = \begin{bmatrix} \mathbf{\Omega}_{RIMU}^{1\_x} & \mathbf{\Omega}_{RIMU}^{1\_y} & \mathbf{\Omega}_{RIMU}^{1\_z} \end{bmatrix}^T \quad (16)$$

According to the inverse operation of matrices and (8), $\mathbf{X}_C$ is obtained as in (17).

$$\mathbf{X}_C = \mathbf{M}_\Omega^{-1} \mathbf{\Omega}_{RIMU}^1 \quad (17)$$

As (6) and (8) indicates, transformational relation between $\mathbf{X}_C$ and transformation matrix $\mathbf{C}$ of IMU_1 is presented as in (18).

$$\mathbf{C}_{RIMU}^1 = \begin{bmatrix} \mathbf{X}_C(1) & \mathbf{X}_C(4) & \mathbf{X}_C(7) \\ \mathbf{X}_C(2) & \mathbf{X}_C(5) & \mathbf{X}_C(8) \\ \mathbf{X}_C(3) & \mathbf{X}_C(6) & \mathbf{X}_C(9) \end{bmatrix} \quad (18)$$

Moreover, the two other NRIMU is taken into account; transformation matrix model of RIMU is obtained as in (19).

$$\begin{bmatrix} \mathbf{\Omega}_{RIMU}^1 \\ \mathbf{\Omega}_{RIMU}^2 \\ \mathbf{\Omega}_{RIMU}^3 \end{bmatrix} = \begin{bmatrix} \mathbf{C}_{RIMU}^1 & \mathbf{0}_{3\times3} & \mathbf{0}_{3\times3} \\ \mathbf{0}_{3\times3} & \mathbf{C}_{RIMU}^2 & \mathbf{0}_{3\times3} \\ \mathbf{0}_{3\times3} & \mathbf{0}_{3\times3} & \mathbf{C}_{RIMU}^3 \end{bmatrix} \begin{bmatrix} \mathbf{\Omega}_{IMU\_1} \\ \mathbf{\Omega}_{IMU\_2} \\ \mathbf{\Omega}_{IMU\_3} \end{bmatrix} \quad (19)$$

Transformation matrix model of accelerometer is similar to that of gyro. Also the method of calibration is the same as gyro.

IV. SIMULATION VERIFICATION

A. Simulation Calibration Data of NRIMU Collection

We designed the following data collection process to ensure the NRIMUs output data for calibration is completely, more specifically arrangement are provided in Table I.

TABLE I.   TABLE TYPE STYLES

State of NRIMU	Motion	Sensor Collected	sampling time
X(U) Z(W) Y(N)	Static	Accelerometer	90s
	Round x-axis 20°/s	Gyro	90s
Y(N) X(D) Z(E)	Static	Accelerometer	90s
	Round x-axis −20°/s	Gyro	90s
Y(U) Z(S) X(E)	Static	Accelerometer	90s
	Round y-axis 25°/s	Gyro	72s
Z(N) X(E) Y(D)	Static	Accelerometer	90s
	Round y-axis −25°/s	Gyro	72s
Z(U) Y(N) X(E)	Static	Accelerometer	90s
	Round z-axis 15°/s	Gyro	120s
Y(S) X(E) Z(D)	Static	Accelerometer	90s
	Round z-axis −15°/s	Gyro	120s

When turntable is static, the accelerometer data used for accelerometer calibration is collected. When turntable is rotating, the gyro data used for gyro calibration is collected. According to the turntable rate denoted in Table I and calibration parameters given in Table II, the simulation output data of NRIMU with zero mean and Gaussian white noise is generated.

Fig. 2 demonstrates the simulation output data of accelerometer and gyro when x-axis of IMU_1 is up.

TABLE II. TABLE TYPE STYLES

		Accelerometer	Gyro
Bias	**X**	21.80324	-51.08025
	Y	6.69792	-21.58649
	Z	55.55956	-31.69460
Scale factors	**X**	-835.6030	647.2877
	Y	-728.1473	645.2037
	Z	-831.8440	647.9611
Misalignment errors	**XY**	-0.00167524	-0.00016699
	XZ	0.00000358	-0.00147865
	YX	0.00042423	0.00007166
	YZ	0.00016856	0.00147072
	ZX	0.00149799	0.00007331
	ZY	-0.00054001	-0.00002994

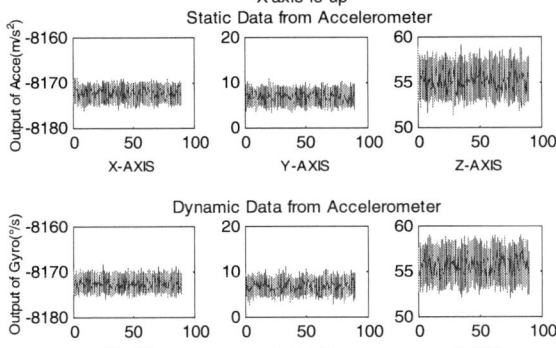

Figure 2. Simulation Output Data of Accelerometer and Gyro, X-axis of IMU_1 is Up

B. Simulation Transformation Matrices Calibration Data Collection

According to the results in Section IIIB，the data collected from three turntable positions used for computation transformation matrices of accelerometer and gyro, and presented in Table III.

TABLE III. TRANSFORMATION MATRICES CALIBRATION ARRANGEMENT

State of RIMU	Motion	Sensor Collected	Sampling Time	Input of RIMU *i*-axis in Theory		
				X-axis	**Y-axis**	**Z-axis**
X(U) Z(W) Y(N)	Static	Accelero-meter	90s	$9.80655\ m/s^2$	$0\,m/s^2$	$0\,m/s^2$
	Round x-axis $20°/s$	Gyro	90s	$20°/s$	$0°/s$	$0°/s$
Y(U) X(E) Z(S)	Static	Accelero-meter	90s	$0\,m/s^2$	$9.80655\ m/s^2$	$0\,m/s^2$
	Round y-axis $25°/s$	Gyro	72s	$0°/s$	$25°/s$	$0°/s$
Z(U) Y(N) X(E)	Static	Accelero-meter	90s	$0\,m/s^2$	$0\,m/s^2$	$9.80655\ m/s^2$
	Round z-axis $15°/s$	Gyro	120s	$0°/s$	$15°/s$	$0°/s$

When turntable with RIMU is static, the accelerometer data used for transformation matrices of accelerometer calibration is collected. When turntable is rotating, the gyro data used for transformation matrices of gyro calibration is collected. According to the input of RIMU *i*-axis in theory and transformation matrices given in Table V, the simulation output data of NRIMU with zero mean and Gaussian white noise is generated.

TABLE IV. THE VALUE IN THEORY AND THE VALUE CALCULATED

	The Value in Theory			**The Value Calculated**		
	X-axis	**Y-axis**	**Z-axis**	**X-axis**	**Y-axis**	**Z-axis**
Acceleration	9.806550	0	0	9.806551	-0.000014	0.000449
	-9.806550	0	0	-9.806549	-0.000014	0.000449
	0	9.806550	0	-0.000002	9.806548	0.000457
	0	-9.806550	0	-0.000002	-9.806552	0.000457
	0	0	9.806550	-0.000001	0.000015	9.806551
	0	0	-9.806550	-0.000001	0.000924	-9.806548

978-1-4799-1215-5/13 $31.00 © 2013 IEEE

Angular Velocity	20	0	0	20.000010	-0.000004	0.000012
	-20	0	0	-19.999989	-0.000004	0.000012
	0	25	0	-0.000007	24.999998	-0.000012
	0	-25	0	-0.000007	-25.000002	-0.000012
	0	0	15	-0.000003	-0.000006	14.999999
	0	0	-15	-0.000003	-0.000006	-15.000001

TABLE V.   TRANSFORMATION MATRICES SIMULATION OF NRIMU

		Accelerometer	Gyro
IMU_1 Transformation Matrices Simulation	Euler Angle for Simulation	$\begin{bmatrix} 109.5° & -0.2° & 90.3° \end{bmatrix}$	$\begin{bmatrix} 109.5° & -0.2° & 90.3° \end{bmatrix}$
	Simulation Value	$\begin{bmatrix} -0.0052 & -0.3338 & 0.9426 \\ -1.0000 & 0.0050 & -0.0038 \\ -0.0035 & -0.9426 & -0.3338 \end{bmatrix}$	$\begin{bmatrix} -0.0052 & -0.3338 & 0.9426 \\ -1.0000 & 0.0050 & -0.0038 \\ -0.0035 & -0.9426 & -0.3338 \end{bmatrix}$
	Calculated Value	$\begin{bmatrix} -0.0065 & -0.3323 & 0.9427 \\ -1.0013 & 0.0065 & -0.0037 \\ -0.0048 & -0.9412 & -0.3337 \end{bmatrix}$	$\begin{bmatrix} -0.0054 & -0.3337 & 0.9422 \\ -1.0001 & 0.0051 & -0.0042 \\ -0.0037 & -0.9426 & -0.3342 \end{bmatrix}$
IMU_2 Transformation Matrices Simulation	Euler Angle for Simulation	$\begin{bmatrix} 0.5° & 0.2° & 0.3° \end{bmatrix}$	$\begin{bmatrix} 0.5° & 0.2° & 0.3° \end{bmatrix}$
	Simulation Value	$\begin{bmatrix} 1.0000 & 0.0053 & -0.0034 \\ -0.0052 & 0.9999 & 0.0087 \\ 0.0035 & -0.0087 & 1.0000 \end{bmatrix}$	$\begin{bmatrix} 1.0000 & 0.0053 & -0.0034 \\ -0.0052 & 0.9999 & 0.0087 \\ 0.0035 & -0.0087 & 1.0000 \end{bmatrix}$
	Calculated Value	$\begin{bmatrix} 0.9998 & 0.0042 & -0.0036 \\ -0.0054 & 0.9989 & 0.0085 \\ 0.0033 & -0.0098 & 0.9998 \end{bmatrix}$	$\begin{bmatrix} 1.0006 & 0.0046 & -0.0041 \\ -0.0046 & 0.9993 & 0.0081 \\ 0.0041 & -0.0094 & 0.9993 \end{bmatrix}$
IMU_3 Transformation Matrices Simulation	Euler Angle for Simulation	$\begin{bmatrix} 108.5° & 0.3° & -29.7° \end{bmatrix}$	$\begin{bmatrix} 108.5° & 0.3° & -29.7° \end{bmatrix}$
	Simulation Value	$\begin{bmatrix} 0.8686 & 0.1615 & -0.4684 \\ 0.4955 & -0.2732 & 0.8246 \\ 0.0052 & -0.9483 & -0.3173 \end{bmatrix}$	$\begin{bmatrix} 0.8686 & 0.1615 & -0.4684 \\ 0.4955 & -0.2732 & 0.8246 \\ 0.0052 & -0.9483 & -0.3173 \end{bmatrix}$
	Calculated Value	$\begin{bmatrix} 0.8686 & 0.1630 & -0.4667 \\ 0.4955 & -0.2717 & 0.8263 \\ 0.0052 & -0.9468 & -0.3156 \end{bmatrix}$	$\begin{bmatrix} 0.8682 & 0.1616 & -0.4683 \\ 0.4951 & -0.2731 & 0.8247 \\ 0.0049 & -0.9483 & -0.3172 \end{bmatrix}$

V.   RESULTS AND DISCUSSION

A.   Results and Discussion of NRIMU Calibration

In (5), adopting the simulation output data and the real value in theory, and calibration parameter is calculated with MATLAB. The calculated value of acceleration and angular velocity in six different positions based on calibration parameter calculated and the simulation output data is denoted in Table IV. For comparing, the theoretical value listed in left side of Table IV, the error range of acceleration and angular velocity calculated reach to 10^{-5}, which meet the demand of inertial navigation computation. The error is mainly caused by white noise of the simulation output data .

B.   Results and Discussion of Transformation Matrices Calibration

Similarly, the simulation output data and the real value in theory in (17), transformation matrices of three NRIMUs is calculated with MATLAB. The transformation matrices of NRIMU simulated test are implemented. Comparing the simulation value and the calculated value appeared in Table V, the error range each to 10^{-4}. The error is caused by white noise.

VI.   CONCLUSION

In the paper, a novel RIMU adopting NRIMU instead of single inertial sensor as the basic composition unit is proposed. The new RIMU is easy to construct and maintenance and the advantage of high-precision and high-reliability is presented. Additionally, stratified calibration according to the novel RIMU is presented in the paper. More importantly, the simulation verified the feasibility of the method of calibration, and this method is benefit for the engineering applications with simple operation, high accuracy.

ACKNOWLEDGMENT

The research of the novel RIMU was a team effort. The authors would like to thank GUAN Lianwu and SHAO Yuping for their assistance in the preparation of this paper.

REFERENCES

[1] Pittelkau, Mark EP . "Calibration and attitude determination with redundant inertial measurement units". Journal of Guidance, Control, and Dynamics. Vol.28, No4, pp.743-752,2005.

[2] Liang Haibo . "Key technique of micro inertial system based on redundant gyroscopes". Harbin : Harbin Engineering University, 2011 .

[3] J.V Harrison, E G Gai. "Evaluating sensor orientations for navigation performance and failure detection". IEEE Transactions on Aerospace and Electronic Systems. Vol.13, No6,pp.631-643, 1977.

[4] E.J Oliveira, et al. "Inertial measurement unit calibration procedure for a redundant tetrahedral gyro configuration with wavelet denoising". Journal of Aerospace Technology and Management . Vol.4, No2, pp.163-168,2012.

[5] Seong Yun Cho, Chan Gook Park . "A Calibration technique for a redundant IMU containing low-grade inertial sensors". ETRI Journal. Vol.27, No4, pp.418-426,2005.

[6] Z.F. Syed, P. Aggarwal, C. Goodall, X. Niu, and N. El-Sheimy. "A new multi-position calibration method for MEMS inertial navigation systems". Measurement Science and Technology. Vol.18, No7, pp.1897 −1907,2007.

A MEMS integrated navigation system based on DSP and FPGA

Jianhui Zeng, Xingzhi Zhang, Kunpeng He, Huiyu Liu

Automation College, Harbin Engineering University, Harbin, China

E-mail: 986893778@qq.com

Abstract—**A scheme for designing a high precision MEMS integrated navigation system based on DSP and FPGA is proposed. The FPGA is used for collecting the multi-channel data and the DSP is for providing the navigation solution. FPGA and DSP can exchange data with each other through a dual-port RAM. The navigation parameters are sent to PC through RS232 interface. The inertial devices of the system use the newly developed MEMS quartz tuning fork gyroscope and MEMS quartz vibrating beam accelerometer. The MIMU/GPS integrated navigation system is fabricated and tested, and the experiment results demonstrate that the EKF algorithm is effective and the system has the high accuracy, strong resistance to disturbance.**

Keywords-integrated navigation; FPGA; DSP; dual port RAM; GPS; EKF

I. INTRODUCTION

MEMS integrated navigation system is widely used in automation, biomedical industry, aerospace and other fields due to its high performance with smaller size and lower price [1]. With the development of micro-electronic technology, the digital signal processing chip (DSP), the programmable logic device such as FPGA and micro inertial sensors have become more mature. The high-speed FPGA multi-channel parallel data acquisition and the ability of time controlling are strong, but the data signal processing capability is relatively weak. In order to obtain a navigation system with high precision, a special digital signal processing chip is used to fulfill the navigation operation. In this paper, a prototype of MIMU/GPS integrated navigation system based on the DSP and FPGA is designed. The latest MEMS quartz tuning fork gyroscope and MEMS quartz vibrating beam accelerometer are used in this system as inertial devices. The attitude error of the system is less than 0.1 °, heading angle error is less than 0.5 °, and positioning accuracy is of 8m using the EKF algorithm.

II. NAVIGATION SYSTEM ARCHITECTUR

The integrated navigation system is mainly composed of MIMU, a GPS receiver, the power module, data sampling module and navigation computer. The hardware block diagram of the integrated system is shown in Fig. 1.

All components are installed in a small and sealed case. The power required for each part of the system is provided by the 24VDC power supply module. The core of the MIMU device is MEMS quartz tuning fork gyroscope and quartz vibrating beam accelerometer.

TABLE I.   TECHNOLOGY PERFORMANCE OF GYROSCOPE

scale factor	25mV/°/s	resolution	≤0.008°/s
Scale factor nonlinearity (1σ)	≤0.05%	bandwidth	≤50Hz
bias	≤0.15°/s	Measurement Range	± 200°/s
Bias stability(1σ)	≤30°/h	threshold	≤0.008°/s
Bias repeatability (1σ)	≤30°/h	Random walk coefficient	$≤0.5°/\sqrt{h}$
Bias G sensitivity	≤0.02°/s/g	Withstand vibration	10g rms; 20~2000Hz

TABLE II.   TECHNOLOGY PERFORMANCE OF ACCELEROMETER

scale factor	50Hz/g	resolution	$≤1×10^{-5}g$
Scale factor repeatability (month,1σ)	≤50ppm	bandwidth	≤400Hz
bias	≤2mg	Measurement Range	± 100g
Bias repeatability (month,1σ))	≤100μg	Natural frequency	900Hz
Second order nonlinear coefficient	≤80μg/g²	threshold	$≤1×10^{-5}g$
Temperature range	-45℃~+85℃	Withstand vibration	20g rms; 20~2000Hz

The three gyroscopes and three accelerometers are installed in three orthogonal axes. The real-time position, speed, accurate location information could be obtained by GPS receiver and then send to the navigation computer by RS232 interface. The AD conversion chip LTC2440 is used for analog-digital conversion because the gyroscopes produce an analogue voltage output. The accelerometer outputs frequency signal [2], so Verilog HDL will be used to write hardware frequency measurement module [3] to obtain the necessary

---

1. National Natural Science Fund of China (51309059).

data and use a timer to achieve gyro and accelerometer synchronous sampling at the same time.

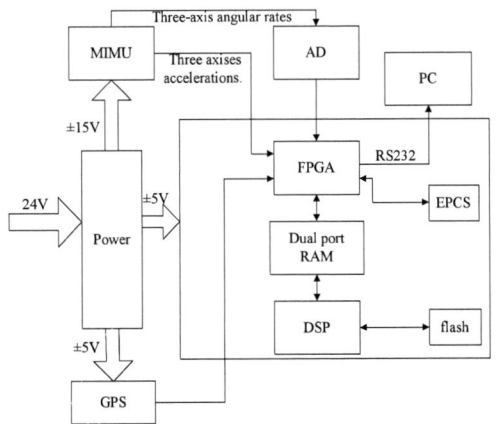

Figure 1. The hardware block diagram of integrated navigation system

In this integrated navigation system, FPGA chip EP3C10 of Altera Company that embedded a soft-core processor is applied to realize the acquisition of the data. Data acquired from the FPGA are sent to the DSP through the dual port RAM by interruption. DSP is used to complete the navigation solution and then send the navigation information to the FPGA. The navigation information is sent to the PC through RS232 interface finally. The design of embedded processing scheme makes this cost less, real-time and high-speed system cost-effective upgrade.

III. THE ALGORITHM OF THE SYSTEM**ERROR! REFERENCE SOURCE NOT FOUND.**

Strap down inertial navigation algorithm and the EKF data**Error! Reference source not found.** fusion algorithm is realized by DSP. The information of three axis angular rates and three axis accelerations collected from the MIMU are used for updating attitude with inertial navigation algorithm. The system could get attitude, position and velocity information more accurately through data obtained from the GPS [4].

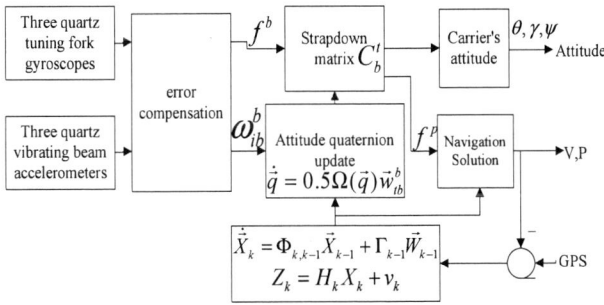

Figure 2. The software block diagram of integrated navigation system

A. System model and state variables

The dynamic system model developed here is a kinematic model for a six DOF rigid body with position and velocity represented in an inertial coordinate frame (Earth fixed), angular velocity and acceleration in a body-fixed frame. The

geographic coordinate frame is chosen as East-North-Up coordinate frame, and the strap down algorithm platform system is set to the geographic coordinate frame. The low accuracy MEMS gyro almost unable to detect the rotational angular velocity of the earth, $\overset{-b}{\omega}_{ip}$ can be seen as part of the noise while calculating attitude rate $\vec{\omega}_{pb}^{b}$.

$$\vec{\omega}_{pb}^{b} = \vec{\omega}_{ib}^{b} + \vec{w}_{\omega} - \vec{b}_{\omega} \qquad (1)$$

$\vec{w}_{\omega}$ —angular rate sensor noise vector, $\vec{b}_{\omega}$ —rate gyro bias vector, $\vec{\omega}_{ib}^{b}$ —rotating angular velocity in the body fixed frame to the inertial frame .

The acceleration that navigation solution needs is also simplified.

$$\overset{-b}{a} = \vec{f}^{b} + \vec{w}_{a} + \mathbf{C}_{p}^{b}\begin{bmatrix} 0 & 0 & g \end{bmatrix}^{T} \qquad (2)$$

$\vec{w}_{a}$ —acceleration sensor noise vector, $\vec{f}^{b}$ —measured acceleration vector in the body fixed frame, $\mathbf{C}_{p}^{b}$ —rotation matrix in the earth fixed frame to the body fixed frame.

This simplification makes the running speed and precision of system improved effectively through the calculation and filtering solution.

The development of the system model will make use of two matrices. The first of these matrices is the rotation matrix as a function of the unit quaternion [5].

$$\mathbf{T} = \mathbf{C}_{b}^{p} = \begin{pmatrix} T_{11} & T_{12} & T_{13} \\ T_{21} & T_{22} & T_{23} \\ T_{31} & T_{32} & T_{33} \end{pmatrix} \qquad (3)$$

The second of these matrices is used in the "strap down" equation, $\dot{\vec{q}} = \frac{1}{2}\boldsymbol{\Omega}(\vec{q})\vec{\omega}_{pb}^{b}$ to relate the body axis angular velocity to the unit quaternion rates.

$$\begin{bmatrix} \dot{q}_{0} \\ \dot{q}_{1} \\ \dot{q}_{2} \\ \dot{q}_{3} \end{bmatrix} = \frac{1}{2} \begin{bmatrix} -q_{1} & -q_{2} & -q_{3} \\ q_{0} & -q_{3} & q_{2} \\ q_{3} & q_{0} & -q_{1} \\ -q_{2} & q_{1} & q_{0} \end{bmatrix} \begin{bmatrix} \omega_{pbx}^{b} \\ \omega_{pby}^{b} \\ \omega_{pbz}^{b} \end{bmatrix} \qquad (4)$$

$\vec{\omega}_{pb}^{b}$ —true rotational rates vector in the body fixed frame.

978-1-4799-1215-5/13 $31.00 © 2013 IEEE

The state variables estimated by the integrated navigation system are the position, velocity, attitude (unit quaternion), and rate gyro biases. The state equations are the derivatives of the state variables. In general, these equations are nonlinear of the state variables and inputs.

$$\dot{x} = \begin{bmatrix} \dot{\vec{P}} \\ \dot{\vec{V}} \\ \dot{\vec{q}} \\ \dot{\vec{b}}_w \end{bmatrix} = \begin{bmatrix} V \\ \mathbf{C}_b^p \vec{a}^b \\ \frac{1}{2}\boldsymbol{\Omega}(\vec{q})\vec{\omega}_{pb}^b \\ \vec{w}_b \end{bmatrix} \quad (5)$$

V—velocity vector in the earth fixed frame, $\vec{w}_b$—noise vector for bias random walks.

While (5) captures the kinematics of a rigid body, it is not in the proper form for state equations. State equations should be written as a function of the states, inputs, and process/disturbance noise.

$$\dot{\vec{X}} = f(\vec{X}, \vec{u}, \vec{W}) \quad (6)$$

Then plugging $\vec{\omega}_{pb}^b$ and $\vec{a}^b$ into (5) gives the state equations as a function of states, the input $\vec{u} = \begin{bmatrix} \vec{\omega}_{ib}^{bT} & \vec{f}^{bT} \end{bmatrix}^T$, and the process noise $\vec{W} = \begin{bmatrix} \vec{w}_\omega^T & \vec{w}_a^T & \vec{w}_b^T \end{bmatrix}^T$.

$$\dot{\vec{X}} = f(\vec{X}, \vec{u}, \vec{W}) = \begin{bmatrix} \dot{\vec{V}} \\ \mathbf{C}_b^p(\vec{f}^b + \vec{w}_a) + \begin{bmatrix} 0 & 0 & g \end{bmatrix}^T \\ \frac{1}{2}\boldsymbol{\Omega}(\vec{q})(\vec{\omega}_{ib}^b + \vec{w}_\omega - \vec{b}_\omega) \\ \vec{w}_b \end{bmatrix} \quad (7)$$

Writing (6) instead of discrete variables,

$$\dot{\vec{X}}_k = \Phi_{k,k-1}\vec{X}_{k-1} + \Gamma_{k-1}\vec{W}_{k-1} \quad (8)$$

Measurement equation:

$$\vec{Z}_k = \mathbf{H}_k\vec{X}_k + \vec{v}_k \quad (9)$$

Where $\mathbf{H}_k$ is the observation matrix, $\vec{v}_k$ is the observation noise. And so the whole mathematical model of the system can be obtained.

B. Linearization of state equation

The process noise vector $\vec{W}_{k-1}$ is assumed to be white noise whose noise covariance matrix is $\mathbf{Q}$. The measurement noise is also assumed to be white noise whose noise covariance matrix is $\mathbf{R}$. The noise of each sensor can be considered as independent and these matrices have numerical diagonal only.

The system model takes the estimated value $\hat{X}(k)$ of the nonlinear system state $X(k)$ as the EKF linear reference point. The nonlinear vector function makes use of Taylor series expansion near the reference point, and the linear equation can be obtained by taking the first item. The linearization of the state equation takes $\hat{X}(k)$ as reference point and the linearization of the measurement equation takes $\hat{X}(k+1|k)$ as reference point.

This linearization results in equations for the following Jacobian matrices,

$$F = \frac{\partial f}{\partial X}, \quad G = \frac{\partial f}{\partial W}, \quad H = \frac{\partial h}{\partial X} \quad (10)$$

The model is a continuous time model. Therefore, some discrete time approximations for implementation as follows should be used.

$$\boldsymbol{\Phi} \cong \mathbf{I} + \mathbf{F}T, \ \boldsymbol{\Gamma} \cong \mathbf{G}T$$

The following formulas will tell us how to calculate $\mathbf{F}$, $\mathbf{H}$ and $\mathbf{G}$.

Calculating the partial derivatives in the elements of $\mathbf{F}$ Gives

$$F = \begin{bmatrix} \boldsymbol{0}_{3\times3} & \boldsymbol{I}_{3\times3} & \boldsymbol{0}_{3\times7} \\ & & F_{Vq} & \boldsymbol{0}_{3\times3} \\ \boldsymbol{0}_{10\times6} & & F_{qq} & F_{qb} \\ & & \boldsymbol{0}_{3\times7} \end{bmatrix} \quad (11)$$

Where

$$F_{Vq} = \begin{bmatrix} F_{Vq0} & F_{Vq1} & F_{Vq2} & F_{Vq3} \\ -F_{Vq3} & -F_{Vq2} & F_{Vq1} & F_{Vq0} \\ F_{Vq0} & F_{Vq0} & F_{Vq0} & F_{Vq0} \end{bmatrix}$$

$$F_{Vq0} = 2(q_0 a_x^b - q_3 a_y^b + q_2 a_z^b)$$

$$F_{Vq1} = 2(q_1 a_x^b + q_2 a_y^b + q_3 a_z^b)$$

$$F_{Vq0} = 2(-q_2 a_x^b + q_1 a_y^b + q_0 a_z^b)$$

$$F_{Vq0} = 2(-q_3 a_x^b - q_0 a_y^b + q_1 a_z^b)$$

$$F_{qq} = \frac{1}{2}\begin{bmatrix} 0 & -\omega_{pbx}^b + b_{\alpha x} & -\omega_{pby}^b + b_{\alpha y} & -\omega_{pbz}^b + b_{\alpha z} \\ \omega_{pbx}^b - b_{\alpha x} & 0 & \omega_{pbz}^b - b_{\alpha z} & \omega_{pby}^b + b_{\alpha y} \\ \omega_{pby}^b - b_{\alpha y} & \omega_{pbz}^b + b_{\alpha z} & 0 & \omega_{pbx}^b - b_{\alpha x} \\ \omega_{pbz}^b - b_{\alpha z} & \omega_{pby}^b - b_{\alpha y} & b_{\alpha x} - \omega_{mx} & 0 \end{bmatrix}$$

$$F_{qb} = -\frac{1}{2}\Omega(\vec{q}) = -\frac{1}{2}\begin{bmatrix} -q_1 & -q_2 & -q_3 \\ q_0 & -q_3 & q_2 \\ q_3 & q_0 & -q_1 \\ -q_2 & q_1 & q_0 \end{bmatrix}$$

Calculating the partial derivatives in the elements of **G**
Gives

$$G = \begin{bmatrix} & 0_{3\times 9} & \\ 0_{3\times 3} & C_b^p & \\ \frac{1}{2}\Omega(\vec{q}) & 0_{4\times 3} & 0_{7\times 3} \\ & 0_{3\times 6} & I_{3\times 3} \end{bmatrix} \quad (12)$$

Calculating the partial derivatives in the elements of **H**
Gives

$$H = \begin{bmatrix} I_{6\times 6} & 0_{6\times 7} \end{bmatrix} \quad (13)$$

C. The system algorithm

The position, velocity, attitude and gyro zero bias of the system at time K can be estimated by using the Runge-Kutta numerical integration algorithm with the value of the acceleration and angular velocity measured at time K-1. Then the value of estimated state variable $\hat{\vec{X}}(k\,|\,k-1)$ that exist in the filtering equation is obtained. Getting the prediction error covariance matrix by calculating **F** and **G**, the covariance will tell us the reliability of state that system estimated.

State variable of the system will diverge with time and the value of P will become larger without other sensor measurements to correct the previous state estimated [6]. With the assistance of GPS, precise position, velocity and azimuth data will be feed backed to the system at a frequency of 1Hz. And then it will provide correction value to the strap down navigation system through the EKF filter. The reduced forecast error covariance P and real error of state estimation will be obtained through update algorithm. Algorithm formulas are as below.

Algorithm of prediction and navigation:

$$\hat{\vec{X}}(k\,|\,k-1) = \hat{\vec{X}}(k-1) + \int_{t_{k-1}}^{t_k} f(\hat{\vec{X}}(k-1),\vec{u}(k-1))dt \quad (14)$$

$$P(k\,|\,k-1) = (I + FT)P(k-1)(I + FT)^T + T^2 GQG^T \quad (15)$$

Update algorithm:

$$\hat{\vec{Z}}(k+1\,|\,k) = H\,\hat{\vec{X}}(k\,|\,k-1) \quad (16)$$

$$K(k) = P(k\,|\,k-1)H^T(HP(k\,|\,k-1)H^T + R)^{-1} \quad (17)$$

$$\hat{\vec{X}}(k) = \hat{\vec{X}}(k\,|\,k-1) + K(k)[\vec{Z}(k) - \hat{\vec{Z}}(k+1\,|\,k)] \quad (18)$$

$$P(k) = P(k\,|\,k-1) - K(k)HP(k\,|\,k-1) \quad (19)$$

In the above algorithm, the system will get navigation information includes the position, velocity and attitude in geographic coordinate system by using the Runge-Kutta numerical integration algorithm without the updated data of GPS. When GPS data is updated, data of inertial navigation system can be corrected through the EKF update algorithm .The system will output position, velocity and attitude information accurately. Flowing chart of the system algorithm is shown in Fig. 3.

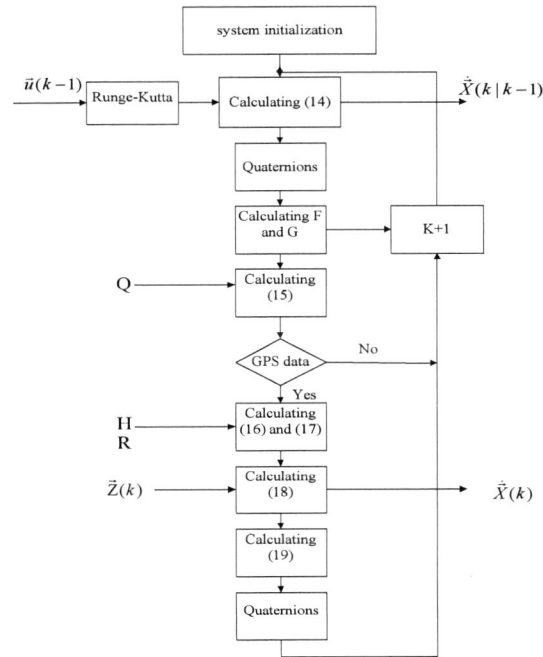

Figure 3. Flowing chart of the system algorithm

978-1-4799-1215-5/13 $31.00 © 2013 IEEE   27

IV. EXPERIMENTS AND RESULT ANALYSIS

The integrated navigation system is shown in Fig.4. In order to test the accuracy of the system, many experiments have been conducted.

Put the system on the three axis rate turntable at normal temperature with a minimum external disturbance to collect the data of gyroscope and accelerometer 30 minutes and the sample frequency is 100Hz. Finally, the correctness of EKF algorithm is verified thought simulation.

As you can see from Fig. 5, the horizontal precision error of the integrated navigation system is less than $0.1^{\circ}(1\sigma)$ and the heading angle error is less than $0.5^{\circ}(1\sigma)$. Since the system temperature will be rise with the sampling time increase, temperature drift of output signal occurs, so temperature compensation is necessary in applications. As shown in Fig. 6, the position error between the system and GPS latitude and longitude is less than 8 meters.

Figure 4. The MEMS integrated navigation system

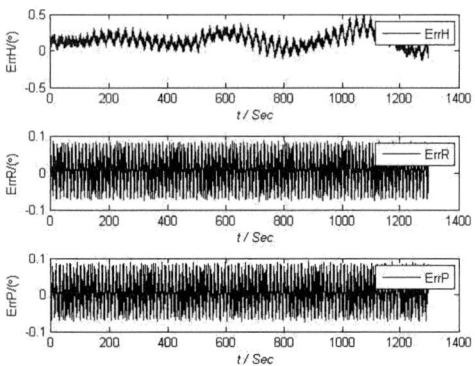

Figure 5. Attitude error of the system

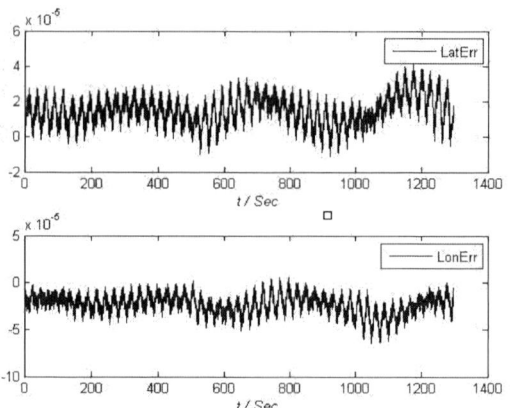

Figure 6. Position error of the system

V. CONCLUSION

In this paper, a integrated navigation system based on DSP and FPGA is proposed. The system has advantages of small size ($105*100*100mm$), low weight ($<1.2kg$), low cost, good real-time performance and high reliability. The system can be applied to self-driving cars, vehicle tracking, intelligent traffic management, intelligent wheelchairs, UAV and so forth.

ACKNOWLEDGMENT

The authors express thanks to my teacher He Kunpeng and Doctor Guan Lianwu, Master Wang Fuchao and Xu Zhenlong, this paper is supported and encouraged by them.

REFERENCES

[1] N. Yazdi, F. Ayazi and K. Najafi. "Micro machined Inertial Sensors", Proceedings of the IEEE, San Diego, Vol. 86, pp. 1640–1659, August 1998.

[2] Randall Jaffe, Ted Aston and Asad M Madni. "Advances in Ruggedized Quartz MEMS Inertial Measurement Units," 2006 IEEE/ION Position, Location, and Navigation Symposium, pp. 390–399, April 2006.

[3] Gao Yanbin, Zhan Junni, He Kunpeng and Wang Tingjun. "Design of a high-precision VBA frequency sampling system," Applied Science and Technology, Vol. 39(3), pp. 61–65, 2012.

[4] Christopher Hide, Terry Moore and Martin Smith. "Adaptive Kalman filtering Algorithms for Integrating GPS and low cost INS," PLANS-2004 Position Location and Navigation Symposium, Monterey. CA. United states, pp. 227-233, April 2004.

[5] Angelo M.Sabatini, "Quaternion-based extended Kalman filter for determining orientation by inertial and magnetic sensing," IEEE Transactions on Biomedical Engineering, Vol. 53, pp. 1346 – 1357, July 2006.

[6] C. Hide, "Integration of GPS and low cost INS measurements," Ph. D. dissertation, University of Nottingham, England, September 2003.

The LPFG temperature characteristic research based on electric heating method

Tao Geng*, Dongdong Zi, Wenlei Yang, Chengguo Tong

Key Laboratory of In-Fiber Integrated Optics, Ministry Education of China, Harbin Engineering University, Harbin, China

gengtao_hit_oe@126.com

Abstract -Under the guidance of the coupled mode theory, we take some mode simulation to the cone area of LPFG produced through melting the cone method. Measured LPFG resonance wavelength curve at different temperatures, find the temperature sensitivity is 62.8pm/°C when temperature below 200°C, and the resonance peak drift and temperature changes in the basic of a linear relationship.

Keywords-component: LPFG; mode simulation; temperature

I. INTRODUCTION

The period of Long period fiber gratings(LPFGs) is relatively long. The phase matching condition meets with the core mode and cladding modes of the same direction to transmission, which features leads to the resonant wavelength and amplitude of the LPFG are very sensitive to changes in the external environment [1-3].

There are several fabrication methods of LPFG, like the amplitude mask method, mechanical microbend method[4], corrosion groove method, local heating method, et al. Local heating method have three different branches: arc discharge heating method, point by point writing method (high-frequency CO_2 laser writing)[5-7], fused method[8-10].

This paper is based on electric heating fused method produced LPFG, then we take mode analysis to the cone area through the COMSOL, and launched the experimental research of temperature sensing characteristics of the LPFG.

II. MODE ANALYSIS OF CONE AREA

As shown in Figure 1, We have established tapering region model, take analysis of a half of the cone area (cone area for O_1 plane is symmetrical, so we just analyze half of the cone area). We set the length of the entire cone area 1600μm, half the length of cone area 800μm. Select O_1, O_2, O_3, O_4, O_5, O_6, O_7, O_8, O_9 nine sections, each section is circular, space is 100μm. We simulated mode of O_1, O_5, O_9 three sections. Diameters of O_1, O_5, O_9 three sections is 65.00μm, 82.58μm, 125.00μm.△S is the largest collapse depth, L=1600μm is collapse area length, Z is horizontal coordinates relative to the plane O_9. $Z(O_1)$=800mm, $Z(O_3)$=600mm, $Z(O_5)$=400mm, $Z(O_7)$=200mm, $Z(O_9)$=0mm.

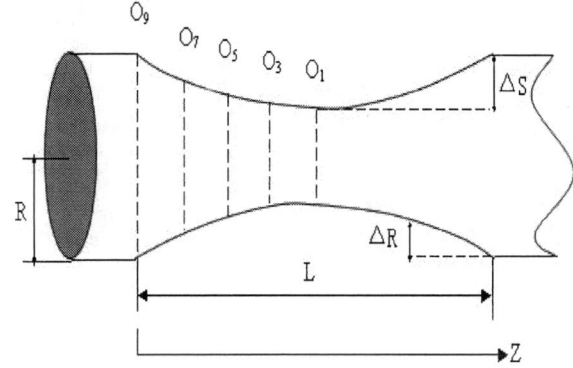

Figure 1. Fiber cone zone structure

The simulation results are shown in Figure 2:

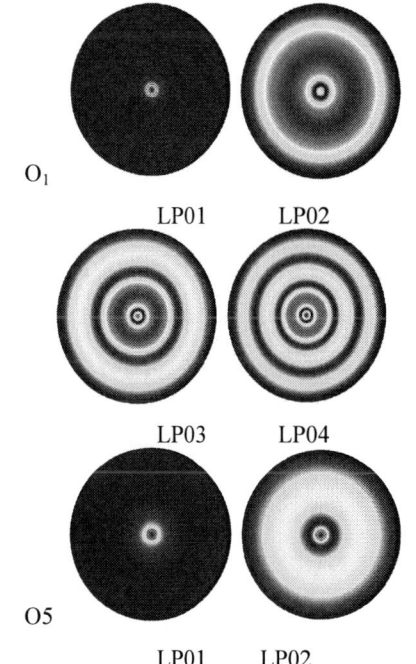

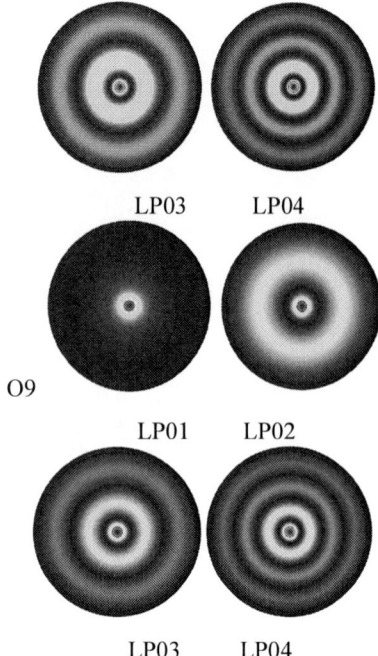

LP03   LP04

O9

LP01   LP02

LP03   LP04

Figure 2.   The change of the mode field section

Contrast O_1, O_5, O_9 sections mode field distribution at the same mode. We can see the energy of the core is diffused outward, the energy is bounded inside the core. After pulling cone O_1 and O_5's energy is diffused from core to cladding. Comparing different mode field, we can find that, with the fused biconical taper region's degree of collapse distribution increases, the fiber's bound ability to the cladding mode is reduced.

Through the mode simulation at O_1, O_3, O_5, O_7, O_9, we find the effective refractive index of each section.

TABLE I.   FIBER CONE ZONE RADIUS PARAMETER OF EACH SECTION

	Cladding diameter/μm	Core diameter/μm
O_1	65.00	4.32
O_3	69.56	4.62
O_5	82.58	5.48
O_7	102.04	6.78
O_9	125.00	8.30

TABLE II.   REFRACTIVE INDEX PARAMETERS OF THE CONE AT EACH SECTION

	LP01	LP02	LP03
O_1	1.461284	1.460681	1.459875
O_3	1.461407	1.460745	1.460053
O_5	1.461818	1.460848	1.460385
O_7	1.462471	1.460912	1.460629
O_9	1.463135	1.460945	1.460765

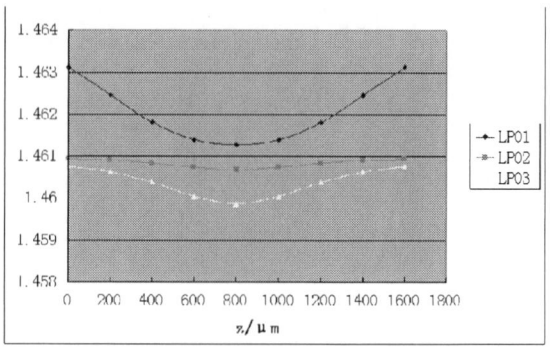

Figure 3.   The curve of cone modes changed with refractive index along the Z axis

As showed in figure 3, as the cone of fiber grating structure changes, the greater the depth of drawing, the smaller the effective refractive index .and we can find that the impact to the refractive index of first-order mod cladding is minimal is minimal, and the effective refractive index of the core changes in the fundamental mode is maximum .

III.   LPFG'S PREPERATION AND TEMPERATURE CHARACTERISTIC RESEARCH

Figure 4.   This experiment used the apparatus shown in Figure 4：

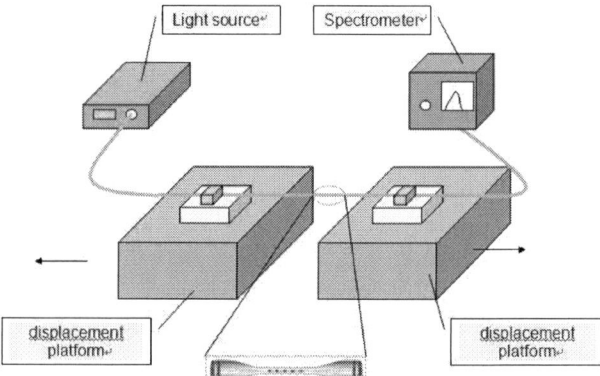

Figure 4 Schematic diagram of electric-heating making device

In this experiment we use the device shown in figure 5, have a research on LPFG temperature characteristics below 200□.

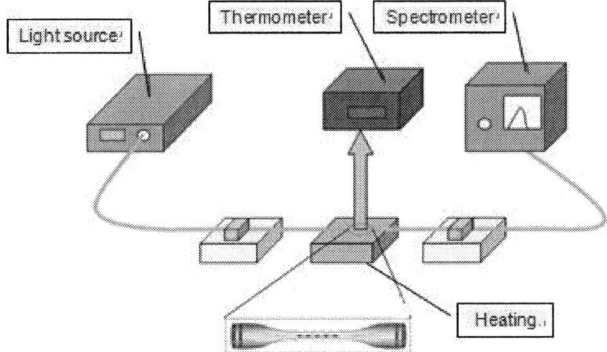

Figure 5.   Experiment device of temperature characteristics

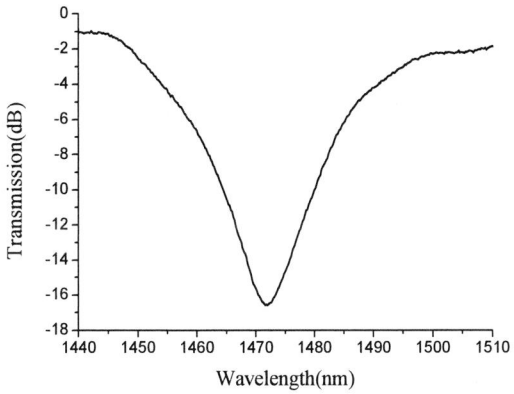

Figure 6.   Temperature characteristic's spectra

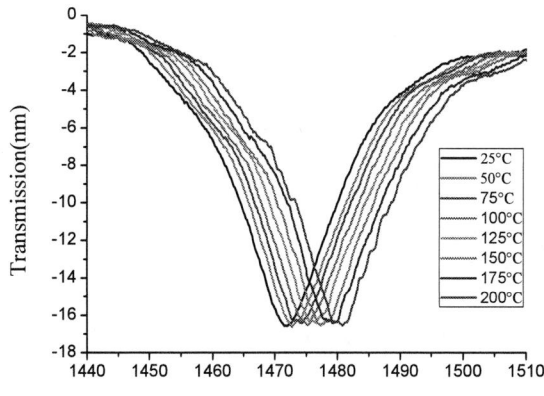

Figure 7 Curve of temperature drift

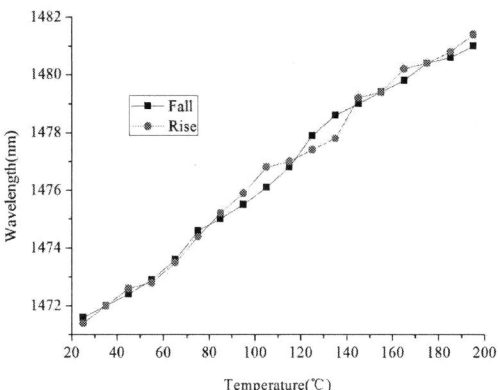

Figure 7.   Curve of resonance peak changed with temperature at heating progress and cooling progress

As shown in figure 6, the cycle of the fiber grating used in the experiment L=1600μm, resonance wavelength is 1472nm, resonant peak depth is -17dB.

Start from 25℃, mark the resonant peak's point every 25 ℃, until the highest temperature 200℃. After the increasing progress, shut down the furnace and mark the resonant peak's point every 25℃ when temperature lowing down. The change of spectra during heating progress is shown in figure 7. This figure record the spectra at 25℃、50℃、75 ℃、100℃、125℃、175℃ and 200℃, with temperature increasing, resonance wavelength drift to longer wavelength, and resonance peak depth increases a little. As shown in figure 8，the relation between resonance peak drift and change of temperature is basically linear, and temperature sensitivity is 62.8pm/℃, the heating progress and cooling progress are basically same.

IV.   CONCLUSION

We take some mode analysis through COMSOL, and find that, change the fiber structure, the mode and energy in the core are both diffusing outward. After that, we have took temperature experiment to LPFG below 200℃. And mark the resonance peak wavelength at different temperature from low to high and from high to low respectively. The resonance peak wavelength is changing linear with temperature, and temperature sensitivity is about 62.8pm/℃ .While in the case of using high-frequency CO_2 in ordinary fiber and photonic crystal fiber to write LPFG, the temperature sensitivity is 52pm/ ℃ and 3.9pm/ ℃ respectively/.In the view of single temperature, the LPFG based on electric heating method has high temperature sensitivity.

ACKNOWLEDGMENT

This work was supported by the National Natural Science Foundation of China (Grants Nos. 61107069, 41174161, 61227013 and 61377084), partially supported by the 111 project (B13015), to Harbin Engineering University. This paper is funded by the International Exchange Program of

978-1-4799-1215-5/13 $31.00 © 2013 IEEE

Harbin Engineering University for Innovation-oriented Talents Cultivation.

REFERENCES

[1] Vikram Bhatia, Ashish M. Vengsarkar. Optical fiber long~period grating sensors. Applied Physics Letters, 1996, 21(9): 692~694p

[2] Xuewen Shu, Lin Zhang, and I. Bennion. Sensitivity characteristics of long-period gratings. Journal of Lightwave Technology, 2002, 20(2): 255~266p

[3] Kersey A. D, Davis M. A, Patrick. Fiber grating sensors. J. of. Lightwave Technology, 1997, 15(8): 1442~1463p

[4] Savin S., Digoneet M. J. F., Kino G. S. et al. Tunable mechanically induced LPFG. Optics Letters, 2000, 25(10): 710~712p

[5] Davis D.D., Gaylod T. K., GLytsis E. N. et al. Long-period fiber grating fabrication CO_2 laser pulses. Electron. Letters, 1998, 34(3): 302~303p

[6] Davis D.D., Gaylod T. K., GLytsis E. N. et al. CO_2 laser-induced long-period fiber gating: spectral charactietics cladding modes and polarization independence. Electron. Letters, 1998, 34(14): 1416-1417p

[7] Yun-Jiang Rao, Yi-Ping Wang, Zeng-Ling Ran and Tao Zhu. Novel fiber optic sensors based on long period fiber gratings written by high-frequency CO_2 laser pulses. J. Lightwave Technolgoy, 2003, 21(5): 1320~1327p

[8] Min-Seok Yoon, Sung-Jae Kim, et al. Development of micro-tapered long-period fiber gratings written in tapered fibers with different diameters for enhancement of strain sensitivity. Proc. of SPIE, 2012, 8421: 1~4p

[9] Min-Seok Yoon, Hyun-Joo Kim, and Young-Geun Han. Long-period fiber grating inscribed in a tapered fiber. Advanced Photonics Congress. 2012,1~2p

[10] Min-Seok Yoon, Sangoh Park, Young-Geun Han. Simultaneous Measurement of Strain and Temperature by Using a Micro-Tapered Fiber Grating. Jouenal of lightwave technology, 2012, 30(8): 1

Influence of Si SBD P$^+$ Ring Junction Depth on ESD Robustness

Jinyu Dong[1], Jinghua Yin[12*], Shiyin Guan[1], Yue Li[1], Shuting Gao[1]

1 School of Applied Science, Harbin University of Science and Technology
2 Key laboratory of Engineering Dielectrics and Its Application, Ministry of Education, HUST, Harbin, China
*yinjinghua1@126.com

Abstract—**As the device size gets smaller and smaller, electrostatic discharge (ESD) has become an important factor affecting the reliability of semiconductor devices. For a Schottky barrier diode (SBD), the P$^+$ ring junction depth has significant impact on ESD robustness. Based on this, the model of a 1A-series SBD chip is established using SILVACO TCAD software in this paper, the ESD failure mechanism and the impact of junction pushing time on P$^+$ ring junction depth, especially on ESD robustness is also investigated by adding HBM mode pulse to anode. The results indicate that the SBD works well when the ESD test level is below 8 kV but failures at the test level of 8 kV because of the over current-induced heat failure; Besides, when time increases from 80 min to 160 min with the increment of 20 min, the junction depth also increases from 1.245 μm to 1.752 μm, at the same time, the current density becomes smaller and the capability of P$^+$ ring sharing electric field with Schottky barrier also gets strengthened respectively, SBD can get HBM-mode ESD robustness range from 5 kV to 20 kV, which means increasing of the junction depth can significantly enhance the HBM ESD robustness of the diode.**

Keywords- Schottky barrier diode; ESD; SILVACO TCAD; P$^+$ ring; junction depth

I. INTRODUCTION

The Schottky barrier diode (SBD) is a kind of single-pole and multiple carrier device working through the metal-semiconductor contact barrier [1]. Compared with the normal PN diode, the forward voltage drop of the SBD is lower since the barrier is lower. In addition, SBD is a multiple-carrier device with no extraction and reverse recovery problem of minority carrier, so its reverse recovery time is much smaller than the PN diode and is suitable for high frequency circuit and ultra high-speed switching circuit [2]. Other applications include rectifier circuit and power protection circuit and so on. While the small reverse voltage and high leakage current are the disadvantages of SBD. Electrostatic discharge (ESD) shows the charge transfer between two objects with different electrostatic potential [3], which can form high voltage, strong electric field and the instantaneous current. Electrostatic discharge can not only cause serious interference and injury to electronic equipment, and may also constitute a potential hazard reducing the operational reliability of electronic devices. Furthermore, with advances in technology and gradually shrinking of the device size, the anti-ESD performance of some devices decreases and ESD failure is more likely to happen. Most papers talk about the reverse characteristics of SBD, only limited publications have

presented the ESD threats of Schottky diodes [4], let alone the papers about the effect of P$^+$ ring junction depth on ESD robustness. Therefore, the ESD properties of Schottky barrier diode are studied in this article and the influence of P$^+$ ring junction depth on that has also been analyzed.

II. DEVICE AND EXPERIMENTAL

In this section, the device (SBD) and the experiment procedure are presented. The process flow and structure of SBD are described in sub-section II-A. The modeling and simulation with SILVACO TCAD is showed in sub-section II-B.

A. Process Flow of the Schottky Barrier Diode and Structure Parameter

Currently, the semiconductor material commonly used to fabricate a SBD are Si[5], diamond[6] and SiC[7] and so on. The mature growth process of the material and the matching technology of the device are the advantages of Si when comparing with other materials. Therefore, a 1-A series SBD with Si substrate is used in this paper.

The actual process flow of the Si Schottky barrier diode is showed in Fig. 1

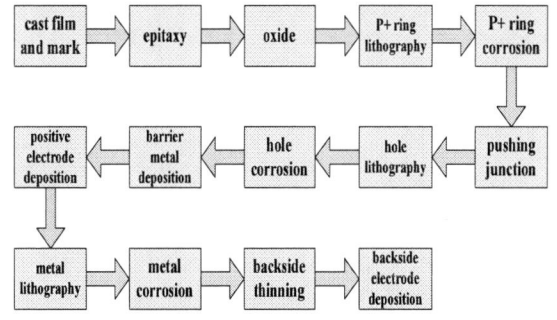

Figure 1. Process flow map of Si-based SBD

Silicon (100) is used as the substrate material. The oxide process lasts for 3h at the temperature of 1000℃. PN junction is formed after that the Boron ions implantation to n-type Si epitaxial layer with the total dose of 2×10^{15} cm^{-2} is performed at 60KeV and the annealing process is done in the Argon ambient at 1050℃ for 80 min. Titanium is deposited on epitaxial layer as the barrier metal to form the Schottky contact

978-1-4799-1215-5/13 $31.00 © 2013 IEEE

after annealing at 470℃. Aluminum deposited on the top and bottom of SBD works as electrode. The cross-sectional view of Si-based Schottky barrier diode is shown in Fig. 2. The Schottky contact is referred as the anode electrode, while the Ohmic contact is defined as cathode electrode. The structure parameters are listed in Table Ⅰ.

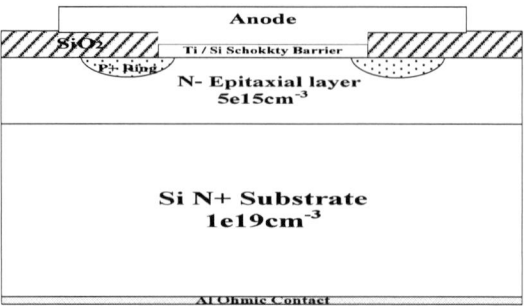

Figure 2. Schematic of Si-based SBD cross-section

TABLE I. STRUCTURE PARAMTERS OF SBD

Structure	Parameter	Value
Substrate	Length	750μm
	Width	750μm
	Thickness	255μm
	Concentration	1×10^{19} cm^{-3}
Epitaxial layer	Thickness	5μm
	Concentration	5×10^{15}cm^{-3}
P+ ring	width	30μm
	Distance from edge	100μm

B. TCAD Simulation Procedure

The experiment in this article is based on SILVACO TCAD instrument.

First, the process simulation framework ATHENA is used to simulate the process flow of the SBD as mentioned in section Ⅱ-A. The structure profile of SBD is obtained as shown in Fig. 3 after the simulation.

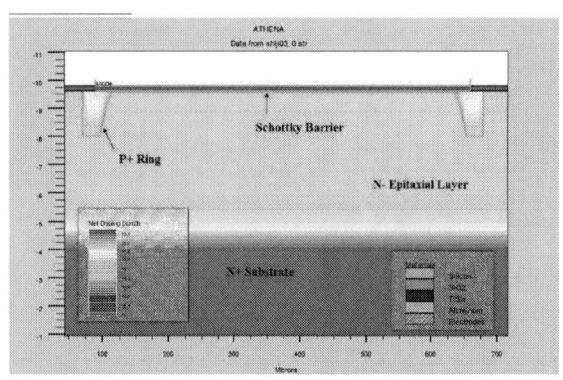

Figure 3. Amplified structure of SBD in ATHENA

In Fig. 3, the frame in the left bottom shows the orders of magnitudes of net doping, and the concentration increases from bottom upward. Thus, it can be seen that concentration in P$^+$ guard ring is relatively higher than the epitaxial layer but lower than the substrate, which is in accord with the design index of parameters. The P$^+$ ring junction depth can be extracted from the simulation results. When time varies from 80 min to 160 min with increments of 20 min, the PN junction depth also changes and gets the values of 1.245 μm, 1.380 μm, 1.504 μm, 1.626 μm and 1.752 μm.

ATLAS is the device simulation module used to simulate electrical, optical and thermal properties of semiconductor devices. First, the SBD structure file obtained from the process simulation is imported into ATLAS module, and a HBM-mode pulse is added to the anode electrode of SBD, then the temperature and electric field distribution are observed to judge whether ESD failure has occurred at SBD. At last, changes of the anti-ESD performance and the junction depth are viewed under the condition of junction pushing time.

III. RESULTS AND ANALYSIS

A. ESD Failure Analysis

Electrostatic discharge failure of SBD is caused by over current-induced heat failure, because the local temperature rise exceeds the melting point of the material in the case of large ESD current in discharge circuit with low impedance, which leads to partial melting of material and the components failing. Taking SBD with junction pushing time of 80 min for instance, 5kV HBM-mode pulse is imported to anode to test the pulse response. According to MIL-STD-883C method 3015.7 HBM model [8], a capacitor C1 with 100pF and a resistance R1 with 1500 Ω are placed in the HBM test circuit. The pulse rises in less than 10 ns and reduces in approximately a range of 120 ns~180 ns. The pulse curve obtained in ATLAS is revealed in Fig. 4

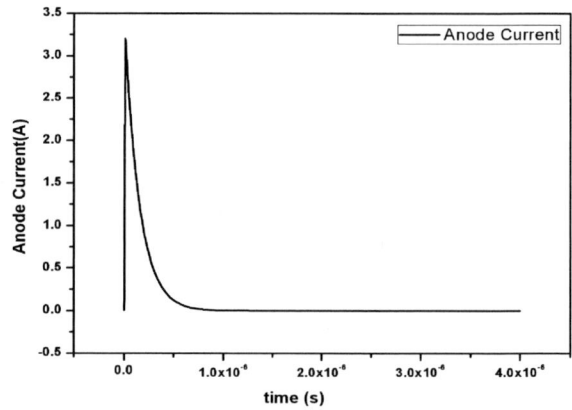

Figure 4. HBM instantaneous pulse current curve

Fig. 4 illustrates the HBM-mode pulse used in the simulation of SBD. The current rises to 3.2 A in nearly 2ns, then it goes down in 170ns and gradually tends to 0 A after 170 ns. The pulse is proven to meet the MIL-STD-883C

978-1-4799-1215-5/13 $31.00 © 2013 IEEE

criterion and it can be used to simulate the ESD robustness of SBD.

Fig. 5 and Fig. 6 illustrate the temperature and electric field distribution after the HBM-mode pulse of 5 kV is added to SBD.

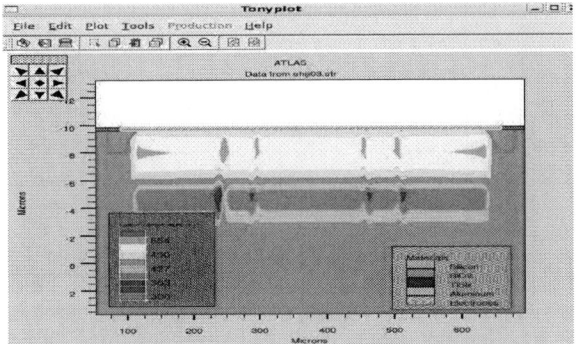

Figure 5. Temperature distribution under the HBM pulse of 5 kV

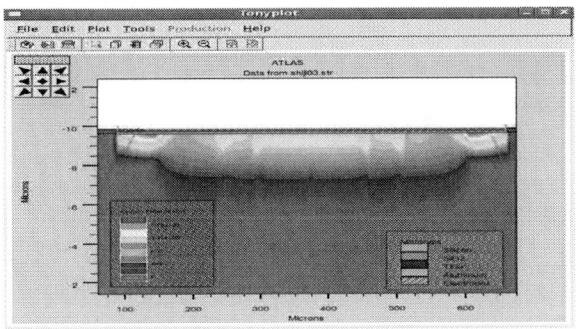

Figure 6. Electric field distribution under the HBM pulse of 5 kV

As shown in Fig. 5 and Fig. 6, the lattice temperature rises to 554 K and the electric field intensity peak arrives at 1.75×10^3 V/cm at the condition of 5 kV HBM pulse. The high temperature areas are mainly located in the epitaxial layer under the barrier area, yet the electric field distributes in both Schottky barrier area and P^+ guard ring, and is higher in P^+ ring. As the introduction of P^+ ring, part of the electric field is assigned to P^+ ring, thus leading to reduction of electric field in Schottky barrier area. In addition, over current-induced heat failure doesn't take place in SBD as a result of temperature being lower than the intrinsic temperature 1688 K of Si material, namely, SBD gets passed through the HBM-mode test of 5 kV.

The simulation results are showed in the following Fig. 7 and Fig. 8 after adding a pulse of 8 kV to SBD.

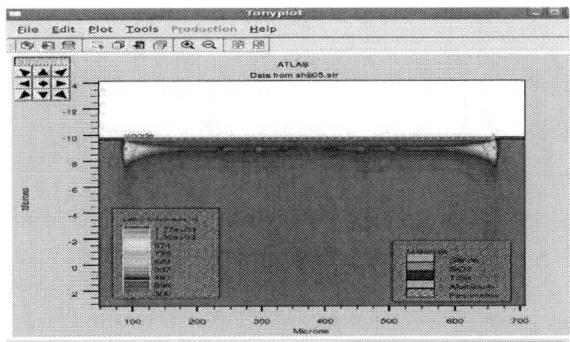

Figure 7. Temperature distribution under the HBM pulse of 8 kV

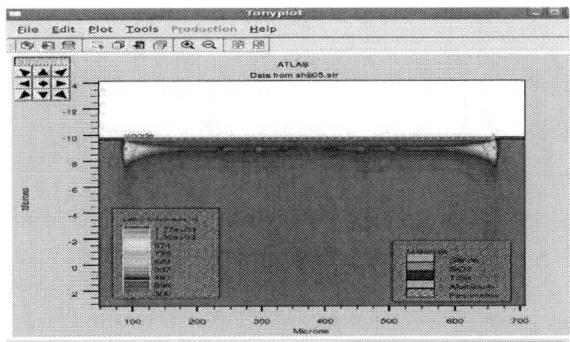

Figure 8. Electric field distribution under the HBM pulse of 8 kV

As shown Fig. 7 and Fig. 8, temperature and electric field distributions have changed in SBD since HBM pulse arrives at 8 kV. The temperature reaches 1770 K and heat distribution mainly focuses on the P^+ protection ring area. The temperature also rises in barrier area and doesn't exceed the intrinsic temperature of Si. Therefore, it can preliminarily be judged that over current-induced heat failure occurs in P^+ ring. Besides, electric field distribution is concentrated in Schottky barrier as well as areas below it. As the P^+ ring loses the capability of sharing the electric field induced by thermal damage, the electric field is wholly added to the barrier area. However, no field-induced failure takes place as the electric field strength is 2.30×10^3 V/cm. Thus, failure occurs in PN junction first because over current-induced heat failure is more likely to occur in the junctions with small areas.

B. Junction Depth Effect on ESD Robustness

As mentioned above, SBD gets through the test of 5 kV HBM pulse but doesn't pass the test of 8 kV, something should be done to improve the ESD robustness of SBD. Due to the main factors affecting SBD failure is power density, which is interrelated with the device material, pulse amplitude and the operative area such as PN junction area, therefore the increase of the junction area can be considered to enhance ESD robustness. And junction depth is one of the factors that have influence on the junction area. Based on this, the junction depth is scanned to study the changes of ESD robustness for SBD.

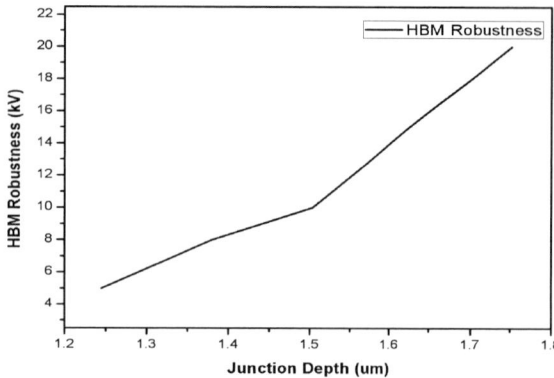

Figure 9. HBM-mode ESD robustness versus junction depth of SBD

The curve that HBM-mode ESD robustness changes almost linearly with junction depth is demonstrated in Fig. 9. The curve can be divided into three parts according to the slope. It can be seen that the slope in part A is larger than that in part B but smaller than that in part C, that is, the stress level increases faster when the junction is larger than 1.5 μm.

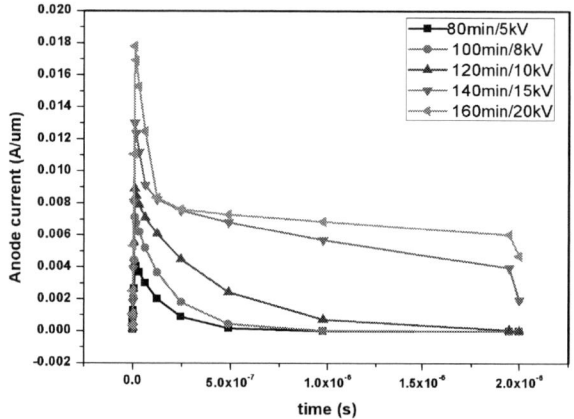

Figure 10. HBM test levels versus different junction pushing time

Fig. 10 reveals the HBM pulse tests of different levels that SBD can pass through. As the device width is 750 μm, the peak pulse current obtained by multiplying 750 μm with the current density are 3.20 A, 5.33 A, 6.67 A, 10.00 A and 13.33 A, which correspond with 5 kV, 8 kV, 10 kV, 15 kV and 20 kV, respectively. The PN junction area increases with the increasing of junction depth. And for the same current, the current density will decrease, which leads to lower over current-induced lattice temperature, smaller possibility of device failure and higher ESD robustness that the device can get.

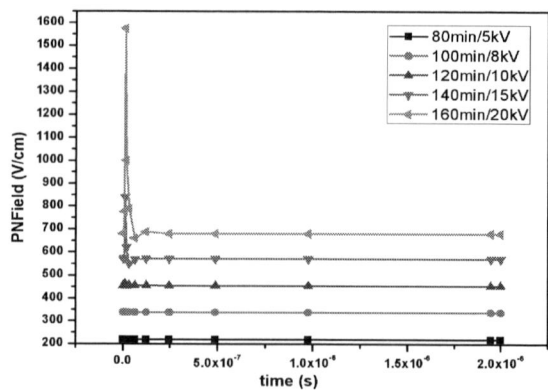

Figure 11. Time-dependent PN junction electric field curves under different test levels

The time-dependent PN junction electric field variation curves under different test levels are also obtained as shown in Fig. 11. It can be seen that the variation trend is consistent with that of the current density. At the same time, the higher the test level is, the larger the electric field in the PN junction becomes. There is a progressive increase in the 20 kV HBM test for SBD with junction pushing time of 160 min. The electric field in Schottky barrier area increases from 800 V/cm to nearly 3000 V/cm with the rise of HBM-mode test level. Compared with PN junction electric field, it can be concluded that the capability of P^+ ring sharing electric field with barrier gets strengthened with the increasing of test stress. The breakdown doesn't happen easily due to higher barrier of PN junction. Thus the ESD robustness of SBD can be improved.

IV. CONCLUSION

In this article, the ESD failure mechanism and the variation of ESD robustness in SBD is analyzed based on different junction pushing time and junction depth. Results is summarized as below.

1. The ESD robustness of the SBD is good when the test level is below 8 kV and turns to bad at the test level of 8 kV because of the over current-induced heat failure.

2. When the test stress is 5 kV, the lattice temperature rises to 554 K and the electric field intensity peak arrives at 1.75×10^3 V/cm. The high temperature areas are mainly located in the epitaxial layer under the barrier area, yet the electric field distributes in both Schottky barrier area and P^+ guard ring, and is higher in P^+ ring.

3. When the test level arrives at 8 kV, the temperature reaches 1770 K and heat distribution mainly focuses on the P^+ protection ring area. The electric field strength is 2.30×10^3 V/cm and mainly distributes in Schottky barrier as well as areas below it.

4. With the increasing of junction pushing time from 80 min to 160 min with increments of 20 min, SBD P^+ guard ring junction depth also increases from 1.245 μm to 1.752 μm. The PN junction area gets increased, which leads to lower over current-induced lattice temperature as well as the strengthened capability of P^+ ring sharing electric field with barrier, and

finally the HBM-mode ESD robustness of SBD gets increased from 5 kV to 20 kV respectively, which means that the increase of P^+ ring junction depth is beneficial for ESD robustness.

ACKNOWLEDGMENT

The authors thank the science and technology research project of Heilongjiang Province Education Department (Grant 11551100) for financial support of this work.

REFERENCES

[1] Zhong Yi, Xu Guolei, and Ou Hongqi et al, "Study on high voltage SBD with improved ESD performance,"[J] Microelectronics, 2012, 42(2): 273.

[2] I. Kang, S. Kim, W. Bahng et al, "Accurate extraction method of reverse recovery time and stored charge for ultrfast diodes,"[J] IEEE Trans on Power Electronics, 2011,1(99):1-4.

[3] Wang Wenshuang, Xu Shaohui and Wang Xiaoqiang, "Analysis of electrostatic discharge model for electronic components," [J] Electronics Quality, 2005:24–25,

[4] S. H. Chen, A. Griffoni and P. Srivastava et al, "HBM ESD robustness of GaN-on-Si Schottky diodes,"[J] IEEE Trans on Device and Material Reliablity, Dec 2012, 12:589-590.

[5] M. Markmann, E. Neufeld and A. Sticht et al, "Excitation efficiency of electrons and holes in forward and reverse biased epitaxially grown Er-doped Si diodes," Appl. Phys. Lett, 2001, 78:210-214.

[6] Bohr Ran Huang, Wen Cheng Ke and Jung Fu Hsu et al, "Successive current-voltage measurements of a thick isolated diamond film,"[J] Materials Chemistry and Physics, 2001, 72(2): 214-217.

[7] F. Roccaforte, F. La Via and S. Di. Franco et al. "Reduction of the power dissipation in silicon carbide Schottky rectifiers by a dual-metal planar structure, " Appl. Phys. Lett, 2002, 81:1125-1129.

[8] A. Tataroglu. "Electrical and dielectric properties of MIS Schottky diodes at low temperatures,"[J] Microelectronic Engineering, 2006, 83(11): 2551-2557.

A novel pre-amplifying and latching comparator

Rongke Ye[1*], Rongbin Hu[2]

1 Sichuan Institute of Solid State Circuits, Chongqing 400060, China
2 Science and Technology on Analog Integrated Circuit Laboratory
No.14 Huayuan Road, Economic and Technological Development Zone, Chongqing 400060, China
yerongke2013@126.com

Abstract—**A novel pre-amplifying and latching comparator is presented, which has several advantages over the traditional one. At first, its preamplifier has three stages, which provides larger gain and wider bandwidth at the same time. Secondly, the outputs of the latch are directly driven by the preamplifier, which will reduce the latching time. Additionally, a pair of capacitors is inserted between the latch and the preamplifier, which reduces the offset of the preamplifier further. The simulated results show that the proposed comparator has an analog bandwidth of 5GHz with sampling rate of 2GHz, which is better than the traditional one.**

Keywords-comparator; ADC; switch; latch; amplifier

I. INTRODUCTION

As a basic structural cell, comparators are necessary for ADC, which decide the performances of the ADC [1-7]. In order to design a high performance ADC [8-11], the first thing we should do is to design a perfect comparator.

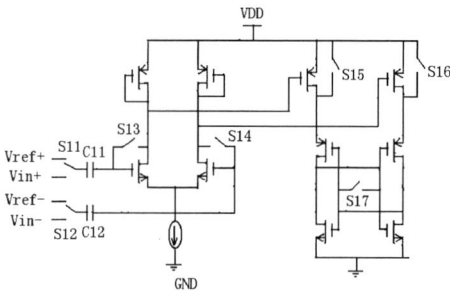

Figure 1. The traditional pre-amplifying and latching comparator

As shown in figure 1, it is the traditional pre-amplifying and latching comparator, which consists of a preamplifier followed by a latch. The preamplifier amplifies the minus difference between the signal and the reference, and the latch latches the compared result of the signal and the reference.

The work process of the comparator consists of two steps. At the first step, the two-throw switches S11 and S12 are thrown to the reference signals Vref+ and Vref- and the two resetting switches S13 and S14 are closed, so the reference signals are sampled onto the capacitor C11 and C12 and the preamplifier is reset. Switches S15 and S16 are closed and S17 is open; the latch is in its latching state. At the second step, two-throw switches S11 and S12 are thrown to the analog signals and switches S13 and S14 are open. The analog signal Vin+-Vin- is compared with the reference Vref+-Vref- which

has been sampled at the capacitor at the first step. The difference is amplifying by the preamplifier. Meanwhile, switches S15 and S16 are open and switch S17 are close. The latch acts as a load for the preamplifier. At the end of the second step, the switches S15 and S16 begin to close and the switch S17 begins to open. The latch begins to latch the state of this time and a new cycle begins.

There are several disadvantages connected to the traditional pre-amplifying and latching comparator. At first, the preamplifier has only one stage which has limited gain and narrow bandwidth. Secondly, the preamplifier doesn't drive the outputs of the latch directly at the first step, which will increase the latching time.

In this paper a newly novel pre-amplifying and latching comparator will be presented. It has several advantages over the traditional one. At first, its preamplifier has three stages, which provides larger gain and wider bandwidth at the same time. Secondly, the outputs of the latch are directly driven by the preamplifier, which will reduce the latching time. Additionally, a pair of capacitors is inserted between the latch and the preamplifier, which reduces the offset of the preamplifier further.

II. THE BLOCK DIAGRAM OF THE PROPOSED COMPARATOR

As shown in figure 2, it is the block diagram of the proposed comparator. In order to suppress common noise such as power supply stir, the comparator is designed to be a fully differential structure. As shown in figure 1, the comparator is symmetric about the middle horizontal line. Because of the differential structure, the signals applied into the comparator are also differential. As shown in figure 2, they are Vref+ and Vref-, and Vin+ and Vin-.

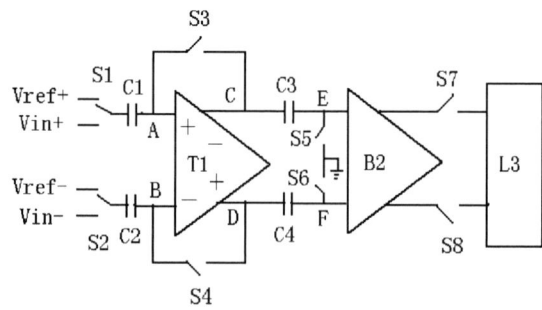

Figure 2. The architecure of the proposed comparator

As shown in figure 2, the proposed comparator consists of a pair of sampling switches S1 and S2, a pair of sampling capacitors C1 and C2, a preamplifier T1, a pair of reset switches S3 and S4, a pair of inter-stage capacitors C3 and C4, a pair of short switches S5 and S6, a buffer B3, and a pair of latch-front switches S7 and S8, and a latch.

The working process of the comparator consists of two steps. At step one, the pair of sampling switches S1 and S2 are thrown to Vref+ and Vref-, respectively, and the reference signals are sampled on the capacitors C1 and C2. The reset switches S3 and S4 are closed, which reset the preamplifier T1. The pair of short switches S5 and S6 is closed, so the capacitors C3 and C4 sample an equal voltage on them. And, the switches S7 and S8 are open; that disconnect the latch from the front circuits.

At the second step, the pair of switches S1 and S2 are thrown to Vin+ and Vin- respectively. Switches S3 and S4, S5 and S6, and S7 and S8, all are open. Because at the first step the reference signals are sampled on capacitors C1 and C2, the voltages at A and B at this step are

$$V_A = V_{in+} - V_{ref+}, \tag{1}$$

$$V_B = V_{in-} - V_{ref-}, \tag{2}$$

respectively. So the signal difference of the inputs of the preamplifier T1 is

$$V_A - V_B = (V_{in+} - V_{in-}) - (V_{ref+} - V_{ref-}). \tag{3}$$

From equation (3), we find that the preamplifier T1 compares the differential analog signal (Vin+-Vin-) with the differential reference signal (Vref+-Vref-) and amplifies their difference. Assuming the gain of the preamplifier is G, we get the voltage difference between the outputs of the preamplifier is

$$V_C - V_D = G(V_A - V_B) \tag{4}$$

where V_C and V_D are the voltage of node C and D, respectively. Because at the first step, the pair of capacitors C3 and C4 has sampled an equal voltage, the voltage difference between node E and F is the same, which is

$$V_E - V_F = G(V_A - V_B). \tag{5}$$

The buffer B2 has no gain characteristic or has a gain of one, so the voltage difference at the outputs of the buffer or the inputs of the latch is the same as given in equation (5). For the latch, we know that the finally settled voltage is related to the original voltage difference between its inputs as below

$$V_t = \Delta V e^{\frac{t}{\tau}}, \tag{6}$$

where V_t is the finally settled value, ΔV is the initial voltage difference between the inputs of the latch, and τ is the time constant of the latch. Adjusting equation (6), we get

$$t = \tau \ln \frac{V_t}{\Delta V}, \tag{7}$$

where t is the settle time of the latch. From equation (7), we see that if ΔV is a minimum, the settle time will be maximal, which means the latch gets into meta-stable situation. If the inputs difference of the preamplifier is ΔV, and its gain is G, we get

$$t = \tau \ln \frac{V_t}{G \Delta V}. \tag{8}$$

From equation (8), we see that the voltage difference at which the meta-stable condition is turned on is reduced by G times if the preamplifier is used.

The function of the buffer is to separate the latch from the preamplifier so that the switch action of the preamplifier won't influence the latch process. The inter-stage capacitors C3 and C4 are used to cancel the offset voltage of buffer B2.

III. Circuit Design

A. The switches and the capacitors

Normally, the switch is made of NMOS transistor because it is much faster than PMOS. But in special case, the combine of NMOS and PMOS is needed, which provides lower pass-through resistance. What kind of capacitor is used depends on the process provided by the foundry. As given below, the transistor-level detail of every block in figure 2 will be given.

B. The Amplifier

As shown in figure 3, it is the transistor-level circuit of the preamplifier T1 and the pairs of switches S3 and S4. As given before, in order to reduce the meta-stability, the gain of the preamplifier should be large. Meanwhile, for the comparator to be high-speed, a large bandwidth of the preamplifier is required.

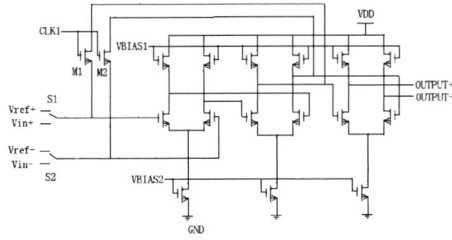

Figure 3. The proposed preamplifier

In order to get these two goals at the same time, the multi-stage structure is adopted. As shown in figure 3, the preamplifier consists of three stages, of which each has a NMOS load. There is no large pole in the signal path, so the bandwidth is very large. Furthermore, because many stages are used, the large gain can be obtained at the same time. The pair of switches S3 and S4 is made of NMOS' M1 and M2, because the NMOS is faster than PMOS transistor. The pair of switches is controlled by the clock signal CLK1.

C. The buffer

As shown in figure 4, it is the transistor-level circuit of the buffer B2, the pair of capacitors C3 and C4, and the pair of switches S6 and S7. The Buffer B2 is composed of two identical source followers, which controlled by the clock signal CLK1. When CLK1 is high, the inputs of the buffer are shorted to a constant voltage and the capacitors C3 and C4 are discharged; when CLK1 is low, the buffer work. The right terminal of capacitors C3 and C4 are connected to the outputs of the preamplifier.

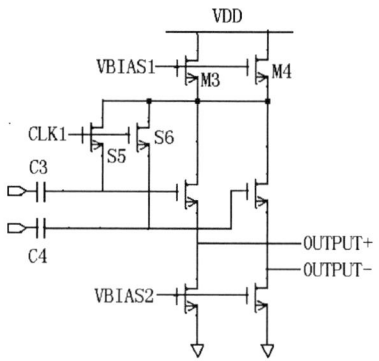

Figure 4. The proposed buffer

D. The Latch

As shown in figure 5, it is the transistor-level circuit of the latch L3 and the pair of switches S7 and S8. When CLK2 is at high voltage level, and CLK1 low, the latch is disabled. The up node G and down node H are connected together by transistor M5.

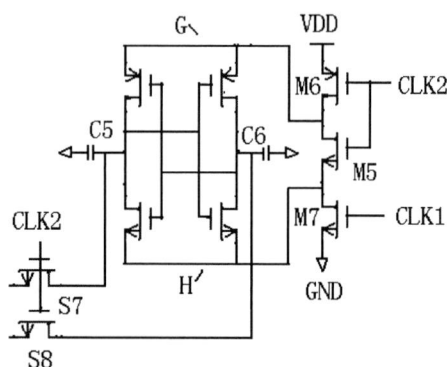

Figure 5. The proposed latch

The preamplifier drives the node capacitors C5 and C6. When CLK 1 is at high voltage level and CLK 2 low, the latch is turned on. The up node G is connected to VDD by PMOS M6 and the down node H to GND by NMOS M7. Switches S7 and S8 are open, which disconnected the latch from the preceding circuit. The latch begins to latch the state at the node capacitors C5 and C6.

IV. THE SIMULATION

As shown in figure 6, they are the simulated waveforms of the proposed pre-amplifying and latching comparator with a difference analog signal of sin wave at the inputs and the difference reference to be 2mV. The sin wave has amplitude of 300mV and frequency of 300MHz. the clock frequency is 1GHz. As shown in figure 5, wave a is the sin wave of the difference analog signal, wave b is the compared result of the comparator.

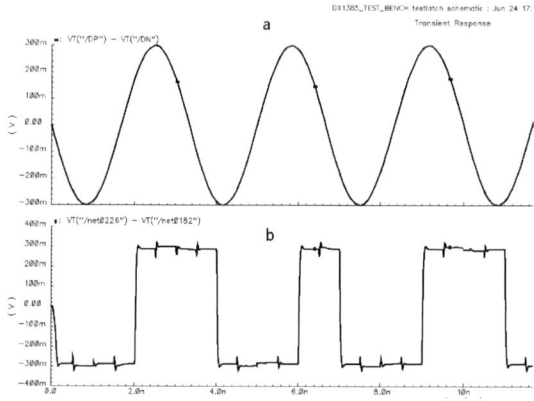

Figure 6. The simualted waveforms

Further simulation show that the proposed comparator has better performances than the traditional one. In table 1, we summarize the performances of the proposed and the traditional pre-amplifying and latching comparator.

TABLE I.   THE SUMMARY OF THE PROPOSED COMPARATOR

Performances	The proposed	The traditional
Process	0.35um CMOS	0.35um CMOS
Analog bandwidth	5GHz	2GHz
Sampling rate	2GHz	1GHz
Power supply	3.3V	3.3V
Power consumption	0.1mW	0.15mW
Area	50um^2	80um^2

V. CONCLUSION

In this paper a newly novel pre-amplifying and latching comparator will be presented. It has several advantages over the traditional one. At first, its preamplifier has three stages, which provides larger gain and wider bandwidth at the same time. Secondly, the outputs of the latch are directly driven by the preamplifier, which will reduce the latching time. Additionally, a pair of capacitors is inserted between the latch and the preamplifier, which reduces the offset of the preamplifier further. The simulated results show that the proposed comparator has an analog bandwidth of 5GHz with sampling rate of 2GHz, which is better than the traditional one.

ACKNOWLEDGMENT

This work was financially supported by the national key laboratory foundation (9140C090104120C09032). The research work was done at Analog Integrated Circuit Laboratory on Science and Technology. The CAD tools and instruments of the laboratory helped the authors a lot. The authors thank RuZhang Li, Guangbin Chen, Yuxin Wang, Dongbin Hu, and other members of Analog Integrated Circuit Laboratory on Science and Technology.

REFERENCES

[1] J. Li and U. Moon "A 1.8 V 67 mW 10-bit 100 Ms/s pipelined ADC using time shifted CDS technique", IEEE J. Solid-State Circuits, vol. 39, no. 9, pp.1468 -1476 2004.

[2] B. Lee , B. Min , G. Manganarom and J. W. Valvano "A 14-b 100-MS/s pipelined ADC with a merged SHA and first MDAC", IEEE J. Solid-State Circuits, vol. 43, no. 12, pp.2613 -2619 2008.

[3] P. Wu , V. Luen and H. Luong "A 1 V 100-MS/s 8 bit CMOS switched-opamp pipelined ADC using loading free architecture", IEEE J. Solid-State Circuits, vol. 42, no. 4, pp.730 -738 2007.

[4] J. Arias , V. Boccuzzi , L. Enriquez , D. Bisbal , M. Banu and J. Barbolla "Low-power pipeline ADC for wireless LANs", IEEE J. Solid-State Circuits, vol. 39, no. 8, pp.1338 -1340 2004.

[5] B. Min , P. Kim , F. W. Bowman III, D. M. Boisvert and A. J. Aude "A 69 mW 10-bit 80-MSample/s pipelined CMOS ADC", IEEE J. Solid-State Circuits, vol. 38, no. 12, pp.2031 -2039 2003.

[6] D. Kurose , T. Ito , T. Ueno , T. Yamaji and T. Itakura "55-mW 200-MSPS 10-bit pipeline ADCs for wireless receivers", IEEE J. Solid-State Circuits, vol. 41, no. 7, pp.1589 -1595 2006.

[7] P. Wu , V. Luen and H. Luong "A 1 V 100-MS/s 8 bit CMOS switched-opamp pipelined ADC using loading free architecture", IEEE J. Solid-State Circuits, vol. 42, no. 4, pp.730 -738 2007

[8] Hu Rongbin, Tang Jie, "A novel bootstrapped switch," in 2012 2nd International Conference on Consumer Electronics, Communications and Networks, CECNet 2012 - Proceedings, p 1545-1547, 2012.

[9] Hu Rongbin, Liu Kun, Liang Cheng, Tang Jie, "A novel sampling and holding circuit for ADC," in MaApplied Mechanics and Materials, v 198-199, p 1241-1245, 2012.

[10] Hu Rongbin, Tang Jie, "A novel full differential double sampling circuit for ADC," in 2012 2nd International Conference on Consumer Electronics, Communications and Networks, CECNet 2012 - Proceedings, p 1548-1551, 2012, 2012

[11] Hu Rongbin, Liu Kun, Tang Jie, "A novel charge pumped clock generator for VLSI," in MaApplied Mechanics and Materials, v 198-199, p 1174-1178, 2012.

A kind of 3-bit flash ADC core

Rongbin Hu[1*], Rongke Ye[2]

1 Science and Technology on Analog Integrated Circuit Laboratory
2 Sichuan Institute of Solid State Circuits, Chongqing 400060, China
No.14 Huayuan Road, Economic and Technological Development Zone, Chongqing 400060, China
Hurongbin2000@126.com

Abstract—**A kind of 3-bit flash ADC core is presented, which adopts CMOS inverter as comparators and PLA encoding scheme, and is very suitable for standard digital CMOS process and easy to be integrated with digital circuits, such as CPU, memory, etc. The optimization of the 3-bit ADC core is given and the simulation shows that the proposed ADC core can achieve 2Gsps sampling rate while consuming only 0.56mW power.**

Keywords-comparator; ADC; inverter; comparator; encoder

I. INTRODUCTION

A high-speed ADC need corresponding high-speed comparators, track and hold circuits [1-7], clock generator [8], and encoders, so someone suggests designing a flash ADC using CMOS invertors as comparators [9]. This method is very suitable for realizing ADCs using standard digital CMOS process. But, from the recent reports, for the ADCs designed by this method, the encoder circuits are too old to match the ultra high speed comparators made by an inverter [10]. For these kinds of encoders, the operation rate is very low. Furthermore, these kinds of encoder will become very large as the resolution of the ADCs increases. For these reasons, we propose a 3-bit flash ADC core, in which a comparator is designed using a CMOS inverter, and the encoder is made using the method suggested by Choudhury, et al. [11].

II. COMPARATOR

The comparator is made of a CMOS inverter, where the threshold voltage of the CMOS inverter is used as the comparing reference voltage of the comparator. As we know, when the input voltage is passing through the threshold voltage, the CMOS inverter will change its state, that is to say, the output changing from the high to low voltage, or vice versa. The ADC designed using this method has many advantages. At first, the reference generator and resistor voltage-divider network can be omitted, because the reference voltages are built-in the CMOS inverters, which is decided by the width-to-length ratio of PMOS gate with respect to that of NMOS gate. Secondly, the comparator made of inverter has only two MOS transistor, which is simpler than the traditional comparator, which means smaller power consumption. At last, because of the digital comparators made of CMOS inverters, the flash ADC can be built on the standard digital CMOS process, which means dramatic cost reduction. Besides, because of the smaller number of MOS transistor, the CMOS inverter is faster than the traditional comparator.

There are some problems for a high resolution flash ADC made of CMOS inverters acting as comparators. At first, as the resolution increases, the inverter at the highest bit has a PMOS or NMOS, which has a width-to-length ratio that is as large as several thousands, which consumes a large chip area because a large number of MOS transistors need to be connected together to realize a so large W-to-L ratio MOS. Secondly, as the resolution increases, the threshold of the inverter at the highest or lowest bit is more and more difficult to control. In order to avoid these kinds of problems, the proposed ADC using this method will have only 3 bits. However, this 3-bit ADC core can be used in multi-stage or pipeline ADC to realize higher resolution (bits).

In order to obtain digital waveforms that are suitable for a following encoder circuit to process, we cascade two inverters to form a comparator. The threshold of comparator is decided by the W-to-L ratio of the PMOS with respect to that of the NMOS of the front inverter. Using the standard digital 0.35um CMOS process PDK and through innumerous theory calculations and simulations, we get seven CMOS inverters whose thresholds distribute equally as shown in table 1 where

$$\beta_0 = \frac{W_p L_n}{W_n L_p} \qquad (1)$$

and W_p, L_p, W_n, L_p are the widths and lengths of PMOS and NMOS of the fist inverter, respectively.

TABLE I. THE COMPARATOR PARAMETERS OF THE ADC

comparator	β_0	Threshold voltage
T1	0.00020416666666	0.4V
T2	0.11666666666666	0.8V
T3	5.02040816326530	1.2V
T4	6.53061224489790	1.6V
T5	20.8163265306032	2.0V
T6	130.612244897914	2.4V
T7	45918.3670235461	2.8V

III. ENCODER

In the process of the traditional encoder design, the outputs of the comparators is made as the inputs of the encoder, the outputs of the ADC as the output of the encoder, then the logic relationships between the inputs and the outputs are enumerated and reduced by algebraic or chart methods. The traditional method has many disadvantages. At first, a large number of logic gates are needed to realize the outputs and inputs logic of the encoder, which will slow down the operation speed of the encoder. Secondly, for high resolution ADC, the input amount will be very large for the encoder. For example, for 8-bit ADC, the input amount is $2^8-1=255$. For such an input amount, using the traditional method, the circuit will be very complicated, having many stages, resulting in low operation speed. Consequently, the encoder will be the bottleneck of the whole ADC.

In the 3-bit ADC core, the encoding scheme proposed by Choudhury, et al, is adopted, which is very suitable for flash ADCs with CMOS inverters acting as comparators, according to our study. The encoding scheme fully takes advantages of PLA network. At first, a pre-encoding circuit composed of AND gates transform the outputs of the comparators into a digital pattern that is suitable for PLA processing. The digital outputs with this pattern are then encoded into the digital codes we want.

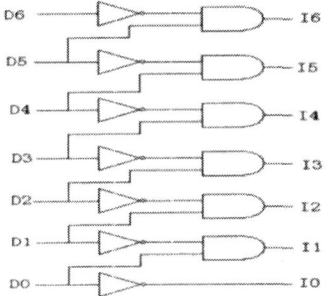

Figure 1. Pre-encoding circuit consisting of AND and NOT gates

The pre-encoding circuit of 3-bit flash ADC core is shown in figure 1. The output of the higher-bit comparator is first inverted, then together with output of the lower-bit comparator dealt by an AND gate. Consequently, there is only one bit with high level in all the outputs of the pre-encoding circuit at any time. For example, when the output bit pattern of the comparators is 0001111, the output bit pattern of the pre-encoding circuit will be 0010000, which is very suitable for PLA processing.

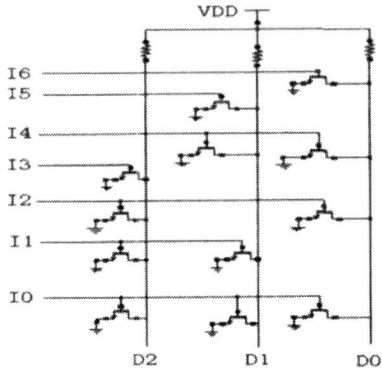

Figure 2. The PLA encoding stage

The PLA encoding circuit is shown in figure 2. After the processing of pre-encoding circuit, only one bit is high level, while all the others are low level, so the only high bit can be used to control corresponding gates of NMOS transistors in the PLA network to realize encoding purpose, as shown in figure 2.

From the analysis above, there are only three stages in the encoding circuit. Furthermore, the amount of the stages won't increase as the resolution of the ADC rises. As shown in figure 2, the last stage is PLA network, which is very fast, because the output level is decided by the voltage division made by a NMOS and a resistor. As a result, the speed of the encoding circuit is decided only by the pre-encoding circuit. As shown in figure 1, the pre-encoding circuit has only two gates in the signal path, so its speed is very fast. Consequently, because of the high ultra-fast speed, the encoding circuit will never be the bottleneck of the ADC. As shown in figure 2, it is the encoding table of the PLA-type encoder.

TABLE II.   THE ENCODING SCHEME

Rang of the input	Outputs of comparators	codes
0.0V to 0.4V	0000000	000
0.4V to 0.8V	0000001	001
0.8V to 1.2V	0000011	010
1.2V to 1.6V	0000111	011
1.6V to 2.0V	0001111	100
2.0V to 2.4V	0011111	101
2.4V to 2.8V	0111111	110

IV. SPEED OPTIMIZATION

A. The signal path

Being a digital circuit, the speed of the ADC is decided by the delay time of every stage on the signal path. The 3-bit ADC core has five stages, which consist of two stages of inverters which form a comparator, two stages of NOT and AND gates which form the pre-encoding circuit, and a stage of PLA encoding network. As was pointed out in the previous section, the PLA stage decides the output by the voltage division of a NMOS and a resistor, which is very fast and has little effect on the ADC performance. As a result, there are only four stages which will have effect on the speed of ADC. In order to speed up the ADC, the delay times of the four stages must be reduced. There are some relationships between these stages. When the width-to-length ratio of the last stage is increased to speed up itself, the load of the precious stage will be heavier, which slow down the speed of the precious stage. At last, we must find a optimal point, at which the ADC is fastest.

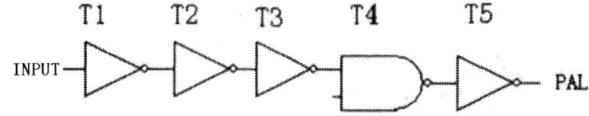

Figure 3. One signal path of ADC

There are a lot of signal paths in parallel. We just need analyze one signal path. As shown in figure 3, it is one signal path of the ADC. From figure 3, it is obvious that one signal path consists of four inverter and a NAND gate, because a AND gate consists of a inverter and a NAND gate.

B. Optimization of T1

As shown in figure 3, the input signal of T1 is the input signal of the whole ADC. In the actual application, there is a driving device in front of the ADC, so the width-to-length ratio of T1 can be prompted to speed up it. But the size of T1 is constrained by the performance of the comparator. For the comparator, the different threshold voltages are obtained by adjusting parameter β_0. In order to save chip area, W_n and W_p are always assigned the minimum size, which is 0.35um in our design.

C. Optimization of T2, T3, T4 and T5

T5 is connected to PAL network. An output of the PAL network is decided by voltage division of a NMOS and a resistor, which is very fast. In order to speed up T4, the minimum transistor size can be used for T5.

To optimize T4, we consider it as a inverter, because when both inputs are high levels, it can be considered as an inverter consisting of two NMOS' in series and two PMOS' in parallel. For two identical NMOS' in series, it is equivalent to a NMOS with double gate length. For two identical PMOS' in parallel, it is equivalent to a PMOS with double gate width. Whatever, T4 can be considered as a inverter. For an inverter, the delay time can be given by

$$\tau_{PHL} = \frac{C_{load}}{K_n(V_{DD}-V_{TN})} \times [\frac{2V_{TN}}{V_{DD}-V_{TN}} + \ln(\frac{4(V_{DD}-V_{TN})}{V_{DD}}-1)] \quad (2)$$

$$\tau_{PLH} = \frac{C_{load}}{K_p(V_{DD}-|V_{TP}|)} \times [\frac{2|V_{TP}|}{V_{DD}-|V_{TP}|} + \ln(\frac{4(V_{DD}-|V_{TP}|)}{V_{DD}}-1)] \quad (3)$$

In equation (2) and (3), τ_{PHL} and τ_{PLH} are the times taken by inverter when the output is falling from VDD to 50%VDD and from 0 to 50%VDD, respectively, at the condition that the input is a ideal step; C_{load} is the load capacitance, K_n, and K_p are the conductance factors of the PMOS and the NMOS, respectively; and V_{TN} and V_{TP} are the threshold of the PMOS and NMOS, respectively. For CMOS digital circuit, the front and the back circuits are correlative, so the expression for τ_{PHL} becomes:

$$\tau_{PHL} = (\frac{a_0 + a_n W_n + a_p W_p}{W_n}) \times (\frac{L_n}{\mu_n C_{ox}(V_{DD}-V_{TN})})$$

$$\times [\frac{2V_{TN}}{V_{DD}-V_{TN}} + \ln(\frac{4(V_{DD}-V_{TN})}{V_{DD}}-1)] \quad (4)$$

And the expression for τ_{PLH} becomes

$$\tau_{PLH} = (\frac{a_0 + a_n W_n + a_p W_p}{W_n}) \times (\frac{L_p}{\mu_p C_{ox}(V_{DD}-|V_{TP}|)})$$

$$\times [\frac{2|V_{TP}|}{V_{DD}-|V_{TP}|} + \ln(\frac{4(V_{DD}-|V_{TP}|)}{V_{DD}}-1)] \quad (5)$$

In equation (4) and (5)

$$a_0 = 2D_{drain}(C_{jswn}K_{eqn} + C_{jswp}K_{eqp}) + C_{int} + C_g \quad (6)$$

$$a_n = K_{eqn}(C_{jon}D_{drain} + 2C_{jswn}) \quad (7)$$

$$a_p = K_{eqp}(C_{jop}D_{drain} + 2C_{jswp}) \quad (8)$$

In equation (4) to (8), C_{jon} and C_{jop} are the diffusion capacitances of the NMOS and the PMOS, respectively; C_{jswn} and C_{jswp} are the zero-bias side capacitances; K_{eqn} and K_{eqp} are equal-division factors; μ_n and μ_p are the electronic and hole mobility, respectively; C_{ox} is the unit capacitance of the gate; W_n and W_p are the gate width of P-type and N-type transistors, respectively; and L_n and L_p are the gate length of P-type and N-type transistors, respectively.

From above, it is obvious that the delay time of a CMOS inverter have a minimum, which can be got using computer simulation.

V. THE SIMULATION

In figure 4, we compare the simulated waveforms of the 3-bit ADC core adopting our proposed encoding scheme (b) to that of the traditional one (a), when the input signal are 50ns skew wave.

From figure 4, it is clear that the output waveforms of the traditional ADC are distorted because of the bad encoding circuit, while the output waveforms of the proposed 3-bit flash core are still standard rectangular waves, which prove our 3-bit ADC core is better than the traditional one.

Further simulation shows that the proposed ADC core can achieve 2Gsps sampling rate while consuming only 0.56mW power. In table 3, a summary of the proposed 3-bit ADC core is given.

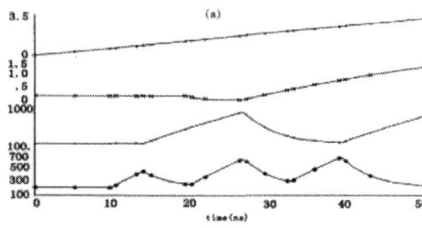

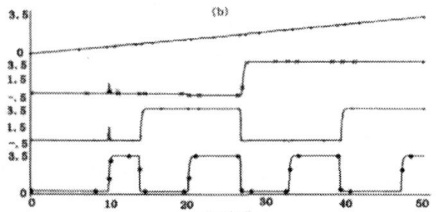

Figure 4.   The simualted waveforms

TABLE III.   THE SUMMARY OF 3-BTT ADC

resolution	3 bits	resolution
Power supply	3.3 V	Power supply
process	0.35um CMOS	process
Sampling rate	2Gsps	Sampling rate
area	0.0465mm2	area
Transistor count	90	Transistor count
Power consumption	0.56mW	Power consumption

VI.   CONCLUSION

A novel 3-bit ADC core is designed which combines new encoding scheme and inverter-comparator technique, solving the problem of encoding circuit as a speed bottleneck when a ultra high speed comparator is used, resulting a overall performance increment. The proposed ADC achieves 2Gsps sampling rate, while consuming only 0.56mW power. Because it combines high sampling rate and low power, the proposed ADC core can be used in pipeline ADC s and multi-stage ADCs to realize high resolution.

ACKNOWLEDGMENT

This work was financially supported by the national key laboratory foundation (9140C090104120C09032). The research work was done at Analog Integrated Circuit Laboratory on Science and Technology. The CAD tools and instruments of the laboratory helped the authors a lot. The authors thank RuZhang Li, Guangbin Chen, Yuxin Wang, Dongbin Hu, and other members of Analog Integrated Circuit Laboratory on Science and Technology.

REFERENCES

[1] H. Kobayashi, M.A. Zin, et al., "High-Speed CMOS Track/Hold Circuit Design", Journal of Analog Integrated Circuits and Signal Processing, Vol.27, pp.161-170, 2001.

[2] S. Shahramian, S. P. Voinigescu, and A. C. Carusone, "A 30-GS/sec track and hold amplifier in 0.13-µm CMOS technology," Proceedings of the Custom Integrated Circuits Conference, pp. 493-496, 2006.

[3] P. Vorenkamp and J. P. Verdaasdonk, "Fully bipolar, 120-Msample/s 10-b track-and-hold circuit," IEEE Journal of Solid-State Circuits, vol. 27, no. 7, pp. 988-992, 1992.

[4] Y. Gai, R. Geiger, and D. Chen, "Noise analysis in hold phase for switched-capacitor circuits," 2008 51st IEEE International Midwest Symposium on Circuits and Systems (MWSCAS), pp. 45-8, 2008/08/10.

[5] Hu Rongbin, Liu Kun, Liang Cheng, Tang Jie, "A novel sampling and holding circuit for ADC," in MaApplied Mechanics and Materials, v 198-199, p 1241-1245, 2012.

[6] Hu Rongbin, Tang Jie, "A novel full differential double sampling circuit for ADC," in 2012 2nd International Conference on Consumer Electronics, Communications and Networks, CECNet 2012 - Proceedings, p 1548-1551, 2012, 2012.

[7] P. J. Lim and B. A. Wooley, "A high-speed track-and-hold technique using a Miller hold capacitance," IEEE J. Solid-State Circuits, vol. 26, pp. 643-65, April 1991.

[8] Hu Rongbin, Liu Kun, Tang Jie, "A novel charge pumped clock generator for VLSI," in MaApplied Mechanics and Materials, v 198-199, p 1174-1178, 2012.

[9] Jincheol Yoo, Daegyu Lee, Kyusun Choi, Tangel A.Future - ready ultrafast 8bit CMOS ADC for system -on - chip applications[J]. ASIC/SOC Conference, 2001. Proceedings. 14th Annual IEEE international 12 - 15 Sept. 2001: 455 - 459.

[10] Shih - Chang Hsia,Wen - ChingLee. A very low - power flashA/D converter based on CMOS inverter circuit[J]. System - on - Chip for Real - Time Applications, 2005.Proceedings. Fifth International Workshop on 20 - 24, July 2005:107 - 110.. 2. Oxford: Clarendon, 1892, pp.68–73.

[11] Choudbury J, Cavanaugh C, Seetharaman G. An efficient encoding scheme for ultra - fast flash ADC[J]. Wireless Communication Technology, 2003. IEEE Topical Cadence Conference on 15 - 17 Oct. 2003:38 - 39.

The Design and Implementation of Oscillograph Based on STM32

Mingxin Song [1]*, Yue Li[1], Yang Yang[1]

College of Applied Science, Harbin University of Science, Harbin, China
songmingxin@126.com

Abstract—This paper reports new oscilloscope with simple structure and low cost. The proposed oscilloscope has utilized the proper combination of two parts: the new 32-bit Cortex-M3 core-based controller and software Keil which can use the C language and convert the C language to assemble language. The proposed oscilloscope displays the waveform on the screen in time longer than one cycle, and the real-time sampling rate is 1Mbyte/s.

Keywords- component; Oscillograph; STM32; low cost

I. INTRODUCTION

In the field of modern electronic measurement and instrumentation, oscilloscope is one of the most commonly measurement instruments[1]. Its main role is to measure and display the measured signal parameters of the waveform. With high-speed development of computer technology, software technology, and applications in measurement technology, the instrument's function and structure are have break the traditional instrument concept in many aspects. A fundamental change has occurred in function and role results in some new methods and theory of test instrumentation [2]. Opened up a new field test and digital oscilloscope is generated in this context. ARM-core digital oscilloscope uses the strong development capability and integration in order to greatly reduce the cost of the system and enhance the flexibility of the system [3]. It breaks through the traditional instrument mode. Oscilloscopes occupy a significant proportion of ultra-small size and flexibility in the market.

Now the majority of our city's high-end oscilloscope needs to be imported. These oscilloscopes which have complex process are difficult to produce and break through. Embedded technology provides the technical foundation for the produce of the new type measurement and control instruments. So in the country developed the low cost oscilloscope with high-performance, simple operation is necessary. Zhou Fu-xiang Chen, De-yi, Liu Pei-gou and Wei Zheng-xia designed a high-performance oscilloscope[4] with the high cost and complex structure. Ding Hongbin, Qin Huibing and Sun Shunyuan designed a low-cost oscilloscope[5]. But the design of software is complex and it cannot work. Purpose of this paper is be able to display the received waveform, and be able to measure the peak-to-peak. The proposed oscilloscope shows that the frequency and amplitude characteristics of waveform. The oscilloscope can obtain the peak of the waveform and frequency. The waveform can be moved up and down on the liquid crystal display. In this paper, the design of oscilloscope has the low cost, simple structure, good portability and the real-time sampling rate of 1Mbyte /s.

II. THE OVERALL DESIGN

Oscilloscope consists mainly of the power supply section, the amplification part, the control part, the acquisition and display of the keyboard function and the chip parts[6,7]. The design receives the small signal. The voltage of small signal which is fed into the oscilloscope ADC is slightly smaller. The signal is not easy to quantitatively choose. We amplify small signals before it is sent to the AD conversion circuit. After AD conversion circuit, data computing and graphics chip perform enable control. Then it entered the start interface to achieve the oscilloscope function. Data calculated by using comparator compares the default value with AD sent. If the AD sent value is higher than the default value, the count will display on the oscilloscope. Then compare the number with the later number to calculate the peak-to-peak.

The core content of the digital oscilloscope is that continuous simulated measured signal is converted to a digital signal[8]. The digital signal is sampled. Sampling methods can be divided into real-time sampling and non-real-time sampling. Both of the sampling methods have their advantages and disadvantages. Real-time sampling can be used to sample the non-periodic signal. Equivalent sampling can obtain a very wide frequency band, but it cannot observe the non-periodic signal. Therefore, the design uses a method that combines the real-time sampling with equivalent sampling. Trigger mode using internal software trigger. This mode can rule out glitches. The glitch is generated by hardware circuit. Internal software trigger has a relatively stable trigger and it is easy to adjust the trigger voltage. Frequency measurement uses precision frequency measurement method. The accuracy of frequency measurement is equal with the signal.

The master chip select STMicroelectronics developed STM32 series. STM32 is a new 32-bit Cortex-M3 core-based controller. The chip integrates 16-channel 12-bit precision A/D converters. The single-chip can complete sample quantization, AD converter, waveform analysis, waveform display and something else. The single-chip also can effectively reduce the complexity of system interface and the difficulty of system development. This greatly improves the stability of the system. We apply STM32 internal keyboard control chip and use color TFT LCD to display. The overall block diagram is shown in Figure 1.

978-1-4799-1215-5/13 $31.00 © 2013 IEEE

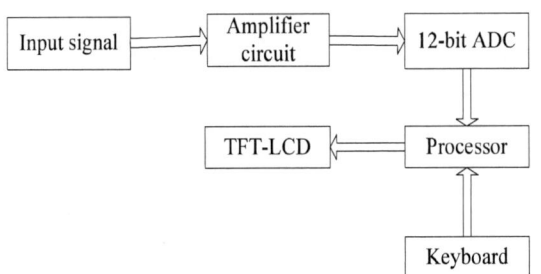

Figure 1.   Overall block diagram

III.   HARDWARE DESIGN

A.   Power Supply

In order to show the complete waveform and the supply of the signal conversion section, the chip needs a stable voltage. The STM32 chip requires 5V. 7805 can convert 12V to 5V. 5V will not hurt the selected chip. The 7805 is a low-cost regulator chip. It can meet the requirements in the design. For example, it can be hot over-protection, short circuit protection, output transistor safe operating area protection and provide a stable 5V. Circuit power application is shown in Figure 2.

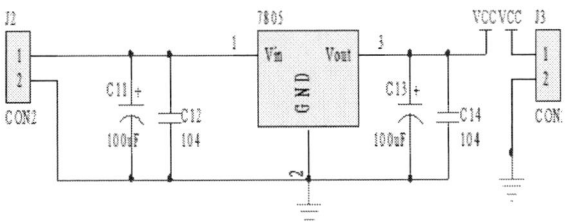

Figure 2.   Design circuit diagram of the power section

The STM32 chip input voltage is 3.3V, so we need to convert the received supply 5V to 3V. Chip AMS1117 is a three-terminal linear regulator circuit. Its regulator adjustment tube is consisted of NPN driven by a PNP. The dropout voltage is defined as: $V_{DROP}=V_{BE}+V_{SAT}$.

In order to ensure the AMS1117 stability and adjustable voltage versions, the output needs to connect a least 22µF tantalum capacitor. For fixed voltage versions, smaller capacitors can be used. Generally, the stability of the linear regulator is reduced as the output current increases. The system power supply circuit is shown in Figure 3.

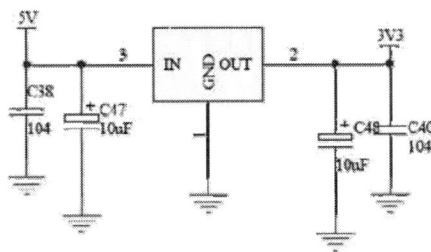

Figure 3.   5V-3.3V circuit

B.   Programmable Amplifier Circuit

This part is designed for display circuit with small-signal waveform and analog-to-digital conversion before data acquiring. The STM32 series chip exist in internal AD converter modules. The module input voltage is 3.3V, so it is necessary to amplify the received signal. Design a voltage follower circuit that stabilizes the received voltage signal before the voltage amplifying. In order to meet the amplification and the voltage follower used in the design, we choose TL074 chip which contains four amplifiers. Programmable amplification circuit uses three amplifiers. The amplifier can not show a negative value. So it is necessary to design a voltage reference. In the waveform display section, if the voltage value is less than the positive input, it will be set below the reference. Part of amplification circuit diagram shown in Figure 4.

The input signal is displayed on the oscilloscope. Need to design an input reference voltage because the voltage can not be negative. The range of the voltage fluctuation must be in the A / D conversion voltage range. Need for pretreated voltage and elevated position of the reference voltage to ensure that the input voltage is positive. In this design we want to obtain the stable waveform, the voltage component must be stable. Then the site will be able to achieve a stable waveform.

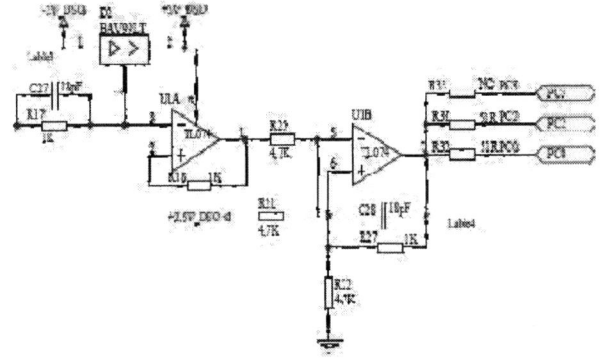

Figure 4.   Part of amplification circuit diagram

C.   Chip Collection and Display Part

Using of STM32F103VE can completely enable chip microcontroller, analog-to-digital conversion chip. It also achieves the numerical calculation, counting and other functions. The chip comes with channel denoising effect which is better than the multiple chips lap denoising effect. In the information acquisition process, the first thing to do is getting the analog circuit converted into a digital circuit. After A/D conversion section, the next is the data calculation and the display portion. STM32F103ZET6 is based on the Cortex-M3 MCU. Users can replace the oscillator (4~6MHz) or 32 KHz oscillator. They have 20 MIPS speed, high-performance cores and strong anti-jamming capability.

ADC section of the STM32F103 series chip is a 12-bit ADC. This ADC is a successive approximation - type analog to digital converter. It has 18 channels. The ADC can measure 16 external and two internal signal sources. Each channel A / D conversion can be performed by the single, continuous, scanning or discontinuous mode. The result of the ADC can be stored in 16-bit data registers by using the left-aligned or right-aligned method. The display section uses 320 × 240 TFT-LCD screen. Using software makes LCD enable to boot start screen. The advantage of TFT-LCD is that it can display a stable and complete waveform. Compared with traditional CRT, TFT-LCD has small size, thin thickness (currently, whole thickness 14.1 inches can be achieved only 5 cm), light weight, low energy consumption (1-10 microwatts/square cm), low operating voltage (1.5V to 6V). It also has no radiation, no flicker and match with a CMOS integrated circuit. The design uses the 320 × 240 LCD display. Waveform display function can be achieved through the connector. This method can meet full waveform display, peak-to-peak calculation. The LCD circuit is shown in Figure 5.

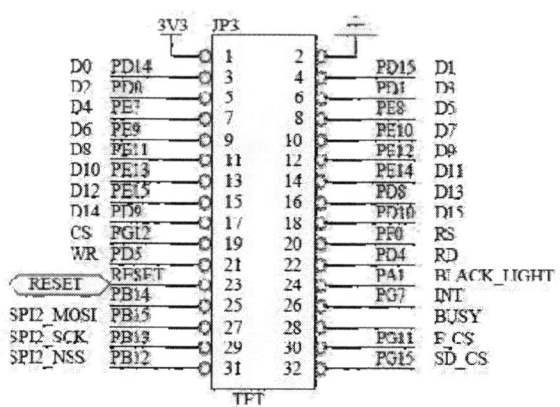

Figure 5. LCD circuit

IV. SOFTWARE DESIGN

A. Application Software

Compilation mode uses the C language, so we use Keil software to convert the C language to assemble language. It includes many features: input source, establish engineering, setting the detailed parameters of the project, and converting the source code into object code. Keil provides a complete development program. It includes a macro assembler, compiler, library management, connectors, and a powerful simulation debugger. Finally put these parts together by using (UVision) integrated development environment.

B. Module Program

In the AD conversion part, we must first set the key state. Three states (help pages, collection status, and suspended state) are controlled by two buttons. Three states can improve the settings of the AD converter. The start state is the suspended interface. Full-speed signal acquisition is a multilayer display. The ADC performs one conversion in single conversion mode. This mode can start by setting ADON bit of ADC_CR2 register (this method only applies to the rules of the channel). It also can start by an external trigger (this method apply to regular channels or injected into the channel). At this time the CONT bit is 0. Use a loop body chooses the operation to be

performed. The control part is controlled by the key part. Jump out and complete the process if the calling function is not established in acquisition part.

At first, judge control part. If established, setting of each key floating-point mode, then prepare for the next key state. Control waveform moves orientation to be more clearly displayed. In the design of the display portion, the number defines access address of LCD drive by use 32-bit addressing mode. After setting the physical coordinates of the LCD, set the start address and end address of the LCD horizontal and the vertical direction, set 80 grids. As follows:

```
void   LCD_SetDisplayWindow(uint16_t   Xpos, uint16_t
Ypos, uint16_t Height, uint16_t Width)
    {
        if (Xpos >= Height)
        {
            LCD_WriteReg(R80, (Xpos - Height + 1)); }
                         else
        {
                LCD_WriteReg(R80, 0);   }
            LCD_WriteReg(R81, Xpos);
                if (Ypos >= Width)
        {
            LCD_WriteReg(R82, (Ypos - Width + 1));}
                         else
        {
                LCD_WriteReg(R82, 0);}
            LCD_WriteReg(R83, Ypos);
        LCD_SetCursor(Xpos, Ypos);}}
```

C. Analysis

Weld according to the circuit diagram, the final overall kind shown in Figure 6. Keyboard Control section have two ways: process control and manual control. This can always collected signal immediately. The oscilloscope includes development board, LCD display, separate buttons, and amplifier. Coupled with the probe will be able to work properly. Physical detection circuit shown in Figure 7. The proposed oscilloscope still have many inadequacies need to be improved. The oscilloscope has the low cost, simple structure, good portability and the real-time sampling rate of 1Mbyte /s.

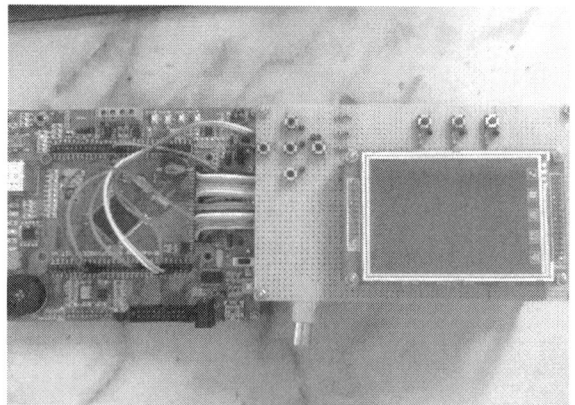

Figure 6. Overall physical

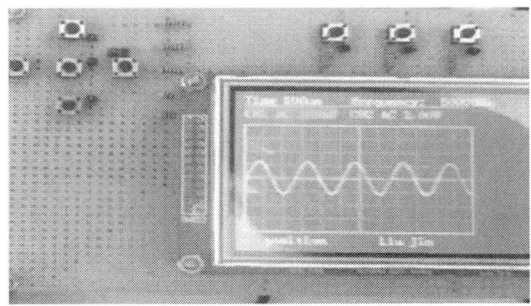

Figure 7. Physical detection circuit

V. CONCLUSION

Easy ARM oscilloscope is based on STM32 series. The oscilloscope displays more than one cycle of the waveform on screen. Non-periodic signal can achieve real-time sampling. Cycle waveform can pan left or right on the screen. Compare the counter value for the peak. At the time the peak appears the register start counting, then timing. After the next peak appears the register stop counting. Application STM32 has its unique advantages. STM32 owns AD conversion module which can save the installed time and space of the channel. The STM32 voltage detection software can achieve the measurement of the peak-to-peak. This software can reduce error-prone when the hardware is structuring. The oscilloscope has the advantages of simple structure, low cost and easy to carry. It also has much space for development with the later improving there will be many new features.

ACKNOWLEDGMENT

This work is supported by the Innovations Research Project of Graduate student of Heilongjiang Province (Grant No. YJSCX2012-103HLJ). The authors would also like to thank the technical support of the student Liu Jin. The authors also thank the open experiment team members of the organization and social support.

REFERENCES

[1] S.K. Pal, A. Kumar, K. Kumawat, "Design and VLSI Implementation of a Digital Oscilloscope," IEEE J. CICN. pp. 473–476, 2012.

[2] Ying-Wen Bai, Hong-Gi Wei, "Loss waveform interval for the data buffering of a multiple-channel microcomputer-based oscilloscope system," TIM. vol 54, NO.1, pp. 45-51, 2005.

[3] D.J. Krause, J.C. Cartledge, C. Laperle, K. Roberts, "Interferometric In-Phase and Quadrature Oscilloscope," JLT. vol 27, NO.14, pp. 5749-5754, Dec.2009.

[4] Zhou Fu-xiang, Chen De-yi, Liu Pei-guo and Wei Zheng-xia. "the design of digital oscilloscope based on STM32," Shanxi Electronic Technology. vol, NO.2, pp. 8-10, 2011.

[5] Ding Hongbin, Qin Huibing and Sun Shunyuan, "the design and implementation of virtual oscillograph sased on STM32," Chinese Journal of Electron Devices. vol 32, NO.6, pp. 1007-1010, Dec.2009.

[6] P. Choi, E. Wyndham, "Fast analog chopper multiplexer for oscilloscope recording of single - shot events," Review of Scientific Instruments. vol 59, pp. 328-331, 1998.

[7] P.D. Hale, T.S. Clement, K.J. Coakley, C.M. Wang, D.C. DeGroot, Angelo P. Verdoni, "Estimating the Magnitude and Phase Response of a 50 GHz Sampling Oscilloscope Using the "Nose-to-Nose" Method," ARFTG. vol 37, pp. 1-8, 2000.

[8] M. Safi-Harb, G.W. Roberts, "70-GHz Effective Sampling Time-Base On-Chip Oscilloscope in CMOS," Solid-State Circuits. IEEE. vol 42, pp. 1743-1757, 2007.

A fabrication method of the micro resonant beam compliant to unfitted thickness

Jiandong Jin *, Mingwei Wang, Yuling Li , Hong Qi

Engineering Center of Chip and Microsystem, The 49th Research Institute of CETC, Harbin, China

mw_wang@163.com

Abstract—**This papers presents a simple bulk micro fabrication method for micro resonant beams based monocrystalline silicon wafer. Experimental results show that the thickness of the micro resonant beams is highly uniform, the length and width of the resonant beams are determined by photolithography, the thickness of the beam, the gap between the beam and Si substrate are determined by the DRIE. The thickness and gap are adjustable in a large range..**

Keywords-resonant beam; anisotropic etching; (111)wafer

I. INTRODUCTION

As a typical application of resonant technology, the resonant silicon pressure sensor offers an order of magnitude greater performance over current technologies, and the resonator provides the highest accuracy pressure sensor at present. The long term accuracy of a silicon micro pressure sensor could reach to 0.01%FS[1]. The silicon MEMS resonant pressure sensor is also qualified with exceptional performance, like anti interference, stability. The long term stability can be up to 0.01% per year. This kind of sensors is imperative in industrial control system, aerospace and atmosphere data processing, etc. Micro resonant beams are the fundamental structures of many devices. Here are four common methods developed to fabricate micro resonant beam..

SOI (Silicon-On-Insulator)wafer can be adopted to make micro beam structures[2]. The device layer is separated from the hold layer by a layer of SiO$_2$, and is etched as the micro beam. The advantage of the resonant beam is that the material the resonant beam is crystal silicon, but the thickness of the beam can't exceed the top layer. Thus, the frequency of the beam is restricted. In addition, SOI wafer is more than ten times expensive than ordinary silicon wafer.

The second technology is that the sacrificial layer method is used to approach the micro vibration beam[3]. In this way, A SiO$_2$ layer is deposited as the sacrificial layer, and the poly silicon layer or nitride layer deposited acts as the beam layer. Due to the beam material is not part of the bulk silicon, and the materials are restricted by the sacrificial material, the shortages are obviously appeared. The mechanical properties, like resonant frequency, residual stress, are less than crystal silicon. The detecting way is limited, too, as the vibration picking piezoresistor can't be made. The structure design is also inflexible, as the thickness of beam and height of the suspending of the beam are limited in narrow range.

The resonant beam can also be made by heavy boron etch-stop technology[4]. Heavy dose of boron is implanted into the beam zone on the wafer, after that, the suspending beam is released in the alkaline wet etching solution. The thickness of beam is defined by the heavy boron doping depth. So it is not easy to make a thicker beam. Moreover, heavy doping may bring larger residual stress, and the long term stability of final device will be influenced.

Depending on Si-Si bonding technology, vibration beam can be made[5]. In this way, two wafers are needed. One wafer is thinned by grinning and polishing from top or bottom side to the thickness the beam needed, and is made to a beam. Meanwhile, a shallow groove is prepared on the other wafer, and a driving vibration frame is made on the same wafer. In this way, the warpage and smoothness is very important, after MEMS process steps, it is difficult to keep the two wafers as smooth and plane as the blanket wafers. If high temperature bonding process adapted, metal related process should be arranged after it. Thus, the flexible of metal connection is restrained. Furthermore, the thickness of the beam made by this way can be uniform, and it is not good for the properties of the beams.

In this paper, a silicon wafer with (111) orientation is used as the main material. Anisotropic wet etching and DRIE(Deep Silicon Reactive Ion Etching) technologies are co-designed. A micro resonant beam made of bulk crystal silicon of the wafer is obtained, and the more crucial point is thickness and suspending height can be adjusted in a large range through MEMS process. document and are identified in italic type, within parentheses, following the example. Some components, such as multi-leveled equations, graphics, and tables are not prescribed, although the various table text styles are provided. The formatter will need to create these components, incorporating the applicable criteria that follow.

II. DESIGN OF THE RESONANT BEAM

Fig.1 shows the schematic of our proposed micro resonant beam. compared with the traditional resonant beam reported in[2]-[5],herein the resonant beam is single crystalline silicon, thickness of which can be adjusted in a large range through DRIE. the cavity beneath the resonant beam is embedded into the silicon substrate by using lateral under-etch.

978-1-4799-1215-5/13 $31.00 © 2013 IEEE

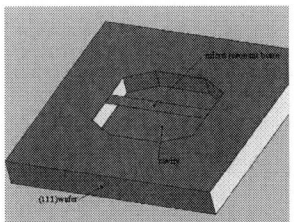

Figure 1. 3-D schematic of our proposed resonant beam structure with front-side fabricated hexagonal embedded-cavity

Fig.2 shows the top-view design schematic for the resonant beam formation from wafer front-side. The resonant beam is aligned<211>orientation for the beam formed by lateral under-etch along<110>and<211>orientation with all the boundary walls as {111}planes. Based on (111)-silicon anisotropic etching theory, hexagonal-shaped cavity can be finally formed by (111)etching stop. It is deserved to point out that to release the beam, must fulfill the relation:

$$l > \sqrt{3}w$$

Where l and w are the length and width of the resonant beam. Angle between the (111)side wall and the bottom is 71°.this angle can calculate on the relationships between the (111)planes.

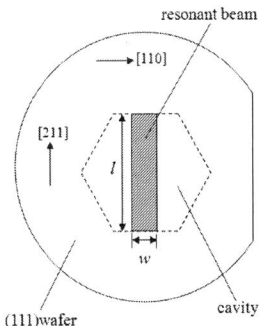

Figure 2. Front –side resonant beam and hexagonal-cavity formation scheme in (111)wafer

The fundamental resonant frequency of the resonant beam can be estimated with Rayleigh-Ritz methods. For the double clamped silicon resonant beam, the fundamental resonant frequencies are[6]:

$$f = \frac{h}{l^2}\sqrt{\frac{E_{Si}}{\rho_{Si}}}$$

Where h is thickness of the resonant beam, l is the length of the resonant beam, E_{Si} is the Yong s modulus of Si, and ρ_{Si} is the density of Si.

III. FABRICATION

The wafers used in the experiment are 4 inch n-type(111) wafers with 0°±1° deviation. The resistivity is 3-8 Ω•cm. The detailed processes are shown in Fig.3 and described as followed:

(a)A 4000Å-thick SiO_2 layer is thermally grown on both sides of the wafer by dry oxidation and wet oxidation.

(b)After thermal oxidization, photolithographic steps are conducted from the front side of the wafer for patterning resonant beam on a photoresist layer. Then buffered HF is used to etch SiO2, with the photoresist as etching mask. DRIE is used sequentially to etched silicon substrate to a designed depth to define the thickness of the resonant beam, with the same photoresist as etching mask.

(c)LPCVD (Low Pressure Chemical Vapor Deposition) nitride is deposited as a hard mask for following TMAH (Tetramethylammonium Hydroxide)lateral under-etch to form a cavity and release the resonant beam. The nitride can protect the sidewalls of the resonant beam from the following anisotropic etch.

(d)The nitride layers at the trench bottom are then dry etched with RIE(Reactive Ion Etching).

(e)DRIE is processed again for an extra depth of the trench with the thermally grown SiO2 as etching mask. where silicon is exposed for further lateral wet etch. The depth is equal to that of the cavity.

(f)Aqueous TMAH is used to laterally under-etch along <110> and <211> orientation that is finally stopped at (111)plane to form the hexagonal cavity.

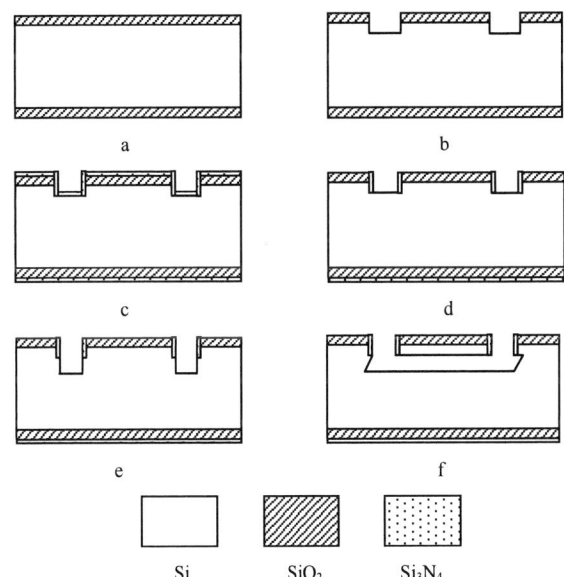

Figure 3. The process flow of micro resonant beam

A fabricated micro resonant beam is shown in Fig.4. The thickness of the beam was 5μm, The length of the beam was 600μm, The width of the beam was 40μm. The fundamental resonant frequencies was 120kHz.

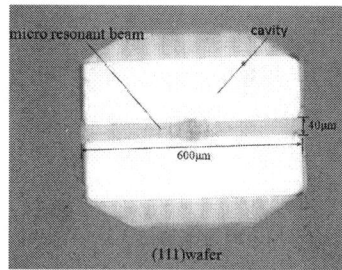

Figure 4. Images of the fabrication resonant beam

IV. DISCUSSION

It is deserved to point that the thickness of the resonant beam are determine by DRIE, The gap between the beam and the substrate are determine also by DRIE. herein the thickness and the gap are adjustable, the thicker of beam is, the resonance frequency higher. and wider of gap, Q value of device higher.

When the thickness of beam is 10μm,fabrication precision of the 10μm-thick is determined by the 10μm-deep front-side DRIE trench etching depth. If conventional backside etching method used, the 10μm beam thickness should be determined by the wafer thickness(450μm) subtracted by the backside etching depth of 440μm. Obviously, our front-sided fabrication method features much higher uniform.

As the surfaces of the (111)wafers deviated slightly from the (111)plane, when the thickness of beam is less than 5000 Å, the application of a TMAH preetching was needed to expose the (111)plane.

REFERENCES

[1] J.C.Greenwood, D.W.Satehell, "Miniature silicon resonant pressure sensor," IEE Proceedings D. Control Theory and Applications, Vol.135, pp.369-372.1988.

[2] Erbe.A, Weiss.C, Zwerger.W, Blick.RH, "Nano mechanical resonant shuttling single electrons at radio frequencies," Phys Rev Lett, Vol.87, pp. 96-106, 2001.

[3] H.S.Ko, C Wliu, C.gau, "Novel fabrication of a pressure sensor with polymer material and evaluation of its performance," J.Micromech. Microeng, Vol.17, pp. 1640-1648,2007.

[4] Kyoichi Ikeda, Hideki Kuwayama, Takashi Kobayashi, et al, "Silicon pressure sensor integrates resonant strain gauge on diaphragm," Sensors and Actuators A, , Vol.21, pp. 146-150,1990.

[5] Kurt Petersen, Farzad Pourahmadi, Joe Brown, "Resonant beam pressure sensor fabricated with silicon fusion bonding," Teeh Digest,6th int Conf Solid-State Sensors and Actuators,Tranducer'91,San Francisco, CA, USA, pp.436-437, 1991.

[6] Minhang Bao, "Analysis and Design Principles of MEMS Devices," Elsevier.Amsterdam, pp.76-79, 2005.

Optical fiber spectroscopy in near infrared spectral range for rapidly discriminating formulated apple vinegar drinks

Bingfang Zhang[1,2], Libo Yuan[1*], Bingxiu Zhang[3]

1 Key Lab of In-fiber Integrated Optics, Ministry Education of China, Harbin Engineering University, Harbin, China
2 College of Science,Northeast Agricultural University,Harbin, China
hljzbf@163.com
3 College of horticulture,Northeast Agricultural University,Harbin, China

Abstract—This paper presents an optical fiber spectroscopy method rapidly discriminating formulated apple vinegar from fermented ones. A collection of 80 apple vinegar drinks samples were considered, 25 of which were formulated apple vinegar drinks and the other 55 were fermented ones. Absorption spectra of these apple vinegar drinks samples were measured by near infrared (NIR) spectrometer. These measured spectral data were processed by means of Principal Component Analysis (PCA), which provided a bi-dimensional map. According to the bi-dimensional map, 25 formulated apple vinegar drinks samples were differentiated from 55 fermented apple vinegar ones. The experimental result showed that the optical fiber spectroscopy method was capable of distinguishing formulated apple vinegar drinks from fermented apple vinegar drinks.

Keywords-optical fiber spectroscopy, apple vinegar, absorption spectra, Principal Component Analysis (PCA)

I. INTRODUCTION

In recent years, the quality and safety of the food has become the focus of public attention. Fruits are rich in sugar, which is the first-class raw material for making vinegar. Compared to food vinegar, fruit vinegar is more abundant in the nutrition. Fruit vinegar contains acetic acid, succinic acid, Malic acid, amino acids, vitamins and other bioactive substances. Fruit vinegar has the functions of softening blood vessel, reducing blood pressure and regulating the acid-base balance of body[1]. Fruit vinegar as a healthy drink, in the United States, Britain, Canada, Southeast Asia and Japan developed very fast. In the United States, the annual output of vinegar is up to 100 million liters, a quarter of which is apple vinegar;In Japan, fruit vinegar consumption accounts for the total amount of vinegar 30%. Fruit vinegar drinks has been seen as "the sixth generation of gold drinks" following the carbonated drinks, water drinks, tea drinks, fruit juice drinks and functional drinks.

Apple vinegar drinks is more popular on the market and it is divided into formulated and fermented two species. Fermented apple vinegar drinks is made by concentrated apple juice after two fermentation (alcoholic fermentation, acetic acid fermentation), which usually need 3 to 6 months, and then mixed with apple juice raw materials. Its nutrition is rich, containing pectin, vitamins, minerals and enzymes, and the acidic component also has bactericidal effect, while the formulated apple vinegar drinks does not contain these compositions. It is made by rice vinegar, edible essence and saccharin, also does not have bactericidal function. Long-term drinking formulated vinegar drinks would cause harm to human body. At present, formulated and fermented apple vinegar drinks are mainly distinguished by appearance, while scientific discriminating method is less.

Absorption spectroscopy is a popular method in conventional analytical chemistry. The near infrared spectra can provide molecular structure, composition and status information of samples. And we can get physical state information, such as density, particle size and polymerization degree. So classification of food based on near infrared absorption spectra has also been studied by many researchers. A.G. Mignani et al.(2004) innovately used absorption spectroscopy and multi-angle scattering measurements for the geographic classification of Italian extra virgin olive oils[2].Anna Grazia Mignani et al.(2009) applied the absorption spectra in both the UV–Vis and NIR spectral ranges to predict important quality indicators of lubricant oils , such as TAN, water, phosphorus content and the JOAP index[3]. Ce Yang et al.(2012) classified the blueberry fruit and leaves based on absorption spectral signatures[4].

Lately NIR transmittance spectroscopy has been studied for discriminating different categories of fruit vinegar such as apple vinegar, aloe vinegar, plum vinegar and so on[5].This paper presents a rapidly method based on NIR absorption spectroscopy to discriminate formulated apple vinegar from fermented ones. The experiment results showed that NIR absorption spectroscopy combined with chemometrics was capable to successfully discriminate formulated apple vinegar from fermented ones, which was a simple, fast and convenient method.

II. SPECTRAL SIGNATURE OF APPLE VINEGAR BY MEANS OF NIR ABSORPTION SPECTROSCOPY

A. Instrumentation

As shown in Figure 1, an instrumentation based optical fiber was used to measure the absorption spectral of the apple vinegar samples in near infrared spectral range. A high powerful and optical fiber-compatible Tungsten-halogen lamp (GuangZhou Electronic Science Technology c.,LTD, LS-3000) served as the source, which spanned the spectral range of 360~2000 nm. A cuvette holder with standard 10mm path length was installed with two collimating lenses (74-UV) for beam collimating and focusing. NIR spectrometer (Ocean Optics, Inc., QUEST) was used as detector. Two multimode optical fiber strands, with a 600 μ m core diameter, were used for connecting the source and the detector to the cuvette holder containing the apple vinegar sample. An empty glass cuvette was used as reference for absorbance evaluation.

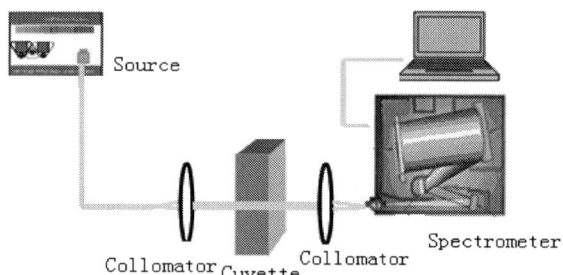

Figure 1. Set-up for absorption spectroscopy by means of optical fiber technology

B. Data Processing

In spectral data processing, multivariate data analysis technique was used to reduce data dimensionality and better extract significant information for sample identification. These acquired spectral data were processed by Principal Component Analysis (PCA)[6-8], which provided the coordinates for identifying the samples in bi-dimensional map. The PCA program was written in Matlab code. The data processing was performed as following:

- A matrix was created by NIR absorption spectral data, and each row in the matrix representing the spectrum of an apple vinegar sample.

- The acquired spectrum was inevitably accompanied by high frequency random noise, sample particle size and light scattering noise, so spectral pretreatment is required to eliminate the noise. This experiment adopted the Multiplicative Signal Correction (MSC), Savitzy-Golay(SG) and Standard Normal Variate Correction(SNV) to eliminate the noise information.

- In order to compress the relevant information, the matrix was processed by PCA, thus created a score-matrix for NIR PCA-data.

- Two linear irrespective functions were extracted, the first two principal components, PC1 and PC2, which were adequate to characterize the sample. Because the high order principal components with loss of information could be disregarded

- Finally, PC1 and PC2 were used to build a bi-dimensional map, each point of which represented one apple vinegar sample. The map was populated by point clusters, and points from same cluster represented the similarity in characteristic.

C. Fingerprinting Formulated Apple Vinegar drinks According to the Spectral Characteristic

The compact and low-cost spectrometric instrumentation was used to obtain apple vinegar sample NIR spectra. Experiments using 3 brands of fermented apple vinegar drinks, and they were 'Yuanchuang,', Letian' and 'Yangshengyuan', as well as 1 brand of formulated apple vinegar drinks contained rice vinegar, edible essence and saccharin. Eighty apple vinegar samples (listed in Table 1) of the 4 brands, were purchased from supermarket, 25 of which were formulated apple vinegar drinks, while the other 55 were fermented ones. Samples of the same brand were belong to different batches. To obtain absorption spectral of the apple vinegar samples, experiments were operated at 20℃. Absorption spectra of all the samples were acquired in the wide 900~1600nm spectral range.

TABLE I. LIST OF INGREDIENTS OF ANALYZED APPLE VINEGAR SAMPLES

Brand	Number of samples	Ingredients
Yuanchuang	30	apple vinegar, apple juice concentrate, food additives, edible essence
Letian	20	apple vinegar, apple juice concentrate, food additives, edible essence
Yangshengyuan	5	apple vinegar, apple juice concentrate, honey, food additives, edible essence
Formulated apple vinegar	25	rice vinegar, edible essence, edible saccharin,

Figure 2 shows the absorption spectral of examples in the NIR spectral range. Different brands of apple vinegar curve are overlapped, and have no obvious difference. So spectral data must be combined with chemometrics method to distinguish the formulated apple vinegar sample.

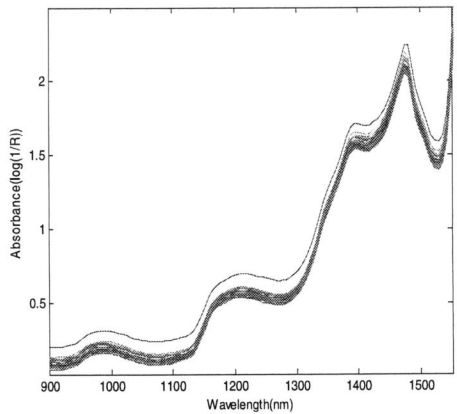

Figure 2. NIR absorption spectra of four apple vinegar varieties

Then these spectral data were processed by Principal Component Analysis (PCA), figure 3 shows the two dimensional map, in which the PC1 axis is the first principal component scores of samples, the PC2 axis of is the second principal component scores. It shows the clustering effects of two types of apple vinegar drinks and gives a qualitative description of characteristic difference between the two types of vinegar drinks. The characteristic difference has provided the feasible basis to distinguish different types of fruit vinegar drinks. In the plot, 25 formulated apple vinegar samples clearly cluster from analyzed 80 apple vinegar samples.

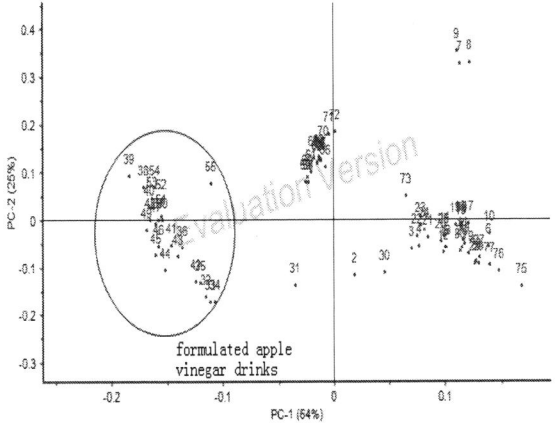

Figure 3. principal component scores of four types apple vinegar

PERSPECTIVES

In this paper a collection of 80 apple vinegar samples was analyzed by means of absorption spectroscopy carried out in the wide 900-1600nm spectral range, and the spectral data were processed by means of Principal Component Analysis. This compact and low-cost instrumentation presented in this paper does not require any preparation of the sample. It is potential to achieve a quick quality control and product classification, and has an important complementarity to the data obtained by conventional chemical analytical techniques.

ACKNOWLEDGMENT

This work was supported by the National Natural Science Foundations of China, under grant numbers 61290314 and 11274077, partially supported by the 111 project (B13015) and Sino-Japan S & T cooperation project (2010DFA-2770), to Harbin Engineering University.

REFERENCES

[1] Ju-xiu Li, Jun-ling Shi. Food Science and Technology,2001,(2):47.

[2] A.G. Mignani, L. Ciaccheri, A. Cimato, G. Sani and P.R. Smith. "Absorption spectroscopy and multi-angle scattering measurements in the visible spectral range for the geographic classification of Italian extra virgin olive oils." Proc. of SPIE Vol. 5271,2004,pp.285-288

[3] Anna Grazia Mignani, Leonardo Ciaccheri et al. "Optical fiber spectroscopy for measuring quality indicators of lubricant oils." Meas. Sci. Technol. 20 (2009) 034011,pp.7.

[4] Ce Yang,Won Suk Lee,et al. "Classification of blueberry fruit and leaves based on absorption spectral signatures." Biosystems Engineering II3,2012, pp. 351-362.

[5] WANG Li, LI Zeng-fang,et al. „Fast Detection of Sugar Content in Fruit Vinegar Using NIR Spectroscopy", Spectroscopy and Spectral Analysis. Vol. 28, No.8, 2008, pp. 1810-1813.

[6] T. Davies, The Principles of Principal Component Analysis, Spectroscopy Word, Vol. 4, No.1, 1992, pp. 23-27.

[7] I.A. Cove and J.W. McNicol, 'The use of Principal Component in the analysis of the near infrared spectra", Applied Spectroscopy, Vol. 39, No. 2, 1985, pp. 257-266.

[8] CHU Xiao-li. Molecular Spectroscopy Analytical Technology Combined With Chemometrics and its Applications.2011.

A study on MEMS acoustic vibration sensor technology

Yonghe Qin[1,2], Yingjie Qiao[1*], Wenjiang Zou[2], Xingyue Xu[2]

1 Institute of materials science and chemical engineering, Harbin engineering university, Harbin, China
2 The 49th research institute of China electronics technology group corporation，Harbin, China
qinyh@139.com

Abstract—by using ANASYS finite element analysis software to analyze the first three order resonance frequencies and strain values of the porous and porous free structure of the acoustic sensing chips, the optimization of the structure indicates: the resonance frequency of the porous structure varies lower relatively, but the stress improves relatively higher, we utilize beam membrane structure, the featured size of the beam thickness is 50 um. We utilize MEMS technology to fabricate acoustic vibration sensor chips on single crystalline silicon, and use second order vibration mode test unit to measure vibration signal, on condition of 5v power supply, the sensitivity of chip is about 26.8mV/g, resonance frequency is 5.19kHz,frequency response range is over 20~2000Hz,the frequency at the point of 3dB is 2.93kHz.

Keywords-MEMS, resonance frequency, acoustic vibration, sensor, single crystalline silicon

I. INTRODUCTION

Acoustic vibration sensor has a wide application in urban road monitoring, it can be used to measure the vehicles and pedestrians traffic accurately. The acoustic vibration sensor manufactured by MEMS technology has the features of high sensitivity, wide range of low frequency, easy to be integrated and manufactured, etc. Through measuring and processing the acoustic vibration signals from the pedestrians and vehicles, extracting effective characteristic parameters, target classification can be achieved, and the intelligent monitoring management of the road traffic can be realized.

II. DESIGH AND MANUFACTURE OF MEMS SENSING CHIPS

A. The principle of MEMS acoustic vibration sensor

The measuring principle of MEMS acoustic vibration sensor is to utilize the Piezoresistive effect of single crystalline silicon, the bigger of the piezoresistive coefficient, on the condition of the same vibration stress, the bigger of the changes of sensitive resistance, and the higher of sensor sensitivity. For the measuring principle, we can use first order vibration mode and second order vibration mode, since first order vibration mode does not have orientation, in order to be suitable to determine the directional position of the target, we use second order vibration mode to measure vibration signals. The working process is that the vibration signal from the outside is delivered to sensitive chip through acoustic coupling material, under the action of the internal mass of the sensitive chip, the elastic beam deformation occurs, and the value of the sensing resistor changes. By measuring the voltage of Wheatstone bridge, the change of sensitive resistance can be measured accurately, through standard dissemination calibration of vibration, to realize the measurement of vibration.

B. Structural design of chips

Sensitive chip is manufactured by using piezoresistive effect principle, the bigger of the piezoresistive coefficient, on the condition of the same stress, the bigger of the change of sensitive resistance, the higher of the sensor output, and the more sensitive of the sensor.The piezoresistive coefficient of silicon is closely relative to conductive type, crystal orientation, and also to doping concentration and temperature, the determining factor is the first two. The symbols of horizontal and vertical piezoresistive coefficient are always the opposite, so they make offset against each other to some degree.Only when piezoresistive coefficient is used reasonably, can be attained the better results.After the structural design has been confirmed, the reasonability of process design plays an key role to the function of the chip.

According to the requirements of micro electronic process design, considering that the overload capacity of the elastic beam should not be lowered during layout design, the structure size of sensitive chip and geometric parameters of resistor strip should make the sensor have higher sensitivity.Fully making use of the joint action of piezoresistive and stress, to make a reasonable layout and design, the whole layout of vibration sensitive chip is shown in figure1.

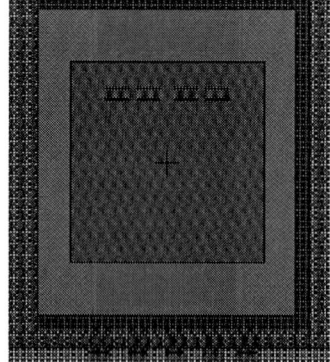

Figure 1. Chip layout design

Identify applicable sponsor/s here. If no sponsors, delete this text box.
(sponsors)

978-1-4799-1215-5/13 $31.00 © 2013 IEEE

C. simulation of chip structure

We use finite element analysis method to design chip structure, and optimize the structure parameters. Because vibration signal tested is weak, and the signal also has transmission loss, the design of chip structure should have higher sensitivity. The resonant calculating value of designed vibration chip is:

$$f = \frac{w}{2\pi} = \frac{1}{2\pi}\sqrt{\frac{k}{m}} = \frac{1}{2\pi}\sqrt{\frac{192Ebh^3}{12l^3 m}} = 1424Hz \qquad (1)$$

The first three resonant frequencies of finite element analysis result are 385, 425, 888Hz,respectively. If estimating by analogical method, the real first order resonance frequency should be over 1420 HZ. From the above analysis, we can conclude that structure size should be reduced to improve frequency response, but it will affect the sensitivity, we should use stress concentrated solution, that under the condition of resonance frequency not be lowered, to improve the sensitivity, and to inhibit the unexpected twist.

On the basis of the above work, through adjusting the structure size, we can reduce mass size, and establish analysis model, they are shown in figure 2 and figure 3.

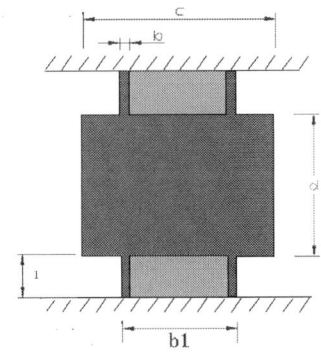

Figure 2.   analysis model 1

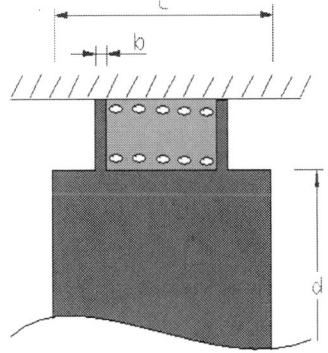

Figure 3.   Analysis model 2

Through the analysis of finite element, the analysis results of analysis model 1, porous free structure are shown as figure 4, the first three order resonant frequencies are 1083,1342, and 2561,respectively, stress is 7.

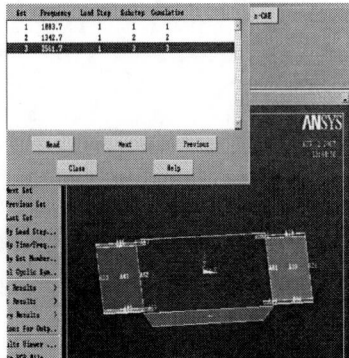

(a)the first three order resonance frequencies analysis results of porous free structure

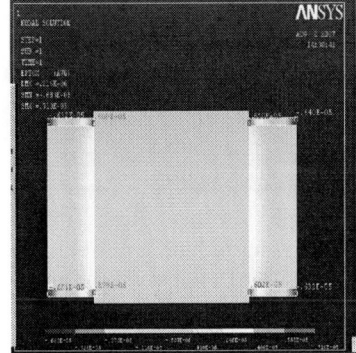

(b)Schematic diagram of every part stress variable in porous free structure

Figure 4. The analysis results of analysis model(porous free)

Analysis mode2, the analysis results of porous structure are shown as figure 5, the first three order resonance frequencies are 758,1181,and 1797.3,respectively, stress is 17.

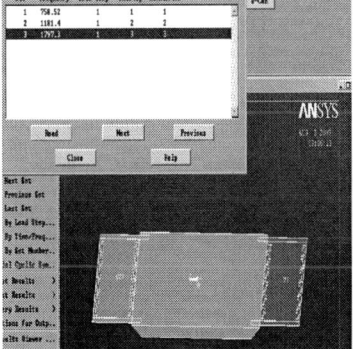

(a)the first three order resonance frequencies analysis results of porous structure.

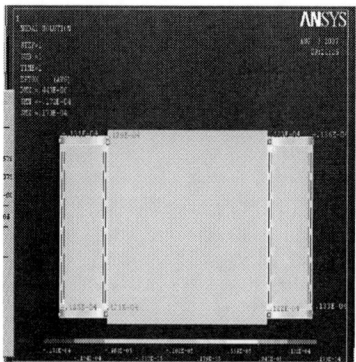

(a)Schematic diagram of every part stress variable in porous structure

Figure 5. analysis results of analysis model

As shown in figure4 and figure5,the frequencies of porous structure lowered slightly, but the stress improved much. After analyzing the results, we adjusted the mass size, and the thickness of the beam membrane, we determined the final structure size of the chip is: outer frame:9 × 7.5mm,inner frame:7 × 5.5mm,mass:4 × 4.5mm,beam width:3.6mm,beam thickness:50 μ m.

D. manufacturing of sensing chip

Choosing n type ＜ 100 ＞ silicon wafer, the diffusion resistance of the sensing resistor strip manufactured is 3～5K Ω ,sheet resistance is 300 Ω .The key technology is to use boron diffusion to manufacture sensing resistor, the doping concentration has impact not only the sheet resistance of the sensing resistor, but also piezoresistance coefficient and severe degree of the piezoresistance coefficient varies with the temperature. The manufacturing process flow of chips is shown as Figure 6.

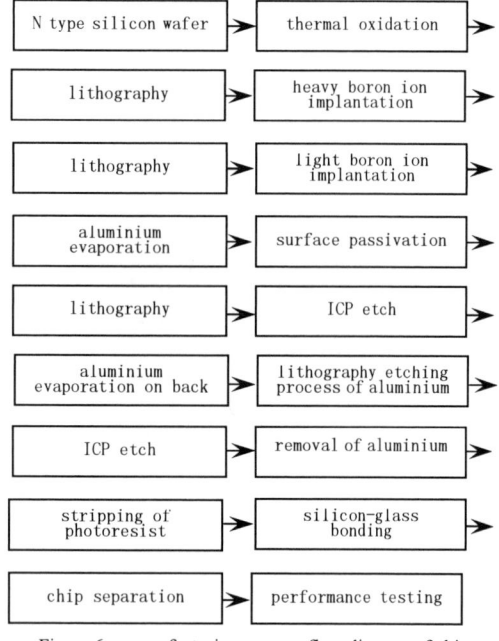

Figure.6   manufacturing process flow diagram of chip

In this paper, we use ion implantation technology as doping method, after theoretical calculation and process simulation, in order to obtain the ideal surface doping concentration and junction depth, we optimized the energy of ion implantation, dosage and annealing condition, the doping concentration of the real measured surface is 1.5e15/cm2 , junction depth is 2 μ m, sheet resistance of sensing resistance is 300 Ω .

Electrostatic bonding is key technology to realize chip packaging and protection, glass material is pyrex7740,EVG501 type wafer bonder is used to fulfil the process, process parameters are as follows: temperature 360 ℃ ,pressure 500N,voltage 1000V, power supply time 15 minutes, after bonding, the sealing strength tested is 9.8MPa.

The mass and beam membrane is fulfilled by adopting ICP dry etch technology. We choose 100 N type silicon wafers, using heavy boron diffusion technology to process upper layer single crystalline silicon, then diffuse the light boron resistance, evaporate aluminium wire, the fulfil the surface passivation, ICP etch, wet erode the upper and lower shell. After finishing the above process, we use static sealing scribing technology to process the chips.

Electrostatic bonding is key technology to realize ship packaging and protection, glass material is pyrex7740, EVG501 type wafer bonder is used to fulfil the process, process parameters are as follows: temperature 360℃,pressure 500N,voltage 1000V, power supply time 15 minutes, after bonding, the sealing strength tested is 9.8MPa.

E. packaging of sensor

The cross sectional view of vibration sensor is shown in figure 7, it is composed of shell, circuit board, sensing chip, potting material. We use co-vibration method to test acoustic vibration sensor, under the action of outer acoustic vibration signals, the sensor vibrated with the transmission medium, by measuring the weak signal, we can measure the vibration signal from the target outside.

The working environment of the acoustic vibration is soil medium, it should be buried under the soil, to meet this requirement, the density of acoustic coupling part of the sensor should be near to the soil density, and have the function of acoustic transmission, and reduce the phenomenon of acoustic reflection and acoustic scattering as little as possible. In this paper, we use acoustic material Polyurethane as acoustic coupling potting material, by doping and modulating polyurethane to meet the requirement of matching soil hardness. The real vibration sensor sample photo is shown in figure 8.

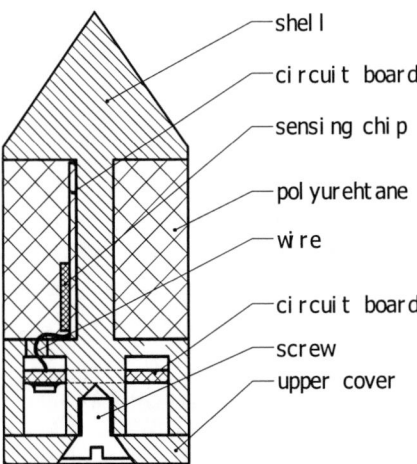

Figure 7. Schematic diagram of sensor structure

Figure 8. The real vibration sensor sample photo

III. TESTING RESULTS AND ANALYSIS

The comprehensive performance of MEMS vibration sensor is tested by B&K4808 type vibration instrument, the testing condition is, environmental temperature 22℃,humidity 30%RH,an atmospheric pressure. At the range of 5～5000Hz, the vibration instrument performs Logarithmic frequency sweep, The acceleration amplitude is about 1g.

A. frequency response test

At normal atmospheric condition, the vibration instrument performs Logarithmic frequency sweep, The acceleration amplitude is about 1g, frequency response curve is shown below, the point of 3db is 2.93kHz.

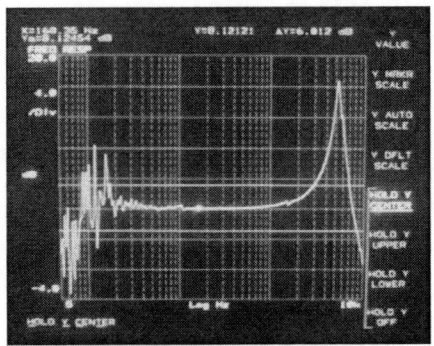

Figure 9. vibration frequency response curve of chip

B. sensitivity test

On the standard condition of measurement, we fix the vibration sensor and standard sensor to the vibration instrument with clamp or hard glue, the sensing axes of standard acceleration sensor and vibration sensor are both parallel to vibration direction. Connecting standard acceleration sensor to charge amplifier, adjusting the sensitivity range of charge amplifier to be able to match standard acceleration sensor, connecting power supply to vibration sensor, setting the working condition of vibration instrument is as follows: frequency 160Hz, acceleration amplitude 1-5g,and recording the output value of vibration sensor, the results is shown as table 1.

TABLE I.   TEST RESULTS OF VIBRATION SENSOR SENSITIVITY

NO.	1g	2g	3g	4g	5g
Z01	26.7	26.6	26.9	26.8	27.0
Z02	26.8	26.8	26.8	26.7	26.9
Z03	26.7	26.9	26.9	26.7	26.8

The test results indicate that the sensitivity of vibration sensor is 26.8mV/g.

C. resolution measurement

Installing the axis of vibration sensor parallel to rotating axis of turntable, powering the system, rotating the turntable one cycle, recording the angle of turntable when DC output is the minimum, at the point of the angle, rotating the turntable 1.8° again, recording the changing value of voltage, repeatedly measuring more than 3 times, the average value is the output variation Δ EP, if the results meet the equation $\dfrac{\Delta E_P}{\Delta E} \times 100\% \geqslant 50\%$,the sensor is qualified.

Δ E is the resolution of sensor, we measured the resolution of sensor is bigger than 5×10^{-4}V.

TABLE II.   MEASUREMENT RESULTS OF VIBRATION SENSOR RESOLUTION

NO	Δ EP	Δ E	Δ EP/ Δ E×100%
Z01	0.33mV	5×10^{-4}V	66%
Z02	0.38mV	5×10^{-4}V	76%
Z03	0.35mV	5×10^{-4}V	70%

D. *Road monitoring test of vibration sensor for vehicles and pedestrians traffic*

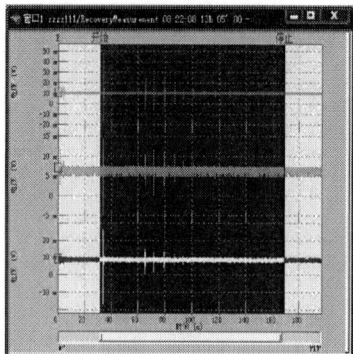

(a)At a certain period of time, the measuring signal of a same signal source

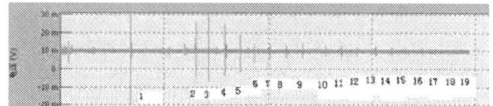

(b)the signal characteristic of a same target but at different distance

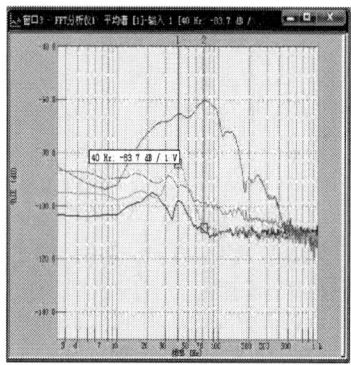

(c)T analysis results

Figure 10. The measuring results of vibration sensor for vehicles traffic

In the vehicle traffic test of vibration, （a）shows the monitoring signal to a same signal source at a certain period of time at a certain roadside. （b）shows the signal characteristics of a same target at different distance. （c）shows the FFT analysis results of featured signal.

It indicates that the vibration sensor has the feature of high sensitivity, and it can measure the targets, such as, vehicles, pedestrians traffic etc, accurately. Its signal amplitude is inversely proportional to transmission distance, featured signal is approximately 40Hz.

IV. CONCLUSION

There are many methods to improve the sensitivity of vibration sensor, in this paper, from the real application of vibration sensor, we propose a kind of beam membrane design of vibration sensing chip, and utilize analysis software to simulation and optimization, we solve the key boron diffusion process technology, silicon-glass bonding process technology and beam membrane ICP dry etch process technology, and

fabricate vibration sensing chip with an excellent performance. In order to meet the application condition of solid soil medium, we adopt performance-excellent acoustic material polyurethane as acoustic coupling potting material to package the vibration sensor, by doping and modulating polyurethane to match the hardness of the soil,in order to reduce acoustic scattering and acoustic reflection phenomenon. The measuring test indicates that the sensor has the features of high sensitivity, wide frequency response, high resolution, etc., the sensor is suitable to be installed in the soil to monitor the vehicles and pedestrians traffic.

REFERENCES

[1] B.L. Jiang, K.T.Young, J.A. Wang, S.G. Zong, "Desian and experiment research on MOEMSlow frequency acoustic sensor," Laser and Infrared, Vol.40,NO10, pp.1096-1100,2010.

[2] A. J. Zuckerwar, G. C. Herring, B. R. Elbing, "Calibration of the pressure sensitivity of microphones by a free-field method at frequencies up to 80 kHz," Journal of the Acoustical Society of America, Vol.119 NO1, pp.320−329,2006.

[3] S. Choi, etc, "A micro-machined piezoelectric hydrophone with hydrostatically balanced air backing," Sensors and Actuators A, 158, pp, 60−71,2010.

[4] B.Xie, et el, "Research and fabrication on Micro silicon bionic vector under acoustic sensor," Journal of sensing technology, Vol.10, NO 19, pp. 2300-2303, 2006.

This page intentionally left blank.

Quantitative Analysis and Identification of liver B-scan Ultrasonic Image Based on BP Neural Network

Fuzhen Zhu[*1], Bing Zhu[2], Peihua Li[3], Zhifang Wang[4], Liqiu Wei[5]

[*1]Electronic Science and Technology Post-Doctoral Research Station, Heilongjiang University, Harbin, China

[1,3,4] Colledge of Electronic Engineering, Heilongjiang University, Harbin, China

[2,5] Colledge of Electronic and Information Engineering, Harbin Institute of Technology, Harbin, China

[*1]zhufuzhen_1978@163.com

Abstract—**This paper aims to provide a method of quantitative analysis and intelligent identification for diagnosing fatty liver by B-scan ultrasonic image. By analyzing textural features of liver B-scan ultrasonic image, some features including angular second moment, entropy and inverse differential moment of gray level co-occurrence matrix are extracted from B-scan ultrasonic liver images, and then feature vectors are formed, which are input BP neural network for training classification and identification. BP neural network classifier can identify normal liver and fatty liver with the accuracy rate 85.71% and 88.89% respectively, at the same time, can identify slight, moderate and serious fatty liver with the accuracy rate 71.42%、 71.42% and 85.71% respectively. The method proposed here for quantitative analysis and diagnosing fatty liver can provide reference for doctor clinical diagnosis, and it will greatly improve the accuracy and efficiency of the fatty liver diagnosis combining clinical experience.**

Keywords-Gray level co-occurrence matrix；Fatty liver；B-scan ultrasonic imaging；BP neural network classifier

I. INTRODUCTION

Currently, the incidence of fatty liver is increasing, while it is not paid more attention by people. Although fatty liver is a benign lesion, it is likely lead to a permanent liver injury if not timely control and treatment, even growing to some irreversible diseases，such as liver fibrosis, liver cirrhosis, etc.

The most reliable way to diagnose fatty liver is liver biopsy in clinic[1], that is liver biopsy cytology, which is hard to be accepted by patients because of its traumatism. Currently, ultrasound imaging technology is a popular method to diagnose fatty liver in clinic. While this way is more dependent on individual vision observation and is lack of specific quantitative criteria. So it is urgent to establish an objective evaluation to provide reference and computer-aided diagnostic tool for doctors to diagnose fatty liver.

Large numbers of data on liver B-scan ultrasonic image home and abroad show that most of studies for diagnosis are based on texture features changes of B-ultrasonic images. Therefore, it is feasible to distinguish liver diseases by quantitative analysis of ultrasound image texture. At the same time, if we establish a computer-aided diagnosis of ultrasound image by way of combining the texture features with image recognition technology, that will be very helpful for doctors diagnosing fatty liver in clinical.

II. TEXTURE FEATURE EXTRACTION OF B-SCAN ULTRASOUND IMAGE

A. Collection of B-scan ultrasound images

In the help of experienced physician, we select 60 frames most representative liver B-ultrasound images (15 frames normal liver images, 45 frames fatty liver images with different degree). Two ultrasonic instruments are used in study. One is GE color Doppler ultrasonic diagnostic apparatus made in US, its model is C358 and transmit and receive frequencies are both 3.3MHz; the other is ALOKA ultrasound diagnostic apparatus made in Japan, its mode is SSD-1400 and transmit and receive frequencies are both 3.5MHz. Sizes of the collected images are 720×576 pixels.

B. Region of interest(ROI) Selection

According to doctors' experiences in diagnosing fatty liver , only the most effective pathological changes region are need to be analysed, and we called them region of interest(ROI). So ROI of each B-ultrasonic image are selected under the guidance of doctors, and principles are as follows:

978-1-4799-1215-5/13 $31.00 © 2013 IEEE

(1) There is only one ROI in each B-scan ultrasonic image.

(2) All pathological changes regions are considered referencing to doctors' experiences and fatty liver B-ultrasonic imaging characteristics, including front layer, middle layer and back layer of images (images are divided into three parts from top to bottom, the front, middle and back layer are of 1/3 each in turn.

(3) ROI is fixed, i.e. same parts lesions in the liver tissue are studied in different images.

(4) Sizes of ROI are fixed, i.e. 89×230 pixels.

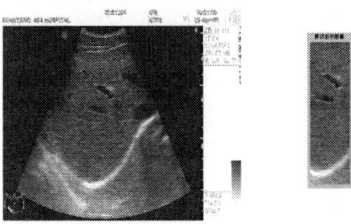

a）Origin image   b）ROI of Selection

Figure 1.   ROI of Liver B-Scan Image

C. Texture feature extraction based on GLCM

1) GLCM calculation of ROI

Image feature extraction is to extract statistical characteristics of ROI for image analysis, which is the base of image recognition. Texture features of ultrasound image include GLCM, statistical feature matrix, Fourier power spectrum, gray scale difference and Laws texture energy, etc[3]. Numerous studies find that GLCM (Gray Level Co-occurrence Matrix) can descript texture of the B-ultrasonic image commendably[4-9]. GLCM is an important method for analyzing image texture features, it is a way to calculate correlation of two points gray which meet a certain distance and a certain direction to reflect some comprehensive information such as direction, space, magnitude changes and so on. Therefore, GLCM can accurately describe texture rough degree and repeating directions.

The definition of GLCM is as follows: Probability of two gray i and j whose position direction is θ, distance is d, denoted as $P(i,j,d,\theta)$. There are four information including distance, angle, gray value and probability in GLCM. Generally, $d=1$, θ are $0°,45°,90°,135°$, because the four directions can representatively describe image grayscale changes. For example, two pixels $I(k,l)=i$ and $I(m,n)=j$, probabilities on four directions are respectively defined as：

$$P(i,j,d,0°) = \#\{[(k,l),(m,n)] \mid k-m\mathord{=}0, \mid l-n\mid\mathord{=}d,$$
$$I(k,l)=i, I(m,n)=j\}$$
$$P(i,j,d,90°) = \#\{[(k,l),(m,n)] \mid\mid k-m\mid\mathord{=}d, l-n=0,$$
$$I(k,l)=i, I(m,n)=j\}$$

$$P(i,j,d,45°) = \#\{[(k,l),(m,n)] \mid k-m\mathord{=}d, l-n=d,$$
$$I(k,l)=i, I(m,n)=j\}$$
$$P(i,j,d,135°) = \#\{[(k,l),(m,n)] \mid k-m=\text{ -}d, l-n=d,$$
$$I(k,l)=i, I(m,n)=j\}$$

Thereinto, # indicates the number of sets elements. We can see that GLCM is symmetric, i.e. $P(i,j,d,\theta)\mathord{=}P(j,i,d,\theta)$.

2) Features parameters derived from GLCM

Texture features parameters derived from GLCM are as follows[10,11]:

(1) Angular Second Moment (ASM)

$$ASM = \sum_{i=0}^{L-1}\sum_{j=0}^{L-1} \hat{P}_{\delta}^2(i,j) \tag{1}$$

ASM is the measure of image gray distribution. ASM is larger when image texture is in great detail and gray scale distribution is even. On the contrary, ASM is less when image gray distributes unevenly and roughness.

(2) Entropy(H)

$$H = -\sum_{i=0}^{L-1}\sum_{j=0}^{L-1} \hat{P}_{\delta}(i,j)\log \hat{P}_{\delta}(i,j) \tag{2}$$

Entropy is the measure of information amount in the image and shows the complexity or non-uniformity of image texture. Entropy is larger when texture is complex, otherwise is less when texture is simple.

(3) Inverse Differential Moment (IDM)

$$IDM = \sum_{i=0}^{L-1}\sum_{j=0}^{L-1} \frac{\hat{P}_{\delta}(i,j)}{1+(i-j)^2} \tag{3}$$

IDM reflects the rule degree of image texture. IDM is larger when texture is regular and easy to describe, otherwise it is less when texture is irregular and uneasy to understand.

In the experiment, we calculate 60 frames typical sample images which we selected，and summarize the range of parameters different types of B-ultrasound images, showing as table 1.

978-1-4799-1215-5/13 $31.00 © 2013 IEEE   63

TABLE I. STATISTICAL RESULT OF TEXTURE FEATURE PARAMETER

			ASM	H	IDM
0 angle	normal liver		0.001748~0.0022235	4.0537~4.2762	0.093702~0.11293
	fatty liver	Slight	0.0015747~0.0017639	4.2806~4.3039	0.086027~0.10076
		moderate	0.0015544~0.0015679	4.3022~4.3211	0.082778~0.096104
		serious	0.0013046~0.0013652	4.3267~4.3539	0.080297~0.081006
90 angle	normal liver		0.0011827~0.0014936	3.9931~4.1128	0.059945~0.064538
	fatty liver	slight	0.001223~0.0014857	4.0894~4.1427	0.053613~0.056678
		Moderate	0.001187~0.001289	4.1179~4.1317	0.052224~0.054165
		serious	0.0010475~0.0011624	4.1338~4.1863	0.037904~0.042215
45 angle	normal liver		0.0011759~0.0014598	4.0043~4.1211	0.09847~0.12859
	fatty liver	slight	0.0012151~0.0012841	4.0747~4.1405	0.075536~0.090885
		Moderate	0.0011624~0.0012624	4.1284~4.1533	0.058428~0.063726
		serious	0.001034~0.001076	4.1648~4.1919	0.031643~0.034215
135 angle	normal liver		0.0011489~0.0013652	4.0453~4.0712	0.057106~0.065424
	fatty liver	slight	0.0011151~0.001257	4.0889~4.1173	0.054353~0.058355
		Moderate	0.0011016~0.0011827	4.1151~4.1346	0.041325~0.045438
		serious	0.0011003~0.0011557	4.1349~4.1537	0.037791~0.042338

From above Table 1, We can see that these statistical parameters are in a certain scopes and rules, i.e. according to the normal liver, mild, moderate, severe fatty liver order, values of energy and inverse moments are decreasing, while values of entropy are increasing.

III. B- SCAN ULTRASONIC IMAGE CLASSIFIER BASED BP NEURAL NETWORK

BP neural network is one of the widely used neural network model. It is the map of input to output by minimizing the cost function[12-14]. To realize the recognition classifier, we use the Matlab software which has a complete neural network toolbox[15-17] , so that we can focuse more attentions on network identification program and network structure, training parameters , and so on.

A. Determination of network structure and parameters

BP network structure mainly includes network layers, nodes of each layers and transfer functions between layers. These factors are determined as following[18-20]:

(1) Determination of network layers. Previous experience is that as long as a sufficiently large number of hidden nodes, single hidden layer neural network can approach any continuous function with arbitrary precision on a bounded domain'. Therefore, layers of BP network are determined as three layers.

(2) Nodes determination of each layers.

□ Input layer nodes number. Generally, BPNN input layer nodes number is determined by input vectors dimensions. In the experiment, feature extraction is done for each B-ultrasonic image, eigenvalues are three statistics of four directions derived from GLCM , i.e. second moment, entropy and inverse moment, total of 12 eigenvalues. Such 12

eigenvalue composes a feature vector to represent a B-mode images. Therefore, number of input layer nodes is 12.

②Output layer nodes number. BPNN onput layer nodes number is determined by expected output vectors dimensions. BPNN classifier goal is to divide liver images into normal liver, mild, moderate and severe fatty liver four categories. So we use a 2-dimensions vector to represent the class attribute, i.e. setting desired output (1,1) is normal liver, (-1，1)、 (1，-1) and(-1，−1) are light, moderate and severe fatty liver respectively. So, number of output layer nodes is 2.

③Hidden layer nodes number. So far, there is no theory to accurately determine the hidden layer nodes. So we determine it by repeated experiments based on empirical formula, i.e. ultimately is 20. Finally, BPNN structure is as Figure 2.

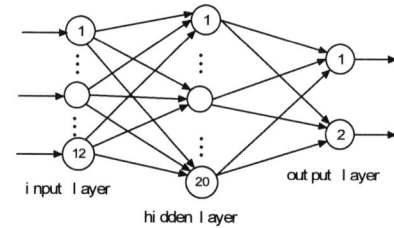

Figure 2. BPNN Structure Sketch Map

(3) Determination of transfer functions and training parameters. Hidden layer is Sigmoid function, input and output layer is line function. Training parameters as follows: expected training error is 0.001, maximum training epoch is 5000, training rule is SCG algorithm.

B. Experiment result

Training samples include 32 training samples and 28

testing samples. Net training is convergent after 795 epochs training and costing time 25 minutes. Recognition interface is shown in Figure 3, when 12 eigenvalues are all input in Fig.(a), then classification result is given in Fig.(b). At the same time, a specific classification result will be given in matlab command line.

Recognition results of 28 testing samples in matlab command line are shown in Table2. Thereinto, error results are marked in italic bold and big class recognition is normal liver and fatty liver; small class recognition is light, moderate and severe fatty liver respectively.

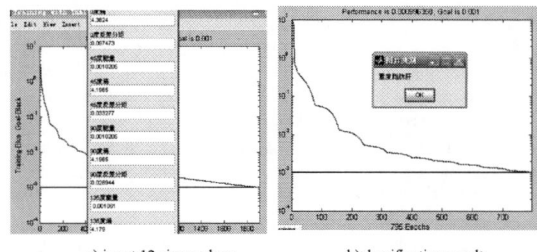

a) input 12 eigenvalues   b)classification result

Figure 3. Recognition Result of BP Network Classifier

TABLE II. TESTING RESULT OF BP NETWORK WITH LEAVING SAMPLES

Sample NO.	Output	small class	big class	Sample NO.	Output	small class	big class
9	0.2820, 0.8796	T	T	39	1.1023, -0.9102	T	T
10	*1.7155, -0.9740*	*F*	*F*	40	0.8512, -0.6356	T	T
11	0.8246, 0.3871	T	T	41	*-1.7323, 0.7629*	*F*	*F*
12	0.3209, 0.9583	T	T	42	*0.8910, 1.1239*	*F*	*F*
13	0.7826，0.5864	T	T	43	1.4209,-0.9413	T	T
14	1.3301, 0.9546	T	T	44	0.8619, -1.0215	T	T
15	0.9358,0.9210	T	T	45	0.9133,-0.8960	T	T
24	-0.3911,1.6954	T	T	54	*1.5010,0.8133*	*F*	*F*
25	*0.8213,0.9985*	*F*	*F*	55	-0.8966,-0.6573	T	T
26	-0.6692,0.7369	T	T	56	-0.3499,-0.7322	T	T
27	*0.1227,-0.8861*	*F*	*F*	57	-0.9812,-0.7908	T	T
28	-1.3021,1.2380	T	T	58	-1.5101,-0.9364	T	T
29	-0.9633,1.0842	T	T	59	-0.7425,-0.6965	T	T
30	-0.5991,1.3948	T	T	60	-0.7932,-0.8610	T	T

Statistical recognition result in Table 2: □ big class recognition result: normal liver correct identification rate is 6/7=85.71%; fatty liver correct identification rate is 40/45=88.89%; □ small class recognition result: correct identification rate of light, moderate and severe fatty liver are respectively 5/7=71.42%、5/7=71.42% and 6/7=85.71%.

IV. CONCLUSION

The paper proposed a method of quantitative analysis and intelligent identification for diagnosing fatty liver by B-scan ultrasonic image. This method mainly include two parts: firstly, texture feature extraction based on gray level co-occurrence matrix was established to form feature vectors of fatty liver B-scan ultrasonic image; then, B-ultrasonic image classifier based on BP neural network was trained. The final trained BP neural network classifier can identify normal liver and fatty liver with the accuracy rate 85.71% and 88.89% respectively, at the same time, can identify slight, moderate and serious fatty liver with the accuracy rate 71.42%、71.42% and 85.71% respectively.

The method proposed here can be used as an objective reference for doctors in diagnosing fatty liver. It will greatly improve the accuracy and efficiency of fatty liver clinical diagnosis combining doctors' experience. It is very meaningful for quantitative analysis and intelligent identification in fatty liver diagnosis.

ACKNOWLEDGMENT

The Paper is supported by the Project of Science and Technology Research of the Education Department of Heilongjiang Provincial, China (No. 12531490) and PhD Start-up Fund of Heilongjiang University.

REFERRENCES

[1] Paul A. Nonalcoholic fatty liver disease[J]. The New England Journal of Medicine，2002，346(16):1221-1231.

[2] Fuzhen Z, Bin W. B-scan ultrasonic image feature extraction of fatty liver based on gray level Co-occurrence matrix. Chinese Journal of Medical Imaging Technology, 2006, 22(2): 287-289.

[3] Zhao H, Bao L, Liang GM, et al. Research on micro-cell image based on synthesized gray level Co-occurrence matrix[J]. Techniques of Automation and Applications, 2004, 23(10):27-33.

[4] Longhao J, Zhongfa Z; Bo L. Study of SAR Image Texture Feature Extraction Based on GLCM in Guizhou Karst Mountainous Region [C]. 2012 2nd International Conference on Remote Sensing, Environment and Transportation Engineering (RSETE), 2012:1-4.

[5] Jiaqiang G, Xiao-ning L. Improved texture feature extraction algorithm based on GLCM[J]. Computer Engineering and Design, 2011, 32(6):2068-2071.

[6] Qing L, XiPing L, LiJun Z, LiMin Z. Image Texture Feature Extraction & Recognition of Chinese Herbal Medicine Based on Gray Level Co-Occurrence Matrix[J]. Advanced Materials Research, 2012: 605－607.

[7] Haralick R M, Shanmugam K, and Dinstein I , et al. Textural Features for Image Classification[J]. IEEE Transactions on Systems, Man and Cybernetics, 1973, SMC-3(6): 610-621.

978-1-4799-1215-5/13 $31.00 © 2013 IEEE

[8] Chanda B, Dutta M D. A note on the use of the graylevel co-occurrence matrix in threshold selection[J]. Signal Processing, 1988, 15(2):149-167.

[9] Steven W, Zucker, Demetri T. Finding structure in Co-occurrence matrices for texture analysis[J]. Computer Graphics and Image Processing, 1980, 12(3):286-308.

[10] David A, Clausi. An analysis of co-occurrence texture statistics as a function of grey level quantization[J]. Canadian Journal of Remote Sensing, 2002, 28(1):45-62.

[11] Xiaofeng W, Hui Z. Visual C++/Matlab Image Processing and Identification Selected Cases[M]. Beijing: Posts & Telecommunications Press, 2004.

[12] Steve Lawrence, Ian Burns, Andrew Back, Ah Chung Tsoi, etc. Neural Network Classification and Prior Class Probabilities[J]. Computer Science, 2012,(7700):295-309.

[13] Zhong Z. Neural Network[M]. Beijing: Higer Education Press, 2009.

[14] Travis A. Jarrell, Yi Wang, Adam E. Bloniarz, Christopher A. Brittin, Meng Xu, J. Nichol Thomson, Donna G. Albertson, David H. Hall, and Scott W. Emmons. The Connectome of a Decision-Making Neural Network[J], science, 2012 , (27) :437-444.

[15] Burks T. F, Shearer S.A, Gates R S, et al. Backpropagation neural network design and evaluation for classification weed species using color image texture[J]. American Society of Agriculture and Biological Engineers. 2000, 43(4):1029-1037.

[16] Dong X, Zheng W. System Analysis and Design—Neural Network[M]. Xi An: Xi Dian University Press, 2003.

[17] Hou Z Y, Dai Q L, Wu X Q, et al. Artificial neural network aided design of catalyst for propane ammoxidation [J]. Applied Catalysis A: General. 1997,161(1-2):183-190.

[18] SuXiang Q, FuXi L , Jian C. Design of Sound Recognition System Based on Modified Neural Network[J]. Applied Mechanics and Materials , 2013: 1178-1181.

[19] DanDan C, Fei L. The Application of BP Neural Network in Internet of Things, Advanced Engineering Forum[C] 2012: 1098-1102.

[20] Changhao X, Zhonghua Y, Bangjun L, Qiufeng Z. SCG and LM Improved BP Neural Network Load Forecasting and Programming Network Parameter Settings and Data Preprocessing[C].   2012 International Conference on Computer Science & Service System (CSSS), 2012: 38-42.

978-1-4799-1215-5/13 $31.00 © 2013 IEEE

Modified BP Neural Network Model is Used for Odd-even Discrimination of Integer Number

LIAN Tongli, XIE Minxiang, XU Jiren, CHEN Ling, GAO Huaihui
Department of Information
Electronic Engineering Institute of Hefei
Hefei, Anhui, China
xujiajun@mail.hf.ah.cn

Abstract—This paper introduced the BP neural network model and the BP algorithm in detail, and points out the BP neural network exists the defects of local optimal tendency of local optimal, slow convergence speed etc. Through the introduction of modified BP algorithm, we can solve the problems existing in the traditional BP algorithm successfully, simulation results for odd-even discrimination of integer number based on MATLAB BP algorithm show that modified BP model compared with BP model, has faster training speed and high study accuracy. Modified BP neural network models is used in practice, as long as it is complementary with effective measures, and we can get satisfactory result completely.

Keywords-BP neural network, modified BP algorithm, odd-even discrimination

Artificial neural network model is simulation and abstraction of basic characterof natural or human neural networks, usually it is interconnected with many simple calculation unit (also called neurons or nodes) and comes into the structure of complex network structure. The neural network has the advantageof self-learning ability, self-organizing, associative memory, distribution parallel information processing, etc. The BP neural network model is considered to be the neural network model which is used widely, but the BP neural network has BP neural network has the defects of local optimal tendency of local optimal, slow convergence speed etc obviously.

I. THE TRAINING PROCESS OF BP NETWORK STUDY

1) Make all starting data of threshold vector $\theta_i(0)$ and power vector $W_{ij}(0)$ are small nonzero random value.

2) Use study sample to train BP network:

Target output is $d_p(p=1,2,...,p)$, and input vector is $X_p(p=1,2,...,p)$.

3) Get the output results of hidden and output layer of BP neural network (suppose incentive function is the function of Sigmoid type):

$$o_{pj} = f_j(\sum W_{ij}o_i - \theta_j)$$

4) Get the error of hidden and output layers after operation:

$$\delta_{pj} = o_{pj}(1 - o_{pj})(t_{pj} - o_{pj}) \quad \text{(output layers)}$$

$$\delta_{pj} = o_{pj}(1 - o_{pj})\sum_k (o_k W_{jk}) \quad \text{(hidden layers)}$$

Among them:: k is the the node number of previous layer of which No.j neuron is on.

t_{pj} is the expected output data of every output node (use binary system to denote).

5) Update threshold values and weights:

$$W_{ji}(t+1) = W_{ji}(t) + \eta\delta_j o_{pj} + \alpha(W_{ji}(t) - W_{ji}(t-1))$$

$$\theta_j(t+1) = \theta_j(t) + \eta\delta_j + \alpha(\theta_j(t) - \theta_j(t-1))$$

Among them: α is constant, which is impact factor of past weights (or threshold values) for the current weights (or threshold values),

η is training speed, which should take larger value as far as possible if system do not have oscillation.

6) Calculation error could meet the given precision index, usually calculte $E \le \varepsilon$. (in the formula, $E_p = \sum(t_{pj} - o_{pj})^2/2$, $E = \sum E_p$, ε takes small real number, and $\varepsilon \le 0.0004$ this section), and if the error meets $E \le \varepsilon$, turn step 7), otherwise turn to step 3).

7) Stop。

II. PROBLEMS OF BP NEURAL NETWORK ALGORITHM

BP algorithm is one of the most effective algorithm accepted currently, and also the basis of pattern identification of BP neural network. But it also exists some shortcomings[1].

1) From mathematics aspect singly, it needs to solve the nonlinear optimization problems, and this will bring about the problem of local minimum value inevitably.

2) Convergence speed of BP training algorithm is slow, and sometimes even repeat thousands of times or more.

3) BP network structure only have forward propagation lines, and does not exist feedback part;

4) To choose hidden layers of the network basically relys on artificial experience value yet;

5) Subsequent learning samples will affect the network to train with the already learned samples, and demand the dimension of single input sample must be equal.

III. MODIFIED BP ALGORITHM

Aiming at the insufficiency of classical BP algorithm proposed in section II , give the corresponding modified algorithm [2].

A. Variable Step Length

BP algorithm is based on the basis of gradient descent method. Usually in dealing with optimization problem, the step length η is a variable searched and gotten from a one-dimension. But when we use classical BP algorithm, η is fixed, because cost function E is a very complicated function, and nonlinear, and it is difficult to get optical step length by means of minimal method. If we calcute y in each iteration time, then computation will be very big. We analyze error curve of BP network, note that a flat fiels exist, and if η is too small in plain area, repetitions will rise, and if η is too large in the region of dramatic change, repetition will increase on the contrary, and this will effect convergence. Therefore, variable step length is a reasonable method, that is to say, we given a step length at first, if error function E decreases after iteration, then step length will multiplied by a constant ψ greater than 1; If error function E increases after iteration, then step length will multiplied by a constant β less than 1, and iterate again (to same direction). Doing it like this do not bring much computation, and can adjust the step length rationally [3].

$$\eta = \eta\psi \quad \psi > 1 \quad \text{when } \Delta E < 0$$
$$\eta = \eta\beta \quad \beta < 1 \quad \text{when } \Delta E > 0$$

Among them：$\Delta E = E(t) - E(t-1)$。

B. Variable Momentum Factor

We introduce a variable momentum factor α in the BP algorithm, and on the one hand, it can accelerate convergence, and can prevent oscillation at the same time[4].

$$W(t+1) = W(t) + \eta(t)d(t) + \alpha\Delta W(t)$$

The NO.3 item is the adjustment of direction of weigh for previous moment in above formula, while modified direction of (t+1) moment is relative to (t+1) moment and (t) moment. So above formula will be converted

$$W(t+1) = W(t) + \eta(t)[d(t) + \alpha\Delta W(t)/\eta(t)]$$
$$= W(t) + \eta(t)[d(t) + \alpha\eta(t-1)d(t-1)/\eta(t)]$$

From form, above formula is similar to conjugate gradient method, but d(t) and $d(t-1)$ are not conjugate actually, but 0 $<\alpha<1$. So when adjusting η, if $\Delta E > 0$, reduce η and make $\alpha = 0$, If $\Delta E > 0$, restore original α.

C. Introducing γ Factor

When training network, if $(t_{pj} - o_{pj}) \neq 0$ and $o_{pj}(1 - o_{pj})$ tends to zero, the second category of local minimum will happen, and it can realize by modifying some network attributes and break away from not sensitive area as soon as possible[5]. We introduce a factor γ in incentive function. Make the network of output become

$$o_{pj} = f_j(Net_{pj}/\gamma_j) = f_j[(\sum W_{ji}\theta_i - \theta_j)/\gamma_j]$$

If system goes into flat area or occurrence of the second category of local minimum in the training process, and decreases weights W and threshold value θ, and thus make the gradient o_{pj} exit zero, and break away from plain area. Usually the second category of local minimum occur easily, and therefore, it can avoid most local minimum, and thus speed up the network convergence speed.

Overall, modified BP training algorithm is expressed as follows:

$$W_{ji}(t+1) = W_{ji}(t) + \eta(t)\sum_p \delta_{pj}o_{pj} + \alpha\Delta W_{ji}(t)$$

If it is ouput layer
$$\delta_{pj} = o_{pj}(1 - o_{pj})(t_{pj} - o_{pj})/\gamma_j$$

If it is hidden layer
$$\delta_{pj} = o_{pj}(1 - o_{pj})\sum_k \delta_k W_{jk}/\gamma_j$$

If $\Delta E < 0$ $\eta(t+1) = \eta(t)\psi, \alpha = \alpha$

If $\Delta E > 0$ $\eta(t+1) = \eta(t)\beta, \alpha = 0$

Among them ψ>1，β<1，$\Delta E(t) = E(t) - E(t-1)$

If there is a local minimum, we can adjust γ_j to solve.

IV. PAY ATTENTION TO SEVERAL QUESTIONS WHEN USING BP NETWORK MODEL

A. BP Network Layers

Neural network with at least a hidden layer (s-type incentive function) and bias and an output layer, can approximate any rational function, and above conclusions theoretically have been given proof. Basically, it become design basis of BP network. Increasing layer can improve precision, and reduce errors, but also makes the network complicated, and study time of network weight increases. Network with only a hidden layer can also improve error precision by increasing the number of neurons in hidden layer, and it is easy for this kind of method to realize than the method with increasing hidden number, and its learning effect is easier to adjust and observe than the method by increasing hidden layer. So usually BP network with only a hidden layer is adopted further[6].

B. Appropriate Node Number of Hidden Layer

Choosing node number of hidden layer is a difficult problem. Because there is no good analytical formula to express, so Eberhart once called it "this is a kind of art" in his writings. Question itself, number of output and input unit and node number of hidden layer is relative directly. If node number of hidden layer is less, it can not come to the demand of study and fault tolerance of network becomes poor, and overmuch hidden nodes will increase training time, and error results may not be ideal. So there must be an optimal hidden nodes and we can consult the inequalities given below [7].

$$k < \sum_{i=1}^n C\binom{n_1}{i}$$

978-1-4799-1215-5/13 $31.00 © 2013 IEEE

Among them, k is sample number，n_1 is node number of hidden layer，n is the node number of input layer，and if i>n

$$, \quad C\binom{n}{i} = 0$$

$$n_1 = \sqrt{n+m} + a$$

Among them：n is the node number of input layer，a is constant from l to 10。M is node number of outputlayer。

$$n_1 = \log 2n$$

Among them, n is the node number of input layer。

C. Normalization of Training Sample Data

Physical quantities which input data of BP neural network use is inconsistent mutually, and numerical difference of same physical quantity is likely to be larger and therefore it is necessary to normalize input data, and avoid that big numerical samples " inundate " small numerical samples. If it is still not enough to normalize every input simple to [0, l]. Because Sigmoid incentive function is saturated in the domain of [0.9, 1.0] and [0.0, 0.1], and fluctuation is extremely smooth, so correct processing method should normalize all input data to the range of [0.1, 0.9]. Therefore use the following formula for normalization:

$$y = \frac{x - x_{min} - 0.01}{x_{max} - x_{min} + 0.01} \times 0.8 + 0.1$$

In above formula, x_{min} and x_{max} is the minimum and maximum of sample data; 0.01 in the numerator and denominator is modification to prevent zero input data appear.

D. Setting Initial Weights

In nonlinear system, the initial weights is very important because it determines whether training can reach global optimal point or not, convergence or not and long learning time or short. If setting too big initial weights, then sum of input after weighted may be in saturated zone of s-type transfer function, and makie its derivative f'() very small, but if f'()$\to 0$, $\Delta w \to 0$, and it makes the whole training process halt almost. Therefore, the ideal situation is: each neuron input value after initial setting appropriate zero, so it ensures that each neuron node can be located where variation is the greatest in their s-type activation function. Usually take initial weights between -1 and 1 randomly.

E. Determining Expected Error

In the training of network, selecting expected error is also critical. But find a suitable numerical value by contrast training, and here the proposed "appropriate", is to aim at node number of hidden layer, because greater expected error value depends on reducing node number of hidden layer and study time to realize. Normally taked as reference, we can use several different expected error value to train network, and finally select one of the errors after considering many factors synthetically.

F. Choosing Sample Characte

The state of the system can be expressed by means of multiple character. So we should take what can reflect its status adequately as character of input layer of the network [8], to distinguish it from other to be identified.

V. MODIFIED BP ALGORITHM BASED ON MATLAB FOR ODD-EVEN DISCRIMINATION OF INTEGER NUMBER

In this section, to demenstrate modified BP algorithm through the simulation of MATLAB toolbox Simulink module and network programming, and it will compared with classical BP algorithm.

For Example, construct BP network of single output, including a hidden layer. Using a variety of different algorithms to train the network and test odd-even discrimination of integer number. Suppose learning samples are p=[0 1 2 3 4 5 6 7 8 9], [t = 1 0 0 0 0 1 1 0 1 1].

Construct a BP neural network which has 10 input layer node, four hidden nodes, and 1 output layer node. With one-dimensional numerical value of p set of as input, and one-dimensional numerical value of t set as output. Use classical BP algorithm, learning algorithm of variable step length, the method of variable momentum factor, fusion algorithm with variable momentum factor and variable step to train the network respectively, after numerous simulation calculation, we can get error curve of network learning as figure 1 - figure 4 below.

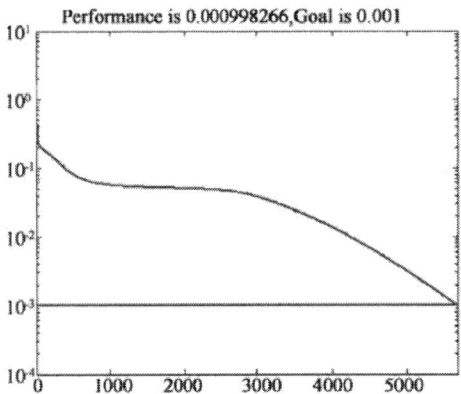

Figure 1. learning error curve of standard BP algorithm

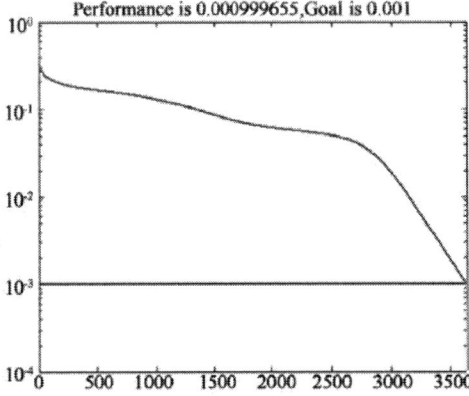

978-1-4799-1215-5/13 $31.00 © 2013 IEEE

Figure 2.   learning error curve of variable momentum factor method

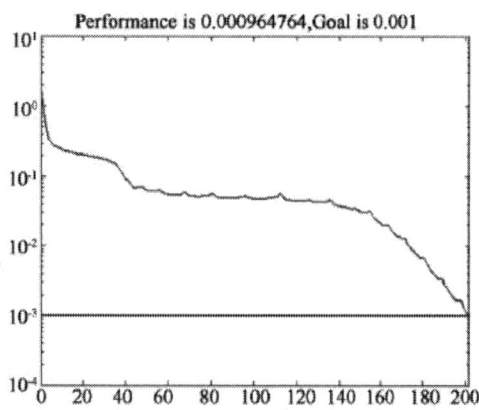

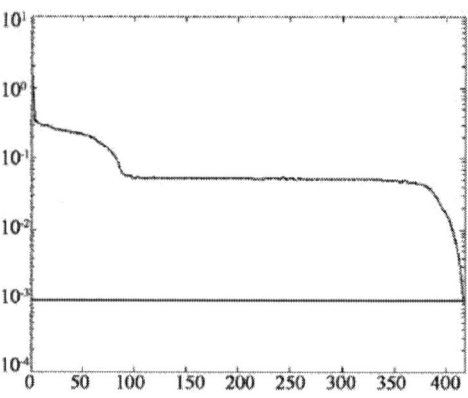

Figure 3.   learning error curve of method combining variable step and introducing γ factor

Figure 4.   learning error curve of fusion algorithm with variable momentum factor and variable step

VI.   CONCLUSION

Through compareing and analysing various simulation results of BP neural network algorithm, we can find under the situation in which convergence rate is the slowest is basic BP algorithm showed as figure 1 in the same precision, after 5562 times learning ceases, From variable momentum method showed as Fig 2, we can find the introduced momentum is almost equal to damping term, and it suppresses the oscillation movement of training process, thus affects the network convergence properties, so learning terminate after 3522 times training, and network study times come down, and learning rate increase, but error convergence rate increased a lot, the method combining variable step and introducing γ factor in Figure 3 has higher convergence rate, and training will be terminate after 211 times study, and enhance above 10 times compared with classic learning algorithm, fusion algorithm with variable momentum factor and variable step in Figure 4 has higher convergence rate, and learning terminates after 408 training, and the simulation data shows that convergence rate of this algorithm is close to fusion algorithm with variable momentum factor and variable step.

From the above analysis, we can find that modified BP network model successfully resolves odd-even discrimination of integer number, and in actual application process, as long as we apply reasonably, obtain satisfactory diagnosis reslut. Modified BP model compared with the BP model, has faster learning rate and higher training accuracy.

[1] Lili Rong, "Building a three-layer BP neural network by fuzzy rules," Proc of the 9th International Conference on Neural Information Processing,2002:452-456.
[2] CUI Wen-bin, ZHANG Yue-wen, and WU Gui-tao etc, "Applying back propagation neural networks to remote monitoring of ocean-going ships," Journal of Harbin Engineering University,2009,(8):935-939.
[3] Rio de Janeiro, "Adaptation of Parameters of BP Algorithm using Learning Automata," VI Brazilian Symposium on Neural Networks, Brazilm,2000,(1):22-24.
[4] LI En-yu, YANG Ping-xian, and SUN Xing-bo, "Improved Algorithm of BP Neural Networks Based on the Activation Function with Four Adjustable Parameters," Microelectronics & Computer,2008,(11):89-93.
[5] Zhang Yun and He Yong, "Study of prediction model on grey relational Bp neural network based on rough set," Machine Learning and Cybernetics,2005,(8):4764-4769.
[6] Zhu Jian-yuan, "Marine diesel engine vibration monitoring based on BP neural network," Mechanical and Electrical Equipment,2008,(3):33-36.
[7] WANG Xiu-ying, "Fault Pattern Recognition Based on Improved BP Network Algorithm," Mechanical & Electrical Engineering Technology,2008,(10):103-105.
[8] LI Zong□kun and ZHENG Jing□xing; ZHOU Jing, "Improved BP model and its application of earth dam monitoring data analysis," Journal of Hydraulic Engineering,2003,(7):111-114.

The effect of incident angle of pumping light to Cholesteric Liquid Crystal

Xiangbao Yin[12], Yongjun Liu[1]*, Lingli Zhang [3]

1 Harbin Engineering University, Harbin, China

2 Heilongjiang University of Science and Technology, Harbin, China

3 Harbin Institute of Technology, Harbin, China

liuyj@hrbeu.edu.cn

Abstract—Optical characteristics research was carried out for the tunable laser of dye doped cholesteric liquid crystal. Based on the characteristic that pitch of the dye doped cholesteric liquid crystal varyed with the incident angle of pumping light, emission characteristics of the tunable laser of dye doped cholesteric liquid crystal were studied. Taking the mixture of laser dye and nematic liquid crystal and left-handed chiral additive as the layer of gain medium, wavelength tunable laser was prepared. The laser sample pumped by 532nm Nd : YAG pulse laser has the following optical characteristics: the wavelength of the emission laser was tuning from 647.38 to 658.11nm with the incident angle of pumping light, reaching 10.73nm.

Keywords-cholesteric liquid crystal; tunable laser; incident angle

I. INTRODUCTION

As early as in 1998, it was reported for the first time that mirrorless lasing was realized with cholesteric liquid crystal(CLC) by V. I. Kopp et al.[1]. in recent years researchers gradually increased,Yuhua Huang[2] vertically put one side of the dye-doped CLC cell near a thermal platform of 40°C, forming a temperature gradient in the cell. Since the temperature would influence pitch, the wavelengths of lasers were tunning from 577nm to 670nm by pumped different position of the cell. L. M. Blino from Russian Academy of Sciences and GciPParrone et al.[3] from Calabria University of Italy adopted a light-tight hollow electrode to replace a ITO electrode in traditional liquid crystal cell, as a lkHz square-wave voltage being applied on two electrodes of the liquid crystal cell, 25nm wavelength tuning range (600nm to 625nm) was obtained. Moreover, some research [4-5] also dropped the mixture of liquid crystal and dye into the pre-made photonic crystal, and obtained tunable laser output with low threshold value.

By summarizing the above mentioned research results, most of the researchers took the influence of temperature, doping concentration of the dye, pressure, etc. On the thread pitch of liquid crystal as the tuning methods, i.e. influencing the optical band gap formed by CLC to reach the target of tuning output laser wavelength. In this article, we studied the tuning laser by varying the incident angle of pumping light.

II. EXPERIMENT

A. Preparation of laser sample

Glass plate with a thickness of 1.1mm was selected as the base plate of laser sample,Evenly apply polyimide (PI) alignment layer on one side of the base plate by spin coating, after treatment of rubbing alignment, make empty cells by arranging the friction direction of two base plates as anti-parallel, and control the thickness of liquid crystal layer by utilizing spacers [6]. The cell gap was 10μm. Nematic LC BHR33200 (n_o=1.508, n_e=1.657, Δn =0.149 at 20°C, 589nm;clearing point is 65°C); left-handed chiral additive S811 is selected as the chiral additive，supplied by Beijing Bayi Space LCD Materials Technology Co .

Ltd.dye4-Dicyanomethylene-2-methyl-6-(4-methylaminos tyryl)-4H-pyrane (DCM) is elected as the laser dye, supplied by American Exciton Company; and UV-2450 UV spectrophotometer made by SHIMUZU of Japan is adopted for transmission spectrum test.The material was a mixture of 72% nematic LC BHR33200,27% left-handed

YAG frequency doubling pulsed laser provided by Beijing Beamtech Optronics Co, Ltd., with a repetition frequency of 1Hz. Pumping laser is focused on the sample by convergent lens after passing energy attenuation plate, there is a specified angle between the face normal of the sample and the pumping light, area of the light spot is

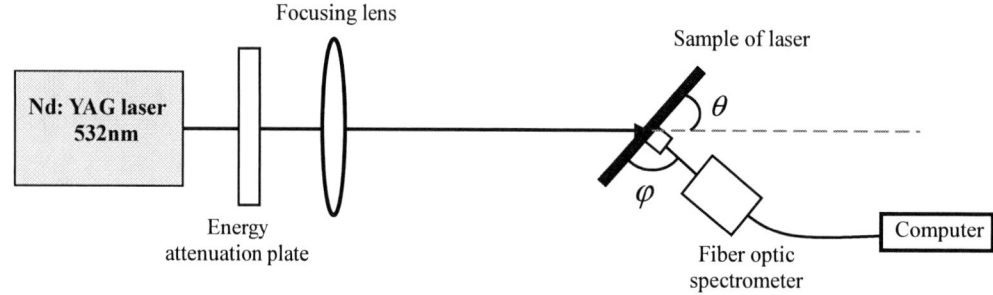

Figure 1. Experimental diagram of the dye doped CLC laser

chiral additive S811 and 1% laser dye DCM, stir well by ultrasonic oscillator after mixing, and inject into the prepared empty sample cells.

B. Emission characteristics experiment of tunable laser varying with the incident angle of pumping light

By mixing nematic LC and chiral additive, CLC can be formed, which has a self-organized periodic helix structure, i.e. periodic reflective index modulation, can be regarded as one-dimensional photonic crystal. If fluorescence spectrum of the dye overlaps photonic band gap of the CLC, narrow linewidth laser can be produced under the excitation of pump light. Adopt Dawa-100Nd:

about $2mm^2$, which will be imported into computer for processing from the spectrometer by emission spectrum received by the fiber optic probe perpendicular to the surface of the sample. The schematic diagram of experimental optical path is shown in Figure1. The emission laser will be produced perpendicular to the surface of the sample. With angle θ between the pumping light and the face normal of the laser sample changing from 90° to 55°, effective refractive index of the LC has corresponding variation, which resulted in the variation of the cavity length of the laser resonant cavity. And wavelength of emission lasers shifted.

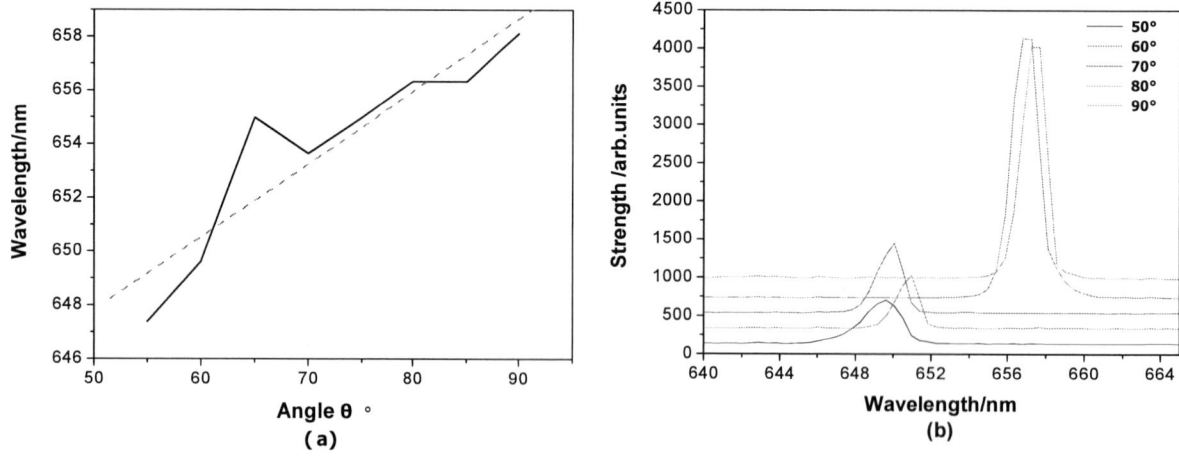

Figure 2. Emission wavelength as a function of incidence

978-1-4799-1215-5/13 $31.00 © 2013 IEEE

III. EXPERIMENT RESULTS AND ANALYSIS

The emission lasers of the sample with liquid crystal layer thickness of 10μm，were studied by changing the angle θ between the pumping light and the surface of the

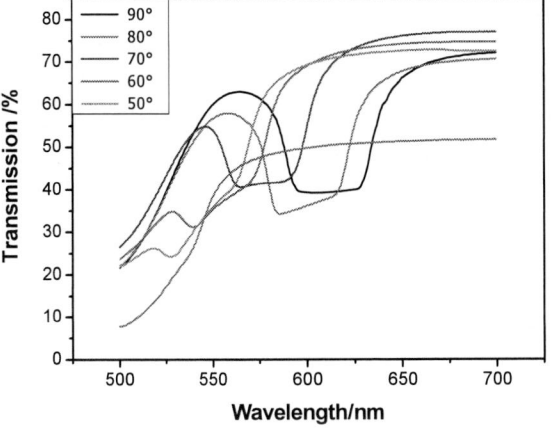

Figure 4.   the transmission spectrum of CLC

laser sample. The results were shown in Figure 2. Figure 3(a)shows that the wavelength changes with the angle of θ and Figure 2(b)shows the tranmission intensity respectively when θ =20°、30°、40° and50°.It shows that wavelength of emission peak appears at 658.11nm when θ=90°, and shifts to the direction of shortwave with decrease of the angle when θ=55°, the shortest wavelength of 647.38nm will appear .The turning range of whe wavelength is from 647.38 to 658.11nm and reaches 10.73nm. Please see Figure 3 for transmission spectrum of the CLC. The factors that results in gradual decrease of wavelength is: since liquid crystal is a sort of material with high bi-refringence, the refractive index of incident light can be periodically modulated by the periodic helix structure of CLC. The intensive Bragg reflection appears for incident light on the surface of CLC, when incident light enters vertically, central wavelength of Bragg reflection is $\lambda_0 = n \cdot p_0$ [7], in which po is the length of CLC pitch, $n = (n_o + n_e)/2$ is the average refractive index of

liquid crystal, central wavelength of Bragg reflection photonic bandgap is determined by Formula $\lambda_0 = n \cdot p_0 \cos\Theta$ in which $\Theta = \sin^{-1}\left[\dfrac{1}{n}\sin\left(\dfrac{\pi}{2} - \theta\right)\right]$. From the above-mentioned formula, we can find that reflection photonic bandgap of cholesteric liquid crystal CLC gradually shifts to the direction of short wavelength with the decrease of θ, appearing "blue shift".

To avoid the impact of pumping light on

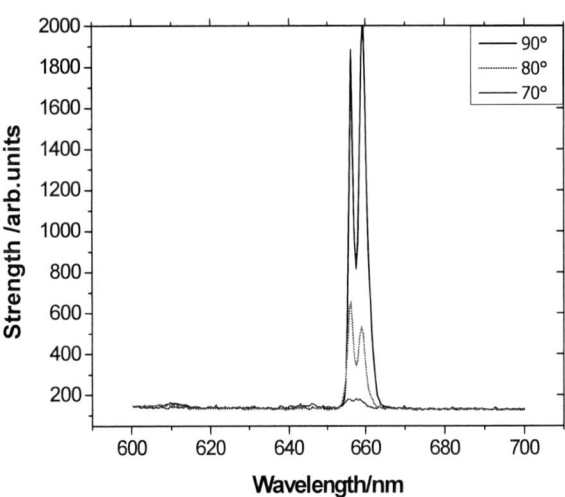

Figure 3.   Emission spectrum of the dye doped CLC laser   at different φ

experimental, select Angle was θ=75°, the wavelength of emission light was studied by varying the angle φ between the probe of fiber spectrometer and the surface of laser sample,Wavelengths of emission light under different φ were given in Figure 4.The result showed that the wavelength of emission light would not change under different φ.

IV. CONCLUSION

Tunable laser can be constructed by utilizing CLC and dye DCM. Regulating incident angle of the laser of CLC,

978-1-4799-1215-5/13 $31.00 © 2013 IEEE

tunable effect of emission wavelength can be obtained. The tuning range of laser emission wavelength with the incident angle of pumping light is from 647.38 to 658.11nm, and reaches 10.73nm. The structure of CLC is easier to realize, more convenient to control, having a good application prospect.In the work, it was found that at a given θ, control angle φ between the probe of fiber optic spectrometer and the surface of laser sample, the reason which the wavelength of emission light does not vary at different φ need to be further researched. Next, the work that decreases the threshold value of laser emission, improves the slope efficiency of laser emission shall be carried out.

ACKNOWLEDGEMENT

This work was supported by the National Natural Science Fund (Approval Number: 61107059, 61077047), China Postdoctoral Science Foundation (Approval Number: 2012M510921) and Heilongjiang Province Postdoctoral Science Foundation (Approval Number: LBH-Z10216)

REFERENCES

[1] V.I.Kopp,B.Fan,H.K.M.Vithana,candA.Z.Genack,"Low-threshold lasing at the edge of a photonic stop band in cholesteric liquid crystals."Opt. Lett, Vol.23, No21, pp. 1707-1709. 1998

[2] YH.Huang ,et al. "Spatially tunable laser emission in dye-doped cholesteric Polymer films". Appl.Phys.Lett, Vol.89, No11, pp. 111106-l. 2006:

[3] L.M.Blinov,GCIPParrone,A.Mazzulla, P.Paglius,et al. "SimPle voltage tunable liquid crystal laser. " Appl.Phys.Lett, Vol.90, No13, pp. 131103-1-31. 2007

[4] Liu Yong-Jun,Sun Wei-Min Liu Xiao-Qi , et al. "Investigation of the tunable laser of one-dimensional photonic crystal with dye--doped nematic liquid crystal defect layer. " Acta Phys.Sin. , Vol.61, No11, pp. 1142111. 2012

[5] Y. Yang,R.Goto,Soichiro Omi,et al. "Highly Photo-stable dye doped solid-state distributed-eedback channeled waveguide lasers by a pen-drawing technique. "Opt. Express, Vol.18, No21, pp. 22080-220891. 2010

[6] Yuhua Huang,Ying Zhou,and Shin-Tson Wu. "Spatially tunable laser emission in dye-doped photonic liquid crystals. "Appl. Phys. Lett, Vol.88, No1, pp. 0111071. 2006

[7] Seiichi Furumi. "Chiral Photonic Band-Gap Liquid Crystals for Laser Applications. "The Chemical Record. Lett, 10, pp. 394–408. 2010:

Generation of a High-Quality Hollow Laser Beam by a Liquid-Core Optical Fiber

Xiaobo HU, Wei GAO[*], Peijing SUN, Shengnan LIU, Xuelian YU, Shaozhi PU

Department of Optoelectronic information science and engineering
Harbin University of Science and Technology
Harbin, China
wei_g@163.com

Abstract—A hollow laser beam (HLB) is produced by a liquid-core optical fiber (LCOF) filled with CS_2. We investigate the dependences of the HLB quality on the laser coherence, the incident angle of the laser beam and the length of the LOCF, obtaining the conditions of high-quality HLBs. The Research results show that the lower laser coherence the better is the quality of HLBs. The quality of HLBs is also influenced by the incident angle of the laser beam and the length of the LCOF. We can achieve high-quality HLBs with longer LCOF when the incident angle of the laser beam is less than 10°. The generated HLBs with high quality can satisfy the needs of different applications.

Keywords-hollow laser beam; liquid-core optical fiber; high quality; coherence; fiber length

I. INTRODUCTION

A hollow laser beam (HLB) is a ring beam with zero central intensity along the beam axis, which has wide applications in atom trapping and guiding, free space optical communication, laser processing and life science because of its special physical properties such as barrel intensity distribution, small dark spot size (DSS) and no heating effect. The generate methods and characteristics of HLBs have been the research focus in the field[1-3].

Many techniques have been used to generate HLBs, such as geometrical optical method, mode conversion and optical holography[4–6]. The experimental arrangements of these methods are usually more complicated. Afterward, some simpler techniques were introduced by using a small hollow fiber, multimode fiber and photonic crystal fiber (PCF)[7-9]. However, the loss of hollow fibers is very large because of its small core contained air[7]; the DSS span of the HLB is confined owing to a small numerical aperture (NA, about 0.3) of multimode fibers[8]; special PCFs are more expensive[1,9]. Recently, a method to generate a HLB by a liquid-core optical fiber (LCOF) was proposed[10], which has advantages of large NA, small bending loss, wide DSS range and so on. Nevertheless, the quality of the HLB generated by this means is not good. Zhao *et al.* presented a method to improve the HLB quality using a rotating ground glass disk to lower the coherence of the incident laser beam[11]. It is bound to reduce the coupling efficiency of laser.

In this paper, we study the effects of the laser coherence and the structure parameters of LCOF on the quality of HLBs, and find that the HLB quality is not only dependent on the incident laser coherence, but also on incident angle of the laser and the length of LCOF.

II. EXPERIMENTAL METHOD

The experimental setup for generating HLB is shown in Figure 1. The laser beam is focused by a lens with focal length of 50mm, and directed into the LCOF filled with CS_2 at a certain angle θ. The LCOF is fixed on the five-dimension adjuster. A CCD is used to record the output beam patterns at about 6mm away from the end of the LCOF.

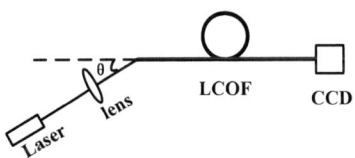

Figure 1.   Experimental setup of generating a HLB by a LCOF

In this experiment, a solid spot which is a Gaussian beam is firstly obtained by adjusting the LCOF input end to achieve maximum output intensity corresponding to 0°. The incident angle of the laser beam can be changed by adjusting the LCOF. A series of HLBs with different sizes can be obtained by increasing the incident angle[10]. To describe or compare different kinds of HLBs, we define DSS as the full-width half-maximum (FWHM) of the radial intensity distribution of the HLB sunken inside, and define W_{HLB} as full-width of $1/e^2$ maximum of radial intensity distribution of the HLB sunken outside, as shown in Figure 2.

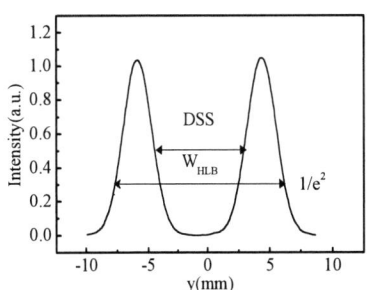

Figure 2. Radial intensity distribution of the HLB

III. RESULT AND DISCUSSION

A. *Effect of the laser coherence on the HLB quality*

In the experiment, three kinds of lasers are used as the laser sources, they are a single-mode narrow-linewidth semiconductor laser at 532nm, a He-Ne laser and a conventional semiconductor laser at 630nm, respectively. The incident angle of the laser beam is 8°, the core diameter and the length of the LCOF are 400μm and 100cm, respectively. The patterns of HLBs and the corresponding radial distribution are shown in Figure 3.

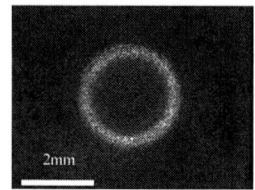

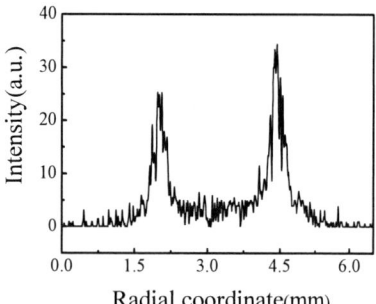

(a) Single-mode narrow-linewidth semiconductor laser at 532nm

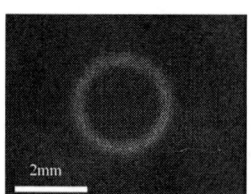

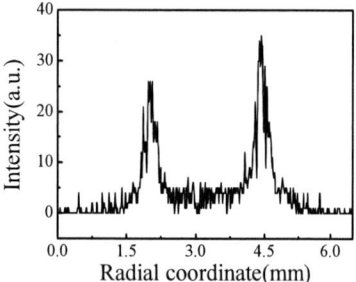

（b）He-Ne laser

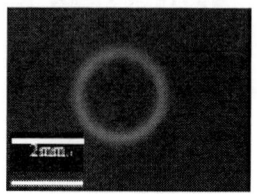

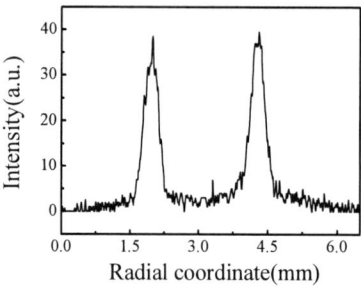

(c) Conventional semiconductor laser at 630nm

Figure 3. HLBs output and the corresponding radial distribution by different laser sources

By comparison, it can be seen that the high-quality HLB can be obtained by using the 630nm-wavelength semiconductor laser. The 532nm-wavelength laser and the He-Ne laser are both single-longitudinal-mode lasers, and the coherence of them are better than that of the 630nm-wavelength semiconductor laser. Thus, the interference will happen in the process of laser beam transmission owing to the high coherence, resulting in a lot of speckle noise. To reduce this speckle noise, the partially coherent optical source should be used to improve the HLB quality.

B. *Effect of the incident angle of laser beam on the HLB quality*

We use the 630nm-wavelength semiconductor laser as the laser source, and investigate the dependence of the HLB quality on the incident angle of the laser beam. The core diameter and the length of LCOF are 100μm and 100cm, respectively. Figure 4 depicts the radial distribution of HLBs for several incident angles.

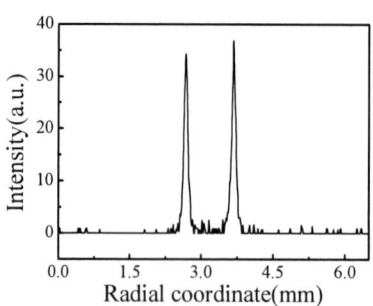

(a) 3°

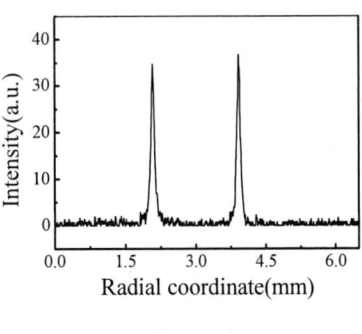

(b) 10°

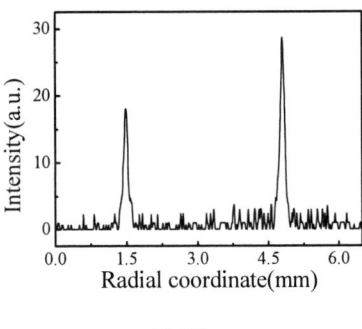

(c) 15°

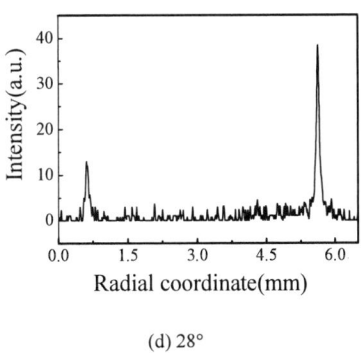

(d) 28°

Figure 4. The radial distribution of HLBs for different incident angles

From Figure 4, we can see that the larger the incident angle gets, the energy distribution of the generated HLB is more uneven. That is to say, the quality of the HLB gets worse as the incident angle increases. The characteristic of asymmetry at the output is kept with the input beam which has an angle with the axis of LCOF leading to the asymmetry of the intensity of the laser beam in the LCOF. This phenomena of asymmetry can be solved by inserting two beams at the incident angles of θ and $-\theta$, respectively[8]. Generally speaking, the HLBs with high-quality can be obtained at the angle of less than 10°, and hence the smaller incident angle should be considered under the same conditions.

C. Effect of the LCOF length on the HLB quality

In this section, we use two lengths of the LCOF to generate HLBs, one is 40cm, the other is 100cm. The incident angle of the laser and the core diameter of LCOF are chosen to be 8° and 400μm, respectively. The generated HLBs and the corresponding radial distributions are depicted in Figure 5.

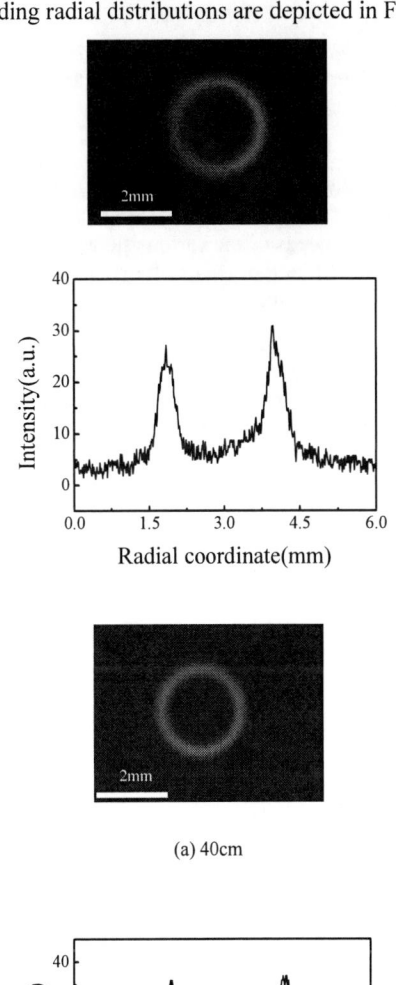

(a) 40cm

(b) 100cm

Figure 5.   HLB output and the corresponding radial distribution by LCOF with different lengths

Figure 5 shows that the radial distribution become smoother as the length of LCOF increases, and the intensity distribution of the HLB spot become more uniform. This indicates that the beam uniformity can be achieved after longer propagation of skew rays in the LCOF. Considering the fiber loss, the length should be chosen as 1m-3m. From Figure 5(a) and 5(b), we also see that the DSS and W_{HLB} are almost the same. This is in agreement with theoretical results in the literature [10].

IV. CONCLUSIONS

We experimentally investigated the influence of laser coherence, incident angle and the LCOF length on the quality of the HLB generated by the LCOF contained CS_2, obtained the conditions of achieving high-quality HLB. Results show that the HLB quality can be improved by using the partially coherent laser source, the incident angle of less than 10° and the fiber length of 1-3m. Therefore, we can obtain high-quality HLBs to meet the needs of the various fields including material processing, optical trapping of free electrons with high efficiency and life sciences [12-15].

ACKNOWLEDGMENT

This work is supported by the Technology of Education the Bureau of Heilongjiang Province, China (Grant No. 12511114).

REFERENCES

[1] M. Y. Zhang, S. G. Li, Y. Y. Yao, B. Fu, and L. Zhang, "A dark hollow beam from a selectively liquid-filled photonic crystal fiber," Chin. Phys. B, vol. 19, 2010, pp. 47103–47108.

[2] Q. G. Sun, K. Zhou ,G. Y. Fang, G. Q. Zhang, and Z. J. Liu, "Hollow sinh-Gaussian beams and their paraxial properties," Opt. Express, vol. 20, 2012, pp. 9682–9691.

[3] Z. Y. Chen and D. M. Zhao, "4Pi focusing of spatially modulated radially polarized vortex beams," Opt. Lett, vol. 37, 2012, pp. 1286–1288.

[4] H. Ito, K. Sakaki, W. Jhe, and M. Ohtsu, "Atomic funnel with evanescent light," Phys. Rev. A, vol. 56, pp. 712–718.

[5] W. L. Power, L. Allen, M. Babiker, and V. E. Lembessis, "Atomic motion in light beams possessing orbital angular momentum," Phys. Rev. A, vol. 52, 1995, pp. 479–488.

[6] N. R. Heckenberg, R. McDuff, C. P. Smith, and A. G. White, "Generation of optical phase singularities by computer-generated holograms," Opt. Lett, vol. 17, 1992, pp. 221–223.

[7] J. P. Yin, H. R. Noh, K. Lee, K. H. Kim, Y. Z. Wang, and W. H. Jhe, "Generation of a dark hollow beam by a small hollow fiber," Opt. Commun, vol. 138, 1997, pp. 287–292.

[8] H. Y. Ma, H. D. Cheng, W. Z. Zhang, L. Liu, and Y. Z. Wang, "Generation of a hollow laser beam by a multimode fiber," Chin. Opt. Lett, vol. 5, 2007, pp. 460–462.

[9] X. B. Zhang, X. Zhu, X. Chen, H. Q. Li, J. G. Peng, N. L. Dai, and J. Y. Li, "A hollow beam supercontinuum generation by the supermode superposition in a GeO2 doped triangular-core photonic crystal fiber," Opt. Express, vol. 20, 2012, pp. 19799–19805.

[10] X. B. Hu, S. N. Liu, W. Gao, D. Sun, H. Y. Zhang, and L. Y. Zhang, "Generation of hollow laser beam by a liquid-core fiber," Chinese J. Lasers, vol. 39, 2012, p. 1105002.

[11] C. L. Zhao, Y. J. Cai, F. Wang, X. H. Lu, and Y. Z. Wang, "Generation of a high-quality partially coherent dark hollow beam with a multimode fiber," Opt. Lett, vol. 33, 2008, p. 1389.

[12] C. J. He, Y. G. Zhang, L. Sun, J. M. Wang, T. Wu, F. Xu, C. L. Du, K.J . Zhu, and Y. W. Liu, "Electrical and optical properties of Nd3+-doped Na0.5Bi0.5TiO3 ferroelectric single crystal," J. Phys. D: Appl. Phys., vol. 46, 2013, p. 245104.

[13] C. J. He, X. D. Fu, F. Xu, J. M. Wang, K. J. Zhu, C. L. Du, and Y. W. Liu, "Orientation effect on bandgap and dispersion behavior of 0.91Pb(Zn1/3Nb2/3)O3-0.09PbTiO3 single crystals," Chin. Phys. B, vol. 21, 2012, p. 054207.

[14] C. J. He, H. B. Chen, L. Sun, J. M. Wang, F. Xu, C. L. Du, K. J. Zhu, and Y. W. Liu, "Effective electro-optic coefficient of (1–x)Pb(Zn1/3Nb2/3)O3–xPbTiO3 single crystals," Cryst. Res. Technol., vol 47, 2012, pp. 610-614.

[15] C. J. He, F. Xu, J. M. Wang, C. L. Du, K. J. Zhu, and Y. W. Liu, "Composition dependence of dispersion and bandgap properties in PZN-xPT single crystals," J. Appl. Phys., vol. 110, 2011, p. 083513.

[16]

Microfluidic Device with Compound Structure

He Zhang[1,*] Xiaowei Liu[1] Li Tian[1] Xiaowei Han[1]

1Key Laboratory of Micro-systems and Micro-structures Manufacturing, Harbin Institute of Technology, Harbin, China
zhanghe.hit@gmail.com

Abstract—**The development of microfluidic chip has become a growing research field. Accompanied by the new polymer materials, the preparation process and the function of the microfluidic chip is becoming diversification. In this paper we propose a novel method for preparing microfluidic amperometric detection chip using the printed circuit board (PCB). The chip is consists of three parts: the printed circuit board substrate with micro-strip electrodes, the polymethyl methacrylate (PMMA) with microchannels and the PDMS as assistant bonding layer. The electrode size can achieve micron level and the detection electrodes which were integrated on the PCB is not easy to fall off or broken. The microfluidic chip would be manufactured numerously while reduce costs by using the mature PCB production process.**

Keywords: microfluidic; amperometric detection; PCB; PMMA; PDMS

I. INTRODUCTION

Over the past decade, microfluidic chip has emerged as an attractive method for chemical and biological analysis using miniaturized systems [1-3]. Recent applications of microfluidic chip-based analytical tools tend toward the integration of multiple unit processes such as sample pretreatment, separation, and detection into a single chip. The electrochemical analysis (EC) detection method especially amperometric detection method comprises very simple instrumentation and integration of microscale electrodes onto a microfluidic chip [4], while maintaining excellent sensitivity and selectivity [5]. As a result, it has been intensively employed as the ideal detection method in microfluidic on-chip separation systems [6].

The chip material is also beginning to play a more important role in the integration of microfluidic chip. Due to the low price and the stability of physical and chemical properties, the high polymer has become the most widely used materials in microfluidic chip. But the most popular high polymer substrate materials such as the poly (methyl methacrylate) (PMMA) and the poly (dimethylsiloxane) (PDMS) are flexible and elasticity, the detection electrodes which were integrated on the high polymer by using hot pressing or lithography are easy to fall off and fracture. So one of the huge challenge with respect to the microfluidic amperometric detection chip is the stability and the yield of microelectrode integrated, which can make amperometric detection inaccurate even impossible.

In this paper, in order to improve the yield of integrated amperometric detection electrode and simplify the production process of high polymer microfluidic chip, we describes a novel compound structure microfluidic amperometric detection chip based on the printed circuit board (PCB). The PCB is used to mechanically support and electrically connect electronic components using conductive pathways. There are quite a few advantages that can be chosen PCB to make micro-strip electrodes, such as: different dielectrics that can be chosen to provide different requirements, production process is mature, low cost of the chip, electrode size can achieve micron level and the detection electrodes which were integrated on the PCB is not easy to fall off or broken.The structure of the chip is shown in Figure 1, the chip consists of three parts: the printed circuit board PCB substrate with micro-strip electrodes, the PMMA coverslip with microchannels and cells and assistant bonding layer.

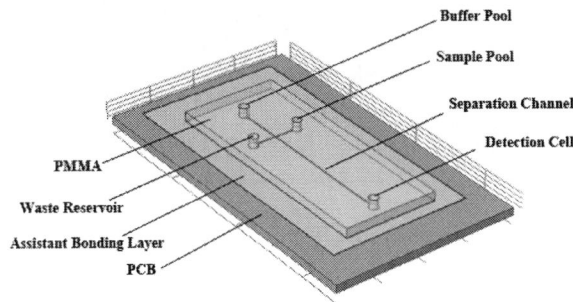

Figure 1. Schematic of the composite amperometric detection chip

II. EXPERIMENTAL PART

A. Instruments and Chemicals.

DH 1722-6A high voltage DC supply (Beijing Dahua radio equipment factory). 3023 x-y recorder (Kyoritsu Electric Shanghai Co., Ltd.). DDB-300 electronic peristalsis pump (Shanghai Letter Instrument Co., Ltd.). CHI630A electrochemical analyzer (CH Instruments Inc.). KQ-SOB ultrasonic cleaner (Kunshan Ultrasonic Instrument Co., Ltd.). VTC-100 spin coating (Oskco Technology Development Co., Ltd.). JGP-800 magnetron sputtering instrument (Shenyang Scientific Instruments Co., Ltd.). BP212 positive photoresist (Baiwan Electronic Science and Technology Center). Acetone, ethanol and NaOH (Shanghai Chemical Reagent Co., Ltd.). Ultrapure water and $Na_2[B_4O_5(OH)_4] \cdot 8H_2O$ (Oriental Institute of Chemical Reagents). All the above chemicals were analytically pure.

B. Manufacture of the PCB electrode

Amperometric detection methods are classified into end-channel, off-channel and in-channel types, depending on the position of the working electrode in a microchannel [7]. End-channel detection is a facile method to measure redox current from analytes that reduces the influence of the capillary electrophoresis voltage and current. Therefore, in this paper the amperometric detection chip has chosen end-channel mode. The design of PCB detection microelectrodes was shown in Figure 2 (a) and the photo of PCB microelectrodes was shown in Figure 2 (b). The driven separation electrode contains four electrodes, which were located at both ends of the injection channel and the separation channel. The End-channel amperometric detection system which used the classic three-electrodes mode was placed in the testing cell at the end of the separation channel. All the microelectrodes produced in the FR-4 epoxy material (the most popular material of PCB) by using electronic printing technology.

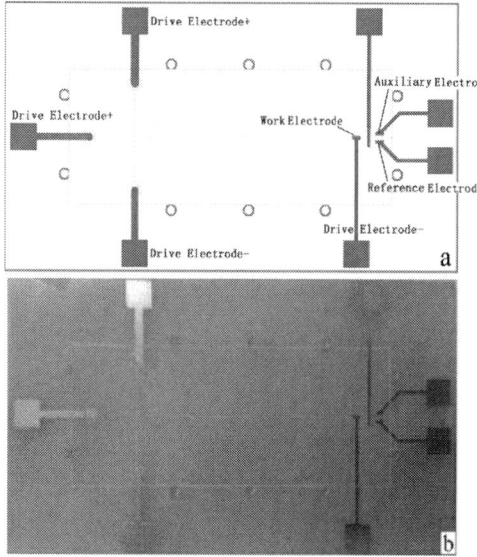

Figure 2.   PCB amperometric detection microelectrodes. (a The design of PCB microelectrodes; b The photo of PCB microelectrodes)

In the three-electrode amperometric detection system, the electrode for different purposes has different materials. In this paper, the working electrode was produced by Au (take advantage of the PCB production process), the reference electrode was produced by Ag and the auxiliary electrode was produced by Pt. In this paper, the PCB amperometric detection electrodes were modification by using lithography and magnetron sputtering technique in order to obtain the electrode of different materials. The process flow of amperometric detection electrode modification was shown in Figure 3.

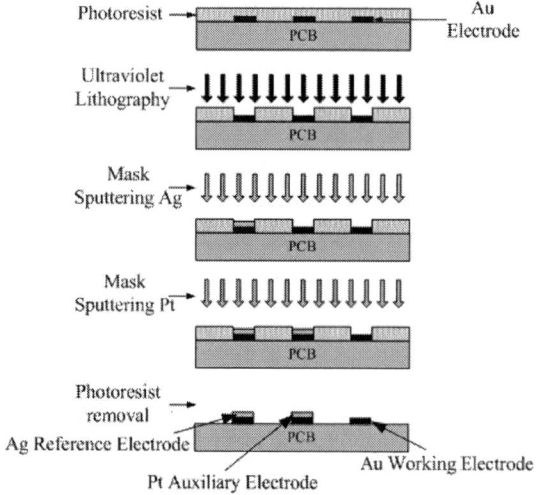

Figure 3.   Process flow of amperometric detection electrode modification

The following is a brief introduction of the PCB electrode modified process, after ultrasonic cleaning by using acetone and ethanol (each for 5min), put the PCB substrate into drying oven (80℃, 10mins). Then PCB substrate was spin-coated with positive photoresist and exposed to UV light through a photolithographic mask.

The process parameters of lithography were shown in Table 1.

TABLE I.   PROCESS PARAMETERS OF LITHOGRAPHY

Process Steps	Parameters
Spin-coated	4500 [rpm/s], 30[s]
Pre-baking	90[□],30[min]
Exposure dose	i-line, 40[mJ/cm2]
Post-baking	90[□],15[min]
Development and Depolymerizing	10[min]
Oven dry	80[□]10[min]

C. Preparation of the PMMA microchannel

The preparation and modification of the microchannels are also very important because of the electroosmotic was determined by the performance of microchannels. A separation microchannel for the specification of 45mm(length) × 75μm(depth) × 150μm(width) and a injection microchannel for the specification of 15mm(length) × 75μm(depth) × 150μm(width) were produced by using hot scribe equipment with independent intellectual property rights of MEMS Center[8], the photo of microchannel was shown in Figure 4 (a) and (b). The property of the microchannels was improved by the means of acetic acid modification. The modification result was shown in Figure 4 (c) and (d). The injection cell, waste pool, buffer cell and detection cell which were produced by using MEMS technology on the PMMA cover slip too.

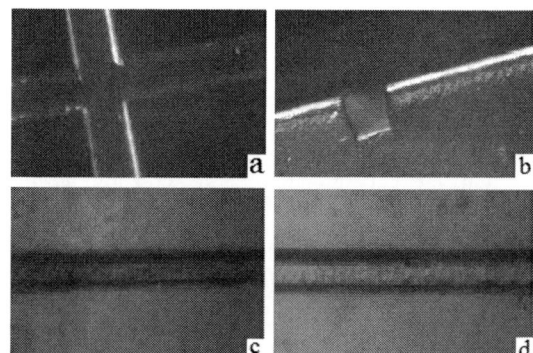

Figure 4.   The photo of microchannel product by mechanical manufacture. (a photo of intercross microchannel; b section plane of microchannel; c the microchannel before modified; d the microchannel after modified)

D. Fabrication of the chip

The bonding of the chip is another important process to the performance of the chip. In this paper, we used PDMS as assistant bond layer because of the PCB substrate and the PMMA coverslip cannot be bonded directly. PDMS is a chemically stable flexible material, it can be poured on the rough surface. The thickness of PDMS membrane which was decided by the speed of spin-coating instrument is very important to the bonding result. If the membrane is too thick, it would block the microchannel; on the other hand, if the membrane is too thin, the chip would leak. The curve of the membrane thickness and the spin-coating speed obtained by experiment was shown in Figure 5, thus we set the speed of the spin-coated between 1500~2000 rad/min to ensure the thickness of membrane about 50μm. After the spin-coating, put the PCB-PDMS chip into the oven (65℃, 20min). When the PDMS cured, removed the PDMS covered on the electrodes carefully by using a scalpel under the microscope. Then we fumigated the PCB substrate by using the organic solvent vapor (chloroform) for 8mins in order to auxiliary bonding. Finally, assembled the PMMA cover slip and the the PCB substrate together by using fixtures and put them into the oven for 10mins. The flow diagram of bonding was shown in Figure 6.

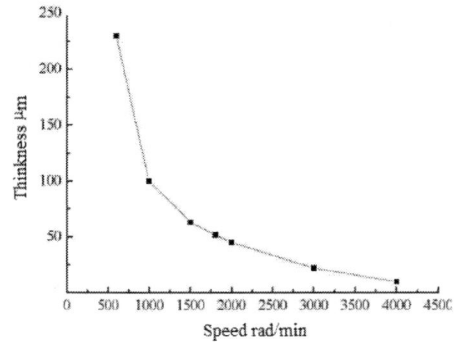

Figure 5. The curve of the membrane thickness and the spin-coating speed

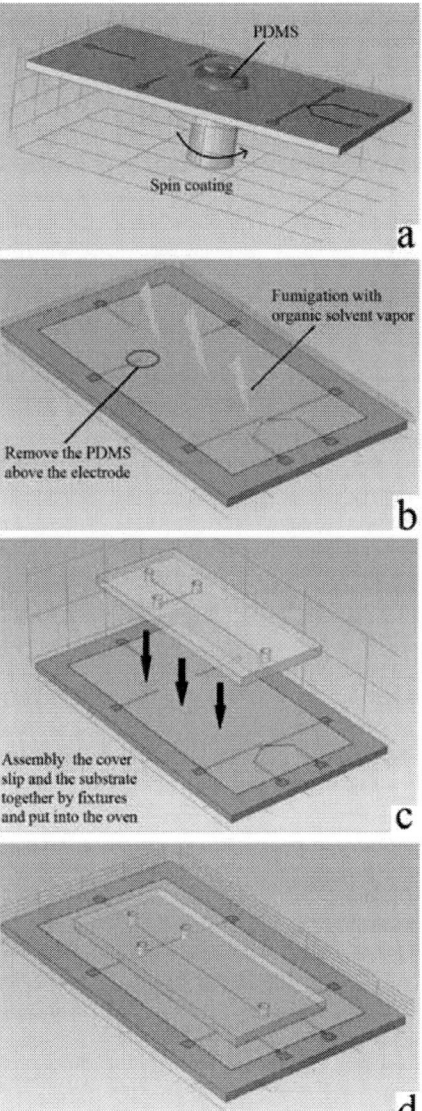

Figure 6. The flow diagram of PCB-PDMS-PMMA compound structure microfluidic bonding

III. RESULTS AND DISCUSSION

The velocity of the electroosmotic flow was measured by using the current method[9]. Before measuring, the surface wetting property of the microchannels were improved by the means of UV modification, the purpose of the modification is to improve the electroosmotic velocity in the microchannel. The result of modification was shown in Figure 7, compared to the unmodified channel, the time of obtain stability velocity reduce by 30 second.

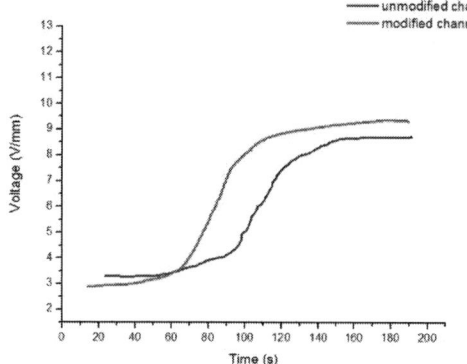

Figure 7. Velocity of the electroosmotic flow

Finally, the compound structure microfluidic amperometric detection chip and characterize PCB electrode performance were tested by using the cyclic voltammetry. The cyclic voltammetry curves were shown in Figure 8. The result of the standard gold disk electrode is presented by red curve and the result of PCB electrode is presented by blue. Due to the different electrode areas, the two electrode oxidation-reduction potential amplitude are varied widely. But the peak of the PCB electrode curve is significantly and symmetric, it shows that the PCB electrode and the standard gold disc electrode have the similar electrochemical properties.

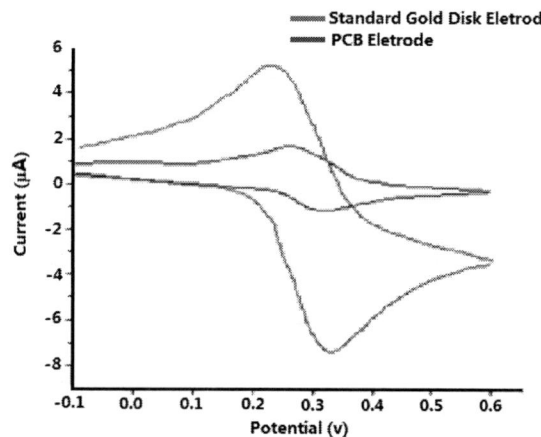

Figure 8. Cyclic voltammetry curves. Reagents: 5mmol/L $K_3Fe(CN)^{6+}$,1mmol/L KLC, scanning frequency: 100mV/s

IV. CONCLUSIONS

In this paper, a novel compound structure microfluidic chip for amperometric detection was proposed and validated. This compound structure microfluidic chip consists of three parts: the printed circuit board PCB substrate with micro-strip electrodes, the PMMA coverslip with microchannels and cells and the PDMS assistant bonding layer. Make the electrodes by using the printed circuit board can improve the yield of integrated amperometric detection electrode and simplify the production process of high polymer microfluidic chip. It should be noted that there is a significant reduction of costs for using the film of PCB as the mask in lithography. Therefore, the PCB-PDMS-PMMA compound structure microfluidic chip for amperometric detection may find technological applications.

V. ACKNOWLEDGMENTS

The authors would like to thank National Basic Research program of China (No. 2012CB934104) and National Science Foundation of China (No. 61071037) for financial support.

REFERENCES

[1] Kenyon, S. M.; Meighan, M. M.; Hayes, M. A. Electrophoresis 2011, 32, 482-493.

[2] Poinsot, V.; Gavard, P.; Feurer, B.; Couderc, F. Electrophoresis 2010, 31, 105-121.

[3] Tran, N. T.; Ayed, I.; Pallandre, A.; Taverna, M. Electrophoresis 2010, 31, 147-173.

[4] Wang, J. Acc. Chem. Res. 2002, 35, 811-816.

[5] Fischer, D. J.; Hulvey, M. K.; Regel, A. R.; Lunte, S. M. Electrophoresis 2009, 30, 3324-3333.

[6] Ghanim, M. H.; Abdullah, M. Z. Talanta 2011, 85, 28-34.

[7] Xu, J. J.; Wang, A. J.; Chen, H. Y. Trends Anal. Chem. 2007, 26,125-132.

[8] Xiaowei Liu, Mingxue Huo, Li Tian, Chinese Patent, 03111185. 3, 2003-03-14.

[9] Xiaohua Huang, Manuel J. Gordon, Richard N. Zare, Anal Chem, 60(1988), 1837-1838.

The Research and Application of Angular Correlation for Helim Optically-pumped Magnetometer

Zong Fabao, Zou Pengyi[*], Chen En, Wang Jingran, Zhang Jin
Hangzhou Applied Acoustics Research Institute, Hangzhou, China
zongfabao@163.com

Abstract—In this paper, the angular correlation of helium optically-pumped magnetometer was studied based on the magnetization vector equation of optical pumping. When the angle θ between optical axis of sensor and external magnetic field vector equals 0°, the resonance signal S_z reaches its maximum. While θ equals 90°, S_z is zero and the magnetometer is unable to work. The vertically-installed single optical axis sensor is suitable for the aeromagnetic survey in high latitude area, but it's unable to work in the low latitude area because of the very small geomagnetic inclination. The horizontally-mounted double orthogonal optical axes sensor was designed. The calculation and measurements showed that the double orthogonal axes sensor is also suitable for low latitude area. The double axes sensor has been mounted on the delta-wing aircraft for the aeromagnetic survey in a West African country (Sierra Leone).

Keywords-optically-pumped; double orthogonal axes sensor; angular correlation; magnetization vector equation;

I. INTRODUCTION

The helium optically-pumped magnetometer is based on the Zeeman Effect of He^4, optical pumping and magnetic resonance [1-3]. The magnetometer usually consists of sensor and resonance detection system. As shown in Fig. 1, the single axis optically-pumped sensor is formed by helium lamp (a), lens (b), polarizer (c), quarter wave plate (d), cell (e), coils (f), light sensor (g).

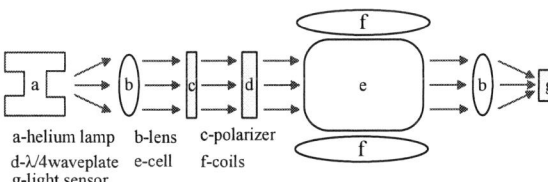

a-helium lamp b-lens c-polarizer
d-λ/4waveplate e-cell f-coils
g-light sensor

Figure 1. The structure of single axis optically-pumped sensor

The metastable helium atoms in the cell can absorb the circularly polarized 1083 nm light from the helium lamp and jump to the sublevel $2^3S_1(m=+1)$ intensively, this process is called optical pumping. Magnetic resonance occurs when the frequency f of alternating magnetic fields from the coils equals the larmor frequency v_L. The resonance causes transitions between zeenam levels; the helium atoms in the cell can absorb the light continuously and the light intensity through the cell will reach its minimum. The resonance detection system always adjusts the frequency of alternating magnetic fields by monitoring the signal of light sensor. When the signal of light sensor reaches its minimum, the magnetic flux density B can be calculated through the equation as follows:

$$B = 2\pi f / \gamma \qquad (1)$$

Where γ is the gyromagnetic ratio of helium atom.

The single axis helium sensor has dead zone [4]. When the angle θ between the optical axis and external magnetic field vector is close to 90 degrees, the magnetometer is unable to work. For the magnetic survey on the ground, the user can prevent the sensor from dead zone, but for the underwater or airborne magnetic survey, the direction of sensor depends on the movement of carrier, it's hard to keep the direction of sensor far from the dead zone all the time.

In this paper, we deduced the expressions of resonance signal S_z and S_x based on the magnetization vector equation of optical pumping. The expressions show that the signal S_z is in proportion to $\cos^4(\theta)$. When $\theta = 0°$, the resonance signal S_z reaches its maximum; when $\theta = 90°$, S_z is zero, the magnetometer is unable to work. In the high latitude area, the geomagnetic inclination is always large, so the single axis sensor which is vertically mounted on the aircraft is satisfied with any heading. But in the low latitude area near the equator, the vertically mounted single axis sensor is unable to work; the horizontally mounted single axis sensor can't keep working in every heading either. The horizontally-mounted double orthogonal axes sensor we designed in this paper can deal with this difficulty effectively; it can work satisfactorily in every heading and has been mounted on the delta aircraft for the aeromagnetic survey in a West African country (Sierra Leone).

II. THEORY OF ANGULAR CORRELATION

In 1957, Bell and Bloom discussed the magnetic resonance detection of optical system with only two Zeennam levels. The magnetization vector equation of optical pumping can be described as Eq. (2). Where τ_{pz} and τ_{px} is the optical pumping time of light along the z and x axis, μ'_{z0} and μ'_{x0} is equilibrium magnetization value along the z and x axis.

978-1-4799-1215-5/13 $31.00 © 2013 IEEE

$$\begin{cases} \dot{\mu}_x = \dfrac{1}{\tau_{px}}\mu'_{x0} - \left(\dfrac{1}{\tau_{px}} + \dfrac{1}{\tau_{pz}}\right)\mu_x \\[2mm] \dot{\mu}_y = -\left(\dfrac{1}{\tau_{px}} + \dfrac{1}{\tau_{pz}}\right)\mu_y \\[2mm] \dot{\mu}_z = \dfrac{1}{\tau_{pz}}\mu'_{z0} - \left(\dfrac{1}{\tau_{px}} + \dfrac{1}{\tau_{pz}}\right)\mu_z \end{cases} \quad (2)$$

When there is constant magnetic field $\vec{H}$ in the cell along the z axis, the Bloch equation as shown Eq. (3) can be used to describe the motion of magnetization vector.

$$\begin{cases} \dot{\mu}_x = -\gamma[\vec{H},\vec{\mu}]_x - \dfrac{1}{T_2}\mu_x \\[2mm] \dot{\mu}_y = -\gamma[\vec{H},\vec{\mu}]_y - \dfrac{1}{T_2}\mu_y \\[2mm] \dot{\mu}_z = -\gamma[\vec{H},\vec{\mu}]_z - \dfrac{1}{T_1}(\mu_z - \chi_0 H_0) \end{cases} \quad (3)$$

Where γ is the gyromagnetic ratio of helium atom, T_1 and T_2 is longitudinal and transverse relaxation time, χ_0 is the paramagnetic susceptibility. For the helium atom, $T_1 = T_2$.

In order to reach the magnetic resonance, we feed the cell with the RF magnetic field $\vec{H}_1$ orthogonal with H_0. Substituting the magnetic field $\vec{H}$ with $\{H_1\cos\omega t, 0, H_0\}$ in Eq. (3), the solution is as follows:

$$\begin{cases} \mu_x = \gamma H_1 S_2 \mu_{z0}\dfrac{\sqrt{1+(\Delta\omega S_2)^2}}{1+(\Delta\omega S_2)^2 + (\gamma H_1)^2 S_1 S_2}\sin(\omega t + \varphi) \\[3mm] \mu_z = \mu_{z0}\dfrac{1+(\Delta\omega S_2)^2}{1+(\Delta\omega S_2)^2 + (\gamma H_1)^2 S_1 S_2} \end{cases} \quad (4)$$

Where $S_1 = T_1$, $S_2 = T_2$ and $\Delta\omega = \omega - \gamma H_0$.

When using the optical method for magnetic resonance detection, the resonance signal S as showed in Eq. (5) is in proportion to the absorption of light in the cell.

$$\begin{aligned} S &= n_1 p_1 + n_2 p_2 \\ &= N_0\dfrac{p_1+p_2}{2} - (n_1-n_2)\dfrac{(n_1-n_2)_0}{N_0}\dfrac{p_1+p_2}{2} \end{aligned} \quad (5)$$

Where n_1 and n_2 is the number of atoms in different sub-level, p_1 and p_2 is probability to energy level transition, $N_0 = n_1 + n_2$ is the sum of atoms. Introduced the symbol $p_z = (p_1+p_2)/2$, $V_z = (n_1-n_2)/N_0$ and $A = (\mu_B g_J)/2$, where μ_B is Lande factor, g_J is Pohl magnetron, the resonance signal can be described as follows:

$$S_z = N_0 p_z - \dfrac{p_z V_z \mu_z}{A} \quad (6)$$

Removing the constant part, the signal S_z can be described as:

$$S_z = k_z p_z^2 \dfrac{1+(\Delta\omega S_2)^2}{1+(\Delta\omega S_2)^2 + (\gamma H_1)^2 S_1 S_2} \quad (7)$$

Where $k_z = V_z \mu'_{z0} S_1 / A$, p_z is in proportion to the intensity of light, $p_z = p\cos^2(\theta) = \dfrac{1}{\tau_p}\cos^2(\theta)$ and θ is the angle between optical axis and external magnetic field H_0. The S_z can be described as follows:

$$S_z = k_z p^2 \dfrac{1+(\Delta\omega S_2)^2}{1+(\Delta\omega S_2)^2 + (\gamma H_1)^2 S_1 S_2}\cos^4(\theta) \quad (8)$$

The Eq. (8) is the angular correlation of helium magnetometer. The resonance signal S_z is in proportion to the $\cos^4(\theta)$. Defining the $K_z(\theta) = S_z / S_{z\,max}$, where $S_{z\,max}$ is the maximum of S_z, then the angular correlation of $K_z(\theta)$ is shown in Fig. 2.

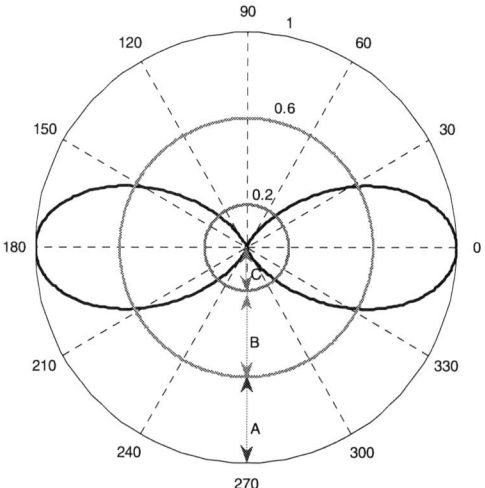

Figure 2. The angular correlation of $K_z(\theta)$

In the practical applications, when $K_z \in (0.6, 1.0]$, the single optical axis magnetometer can work well; when $K_z \in (0.2, 0.6]$, the magnetometer can work but the performance is reduced sharply with K_z; when $K_z \in [0, 0.2]$, the magnetometer is unable to work. These can be shown in Tab. I.

The geomagnetic inclination is always small in the low latitude area near the equator and the geomagnetic field is nearly parallel to the horizontal plane. The single optical axis sensor must be mounted horizontally in this area, but the dead zone is nearly 50% of heading. In the aeromagnetic survey, the surveying lines can be predesigned to avoid the dead zone. But when changing the surveying line, it is unavoidable that the magnetometer will enter the dead zone and it will be unable to work.

TABLE I.　THE INFLUENCE OF K_z ON SINGLE OPTICAL AXIS HELIUM MAGNETOMETER

Region	K_z	Heading(n is integer)	Scale	Working	Performance
A	0.6~1.0	(-28°~28°)+ n×180°	31%	Yes	Unchagned
B	0.2~0.6	(28°~48°) +n×180° (132°~152°) +n×180°	22%	Yes	Reduced
C	0~0.2	others	47%	No	--

III.　DOUBLE ORTHOGONAL OPTICAL AXES SENSOR

In order to cope with the difficulty of aeromagnetic survey in low latitude area, we designed the double orthogonal optical axes sensor as showed in Fig. 3.

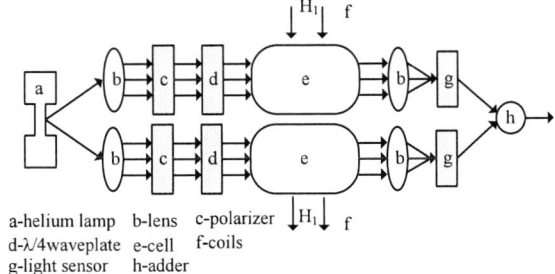

a-helium lamp　b-lens　c-polarizer
d-λ/4waveplate　e-cell　f-coils
g-light sensor　h-adder

Figure 3 The structure of duble orthogonal optical axes sensor

The double orthogonal axes sensor consists of two orthogonal optical axes. Double optical axes share a helium lamp (a), the lenses (b), polarizer (c), quarter wave plates (d), cells (e), and light sensors (g) are all individual, double RF coils(f) are series, the adder (h) is commonly achieved by paralleling two light sensors. The resonance detection system adjusts the center frequency of RF magnetic field in real time by monitoring the summed resonance signal and calculates the magnetic flux density B.

Supposing the parameter of double orthogonal axes is identical, then the angular correlation of double axes sensor can be described as:

$$
\begin{aligned}
S''_z &= S_{z1} + S_{z2} \\
&= K_c(\cos^4(\theta) + \cos^4(\theta + \frac{\pi}{2}))
\end{aligned} \tag{9}
$$

Where K_c is irrelevant to θ. After normalization, the angular correlation of S''_z is shown in Fig. 4.

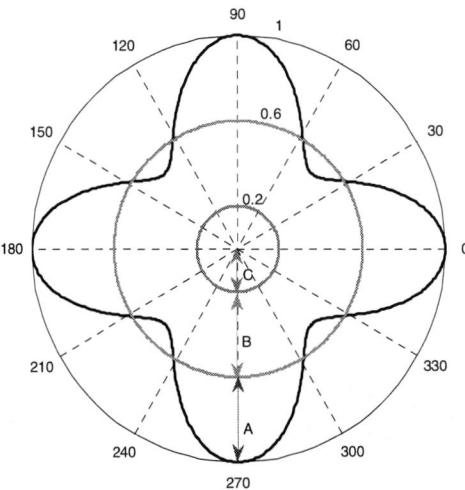

Figure 4 The angular correlation of duble orthogonal optical axes sensor

The minimum of resonance signal of double axes sensor is 0.5, so the magnetometer can work in every heading. The heading that the performance is not reduced is (-31°～31°)+n×90°, it is 69% of all heading.

In order to verify the angular correlation of double axes sensor, experiment had been done in the Xiang Shan Weak Magnetic Laboratory of National Institute of Metrology, China. At first, we adjusted the current of three-dimensional Helmholtz coils to offset the vertical component of earth magnetic field and generated a horizontal field with 35000nT along the north. Then we put the double axes sensor horizontally at the center of Helmholtz coils and let one of the axes along the north. Manually adjusting the frequency of RF field and let the magnetometer in open-loop mode caught the peak resonance signal. At last, we turned the direction of sensor with step 10 degrees and recorded the corresponding amplitude of resonance signal until the sensor had been rotated a circle. The experiment results are showed in Fig. 5.

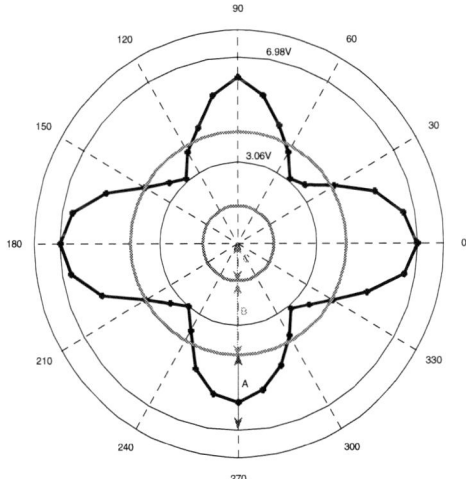

Figure 5 The experiment results

978-1-4799-1215-5/13 $31.00 © 2013 IEEE

The Fig. 4 and Fig. 5 all demonstrated that there is no dead zone (Region C) for the horizontally-mounted double orthogonal axes sensor in every heading at low latitude area. Because of the difference between two axes of sensor, when $\theta = 0°$, the resonance signal is 6.98 V; when $\theta = 90°$, the resonance signal is only 6.14 V. The minimum of resonance signal is 3.06 V, 44% of the maximum.

As shown in Fig. 6, the aeromagnetic survey system consists of a helium magnetometer with horizontally-mounted double axes sensor, a vector magnetometer, a GPS, an altimeter, a magnetic compensator, a data logger and the delta-wing aircraft.

Figure 6 The aeromagnetic survey system

The aeromagnetic survey system with horizontally-mounted double axes sensor had been carried out more than ten thousand kilometers in Sierra Leone, a West African country near the equator. Part of the aeromagnetic data is shown in Fig. 7. The upper dotted curve shows the magnetic field before compensation, the solid curve shows the magnetic field after compensation. After compensation, the peak noise of data is less than 0.3nT; the dynamic noise [7] after forth order derivative is 0.005nT, less than 0.08nT, meets the demand of A-standard.

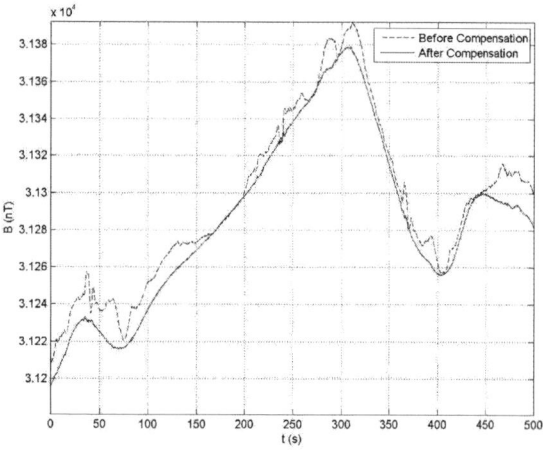

Figure 7 Part of the aeromagnetic data

IV. CONCLUSIONS

In this paper, we studied the angular correlation of tracking mode optically-pumped magnetometer based on the magnetization vector equation. In order to cope with the difficulty of aeromagnetic survey in low latitude area, we designed the horizontally-mounted double orthogonal optical axes sensor. The theoretical simulation and experiments show that there is no dead zone (Region C) in the horizontally-mounted the double axes sensor, the heading in which the magnetometer works well is 69% of all directions. The magnetometer with double axes sensor had been mounted on the delta aircraft for the aeromagnetic survey in West African country (Sierra Leone).

ACKNOWLEDGMENT

This work was supported by the National High Technology Research and Development Program (2007AA09Z319 and 2012AA8112005). The authors are grateful to Professor Wu Wenfu and all the members of group.

REFERENCES

[1] Dmitry Budker, Michael Romalis, "Opitical magnetometry," Nature Physics, VOL.3, pp. 227-234, 2007.

[2] Xiao Jianhua, "Understanding of the principle of optical pumping and magnetic resonance from oscillograph show,".Journal of Sichuan Normal University, VOL.19, NO5, pp. 103-107,1996.

[3] Zhang Zhenyu, "Research on Optically Pumped Helium Magnetic Measurement Technology," JILIN UNIVERSITY, 2012.

[4] Chen Zhunian, "The Angler Dependence of Resonance Signal for Optically Pumped Helium Magnetometry," Acta Metrologica Sinica, VOL.15,NO4, pp. 265-268,1944.

[5] Zhining Guan, "Geomagnetic Field and Magnetic Exploration," Geological Publishing House, 2005.

[6] H.M., B.M. and Г.B., "Atomic Magnetometer," Underwater Weapon Editing Room, 1981.

[7] DZ/T 0142-2010, "Criterion of Aeromagnetic Survey," Ministry of Geology and Mineral Resources of P.R. China, 2010.

A double-ring Mach-Zehnder interferometer for highly sensitive temperature sensing

Xiaoqi Liu, Yundong Zhang*, Xuenan Zhang, Ping Yuan

National Key Laboratory of Tunable Laser Technology, Institute of Opto-Electronics, Harbin Institute of Technology, Harbin,
China
ydzhang@hit.edu.cn

Abstract—We propose a double-ring resonators (DRR) Mach-Zehnder (M-Z) interferometer for highly sensitive temperature sensing. We theoretically calculate the sensitivity of the configuration as a temperature sensor. The sensitivity of our configuration can achieve $481.24/^{\circ}C$, which is enhanced by $140/^{\circ}C$ compared to that of single-ring resonator (SRR) M-Z interferometer. Furthermore, we discuss the relationship between the sensitivity and the temperature detection range of the proposed interferometer. This proposed structure enables highly sensitive, compact and stable temperature sensors.

Keywords- resonator, Mach-Zehnder interferometer, sensor

I. INTRODUCTION

Optical ring resonators have attracted considerable attention due to their applications as compact and sensitive biological and chemical sensors, as well as filters, buffers and modulators [1-5]. In the past few years, various sensors based on optical ring resonators have been typically developed, for example, grating-coupled waveguide sensors [6], planar optical-waveguide sensors [7], directional coupler sensors [8] and micro resonator sensors [9]. Recently, sensors based on all-fiber interferometer have been used for temperature measurement, due to high sensitivity, simple fabrication process and unlimited measuring of wavelength range. Also, it has been proved that the sensitivity of all-fiber interferometer can be significantly enhanced by introducing both normal dispersion and abnormal dispersion structure into it [10]. Optical devices consisting of a resonator coupled to M-Z interferometer are quite useful in tailoring the resonance line shape for the multiple-beam interference between the circulating optical waves in the resonator and the reference arm [11] therefore holding potential applications in sensing [12]. The coupled-resonator-induced transparency structure has been proposed and extensively discussed [13, 14] and results have implied this structure can be potentially used as an ultra-sensitive sensor. By coupling the coupled-resonator-induced transparency structure to the M-Z interferometer, the sensitivity of the interferometer can be enhanced obviously. With the phase bias introduced into the reference arm, the normalized output intensity exhibits asymmetrical resonance line shape due to phase mismatch between the two arms of the interferometer. Such rapid change of the intensity in a narrow frequency range greatly enhanced the temperature sensitivity of our device thus this scheme enables highly sensitive, compact and stable temperature sensors.

II. THEORETICAL ANALYSIS

The schematic diagram of the proposed DRR M-Z interferometer is illustrated in Fig.1. (a). Two fiber ring resonators named R_1 and R_2 are coupled through one waveguide directional coupler and the second ring R_2 is side coupled to one arm of a balanced M-Z interferometer.

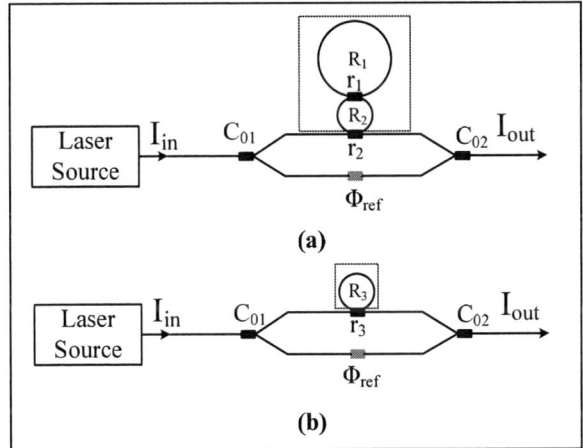

Figure 1. (a) A DRR M-Z interferometer. (b) A SRR M-Z interferometer.

The two arms of the interferometer are balanced so that the optical path difference between the upper and lower arms is caused only by the DRR structure which is used as the temperature sensing unit. r_1 and r_2 are the amplitude reflecting coefficients and are related to the corresponding amplitude coupling coefficient k by the relationship $r_i^2 + k_i^2 = 1$, $i = 1, 2$. In the reference arm, a phase shifter is introduced to produce a phase bias which makes the output intensity exhibits a sharp asymmetrical line shape around the resonance wavelength. Narrowband light source of wavelength $\lambda = 1550nm$ is launched into the input port of the M-Z interferometer where it splits into the two arms of the interferometer by the first coupler C_{01}. Past the structure, the lights are recombined at the second coupler C_{02} and interfere with the reference light through the coupler C_{02}. C_{01} and C_{02} are all 3dB couplers. The reflected and transmitted electric fields, respectively, for the lossless coupling of light to the first ring resonator are:

$$E_2 = r_1 E_0 + i k_1 E_1 , \tag{1}$$

$$E_3 = i k_1 E_0 + r_1 E_1 . \tag{2}$$

A $\pi/2$ phase shift occurs upon transmission across the coupler, and we have ignored the mode function describing the spatial field distribution. Propagation of E_3 around the ring of length L_1 having loss coefficient α_1 yields:

$$E_1 = a_1 \exp(i 2\pi \nu t_1) E_3 , \tag{3}$$

where $t_i = n_{eff} L_i / c$, $i = 1, 2$ is the single-pass transit time, L_i is the length of the $R_i , (i = 1, 2, 3)$, $a_i = \exp(-\alpha_i L_i / 2), i = 1, 2$ is the attenuation factor, ν is the frequency detuning, n_{eff} and c are the effective refractive index of the waveguide and the light velocity in vacuum. Substitution of the equation (3) to the equation (2) can deduce that:

$$\frac{E_2}{E_0} = \frac{r_1 - a_1 \exp(i 2\pi \nu t_1)}{1 - a_1 r_1 \exp(i 2\pi \nu t_1)} = \tau_1 . \tag{4}$$

Similar to equation (1) and equation (2), the transfer equations for the ring $Ring_2$ are given by:

$$E_6 = r_2 E_4 + i k_2 E_5 , \tag{5}$$

$$E_7 = i k_2 E_4 + r_2 E_5 , \tag{6}$$

the complex transmission coefficient of the DRR structure then can be derived as follows:

$$\frac{E_2}{E_1} = \frac{r_2 - a_2 \tau_1 \exp(i 2\pi \nu t_2)}{1 - a_2 r_2 \tau_1 \exp(i 2\pi \nu t_2)} = \tau_2 . \tag{7}$$

The field transmission amplitude is defined as $Tr = |\tau_2|$ and the effective phase shift of the transmitted light is described as $\Phi = \arg(\tau_2)$. As the two arms of the M-Z interferometer are balanced, the light in the sensing arm acquires an extra optical phase Φ while in the reference arm the extra phase is introduced by the phase shift Φ_{ref}. Therefore, the normalized output intensity of the DRR M-Z interferometer can be derived as follow:

$$\frac{I_{out}}{I_{in}} = \left| \frac{E_{out}}{E_{in}} \right|^2 = \frac{1}{4} [1 - 2Tr \cos(\Phi - \Phi_{ref}) + Tr^2] . \tag{8}$$

In Fig.2, we present the normalized output intensity transmission spectra near the resonance which are calculated by using equation (8). We set the parameters as follows: $r_1 = 0.1$, $r_2 = 0.95$, $L_1 = 3m$, $L_2 = 1.5m$, $a_1 = a_2 = 0.95$, $n_{eff} = 1.468$, $c = 2.99792 \times 10^8 \, m/s$.

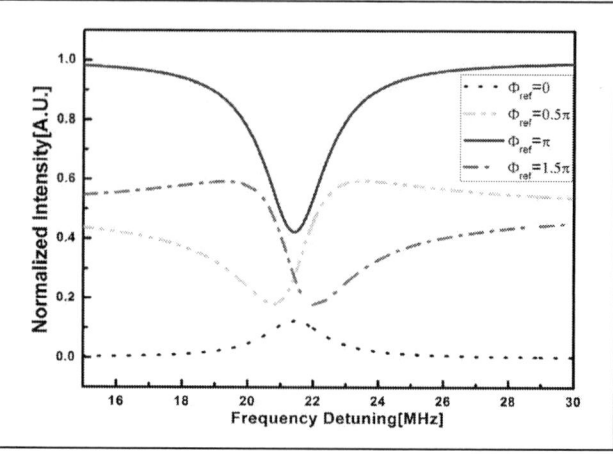

Figure 3. $\Phi_{ref} = 0$, $\Phi_{ref} = 0.5\pi$, $\Phi_{ref} = \pi$, and $\Phi_{ref} = 1.5\pi$ respectively.

The spectra of output transmissions exhibit different properties with different phase bias in the reference arm. We vary the phase shift from $\Phi_{ref} = 0$ to $\Phi_{ref} = 2\pi$. As we can see in Fig.2, the transition from a symmetric Lorentzian line shape to an asymmetric Fano line shape is realized by changing the phase shift Φ_{ref} . Thus it is convenient to produce tunable Fano resonances. For $\Phi_{ref} = 0.5\pi$ and $\Phi_{ref} = 1.5\pi$, the transmittances of the output exhibit asymmetrical resonance line shapes due to phase mismatch between the sensing arm and the reference arm. Such rapid changes of the intensities in the very narrow frequency ranges are useful for the highly sensitive detection. In this paper, the phase bias is set to be 0.5π to produce sharp asymmetric line shapes.

III. SYSTEM SENSITIVITY

When the DRR M-Z interferometer is applied as a sensing system, we choose the DRR structure as the sensing part which is in the dashed square frame in Fig.1. The other parts of the M-Z interferometer are kept isolated from temperature variation. Thus the temperature variation in the sensing region can be easily acquired by analyzing the intensity variation in the interference signal. The temperature sensitivity of the DRR structure M-Z interferometer can typically be defined as:

$$S = \frac{\partial(I_{out}/I_{in})}{\partial T} = \frac{\partial(I_{out}/I_{in})}{\partial \nu} \cdot \frac{\partial \nu}{\partial \varphi} \cdot \frac{\partial \varphi}{\partial T} = S_1 \cdot S_2 \cdot S_3 . \tag{9}$$

In equation (9), S_1 is the changing rate of the normalized output intensity with the change of the frequency detuning ν . As S_1 relies on the structure of the system, we define it as the device sensitivity; S_2 is the changing rate of the frequency detuning ν with the single-pass phase shift φ ; S_3 is the influence of temperature change T on the single-pass phase shift φ , which is determined mostly by waveguide's thermal

---

The research is supported by the National Natural Science Foundation of China (NSFC) (No. 61078006 and No. 61275066) and National Key Technology Research and Development Program of the Ministry of Science and Technology of China (No.2012BAF14B11).

978-1-4799-1215-5/13 $31.00 © 2013 IEEE

expansion coefficient $(dl/dT)/l = 5 \times 10^{-7}/°C$ and thermo-optic coefficient $dn/dT = 1 \times 10^{-5}/°C$. By taking integral transformation to the two coefficients, we can derive that $l = l_0 \exp[(5 \times 10^{-7})\Delta T]$, $n = \int_0^{\Delta T} 1 \times 10^{-5} dT = n_0 + (1 \times 10^{-5})\Delta T$. For the reasons above, we can induce that when the structure is fixed, $S_2 \cdot S_3$ which is a constant value then can be defined as the fiber sensitivity. So in this paper, we just discuss the device sensitivity.

IV. SIMULATIONS AND DISCUSSIONS

In this article, we set the circumferences of the fiber rings R_1 and R_2 are $L_1 = 3m$, and $L_2 = 1.5m$ and the initial temperature is $20°C$. For comparison, the simulation results of the device sensitivity of the DRR M-Z interferometer and the SRR M-Z interferometer are shown respectively in the Fig.3.

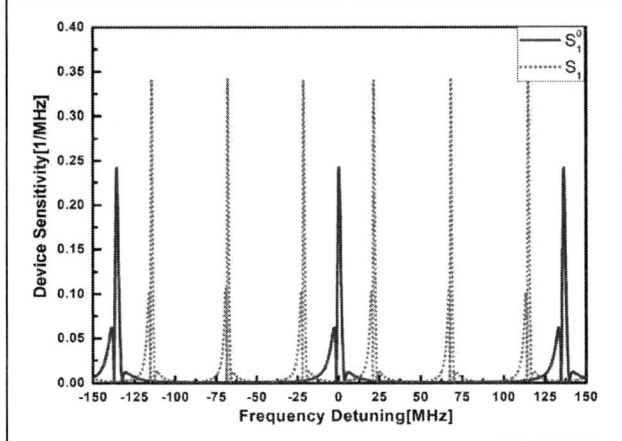

Figure 4. The device sensitivity of the DRR M-Z interferometer S_1 and that of the SRR M-Z interferometer S_1^0.

We can see clearly from the Fig.3, the device sensitivity of The DRR M-Z interferometer S_1 can achieve almost $0.3496/MHz$ while that of the SRR M-Z interferometer S_1^0 is only $0.2481/MHz$. This means by introducing the DRR structure, the device sensitivity of the interferometer exceeds that of the SRR M-Z interferometer by almost $0.1/MHz$. As the temperature sensitivity S is equal to $S_1 \cdot S_2 \cdot S_3$ and the fiber sensitivity $S_2 \cdot S_3$ is a constant value when the structure is fixed, the device sensitivity of the DRR structure M-Z interferometer $S_1 = 0.3496MHz$ correspondingly means the temperature sensitivity of the proposed interferometer is $481.24/°C$. Similar to that, the device sensitivity of the SRR M-Z interferometer $S_1^0 = 0.2481/MHz$ correspondingly means the temperature sensitivity of this interferometer is $341.63/°C$. Thus we can deduce that the temperature sensitivity S of the proposed structure is enhanced by $140/°C$ compared to that of the SRR M-Z interferometer.

By theoretical analysis and simulation of the DRR M-Z interferometer, we find the sensitivity of our structure increase sharply with the lengths of the R_1 and the R_2 while the sensitivity decrease slightly with the increase of the amplitude reflecting coefficient r_1 when the lengths of the two rings are fixed. The simulation results are shown in Fig.4 and Fig.5 respectively.

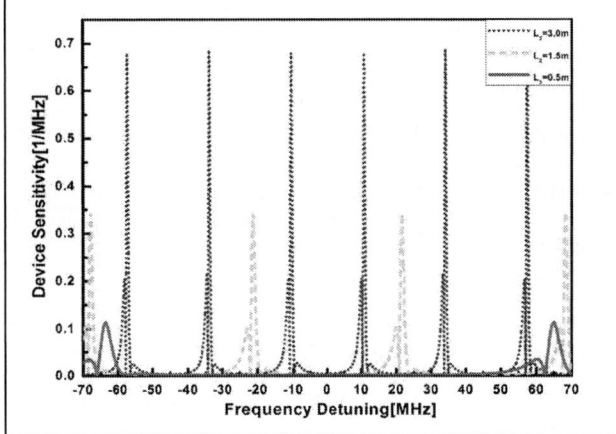

Figure 5. The device sensitivity of the DRR M-Z interferometer with different lengths of the R_1 and the R_2, $L_1 = 2L_2$.

The changes in normalized output intensities with the temperature variation for different ring lengths when the amplitude reflecting coefficients $r_1 = 0.01$ is fixed are shown in Fig.6. The curves are all asymmetric line shapes: they first abruptly rise up to the maximum values and then slowly decrease to the minimum values with increasing temperature. The sensitivity increases with the increase of the R_1 length, yet the linear measurement range narrows down. For $L_1 = 3m$, the linearity is much better than for $L_1 = 0.5m$, however, the temperature detection range shrinks from $0.09°C$ to $0.02°C$.

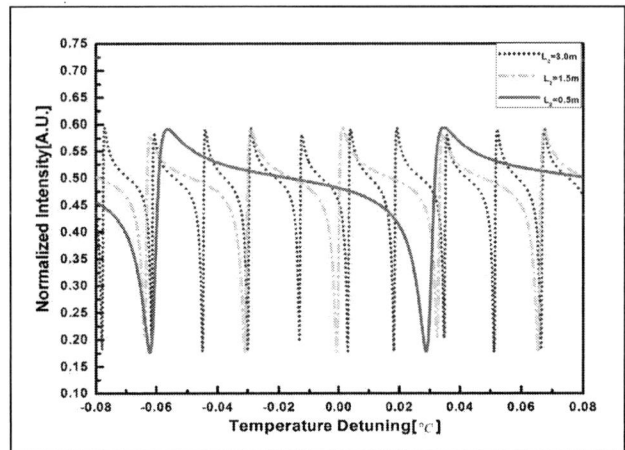

Figure 6. Changes in normalized output intensity as a function of the temperature for different lengths of the R_1 and the R_2, $L_1 = 2L_2$.

From the analysis and simulations above, we can easily summarize that the temperature detection range and the sensitivity of the proposed temperature sensor can be controlled by setting the structural parameters deliberately. We can choose appropriate ring length and amplitude reflecting coefficient to ensure the measurement range and the sensitivity according to actual condition [15-18].

V. CONCLUSION

In summary，we propose a DRR M-Z interferometer for highly sensitive temperature sensing. We theoretically calculate the sensitivity of the configuration as a temperature sensor. It is enhanced by $140/^{\circ}C$ compared to that of the SRR M-Z interferometer. The simulation results indicate that the temperature detection range and the sensitivity are trade-off and can be easily set by choosing the structural parameters deliberately. The proposed structure enables highly sensitive, compact and stable temperature sensors.

ACKNOWLEDGMENT

The research is supported by the National Natural Science Foundation of China (NSFC) (No. 61078006 and No. 61275066) and National Key Technology Research and Development Program of the Ministry of Science and Technology of China (No.2012BAF14B11).

REFERENCES

[1] C. J. Wang, "Fiber Loop Ringdown-a Time-Domain Sensing Technique for Multi-Function Fiber Optic Sensor Platforms: Current Status and Design Perspectives," Sensors 9, 7595-7621(2009).

[2] G. B. Hocker, "Fiber-optic sensing of pressure and temperature, " Appl. Opt. 18, 1445-1448(1979).

[3] F. Xia, L. Sekaric, and Y. Vlasov, "Ultracompact optical buffers on silicon chip, " Nat. Photon. 1, 65–71(2007).

[4] G. T. Reed, G. Mashanovich, F. Y. Gardes, and D. J.Thomson, "Silicon optical modulators" Nat. Photon. 4, 518–526(2010).

[5] C.Y. Chao and L. J. Guo, "Design and optimization of microring resonators in biochemical sensing applications, " J. Lightwave Technol. 24, 1395–1402(2006).

[6] R. Hornath, H. C. Pedersen, N. Skivesen, D. Selmeczi, and N. B. Larsen, "Opticalwaveguide sensor for on-line monitoring of bacteria," Opt. Lett. 28, 1233-1235(2003).

[7] T. Okamoto, M. Yamamoto, and I. Yamaguchi, " Optical waveguide absorption sensor using a single coupling prism, " J. Opt. Soc. Amer. A, Opt. Image Sci. 17, 1880-1886(2000).

[8] B. J. Luff, R. D. Harris, J. S. Wilkinson, R. Wilson, and D. J. Schiffrin, "Integrated-optical directional coupler biosensor, " Opt. Lett. 21, 618-620(1996).

[9] C. Y. Chao and L. J. Guo, "Biochemical sensors based on polymer microrings with sharp asymmetrical resonance," App. Phys. Lett. 83, 1527-1529 (2003).

[10] J. F. Wang, Y. D. Zhang , X. N. Zhang, H. Tian, H. Wu, Y. X. Cai , J. Zhang, and P.Yuan, "Enhancing the sensitivity of fiber Mach–Zehnder interferometers using slow and fast light, " Opt. Lett. 36, 3173-3175(2011).

[11] Y. Lu, J. Q. Yao, X. F. Li, and P. Wang, "Tunable asymmetrical Fano resonance and bistability in a microcavity-resonator-coupled Mach-Zehnder interferometer," Opt. Lett. 30, 3069-3071(2005).

[12] C. Y. Chao and L. J. Guo, "Design and optimization of microring resonators in biochemical sensing applications," J. Lightwave Technol. 24, 1395-1402(2006).

[13] M. Terrel, M. J. F. Digonnet, and S. H. Fan, "Ring-coupled Mach-Zehnder interferometer optimized for sensing," Appl. Opt. 48, 4874-4879(2009).

[14] Y. D. Zhang, X. N. Zhang, Y. Wang, R. D. Zhu, Y. L. Gai, X. Q. Liu, and P. Yuan, " Reversible Fano resonance by transition from fast light to slow light in a coupled-resonator-induced transparency structure,"Opt.Express.21,8572-8577 (2013).

[15] C. J. He, Y. G. Zhang, L. Sun, J. M. Wang, T. Wu, F. Xu, C. L. Du, K. J. Zhu, and Y. W. Liu, "Electrical and optical properties of Nd3+-doped Na0.5Bi0.5TiO3 ferroelectric single crystal," J. Phys. D: Appl. Phys. 46, 245104 (2013).

[16] C. J. He, X. D. Fu, F. Xu, J. M. Wang, K. J. Zhu, C. L. Du, and Y. W. Liu, "Orientation effect on bandgap and dispersion behavior of 0.91Pb(Zn1/3Nb2/3)O3-0.09PbTiO3 single crystals," Chin. Phys. B 21, 054207 (2012).

[17] C. J. He, H. B. Chen, L. Sun, J. M. Wang, F. Xu, C. L. Du, K. J. Zhu, and Y. W. Liu, "Effective electro-optic coefficient of (1−x)Pb(Zn1/3Nb2/3)O3–xPbTiO3 single crystals," Cryst. Res. Technol. 47, 610-614 (2012).

[18] C. J. He, F. Xu, J. M. Wang, C. L. Du, K. J. Zhu, and Y. W. Liu, "Composition dependence of dispersion and bandgap properties in PZN-xPT single crystals," J. Appl. Phys. 110, 083513 (2011).

978-1-4799-1215-5/13 $31.00 © 2013 IEEE

Carbon Nanotube-Based Printed Antenna for Conformal Applications

Yu-Ming Wu, Xin Lv

Beijing Key Laboratory of Millimeter Wave and Terahertz
Technology, Department of Electronic Engineering, School
of Information and Electronics
Beijing Institute of Technology
Beijing, China
wuyuming@bit.edu.cn, lvxin@bit.edu.cn

Beng Kang Tay, Hong Wang

Division of Microelectronics, School of Electrical and
Electronic Engineering, College of Engineering
Nanyang Technological University
Singapore
EBKTAY@ntu.edu.sg, EWANGHONG@ntu.edu.sg

Abstract—**In this paper, a novel printed antenna on the flexible substrate working at Gigahertz frequency is proposed based on the conductive carbon nanotube (CNT) ink. By advanced fabricating technique, the CNT ink is especially prepared and characterized so that a low-cost fabrication process has been developed for such a CNT printed antenna. The electromagnetic radiation properties have been investigated both theoretically and experimentaly. This antenna has many advantages such as flexible, conformable, and portable but durable in severe enviroment. Because it can be conveniently integrated into the clothes and easy to carry, it has important applications in the wireless body area network.**

Keywords-carbon nanotube; printed conformal antenna; inkjet printing; flexible electronics

I. INTRODUCTION

With the emergence and further development of nanofabrication technique, the advanced nanomaterials like carbon nanotube(CNT) has started to play an important role as the new candidate for constituting the antenna and RF devices especially adaptable for more complicated and extreme environment. CNT and its composite has arose numerous research interests because they have unique premier electrical properties, such as half of the density as that of copper, ten times of the thermal conductivity as copper. Its metallic or semi-conductive properties can be controlled and changed through the preparation or growth of either single-wall or multiple-wall samples. In addition, carbon nanotube and its composite are promising because they have been adopted for aerospace and RFID system applications for their improved performance such as high flexibility and conformability, reduced weight while reliable and durable in compared with the conventional metals like copper [1, 2]. CNT can be inkjet printed on the flexible substrate like paper, polymer, film and even textile with high advanced integration capability, which make it suitable for the flexible electronics, sensors and "smart skin" in near future [3-5]. Because it is bendable and shaped without any damage, the flexible electronics are more reliable in providing better circuit performance than the extensively adopted conventional silicon microelectronics. In addition, the flexible antenna design has the advantages of light weight, conformal and wearable capabilities, which are helpful to be conveniently integrated into the clothes or to carry [6, 7]. These properties enable the feasibility of embedding the components such as the antennas [8], integrated circuits, memory, batteries and sensors to a paper module to form a complete and promising system. Moreover, such an antenna is suitable for fast fabrication due to direct write other than traditional metal etching, so it is considered as green, cost and energy-efficient.

II. CNT INK PREPARATION AND CHARACTERIZATION

In this part of work, the printed antennas on the paper is completed under the help of the inkjet-printing technology. In previous study, the silver conductive ink has been used to print the antenna and Radio Frequency Identification (RFID) tag[7, 9], but we innovatively applied CNT for the ink to print the monopole antenna.

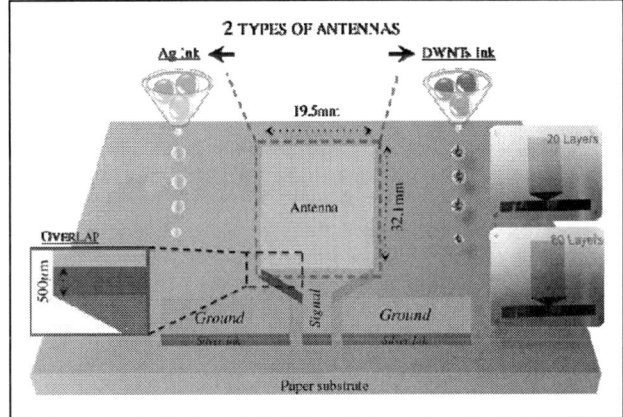

Figure 1.   Schematic of antenna printed by inkjet printing process.

As shown on Figure 1, two types of conductive ink have been used to fabricate the antenna: silver(Ag) nanoparticles and double-wall carbon nanotubes (DWCNTs). For the former, the conductive Ag nanoparticles are with an average particle size of 20-25 nm. The electrical performances (R, σ, |Sij|) and the thickness of Ag metallization were measured using the Agilent analyzers (4142B and HP8510C) and Veeco Dektak 3ST Stylus profilometer. The detailed results are listed on Table I.

In order to form a suitable printable ink for ink-jet printing process, the functionalized DWCNT as stable DWCNT in dimethylformamide (DMF) was especially prepared. The radiation part of the antenna is printed by multiple layers of

DWCNTs films, while the interconnection part with testing cable is printed by the Ag ink for a good electrical connection. An overlap region of both DWCNT and Ag with the length of 500 μm is designed to enhance the contact between the DWCNTs films and the feeding line. The concentration of DWCNTs can be controlled by the number of layers during the inkjet process.

TABLE I.   DC AND MICROWAVE PERFORMANCES OF AG INK AND THE DIELECTRIC PROPERTIES OF PAPER

Ag Ink		Paper Substrate	
Thickness	850 nm	*Thickness*	250 μ m
Resistance	4.68 Ω	*Tan λ*	0.02
Conductivity	1.4×10^6 S/m	*Permittivity*	3.5
S21 at 10 GHz	-1.3 dB		

III.   ANTENNA DESIGN CONSIDERATIONS AND MEASUREMENT RESULTS

A.   Design of CNT Printed Antenna

A coplanar waveguide (CPW)-fed printed antenna is adopted for its simple structure to realize. The CPW type of feeding is used because it can provide low dispersion, low radiation leakage in comparison with the microstrip line type feeding [10]. In addition, the fabrication only involves the processing on single side of the substrate, and there's no via hole, hence make it convenient to be integrated with other devices. The design,simulation and optimization are completed by FEM-based Ansoft HFSS 3-D full-wave electromagnetic solver. The actual thickness of CNT conductive ink and the permittivity, thickness, loss of the paper substrate have been taken into account. In addition, The CNT ink can be printed in multiple layers to obtain tuneable and variable conductivities for controlling purpose. As the layer increases, an increase in the conductivity of CNT is expected. For the initial design, the schematic is given in Figure 2 and the corresponding parameters and their dimension are listed in Table II. It is worth noting that in addition to the antenna printed by CNT ink, the CPW feeding part is printed by the silver ink instead of the CNT ink (i.e., the base part). This is for better connection between the antenna and the SMA connector in the measurement. The CNT coplanar waveguide feeding part is more fragile during the fixing and more likely to be influenced in the performance. The CNT's conductivity is initially taken as 140 S/m, this value can be improved by more condensed layers and corresponding parametric study on the variable conductivity's effects has been done. As an improved conductivity case study, the antenna with the same size however printed fully by the silver ink is also included in our study, where the silver ink conductivity is taken as 1.4e6 S/m.

for the antenna simulation, a lumped port excitation is set by a solid PEC bridge connecting both ground plane strips across the feeding strip in the middle (for the details, see Figure 2). A parameter study is completed by including the cases in which the antenna radiation parts are made from several CNT samples with increasing conductivities from 1.4×10^2 to 1.4×10^6 S/m. Figure 3 shows the simulation results of the CNT

printed antenna with the constant dimension but different CNT conductivity values. It can be seen that our designed CNT printed antenna gives -18.3dB return loss at 4.8 GHz, with 0.9 (4.42-5.32) GHz bandwidth. It can also be found that CNT's conductivity's value has a significant impact on the antenna performance. In general, with higher conductivity, the antenna of the same size can provide wider bandwidth and smaller S11 at resonance. Silver ink with conductivity 1.4e6 S/m can be considered as the case where CNT has an improved conductivity. A return loss of -30.4dB can be obtained at 5.05 GHz. In future, the CNT's conductivity can be improved to achieve better results. The detailed simulation shows that multiple resonance frequency at lower frequency range (2.5-4.0 GHz) start to appear when the conductivity of CNT is better than 1.4×10^3 S/m.

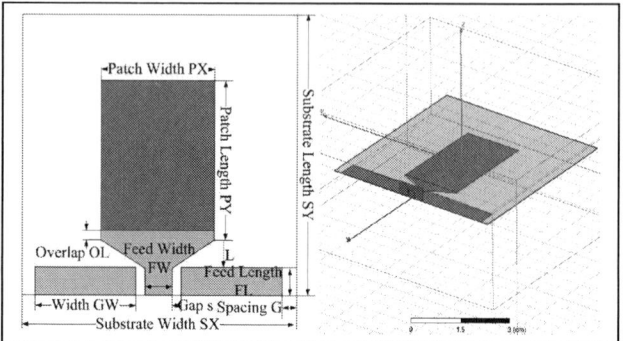

Figure 2.   The schematic of the CNT printed antenna on paper substrate and the simulation including the demonstration of the lumped port setting.

TABLE II.   DIMENSION AND SIZE OF CNT PRINTED ANTENNA INKJET PRINTED ON PAPER SUBSTRATE

Paper Substrate Parameters	εr	tanδ	Substrate Thickness	Ink Thickness
	3.5	0.02	250 μ m	850 nm
CPW printed antenna Dimension Parameters	*S-FW-S*	*FL*	*G, OL*	*SX×SY*
	0.037-0.59-0.037 mm	0.59 mm	0.26, 0.05 cm	5.8×5.8 cm
	PX	*PY*	*GW*	*L*
	1.074 cm	4.978 cm	2.503 cm	0.4 cm

B.   Characterization and Measurement of the CNT Printed Antenna

The antenna measurement is performed in the microwave anechoic chamber. Agilent N5230A PNA-L network analyzer (10 MHz -20 GHz) is used for the return loss measurement.

As the contact pad of antenna are made of silver ink on photopaper substrate, no high-temperature process such as soldering is allowed. This in turn increases the difficulty of obtaining reliable measurement results. For more reliable connection, the SMA connector is soldered to a small piece of solid PCB board. And then the antenna is fixed on top of using three pins. The entire adaptor is fixed onto a plastic foam substrate using double-sided tape. Silver paste is used to enhance the contacts between pins of SMA connector and pads of antenna. The detailed fixture part and connection during the measurement is shown in Figure 4. A multi-meter is used to

978-1-4799-1215-5/13 $31.00 © 2013 IEEE

check the connection between SMA terminals and contact pad of antennas.

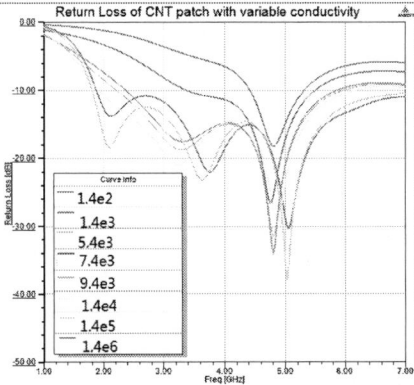

Figure 3.   CNT conductivity's effects in simulation

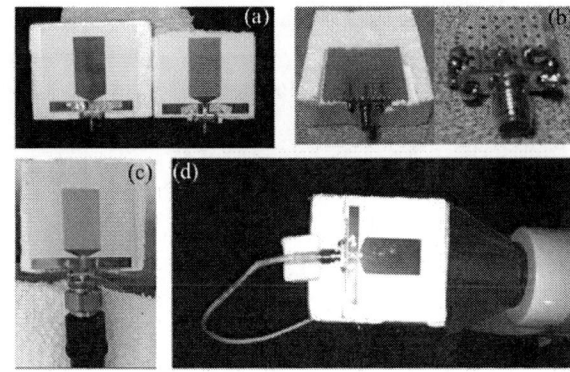

Figure 4.   The antenna connection during the measurement a) the antennas connected to SMA without the fixture part b) the fixture part for connecting the antenna to SMA c) antenna connected during return loss measurement d) antenna connected during pattern measurement.

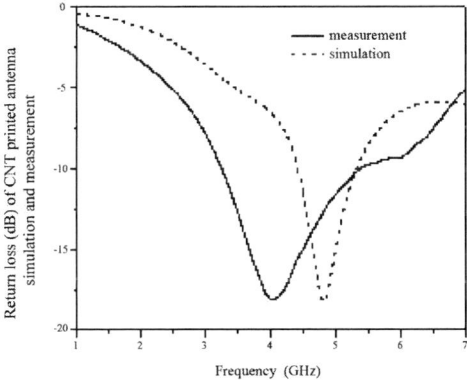

Figure 5.   Results comparison between simulation and measurement for CNT printed antenna.

A comparison between the simulation and measurement is shown in Figure 8.   The solid curves show the results of measurement while the dashed curves illustrate those of simulation. The practical measurement gives bigger return loss at resonance and slightly parallel shift the resonance frequency to lower frequency compared with the simulation. The discrepancies exist here may arise from the connection in the practical measurement. In future, the results can be improved by adopting more reliable fixturer for better connection.

IV.   CONCLUSION

The CNT printed antenna design on the flexible substrate working at <6 GHz has been demonstrated both theoretically and measurementally. A corresponding low-cost fabrication process has been developed for such a CNT printed antenna.The test results verified that a low-cost fabrication process can be used for such a CNT printed antenna design with good performance.In future, very conductive (conductivity > 10^4 S/m) CNT film will be fabricated. In addition, as an extension investigation, the conformal CNT printed antenna array will be designed, fabricated and characterized.

ACKNOWLEDGMENT

This work was supported by nanoantenna and array project. The authors wish to acknowledge Dr. Hong Li, Dr.Wai Leong Chow and Dr. Congxiang Lu for the helpful discussions on some research topics in the fabrication.

REFERENCES

[1]   Mehdipour, A., et al., Carbon nanotube composites for wideband millimeter-wave antenna applications. IEEE Transactions on Antennas and Propagation, 2011. 59(10): p. 3572-3578.

[2]   Mehdipour, A., et al. Carbon-fiber nanotubes for X-band conformal antenna applications. in Proceedings of the International Symposium on Antennas and Propagation (APS/URSI), Toronto, Canada, July 2010.

[3]   Lee, H., et al. A novel highly-sensitive antenna-based "smart skin" gas sensor utilizing carbon nanotubes and inkjet printing. in Proceedings of the International Symposium on Antennas and Propagation (APS/URSI), Spokane, Washington, USA on July 3-9, 2011.

[4]   Lee, H., et al., Carbon-nanotube loaded antenna-based ammonia gas sensor, IEEE Transactions on Microwave Theory and Techniques, 2011. 59(10): p. 2665-2673.

[5]   Vyas, R., et al., Inkjet printed, self powered, wireless sensors for environmental, gas, and authentication-based sensing. Sensors Journal, IEEE, 2011. 11(12): p. 3139-3152.

[6]   Bajwa, H., et al. Nanostructured conformable patch antenna array. in Proceedings of the, 2010 International Conference on Information and Emerging Technologies (ICIET). 14-16 June 2010,Karachi,India.

[7]   Zhou, Y., et al., Polymer-carbon nanotube sheets for conformal load bearing antennas. Antennas and Propagation, IEEE Transactions on, 2010. 58(7): p. 2169-2175.

[8]   Shaker, G., et al., Inkjet printing of ultrawideband (UWB) antennas on paper-based substrates. IEEE Antennas and Wireless Propagation Letters, 2011. 10: p. 111-114.

[9]   Rida, A., et al., Conductive inkjet-printed antennas on flexible low-cost paper-based substrates for RFID and WSN applications. IEEE Antennas and Propagation Magazine , 2009. 51(3): p. 13-23.

[10]   Liu, H.-W., C.-H. Ku, and C.-F. Yang, Novel CPW-fed planar monopole antenna for WiMAX/WLAN applications. IEEE Antennas and Wireless Propagation Letters, 2010. 9: p. 240-243.

Modeling and analysis of analog single event transients in an amplifier circuit

Yongsheng Wang[1]*, Wenjuan Wang [1], Yunfei Du[1], bei Cao[12]

1 Micro-electronic Department, Harbin Institute of Technology, Harbin, China
2 Electronic Science and technology Post-Doctoral Research Center, Heilongjiang University, Harbin, China
yswang@hit.edu.cn

Abstract—**Simulations are used to characterize the single event transient current and voltage waveforms in the common source amplifier with resistive load implemented using SMIC 0.18μm technology. The simulations are all carried out using the circuit and device-level mixed mode with TCAD tools. Results show that the amplitude of the current pulse is depending on the input voltage of the circuit. And when the LET of the incidence particle is higher enough the transient current waveforms do not obey the classical double exponential model.**

Keywords-single event transient; pulse shape; pulse amplitude; Common-Source amplifier.

I. INTRODUCTION

In recent years，a new kind of Single-event Effects which is called the Single-event Transient in analog circuits(ASET) has been received an increasing concern. It has been found that the generation, transportation and collection of the ionizing charges caused by the single particle in circuits and devices is totally applicable in analog circuits, which mainly contain the op-amps and comparators. SETs are temporary variations in the output voltage of a circuit due to the passage of a heavy ion through a sensitive device when the input differential voltage is a certain value [1]. The heavy ions move through the sensitive device in analog circuits and generate movable charges which may cause the drift of the voltage at the strike node and this will cause a large offset in the output voltage.

Several fatal errors have already been observed in flight [2]. Because of the large number of linear components that are used in spacecraft and a very small SET may cause a very large degradation in analog circuits, this phenomenon is becoming a very significant problem. But the radiation effects caused by heavy ions in a circuit or device are dependent on bias conditions, linear energy transfer (LET) and strike location [3]. In this paper, the bias conditions and LET are concerned in the single-transistor common-source(CS) amplifier with resistive load.

Two kinds of methods to analyze the radiation effects are mainly used today: heavy-ion experiments and by technology computer aided design (TCAD) simulation [4], [5]. Heavy-ion experiments are accurate, but they are too expensive to study every possible heavy-ion induced response that may occur in space environment. While, the TCAD simulation tools are very

useful, fast and cost-efficient in analyzing SETs in small circuits. On the other hand, this technique is very helpful in understanding the mechanisms caused by single particle through the device and circuit-level mixed-mode simulation for small circuits.

The TCAD simulation tools are used in this paper to evaluate the influence of the bias conditions and LET of the incident particle on the SET effect in a single-transistor CS amplifier with resistive load.

II. MODELING OF THE DEVICE

The device and circuit-level mixed-mode simulations are used to obtain the shape, amplitude and duration of the transient current generated by the interaction of incident particle and a sensitive transistor in circuits. The 3D TCAD device models are calibrated to the SPICE models which are a 180 nm technology. The simulations reported in this work are carried out on a single-transistor common source amplifier with resistive load and the n-channel transistor is modeled in TCAD, as shown in Figure.1.

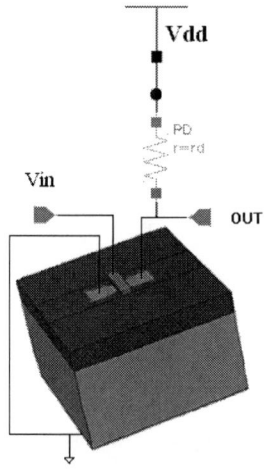

Figure 1. Mixed-mode of cs-amplifier with resistive load

First, the electric characteristics of the 3D n-channel transistor are fitted with the SPICE models. Figure 2 and figure

---

The work is supported by the Research Foundation of Education Bureau of Heilongjiang Province under Grant No. 12531479

978-1-4799-1215-5/13 $31.00 © 2013 IEEE

3 are the calibrated $I_d\_V_d$ and $I_d\_V_g$ curve of the TCAD model and the spice model.

The length and width of 3D n-channel transistor are 0.4μm and 1μm respectively. The junction depth (Xj), thickness of the gate oxide (Tox) and deposition of the channel (NCH) is obtained from the model library. And the Xj is 170nm, the Tox is 0.0038μm. The outer areas of the transistor are SiO₂ isolation layer and the depth of this area is 0.4μm which is an experience value.

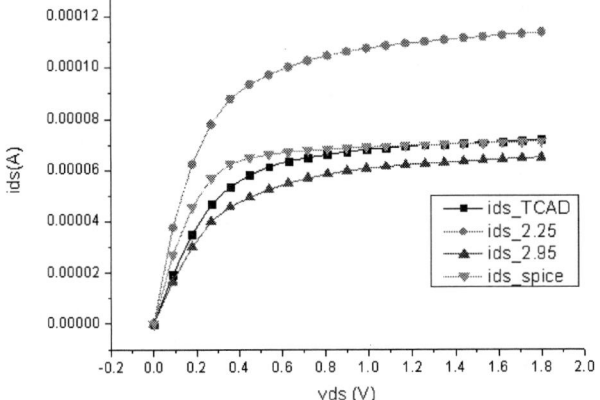

Figure 2.   Id_Vd curve of the TCAD model and SPICE model.

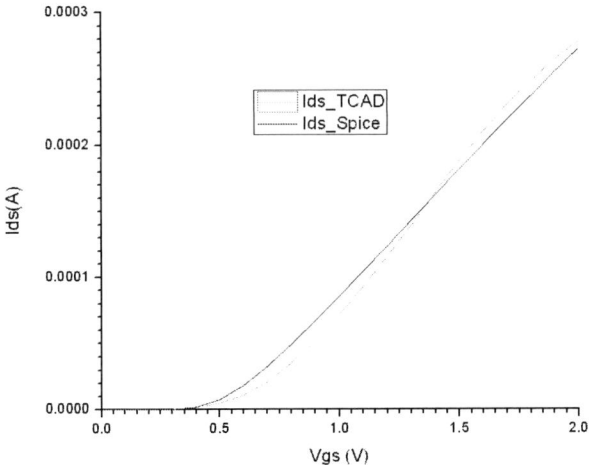

Figure 3.   Id_Vg curve of the TCAD model and SPICE model.

To get the calibrated curve the deposition of the ldd region is changed. In figure 2, the red curve is the original one with the deposition of the ldd region is 2.25e18 which is a little smaller; the blue line show that when the deposition of the ldd region is 2.95e18 which is a little larger. When the deposition of the ldd region is 2.38e18 the TCAD curve is well fitted with the characteristics of the spice model and which is the black one. Figure 3 shows the i_d-v_g curve of the TCAD model and spice model. From these two curves we can see that the 3D TCAD device model is fitted well enough with the spice model.

As shown in figure 1, the n-channel transistor is the 3D TCAD device model while the resistor is spice model and it is 10kΩ. The electric characteristic of this circuit is simulated as shown in figure 4. And when the input voltage of the circuit is

between 0.6V and 1.2V the n-channel transistor is worked in the saturation region, and the amplifier works in linear region.

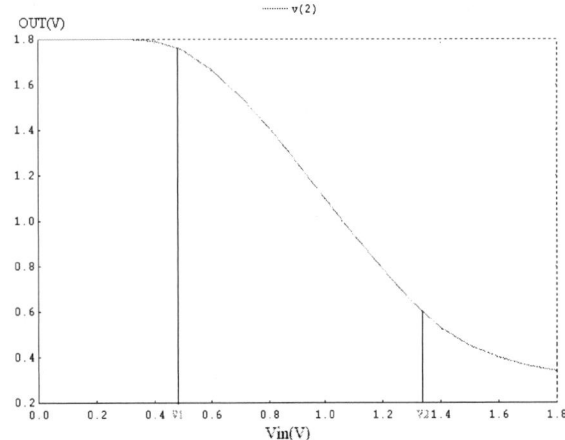

Figure 4.   Input and output characteristics of the CS amplifier.

III.   THE SET EFFECTS IN THE CS-AMPLIFIER

The simulation results reported in this work were carried out on the single-transistor common source amplifier with resistive load with the n-channel transistor modeled in TCAD, as shown in figure1. This NMOSFET was the device that was irradiated. And the model of the heavy ion strike used in these simulations is the model which is adapted to the heavy ions used in the experiments. The heavy ion model used in this circuit contains the radiation energy(0.1 MeV/(mg/cm²))、 incidence range(0.5μm) and the radius(0.1μm). When the input voltage of this circuit is 0.9V and the incidence angle is vertical of the heavy ion, the current and voltage waveforms from mixed-mode TCAD simulations of the CS-amplifier of figure1 are shown in figure 5. From this figure we can see that when at a low LET of 0.1 MeV/(mg/cm²), the current pulse has a double exponential shape. The transient current duration is almost 0.1ns.

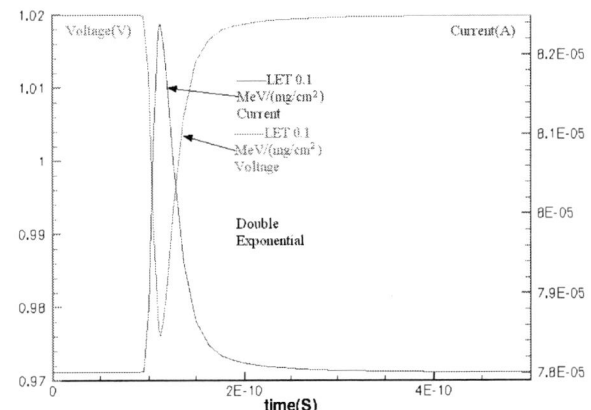

Figure 5.   Transient current and voltage wave form of the circuit.

We know that the SET radiation effects can be affected by a list of factors including the bias conditions, LET and strike location. In this paper the first two factors are concerned. From figure 4, we can see that when the input voltage of the circuit is between V1 and V2 the n-channel transistor is worked in the saturation region. Four different values of the input voltage from V1 to V2 were chosen to analysis the effect of bias conditions on the SETs which are 0.8V, 0.85V, 0.9V and 1.0V. The energy of the incidence particle was 0.1 MeV/(mg/cm^2) and 0.5 MeV/(mg/cm^2). The strike location, the angle of incidence and other conditions were kept constant while the bias condition was changing, Figure 6 and figure 7 illustrate the relationship between the bias condition and the transient current pulses obtained for the same ion used in the simulations.

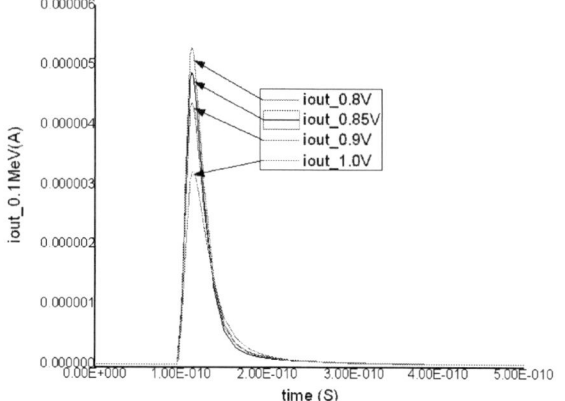

Figure 6. Relationship between the bias conditions and the transient current when the incidence energy is 0.1 MeV/(mg/cm^2).

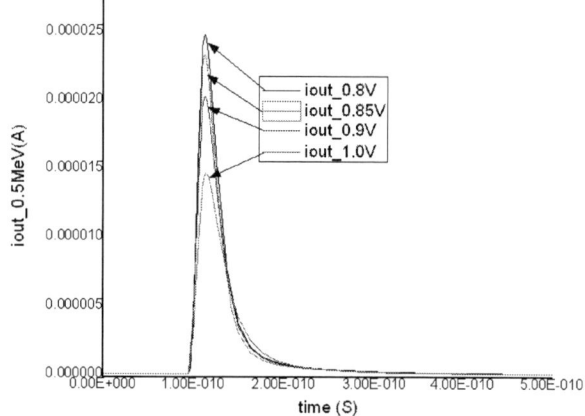

Figure 7. Relationship between the bias conditions and the transient current when the incidence energy is 0.5 MeV/(mg/cm^2).

Figure 6 describes the relations between the bias condition and the transient current pulses with the energy of the incidence particle 0.1 MeV/(mg/cm2). And figure 7 illustrates the relationship obtained for the energy of the incidence particle 0.5 MeV/(mg/cm^2).

It is known that when the input voltage of the circuit changes the current flows through the path changes as well. So for the sake of getting the effects of the bias conditions on the transient current intuitively the initial value of the current was adjusted to 0A as shown in figure 6 and 7. From these two figures we can see that for these two kinds of energy used for the radiation test the current pulses show a uniform trend. As the input voltage increases the amplitude of the transient current pulse decreases. The reason for this phenomenon is that as the input voltage increases the output voltage of a transistor worked in saturation region decreases as shown in figure 4, in turn the electron collected by the drain terminal of the transistor reduces, as a result the amplitude of the transient current decreases.

In these above two figures, when at a low LET of 0.1MeV/(mg/cm^2) and 0.5MeV/(mg/cm^2), the current pulse has a double exponential shape. But when the LET of the incidence particle reaches a right value or higher than this one, the current pulse will not obey the double exponential shape.

The transient current at different LET was studied to get the influence of different LET on the transient current pulse.

Figure 8 shows the heavy ion induced transient current for different LET of the incidence particle, as obtained using 3D cylindrical simulation. All these curves here listed were all obtained when the circuit worked at an input voltage of 0.8V. The figure indicates that, as expected, the current pulse at the LET of 1MeV/(mg/cm^2) according with the double exponential shape. At an LET of 5MeV/(mg/cm^2) or higher, however, the current pulse has a high peak current for a few picoseconds, followed by a distinct "plateau" region where the current is relatively constant for a much longer time than that of the initial prompt peak response. Figure 9 shows the output voltage LET of 1MeV/(mg/cm^2) (the blue line) and 5MeV/(mg/cm^2)(the red line). A "plateau" in the single event current pulse following the prompt response has also previously been observed in [6] for both bulk and SOI CMOS processes for digital SETs.

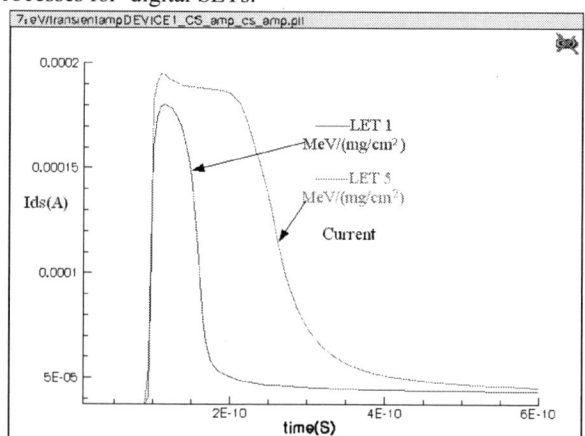

Figure 8. Transient current at different LET.

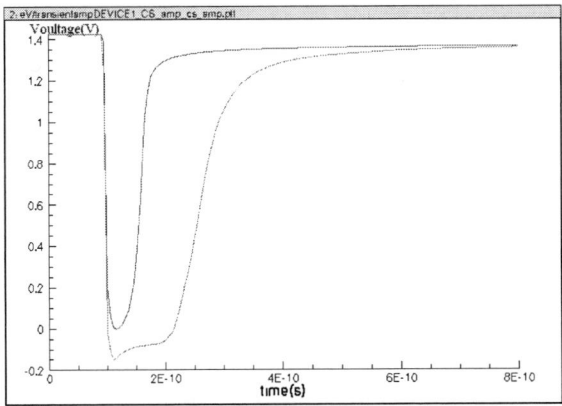

Figure 9. Output transient voltage at different LET.

When the LET of the incidence particle is higher enough, the amplitude of the transient current pulse and the output transient voltage pulse remains constant as shown in figure 10 and figure 11. The transient current pulses (1), (2), (3) and (4) were obtained separately at the LET of $0.5\text{MeV}/(\text{mg}/\text{cm}^2)$、$1\text{MeV}/(\text{mg}/\text{cm}^2)$、$5\text{MeV}/(\text{mg}/\text{cm}^2)$ and $10\text{MeV}/(\text{mg}/\text{cm}^2)$. In this figure we can see that, when the LET is higher enough the high peak current remains constant while the length of the "plateau"region increases.

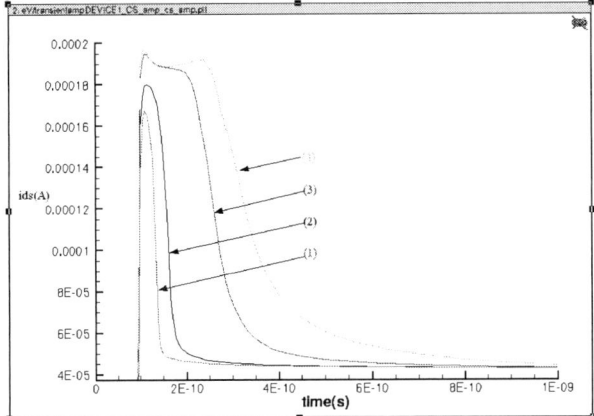

Figure 10. Transient current at different LET

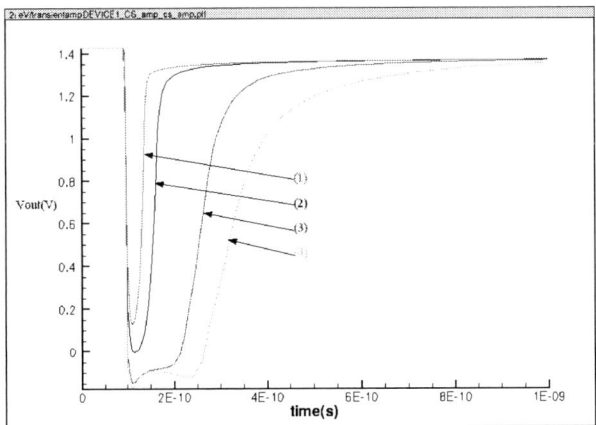

Figure 11. Output transient voltage at different LET.

For above two figures, the input voltage of the n-channel transistor was 0.8V and the initial output voltage was remained a constant. So when the LET of the incidence particle exceeds the threshold, although the amount of electron-hole pairs generated in device increases the electron collected at the drain remains unchangeable. Thus for more electrons the circuit needs more time to totally collecting all these movable electron-hole pairs.

IV. CONCLUSIONS

The Single event transient effect on the CS amplifier with resistive load is studied using a combination of device and circuit simulations in this paper. From these simulation results it can be concluded that the transient current pulse is relied on the bias conditions and LET of the incidence particles. As the input voltage of the CS amplifier increases, the amplitude of the transient current pulse decreases. And the SET waveforms reveal that although low LET pulses($\sim 1\text{MeV}/(\text{mg}/\text{cm}^2)$) still have a classical double exponential waveform in the SMIC 0.18μm technology, for higher LETs($>5\text{MeV}/(\text{mg}/\text{cm}^2)$) the current pulses have a plateau region in addition to the double exponential waveshape. In addition, when the LET exceeds the threshold the peak amplitude of the current pulse remains a constant and the length of the "plateau" region increases.

REFERENCES

[1] S.Buchner and D.McMorrow,"Single-Event Transients in Bipolar Linear Integrated Circuits," IEEE Transactions on Nuclear Science,Vol.53,no.6,pp.3079-3102,Dec.2006.

[2] R. Ecoffet, S. Duzellier, P. Tastet, C.Aicardi, and M. Labrunee, "Observation of heavy ion induced transients in linear circuits," in IEEE Radiation Effects Data Workshop, 1994, pp. 72 -77.

[3] R. Koga, S. D. Pinkerton, S. C. Moss, S.LaLumondiere, S.J. Hansel, K. B. Crawford, and W. R. Crain, "Observation of single event upsets in analog microcircuits," IEEE Trans. Nucl. Sci., vol. 40, pp.1838-1844,1993.

[4] P.E.Dodd,M.R.Shaneyfelt,J.A.Felix,and J.R.Schwank,"Produc- tion and propagation of single-event transients in high-speed digital logic ICs,"IEEE Trans.Nucl.Sci.,vol.51,no.6,pp.3278–3284,Dec.2004..

[5] V. Ferlet-Cavrois, P. Palliet, M. Gaillardin, D. Lambert, J. Baggio, J. R. Schwank, G. Vizkelethy, M. R. Shaneyfelt, K. Hirose, E. W. Black-more, O. Faynot, C. Jahan, and L. Tosti, "Statistical analysis of thecharge collected in soi and bulk devices under heavy Ion and proton ir-radiation—implications for digital SETs,"IEEE Trans.Nucl.Sci.,vol.51,no.6,pp.3242–3252,Dec.2006.

[6] V.Ferlet-Cavrois,P.Palliet,M.Gaillardin,D.Lambert,J.Baggio,J. R.Schwank,G.Vizkelethy,M.R.Shaneyfelt,K.Hirose,E.W.Black-more,O.Faynot,C.Jahan,and L.Tosti,"Statistical analysis of the charge collected in soi and bulk devices under heavy Ion and proton ir-radiation—implications for digital SETs,"IEEE Trans.Nucl.Sci., vol.51,no.6,pp.3242–3252,Dec.2006.

Design of the Photo-ionization Detector of Total Hydrocarbon

Xinping Dong, Peng Zhang, Zhenqi Zhao, Hui Wang, Shouchen Chai

The center of the physical sensor research

CETC NO.49th research institute

Harbin, China

zhaozhenqi1987@163.com

Abstract—In order to solve problem that the traditional gas detector are unable to detect Volatile Organic Compounds(VOC), the photo-ionization gas detector that can detect the VOC effectively by using the photo-ionization have been designed. The principle, structure, technology for ultraviolet (UV) light and ionization chamber and small-signal processing of the photo-ionization gas detector are introduced. Especially the detector incorporates an innovating ionization chamber, which can effectively improve the ionization efficiency and sensitivity. The instrument can effectively detect total hydrocarbon gas, has the advantages of small volume, stable working state of high sensitivity and the smallest error.

Keywords-Photo-ionization; Hydrocarbon gas; Detector; Small-signal processs

I. INTRODUCTION

Along with the development of the science, new materials, more and more applied to automobile, shipbuilding, railway and aerospace. But these new materials bring baneful influence while provide a convenience traffic environment, the Volatile Organic Compounds(VOC) will appear due to long time usage of new materials. The VOC mainly damage to people's respiratory system and nervous system[1]. And due to the complexity of composition and concentration is extremely low, the traditional detector is hard to detect total hydrocarbon gas, so the study of total hydrocarbon gas detection method is very necessary and urgently needed.

Recently, photo-ionization detection technology got great development, compared with the traditional detection technique, photo-ionization technique has the advantage of portable, high sensitivity, high accuracy, fast response, and continuous using. The detector based on the photo-ionization detection technology can be used in monitoring organic gas and play an important role[2-3].

II. PRINCIPLE OF PHOTO-IONIZATION DETECTOR

Following UV exposure, the organic gas was ionized, produced positive ions and electrons, while the basic ingredients such as N_2, O_2 in the air were not ionized. The resulting ions will cause a current flow between two electrodes disposed inside the chamber. An electrometer is used to measure the current. The current measurement can be converted into concentration in parts per million (ppm) of the organic gas based on the flow rate of the gas stream[4-5]. Principle is shown in figure 1.

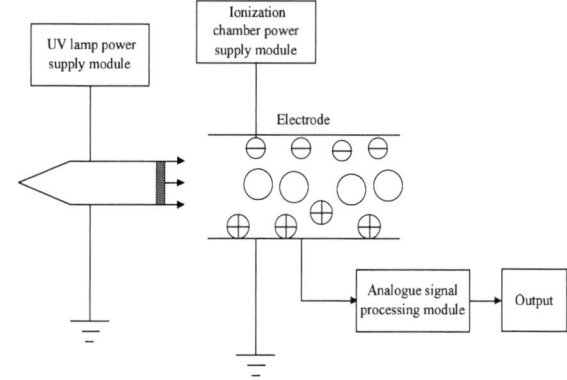

Figure 1. Principle of photo-ionization detector

Photo-ionization signal could be presented by the following equation [6]:

$$I = j_0 \cdot N_A \cdot C \cdot V \cdot \delta_i \tag{1}$$

—Luminous flux density; N_A—Loschmidt constant （ 2.69×10^{25} gram-atom/m³ ） ; C—Concentration ; V—Volume of the ionization chamber ; δ_i—Effective ionization section

III. STRUCTURAL DESIGN

The detector consists of UV lamp, UV lamp holder, ionization chamber, air pump, circuit boards, gas filter, etc. Arranging component according to the requirements of the corresponding size; Electromagnetic shielding on the ionization chamber and the holder used to avoid the electromagnetic interference between each other.

A. UV Lamp Design

Typically, a gas discharge UV lamp is used as the high-energy photon source. This lamp produces photons with photon energy of 10eV and above, corresponding to the wavelength of 100 nm or so, in the vacuum UV wavelengths.

B. Ionization chamber Design

In ionization chamber, the UV photons collide with and ionize volatile gas molecules having ionization potentials below the energy of the photons, creating ions and electrons. Electrode is biased to a high voltage to attract negatively charged particles (electrons) and repel positively charged particles (ions), and electrode is grounded to collect the positively charged particles (ions). The movement of the ions to electrode 22 produces a current, from which the concentration of the volatile gas can be determined. So the ionization chamber is an important component of the detector used to be signal conversion[7].

IV. CIRCUIT DESIGN

Circuit consists of UV lamp power supply module, ionization chamber power supply module and analogue signal processing module.

A. UV lamp power supply module

UV lamp power supply is high frequency (100 kHz) and high voltage (1500V).In order to get stable UV lamp output intensity, the voltage and frequency should be stability. In order to avoid high energy usage and the excitation signal unstable, the common emitter of push-pull self-excited oscillation circuit is used. It uses fewer devices, and need not too much filtering, the frequency shift circuit, can greatly improve the stability of the 1500V AC signal output, and effectively solve the problem of the driver circuit power supply consumption. Circuit diagram as shown in the figure 2:

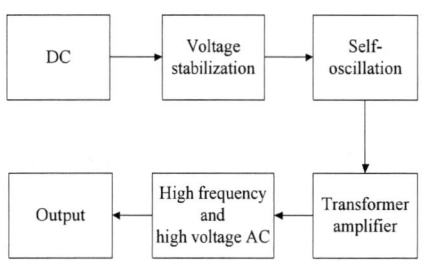

Figure 2. Block diagram of UV lamp power supply

B. Ionization chamber power supply module

The ionization chamber power supply is 150V AC power supplies, in order to get stable ionization chamber signal output, the voltage and frequency should be stability, thus improve the long-term working stability of the detector.

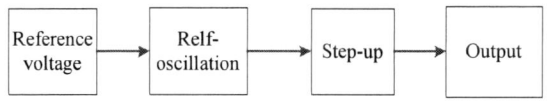

Figure 3. Block diagram of ionization chamber power supply

C. Analogue signal processing module

Due to the ionization chamber is the high impedance source, signal is very weak, and so the signal detection circuit should be high input impedance, high gain and low noise. Transforming impedance of ionization chamber by high-impedance circuit and the phase-sensitive detection circuit is adopted to improve the signal amplification.

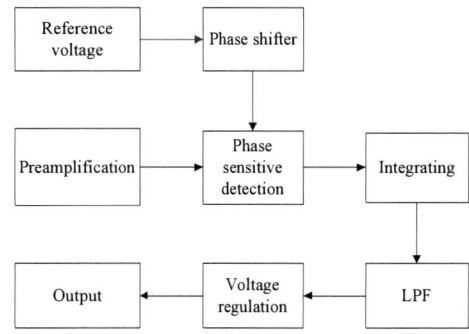

Figure 4. Block diagram of analogue signal processing

V. DETECTOR PERFORMANCE TEST AND INDICATORS

A. Error and the Linearity

After calibrating the two detectors, measure the output voltage when the concentration is 0 PPM, 25 PPM, 50 PPM, 75 PPM and 100 PPM, and mapping the output characteristic curve of the detectors. The measurement data is shown in table 1 and the output characteristic curve is shown in figure 5.

TABLE I.   OUTPUT VOLTAGE OF THE TWO DETECTORS

	0ppm	25 ppm	50 ppm	75 ppm	100 ppm
1	-0.05V	1.25V	2.52V	3.75V	4.93V
2	0.06V	1.29V	2.58V	3.77V	5.02V

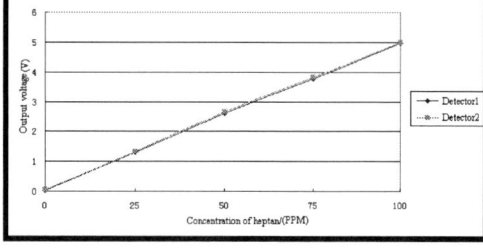

Figure 5. Output characteristic curve

From the table 1 can be seen that the maximum error of the output voltage is 70mV and 80mV, and corresponding to the concentration of error of 1.4PPM and 1.6PPM respectively.

B. Resolution measurement

Measure the output voltage fluctuations within a certain time (30s), the voltage fluctuation range is approximately less than 20mV, and less than half of the 50mV, which the concentration of heptane gas is 1PPM corresponding to, can be considered that the resolution of the detector is 0.5 PPM. The detection range is 0 to 100ppm.

C. Response time measurement

Place the detector under normal atmospheric conditions, and provide the detector 100 PPM of the concentration heptane gas quickly, measure output voltage waveform by oscilloscope, and the response time is received. The response time of the detector measurement result is shown in figure 6.

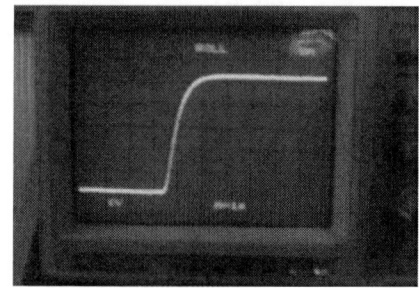

Figure 6. Response time measurement

VI. CONCLUSION

The instrument can effectively detect total hydrocarbon gas, has the advantages of small volume, stable working state of high sensitivity and the smallest error. Although the invention has been described with reference to particular embodiments, the description is only an example of the invention's application and should not be taken as a limitation. The method of test must lucubrate on testing the reliability and stability of the invention with a long-term use.

REFERENCES

[1] Han D Q,Ma WY,Chen D Y.Chromatographia,2007,66(11/12);899~904

[2] Zhang fan,Wei QingNong,Peng FuMin,Zhang Wei.The Overview of Development of Photoionization Technique[J].Modern Scientific Instruments 2007(2):8~18.

[3] Freedman A N.Photoionization detector response[J].Chromatography A,1982,236(1):11~15.

[4] Mergemeier S,Ebner I,Scholx F.Basic experimental studies on the operation of photoionization detectors[J].Fresenius'Journal of Analytical Chemistry,1998,361(1):29~33.

[5] Gerard Gremaud.Wentworth WE,Zlatkis Aetal.Windowsless pulsedMicharge photoionization detector application to qualita-tire analysis of volatile organic compounds,J,Chromatogr.A 1996-724,235-250.

[6] Verner P.J Chromatogr,1984,29(300):249.

[7] Fan QiuSheng,Xue ChenYang,Liang Ting,etc.Design of sensor for hydrazine gas based on photo-ionization principle[J].Transducer and Microsystem Technologies,2009(7):62~65.

Numerical Analysis of Pulse Laser Deformation on GaAs

Haijiao Zhou[1], Wenjun Sun[1]*, Zhong Meng[2], Zhongyang Liu[1]

1 Key Laboratory of Photonic and Electronic Bandgap Material(Ministry of Education); Heilongjiang Key Laboratory of Advanced Functional Materials and Excited State Processes, School of Physics and Electronic Engineering, Harbin Normal University,Harbin ,China
2 Key Laboratory of Airborne Optical Imaging and Measurement, Changchun Institute of Optics, Fine Mechanics and Physics, Chinese Academy of Sciences,Changchun,China
*:swjgood0139@126.com

Abstract—Two-dimension axis symmetric physical model of Gaussian distribution pulse laser irradiation for semiconductor material was established by software COMSOL Multiphysics. The temperature field distribution of semiconductor GaAs was analyzed under irradiation of nanosecond pulse laser with wavelength of 1064nm by means of solving the thermal conduction equations, in consideration of the change of the thermal physical parameters of GaAs along with temperature. The influences of three kinds of heat exchange mode on temperature field of GaAs were compared. Calculation results matched with related experiments, which showed the built physical model is scientific, at the same thermal convection and thermal radiation could be ignored for calculating the temperature field of semiconductor materials under nanosecond pulse laser irradiation.

Keywords-pulse laser irradiation; semiconductor material;

temperature field; damage threshold.

I. INTRODUCTION

GaAs has become an important semiconductor optoelectronic material because of its direct bandgap structure, high photoelectric conversion efficiency and high electron mobility. Thermal deformation of GaAs caused by temperature gradient when pulse laser irradiation GaAs material, what is due to the thermal absorption and the thermal conduction of semiconductor materials. Thermal deformation effect the device performance and even damage the device. So it is a significant research topic to explore and solve the changes of temperature on the surface of GaAs. The thermal conduction equation of pulse laser irradiation samples can be solved in two ways: analytic solution method [1-2] and numerical solution method [3] under the certain boundary conditions. It is need to consider the thermal conduction, the thermal convection and the thermal radiation of material surface while solving the problem of pulse laser irradiation GaAs material, as well as the change of semiconductors materials thermal physical parameters along with temperature must be considered.

In this paper, a physical model of Gaussian distribution pulse laser irradiation for semiconductor GaAs was established by COMSOL Multiphysics[4], considered

comprehensively three kinds of thermal exchange mode (the thermal conduction, the thermal convection, the thermal radiation) influences on temperature distribution while pulse laser interaction with semiconductor material and the change of thermal physical parameters along with temperature for GaAs. The thermal conduction equation of the Gaussian distribution pulse laser irradiation GaAs material was solved, relatively exact temperature distribution was obtained, and the damage mechanism was analyzed.

Ⅱ. PHYSICAL MODEL

Two-dimension axis symmetric model of pulse laser irradiation GaAs material was shown in Figure1. Coordinate origin locate at the center of the surface of the GaAs sample, it is the pulse laser incidence center. r and z axes indicate the radial directional and longitudinal direction of the sample respectively. h and b are the thickness and width of the sample, a is beam waist radius of Gaussian distribution pulse laser. Thermal conduction equation is given by equation (1), q is heat source that comes from equation (2), equation (3) and equation (4) shows the physical model of pulse laser. Equation (3) is the space distribution model of pulse laser. And equation (4) is the time distribution model of pulse laser.

$$\rho c \frac{\partial T(r,z,t)}{\partial t} = \frac{1}{r}\frac{\partial}{\partial r}(rk\frac{\partial T(r,z,t)}{\partial t}) + \frac{\partial}{\partial r}(k\frac{\partial T(r,z,t)}{\partial t}) + q \tag{1}$$

$$q = \alpha I an(t) \tag{2}$$

$$I = (1-R)I_0 \exp(-r^2/a^2)\exp(-\alpha z) \tag{3}$$

$$an(t) = \begin{cases} 1 & (n-1)\Gamma < t < (n-1)\Gamma + \tau \\ 0 & (n-1)\Gamma + \tau < t < n\Gamma \end{cases}$$

$$n = 1,2,3\ldots\ldots \tag{4}$$

In above formulas, ρ, k and c indicate the density, the thermal diffusivity and the thermal capacity of the GaAs respectively. α is a absorption coefficient without considering the nonlinear absorption of GaAs for Nd: YAG pulse laser with wavelength of 1064nm. an(t) is the time distribution function of pulse laser. R is reflectivity of sample surface, I_0 is the peak power of the incident pulse laser, τ is pulse duration.

978-1-4799-1215-5/13 $31.00 © 2013 IEEE

r is the distance from the surface of material to the incident pulse laser center. Γ is pulse period.

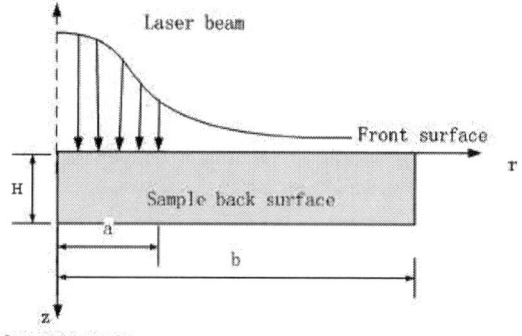

Figure1. The model of pulse laser irradiation GaAs

Initial conditions:

$$q_0 = h(T_0 - T^{'}) \tag{5}$$

$$T\big|_{t=0} = T_0 = 293.15K \tag{6}$$

$$\frac{\partial T}{\partial r}\bigg|_{r=b} = 0 \tag{7}$$

$$\frac{\partial T}{\partial Z}\bigg|_{z=H} = 0 \tag{8}$$

In equation (5): h is heat transfer coefficient on the surface of GaAs material. T ' is the temperature of the GaAs material, T_0 with value of 293.15K is the environment temperature around the GaAs material. Side surface of the sample is adiabatic, that need to meet equation (7). Lower surface of the sample is adiabatic, that need to meet equation (8). There are free convection between the upper surface and the air and materials irradiate to the thermal surroundings, need to meet the following condition:

$$K\frac{\partial T}{\partial z}\bigg|_{z=0} = h(T_0 - T^{'}(r,0,t)) + \varepsilon\sigma(T_0^4 - T_0^4(r,0,t)) \tag{9}$$

In equation (9): ε with a value of 0.1 is emissivity of the material surface. σ with a value of $5.67e\text{-}8\,W/(m^2 \cdot K^4)$ is a Stepan constant.

Table 1 shows thermal physical parameters of GaAs material in the process of calculation. Pulse laser wavelength is λ=1064nm, pulse duration is τ=16ns, pulse period is Γ=0.1s. pulse laser beam waist radius is a=0.04cm. The radius is r=03cm, the thickness along z axis is H=0.0005cm of GaAs material. Figure 2 shows the mesh of the model.

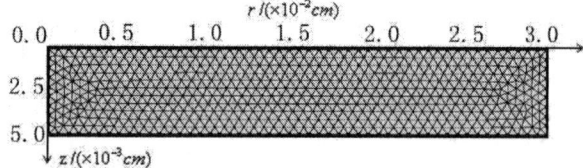

Figure2. Grid division of physical model

TABLE□. THERMAL PHSICAL PARAMETERS OF GaAs[5-9]

Parameter	Value or expression
Thermal conductivity k /(W·cm·K^{-1})	$0.425(300/T)^{1.1}$
Density ρ /(kg·m^{-3})	5316
Heat capacity C/(J·g^{-1}·K^{-1})	$0.307+7.25*10^{-5}T$
Energy gap Eg/eV	1.575-0.15*T/300
Coefficient of heat transfer h/ (W·m^{-2}·K)	100
Decomposition temperature /K	800
Thermal radiation rate / ε	0.1
Laser radius a/cm	0.04
Initial temperature T_0/K	293.15
Reflectivity R	0.4
Absorption coefficient α /cm^{-1}	10*exp(1.49*(1.17+1.43-Eg-1.38))

□. CACULATION RESULTS AND DISCUSSIONS

A. Temperature distribution of GaAs material

According to the analysis above, the temperature distribution of pulse laser irradiation GaAs material was calculated by software COMSOL Multiphysics. Figure 3 shows the temperature distribution curve of material surface along the radial direction when pulse duration is 16ns, pulse period is 0.1s, peak power density of incident pulse laser is 2.2×10^9W/cm^2, irradiation time is 3×10^{-8} s. The maximum temperature value of the pulse laser radiation center is 1154K. This value matches well with related experiments [5]. At the same time, this temperature value is higher than the decomposition temperature of 800K for GaAs material. And this temperature value is the same as the decomposition temperature in related experiments [6] in which semiconductor GaAs was analyzed under irradiation of nanosecond pulse laser with wavelength of 1064nm, which means the GaAs material has been decomposed. The decomposition damage time of GaAs material is very short (3×10^{-8}s approximately), when the pulse duration is 16ns, pulse period is 0.1s, and peak power density of incident pulse laser is $2.2 \times 10^9 \text{W/cm}^2$.

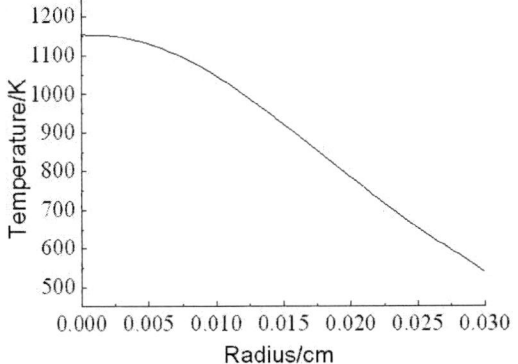

Figure3. Temperature distribution along radius

Figure 4 shows the temperature distribution curve of the radiation center along longitudinal direction when pulse duration is 16ns, pulse period is 0.1s, peak power density of incident pulse laser is $2.2 \times 10^9 \text{W/cm}^2$, irradiation time is 3×10^{-8}s. We found that the temperature value of the radiation center reaches the damage temperature firstly, that is to say, the damage of GaAs material initiates from the radiation center. According to Figure 3 and Figure 4, the maximum temperature value is appearing at the position of (r=0, z=0).

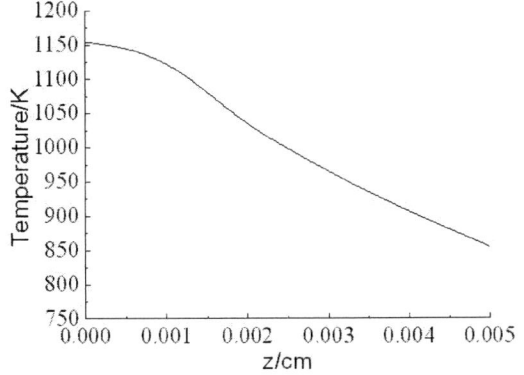

Figure 4. Temperature distribution of the irradiation center along z axis

B. The influences of the change of pulse duration on the temperature field

Figure 5 shows the temperature distribution curve of the maximum temperature of GaAs surface radiation center at different pulse duration, and pulse period is 0.1s, peak power density of incident pulse laser is $2.2 \times 10^9 \text{W/cm}^2$, irradiation time is 3×10^{-8} s. The maximum temperature of irradiation center increases along with the increase of pulse duration. The bigger the pulse interval is, the easier damage temperature of GaAs materials reaches when pulse period, peak power density of incident pulse laser, irradiation time are all unchanged. This calculation result is consistent with the result of the related experiment [10].

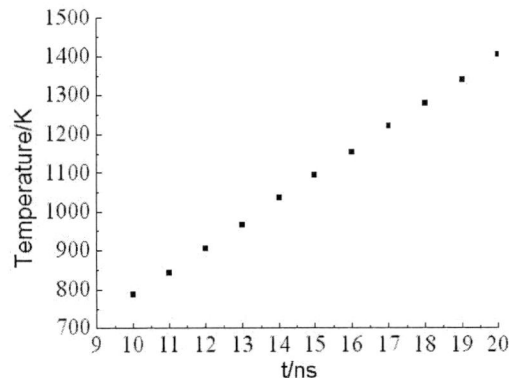

Figure5. Temperature vs pulse duration

C. Damage threshold of GaAs material under pulse laser irradiation

Usually the longer the time of laser irradiation is, the smaller decomposition damage power density is. It is because that when the same energies were calculated, the bigger the power density of pulse laser is, the shorter the time requires, which is consistent with statement of related experiment [11]. And the calculation result is lower than decomposition temperature when pulse duration is 16ns, pulse period is 0.1s, peak power density of incident pulse laser is $2.1 \times 10^7 \text{W/cm}^2$, irradiation time is 0.1s. That is to say, when pulse duration is 16ns, pulse period is 0.1s, the pulse laser damage threshold is $2.1 \times 10^7 \text{W/cm}^2$ in this theoretical model. The damage threshold of this paper was compared with other damage threshold of related literature [5][12] as the follow table 2. We can find that these damage thresholds are basically similar, which proved this calculation result is correct.

TABLE. DAMAGE THRESHOLD OF GaAs

this paper	reference[5]	reference[12]
$2.1 \times 10^7 \text{W/cm}^2$	$2.6 \times 10^7 \text{W/cm}^2$	$4.2 \times 10^7 \text{W/cm}^2$

D. The influence of three kinds of heat exchange mode on temperature field of GaAs

Figure 6 shows the temperature distribution curve of the material surface radiation center when pulse duration is 16ns, pulse period is 0.1s, peak power density of incident pulse laser is $2.2 \times 10^9 \text{W/cm}^2$, irradiation time is 3×10^{-8}s. Curve (a) shows the temperature distribution of the material surface radiation center only considering the thermal conduction; Curve (b) shows the temperature distribution not only considering the thermal conduction but also considering the thermal convection and the thermal radiation. From the figure, we can find that the temperature value changes a little under the two different conditions. That is to say, when we calculate the temperature field of nanosecond pulse laser irradiation for semiconductor GaAs material, the influence of the thermal convection and the thermal radiation could be ignored.

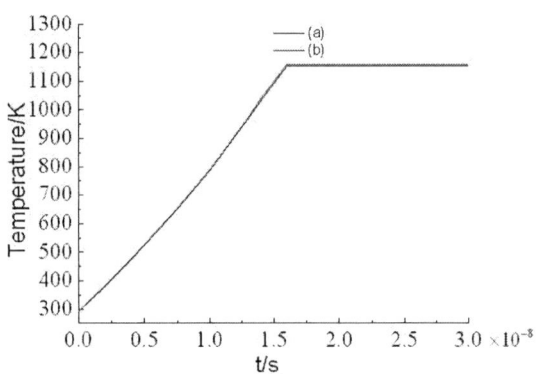

Figure6. Influences of thermal convection and thermal radiation on temperature

E. The influences of the thermal physical parameters of GaAs on the temperature field

Figure 7 shows the temperature distribution curve of material surface radiation center when pulse duration is 16ns, pulse period is 0.1s, peak power density of incident pulse laser is $2.2 \times 10^9 W/cm^2$, irradiation time is 3×10^{-8}s. Curve (a) and Curve (b) show the change of material surface irradiation center temperature without considering the change of thermal physical parameters of GaAs material along with temperature and under considering the change of thermal physical parameters of GaAs material along with temperature respectively. Curve (b) is more close to experiment result, so the change of the thermal physical parameters of GaAs material along with temperature must be considered when we calculate the temperature field of pulse laser irradiation GaAs material [5].

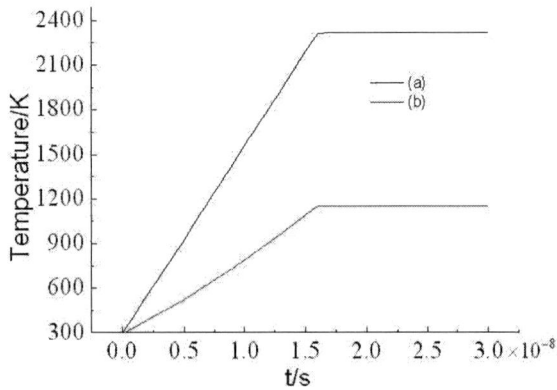

Figure7. Influences of thermal physical parameters of GaAs on temperature field

□. CONCLUSIONS

Two-dimension axis symmetric physical model of Gaussian distribution pulse laser irradiation for semiconductor material was successfully established by software COMSOL Multiphysics. The temperature distribution of semiconductor

GaAs material was numerically calculated under irradiation of nanosecond pulse laser. And this paper obtained the relationship between the maximum temperature of surface radiation center and the pulse duration. Numerical results are basically consistent with the related experiment results, which show the built physical model is scientific. And the thermal convection and the thermal radiation could be ignored under used pulse power density in this paper. At the same time, the influences on the temperature field by the change of the thermal physical parameters of GaAs material along with temperature were calculated. The established model in this paper also applied to calculate the heat effect of nanosecond pulse laser irradiation for other semiconductor materials.

Acknowledgment

This work is supported by the Natural Science of Foundation of Heilongjiang Province (No.F201202);the Education Department Key Teacher Project of Heilongjiang Province (No. 1251G031) and the Opening Project of Key Laboratory of Photonic and Electronic Bandgap Materia(Ministry of Education); Heilongjiang Key Laboratory of Advanced Functional Materials and Excited State Processes, School of Physics and Electronic Engineering, Harbin Normal University.

REFERENCE

[1] Mingqiang Liu, Bincheng Li, "Analysis of temperature and deformation fields in an optical coating sample,"J. Acta Physica Sinica, vol.57,No6, pp.3402-3409,2008.

[2] Bincheng Li, Yapei Yang,"Theory of surface thermal lens signal in optical coating with CW modulated top-hat beam excitation,"J.Acta Physica Sinica, vol.55,No9, pp.4673-4677, 2006.

[3] Shaofeng Guo, Qisheng Lu, Xiang-ai Cheng, Xuewen Zeng, Jin Li , Qianrong Chen,"Rotationeffects of intensive laser,"J. Chin. J.Lasers, vol.31,No2, pp.145-148,2004.

[4] Meifang Bao, Zhiyu Qian,Weitao Li , Di Xiao, Yuyang Wang, Lu Qian,"Biological Tissue's Temperature Field During the Laser-Induced Interstitial Thermotherapy,"J.Acta Photonica Sinica, vol.40,No5, pp. 718-721, 2011.

[5] Haifeng Qi,"Investigation on damag process of GaAs induced by 1064nm continuous laser and nanoscale pulse laser," D.ShanDong University,April,2008.

[6] Juan Li,Wenjun Sun,Jingnan Sun,Liping Zhao,Mengyang Li, Hongwu Zhi,"Numerical analsis of CW laser damage in GaAs,"J.Acta Photonica Sinica ,vol.41,No5, pp.571-574, 2012.

[7] J. R .Meyer, M. R. Kraer, and F.J. Bartoli,"Optical heating in Semiconductors:Laser damage in Ge , Si, Insb, and GaAs ,"J. J Appl Phys,vol.51,No10, pp.5513-5522,1980.

[8] Yongfu Li, Haifeng Qi, Qingpu Wang,"532nmCW laser induced damage of GaAs material,"J.Chinese Journal of Quantum Electronics,vol.24,No5, pp.625-629,2007.

[9] Amit Garg, Avinashi Kapaoor,K.N.Tripathi, "Laser-induced damage studies in GaAs," J. Optics & Laser Technology,vol.35,No1, pp. 21-24,2003.

[10]Haifeng Qi, Qingpu Wang, Yongfu Li,"Thermal procee and surface damage of GaAs induced by 532nm continuous laser,"J.Applied Surface Science,vol.254,No5, pp.1373-1376,2007.

[11] Jianjun Zhao, Jin Liu, Chunrong Song. Yanxiong Niu,"Thermal and Mechanical Damage in InSb (pv) Detector Induced by Repetitive Pulse Laser,"J. Journal of Ordnance Engineering College,vol.18,No5, pp.23-26,2006.

[12]Yongfu Li,"strength pulse laser induced damage mechanism of GaAs Material,"D.ShanDong University,April,2007.

Add-drop ring resonator coupled Mach-Zehnder interferometer for highly sensitive sensing

Xiaoqi Liu, Yundong Zhang*, Xuenan Zhang, Ping Yuan

National Key Laboratory of Tunable Laser Technology, Institute of Opto-Electronics, Harbin Institute of Technology, Harbin, China
ydzhang@hit.edu.cn

Abstract—We propose an add-drop fiber ring resonator (AFRR) coupled Mach-Zehnder (M-Z) interferometer for highly sensitive sensing. Theoretically it is shown that the sensitivity of such an interferometer can be much higher than that of a single fiber ring resonator (SFRR) coupled M-Z interferometer as a result of the dispersion properties of the add-drop ring resonator. The sensitivity of our configuration can achieve 49.5 , which is enhanced by 30.06 compared to that of the SFRR coupled M-Z interferometer. This proposed configuration enables highly sensitive, compact and stable sensors.

Keywords- resonator;interferometer;sensor

I. INTRODUCTION

In modern industries, biological and manufacturing field, there has been great potential demand on measuring and detecting minor changes in the environment, which inquires the development of highly precise as well as safe sensing techniques [1]. Of these techniques, devices based on optical ring resonators have attracted considerable attention due to their simplicity, compactness, and stability. In the past few years, various sensors based on optical ring resonators have been typically developed, for example, grating-coupled waveguide sensors [2], planar optical-waveguide sensors [3], directional coupler sensors [4] and micro resonator sensors [5]. In the past few years, a number of schemes have been proposed to enhance the sensitivity of the interferometers, such as using photonic crystal structures to minimize the size of on-chip devices, utilizing the dispersive property of semiconductors to enhance the spectral sensitivity of interferometers[6], making use of slow light media to enhance the resolution of Fourier transform interferometer, and exploiting fast light media or slow light structures to increase the rotation sensitivity of a Sagnac interferometer[7]. Resonators coupled to M-Z interferometer are quite useful in tailoring the resonance line shape for the multiple-beam interference between the circulating optical waves in the resonator and the reference arm [8] therefore holding potential applications in sensing [9].

In this letter, we propose the AFRR coupled M-Z interferometer with a compact size which enables the simultaneously producing fast light and slow light. In addition, we show that the sensitivity of the M-Z interferometer would be enhanced by combining slow light with fast light together. Also, by theoretical calculation, we obtain that the dispersion sensitivity of the AFRR coupled M-Z interferometer can achieve 49.5 . For comparison, we also theoretically calculated

the dispersion sensitivity of the SFRR coupled M-Z interferometer which is 19.44 . The simulation results show that the dispersion sensitivity of our configuration is enhanced by 30.06 compared to that of the SFRR coupled M-Z interferometer. This scheme may provide a convenient way to enhance the sensitivity of the M-Z interferometer.

II. THEORETICAL ANALYSIS

Recently it has been proved that the sensitivity of an optical fiber M-Z interferometer can be enhanced by introducing both normal dispersion and abnormal dispersion structure into it, like the nested fiber ring resonator. But it was difficult to insure the stable resonance of both the inner ring and the outer ring in the nested fiber ring resonator. Here we adopt the add-drop ring resonator to produce the fast and slow light for the simplicity to construct in the experiment and for the good temporal stability.

Fig.1 (a) gives the schematic of the balanced M-Z interferometer coupled with the add-drop resonator which consists of a single fiber ring resonator and two waveguide directional couplers. r_1 and r_2 are the amplitude reflecting coefficients and are related to the amplitude coupling coefficients t by $r_i^2 + t_i^2 = 1$, $i = 1, 2$. The output light of resonator's port one and two interfere with each other at $C3$ and then interfere with the reference light through the coupler $C4$, which are all 3dB couplers together with the coupler $C0$. The fiber length of the two arms of the M-Z interferometer are made equal, then the optical path difference between the upper and lower arms through the M-Z interferometer is caused only by the AFRR structure. The schematic of the SFRR coupled M-Z interferometer was shown in Fig.1 (b).

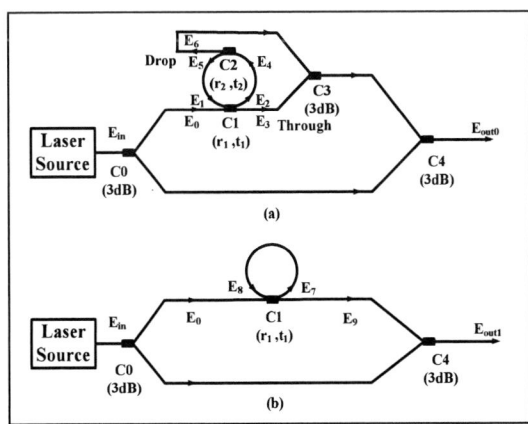

(a)

(b)

The research is supported by the National Natural Science Foundation of China (NSFC) (No. 61078006 and No. 61275066) and National Key Technology Research and Development Program of the Ministry of Science and Technology of China (No.2012BAF14B11).

Figure.1.(a) M-Z interferometer coupled with an AFRR, (b) M-Z interferometer coupled with a SFRR.

Figure.2. The effective phase shift versus the round-trip phase shift of the two outputs of the AFRR

To illustrate the dispersion properties of the outputs of the add-drop resonator, we set the parameters as follows: the perimeter l of the resonator is $1m$, the attenuation factor a is equal to 0.98, and $r_1 = r_2 = 0.9$. In addition, the effective refractive index of the fiber is 1.468 and the wavelength of the input light is $1550nm$ which is compatible with other fiber systems. By using the transfer matrix theory [10, 11], we can find that the input and two outputs are related by the complex amplitude transmission as follows:

$$\tau_1 = \frac{E_3}{E_0} = \frac{r_1 - r_2 a e^{i\phi}}{1 - r_1 r_2 a e^{i\phi}}, \qquad (1)$$

$$\tau_2 = \frac{E_6}{E_0} = \frac{-t_1 t_2 \sqrt{a} e^{i\phi/2}}{1 - r_1 r_2 a e^{i\phi}}. \qquad (2)$$

The field transmission intensities of the two ports are defined as $T_1 = |\tau_1|^2$ and $T_2 = |\tau_2|^2$. The effective phase shifts of the transmitted lights are described as $\phi_1^{eff} = \arg(\tau_1)$ and $\phi_2^{eff} = \arg(\tau_2)$ respectively, so we can explicitly express the effective phase shifts as follows:

$$\phi_1^{eff} = \arg(\tau_1)$$
$$= \pi + \phi + \arctan \frac{r_1 \sin \phi}{r_2 a - r_1 \cos \phi}, \qquad (3)$$
$$+ \arctan \frac{r_1 r_2 a \sin \phi}{1 - r_1 r_2 a \cos \phi}$$

$$\phi_2^{eff} = \arg(\tau_2)$$
$$= \pi + \phi / 2 + \arctan \frac{r_1 r_2 a \sin \phi}{1 - r_1 r_2 a \cos \phi}. \qquad (4)$$

The simulation results based on the transfer matrix theory are shown in Fig.2. The two outputs all undergoes the rapid phase shift around the resonance region, but has the opposite variation direction, which are corresponding to the abnormal dispersion and the normal dispersion. Therefore we can combine the two outputs to enhance the phase sensitivity of the resonator coupled M-Z interferometer.

As shown in Fig.2, we can find that the effective phase shifts of the two outputs have opposite changing trends at resonance corresponding to anomalous and normal dispersion, respectively. Optical resonators can be analogous to atomic resonances, and the effective phase shift ϕ_i^{eff}, $i = 1, 2$ imparted to light transmitted across a structure is analogous to the polarizability of an atom. Therefore, the dispersive response of such structures is contained in the derivative of the effective phase shift $d\phi_i^{eff} / d\phi$, $i = 1, 2$. $d\phi_i^{eff} / d\phi > 0$ corresponding to normal dispersion and referred to as slow light, $d\phi_i^{eff} / d\phi < 0$ corresponding to anomalous dispersion and referred to as fast light.

The derivative of the effective phase shifts of the two outputs of the AFRR coupled M-Z interferometer can be express as follows:

$$\frac{d\phi_1^{eff}}{d\phi} = 1 + \frac{r_1 r_2 a \cos \phi - r_1^2}{r_1^2 - 2 r_1 r_2 a \cos \phi + r_2^2 a^2} + \frac{r_1 r_2 a \cos \phi - r_1^2 r_2^2 a^2}{1 - 2 r_1 r_2 a \cos \phi + r_1^2 r_2^2 a^2}, \qquad (5)$$

$$\frac{d\phi_2^{eff}}{d\phi} = \frac{1}{2} + \frac{r_1 r_2 a \cos \phi - r_1^2 r_2^2 a^2}{1 - 2 r_1 r_2 a \cos \phi + r_1^2 r_2^2 a^2}. \qquad (6)$$

As we can see in Fig.3, for the through port, near the resonance $d\phi_1^{eff} / d\phi = -41.5 < 0$ is the anomalous dispersion and referred to as fast light; for the drop port, near the resonance $d\phi_2^{eff} / d\phi = 7.98 > 0$ is the normal dispersion and referred to as slow light. Therefore, this add-drop resonator configuration with a compact size enables the simultaneously producing fast light and slow light.

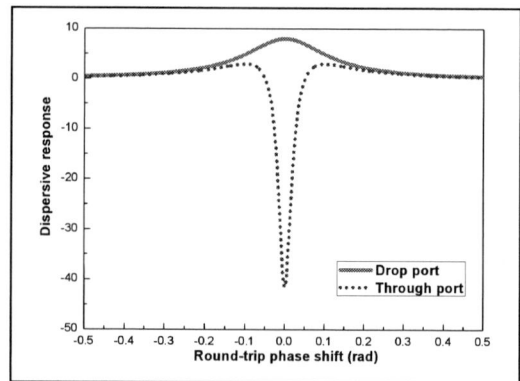

Figure.3. The dispersive response of the two outputs of the AFRR

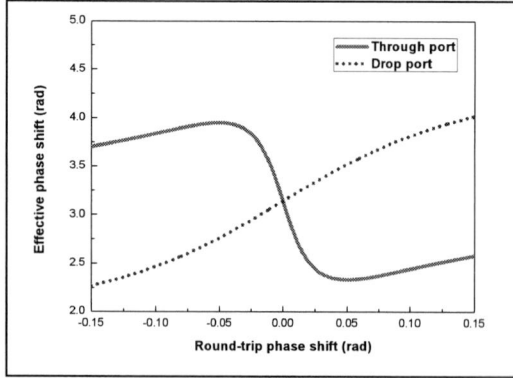

978-1-4799-1215-5/13 $31.00 © 2013 IEEE

The sensitivity of the ring coupled M-Z interferometer to any measured variable ε can typically be decreased in terms of the quantity $d\Delta\Phi/d\varepsilon$, where ε is the measured variable. The sensitivity of our interferometer is defined as:

$$S = \frac{d\Delta\Phi}{d\varepsilon} = \frac{d\Delta\Phi}{d\phi} \cdot \frac{d\phi}{d\varepsilon} = S_1 \cdot S_2, \qquad (7)$$

where ϕ is the round-pass phase shift of the ring. $S_1 = d\Delta\Phi/d\phi$ is the dispersion sensitivity, $S_2 = d\phi/d\varepsilon$ is the rate of change of ϕ with respect to the measured variable ε. We know how a measurand affects the phase of the signal traveling through a ring for a wide range of measurands. For biochemical sensors, the interaction of the waveguide mode's evanescent field with a dissolved analyte changes the effective index n_{eff} of the waveguide mode, thus changing ϕ [9]. If the measurand is a rotation Ω, then $S_2 = d\phi/d\varepsilon = d\phi/d\Omega$ is a constant determined by the wavelength [12]. If the measured variable is temperature T, then $S_2 = d\phi/d\varepsilon = d\phi/dT$ is determined mostly by the thermal expansion coefficient and thermo-optic coefficient. As the fiber used in the M-Z interferometer is the same, thus the result of S_2 is the same for the same measured variable. Thus the sensitivity S will be enhanced while the dispersion sensitivity S_2 is increased. However, if a strong dispersive slow light structure is exploited, the sensitivity of such an interferometer can be greatly enhanced.

The dispersion sensitivity of the SFRR coupled M-Z interferometer is given by:

$$S_1 = \frac{d\Delta\Phi}{d\phi} = \frac{d\varphi^{eff}}{d\phi}, \qquad (8)$$

where $\Delta\Phi$ is the phase difference of the two arms, and here it is equal to the effective phase of the ring resonator φ^{eff}.

The dispersion sensitivity of the AFRR coupled M-Z interferometer is given by:

$$S_1 = \frac{d(\varphi_1^{eff} - \varphi_2^{eff})}{d\phi} = \frac{d\varphi_1^{eff}}{d\phi} - \frac{d\varphi_2^{eff}}{d\phi}, \qquad (9)$$

where φ_j^{eff} ($j = 1, 2$) are the effective phase shifts of the two outputs, and $\varphi_1^{eff} - \varphi_2^{eff}$ is equal to the phase difference of the two arms. As the derivatives $d\phi_1^{eff}/d\phi$ and $d\phi_2^{eff}/d\phi$ are opposite in sign, then the subtraction of $d\phi_1^{eff}/d\phi$ and $d\phi_2^{eff}/d\phi$ will increase the dispersion sensitivity S_1 of the AFRR coupled M-Z interferometer. Therefore we can combine the two outputs to enhance the phase sensitivity of the resonator coupled M-Z interferometer. For comparison, the simulation results of the dispersion sensitivity of the AFRR coupled M-Z interferometer and the SFRR coupled M-Z interferometer are shown in the Fig.4.

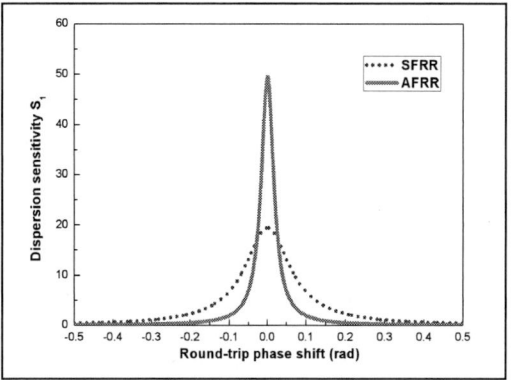

Figure.4. The dispersion sensitivity of the AFRR coupled M-Z interferometer and that of the SFRR coupled M-Z interferometer

As we can see clearly from the Fig.4, the sensitivity of our configuration can achieve 49.5, which is enhanced by 30.06 compared to that of the traditional single-bus ring resonator which is 19.44. This means the performance of the proposed AFRR coupled M-Z interferometer structure is better than that of the SFRR coupled M-Z interferometer.

III. CONCLUSION

In summary, we propose an AFRR coupled M-Z interferometer with high sensitivity. Through theoretical analysis of the transmission characteristics, we demonstrate the proposed structure enables the simultaneously producing superluminal and slow light. Therefore we can combine the two outputs to enhance the phase sensitivity of the resonator coupled M-Z interferometer. Also, the simulation results show that the dispersion sensitivity of the proposed configuration can achieve 49.5, which is enhanced by 30.06 compared to that of the SFRR coupled M-Z interferometer. Thus the proposed AFRR coupled M-Z interferometer structure enables highly sensitive, compact and stable sensors [13-16].

ACKNOWLEDGMENT

The research is supported by the National Natural Science Foundation of China (NSFC) (No. 61078006 and No. 61275066) and National Key Technology Research and Development Program of the Ministry of Science and Technology of China (No.2012BAF14B11).

REFERENCES

[1] C. J. Wang, "Fiber Loop Ringdown-a Time-Domain Sensing Technique for Multi-Function Fiber Optic Sensor Platforms: Current Status and Design Perspectives," Sensors 9, 7595-7621(2009).

[2] R. Hornath, H. C. Pedersen, N. Skivesen, D. Selmeczi, and N. B. Larsen, "Opticalwaveguide sensor for on-line monitoring of bacteria," Opt. Lett. 28, 1233-1235(2003).

[3] T. Okamoto, M. Yamamoto, and I. Yamaguchi, " Optical waveguide absorption sensor using a single coupling prism, " J. Opt. Soc. Amer. A, Opt. Image Sci. 17, 1880-1886(2000).

[4] B. J. Luff, R. D. Harris, J. S. Wilkinson, R. Wilson, and D. J. Schiffrin, "Integrated-optical directional coupler biosensor, " Opt. Lett. 21, 618-620(1996).

[5] C. Y. Chao and L. J. Guo, "Biochemical sensors based on polymer microrings with sharp asymmetrical resonance," App. Phys. Lett. 83, 1527-1529 (2003).

[6] Y. Cai, Y. Zhang, C. Yang, B. Dang, J. Wang, and P. Yuan,Opt. Express 17, 22254 (2009).

[7] Y. D. Zhang, H. Tian, X. N. Zhang, N. Wang, J. Zhang, H. Wu, and P. Yuan, "Experimental evidence of enhanced rotation sensing in a slow-light structure," Opt. Lett. 35, 691-693 (2010).

[8] Y. Lu, J. Q. Yao, X. F. Li, and P. Wang, "Tunable asymmetrical Fano resonance and bistability in a microcavity-resonator-coupled Mach-Zehnder interferometer," Opt. Lett. 30, 3069-3071(2005).

[9] C. Y. Chao and L. J. Guo, "Design and optimization of microring resonators in biochemical sensing applications," J. Lightwave Technol. 24, 1395-1402(2006).

[10] J. E. Heebner, V. Wong, A. Schweinsberg, R. W. Boyd, and D. J. Jackson.Optical transmission characteristics of fiber ring resonators[J]. IEEE J Quantum Elect, 40, 726-730(2004).

[11] A.Yariv,"Critical coupling and its contral in optical waveguide-ring resonator systems," IEEE Photon. Technol.Lett.14, 483-485 (2002).

[12] H.Lefevre, The Fiber-Optic Gyroscope(ARTECH,1993),Chap.11

[13] C. J. He, Y. G. Zhang, L. Sun, J. M. Wang, T. Wu, F. Xu, C. L. Du, K. J. Zhu, and Y. W. Liu, "Electrical and optical properties of Nd3+-doped Na0.5Bi0.5TiO3 ferroelectric single crystal," J. Phys. D: Appl. Phys. 46, 245104 (2013).

[14] C. J. He, X. D. Fu, F. Xu, J. M. Wang, K. J. Zhu, C. L. Du, and Y. W. Liu, "Orientation effect on bandgap and dispersion behavior of 0.91Pb(Zn1/3Nb2/3)O3-0.09PbTiO3 single crystals," Chin. Phys. B 21, 054207 (2012).

[15] C. J. He, H. B. Chen, L. Sun, J. M. Wang, F. Xu, C. L. Du, K. J. Zhu, and Y. W. Liu, "Effective electro-optic coefficient of (1−x)Pb(Zn1/3Nb2/3)O3–xPbTiO3 single crystals," Cryst. Res. Technol. 47, 610-614 (2012).

[16] C. J. He, F. Xu, J. M. Wang, C. L. Du, K. J. Zhu, and Y. W. Liu, "Composition dependence of dispersion and bandgap properties in PZN-xPT single crystals," J. Appl. Phys. 110, 083513 (2011)

The application of auto correlation detection technology in magnetic targets searching

Li Yuxiang, Sun Weimin* , Wang Shuai
Harbin Engineering University,Harbin, China
sunweimin@hrbeu.edu.cn

Abstract—Forward calculations and inverse calculations for magnetic targets are two important components of magnetic data processing. Detection of magnetic targets is the premise of parameter inversion. By done some studies on the correlation-detecting principle of magnetic data, we obtained the detection method of magnetic targets searching. We also analyzed some related factors that affecting the detection performance, such as signal-to-noise the ratio, measuring height, measuring range and so on.

Keywords- Forward and Inverse; correlation-detecting; magnetic targets;

I. INTRODUCTION

Locating the positions and getting the magnetic moments of magnetic targets are the primary missions for magnetic data inversion. In order to locate the magnetic targets, we must judge whether the magnetic measurement curves contain magnetic targets. When the magnetic measurement curves containing magnetic targets is sure, we can extract the target parameters further. To judge whether the magnetic measurement curves containing magnetic targets, we study the correlation-detecting technology of magnetic objects searching. Assuming that the airplane doing roundtrip at a fixed-height flight plane, A roundtrip is defined as the forward and reverse measurement of a measurement, and we do correlation calculation on these two measurement data. Basing on the characteristics of the strong correlation of signal and the weak correlation of noise, we detect magnetic targets contain in the magnetic measurement curves.

II. ESTIMATING THE CORRELATION FUNCTIONS AND THE CORRELATION COEFFICIENTS

Correlation function is the numerical characteristics of a stochastic process. After the stationary stochastic process(x_i, y_i),the estimation of the correlation function $\hat{R}_{xy}$ for each discrete states can be expressed by the following formulation[1]:

$$\hat{R}_{xy}(k) = \sum_{i=1}^{n-k} x_{i+k} y_i \quad k \geq 0$$

This work was supported by the Fundamental Research Funds for the Central Universities

The correlation coefficients are the standardized random variables, and it also can be the correlation values or covariance-values of the stochastic process. The estimated values of the correlation coefficient for the stationary stochastic process(x_i, y_i) can be expressed by the following formulation

$$\hat{\rho}_{xy} = \frac{\sum_{i=1}^{n}(x_i - \bar{x})(y_i - \bar{y})}{\left(\sum_{i=1}^{n}(x_i - \bar{x})^2\right)^{1/2}\left(\sum_{i=1}^{n}(y_i - \bar{y})^2\right)^{1/2}}$$

$\hat{\rho}_{xy}$ are values between 0-1, it can be considered as the normalization of the correlation function.

III. STATISTICAL CALCULATION OF THE DETECTION PROBABILITY

Assuming that the received data only contains white Gaussian noise, the probability density function of the received signal is Gaussian distribution when there is no signal.[2,3]

H0: $\quad p_0(x) = \frac{1}{\sqrt{2\pi}\delta_n}\exp(-\frac{(x-\mu_0)^2}{2\delta_n^2})$, $\mu_0 = 0$

The probability density function of the received signal is still Gaussian distribution when there is signal, just the mean values changing.

H1: $\quad p_1(x) = \frac{1}{\sqrt{2\pi}\delta_s}\exp(-\frac{(x-\mu_s)^2}{2\delta_s^2})$

where the μ_s is the mean of correlation calculation.

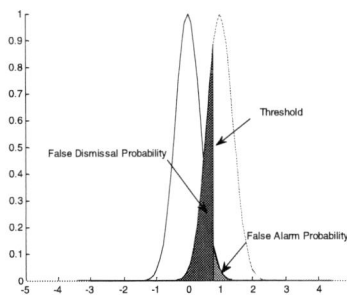

Figure 1. The relationship between false alarm probability and threshold

Under the condition of a given false alarm probability, threshold can be calculated by the following formula:

$$P(D_1 \mid H_0) = \int_{R1} p_0(y)dy = \int_{\gamma}^{\infty} \frac{1}{\sqrt{2\pi}\sigma} \exp(-\frac{y^2}{2\sigma^2})dy = \text{Pfa}$$

Judgment Rule: if the correlation coefficient y> , we select H1 and judge it containing targets; otherwise we select H0 and judge it containing no targets.

Basing on the observed quantity y, the detection probability can be calculated by the following formula:

$$P(D_1 \mid H_1) = \int_{R1} p_1(y)dy = \int_{\gamma}^{\infty} \frac{1}{\sqrt{2\pi}\sigma} \exp(-\frac{(y-\mu)^2}{2\sigma^2})dy = \text{Pde}$$

When the threshold γ is given, we can calculate the detection probability of judging whether the target exists after a measurement.

IV. SIMULATION RESULTS

A. With the same SNR, changing the height z

Simulation conditions: SNR=0.5, false alarm probability Pfa=0.001, 1000 times of simulation, height of 600 meters, we calculated that the threshold is 0.0630 and the detection probability Pde=1; when the height is 1200 meters, the threshold is 0.0626 and the detection probability Pde=0.9791.[4]

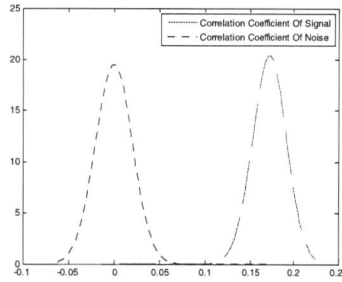

Figure 2.   The correlation coefficient when the SNR= 0.5 and z= 600m

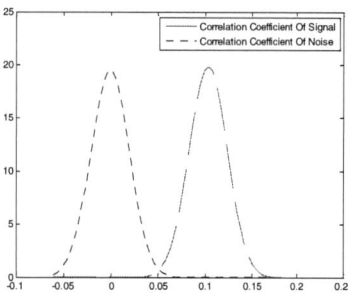

Figure 3.   The correlation coefficient when the SNR= 0.5 and z= 1200m

Comparing figure 2 and figure 3, we can see that: under the condition of fixed signal-to-noise ratio, with the height increasing, the distance between the distribution curves of correlation coefficient for signal and noise is smaller, and its

detection probability is also reduced. The reasons for the above situation: with the height increasing, the anomaly range of the magnetic measurement curves is lager, and the range affected by noise increasing. Therefore, we can increase the measuring range approximately.

With the height of 1200 meters, we calculate that the threshold is 0.0622 and the detection probability Pde=1 when doubling the measuring distance.. It can be seen that with the measurement distance increasing, the distance between the distribution curves for correlation coefficient of signal and noise increases, therefore the detection probability also improves.

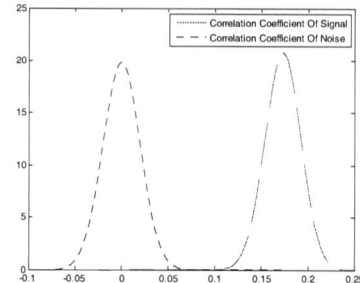

Figure 4.   The correlation coefficient of x=2400m, SNR= 0.5 and z= 1200m

B. With the same height, changing the SNR

Table 1 shows the detection threshold and the detection probability of the detection threshold when the measured height is 600 meters and the false alarm probability is 0.001. For each measurement, we can get the detection probability by calculating the correlation coefficient.

TABLE I.   THE DETECTION THRESHOLD AND THE DETECTION PROBABILITY WHEN THE FALSE ALARM PROBABILITY IS 0.001

SNR	1	0.8	0.5	0.3
Threshold	0.0902	0.0888	0.0901	0.0893
Detection probability	1	1	0.9981	0.256

Therefore, it can be concluded that: the detection performance is affected by some factors, such as SNR, measuring height, measuring range and so on. We can improve the detection performance by enhancing SNR, reducing the measuring height, expanding the measuring range and so on.

By analyzing the simulation results, we obtain the factors affecting the detection performance. We can improve the detection performance by enhancing SNR, reducing the measuring height, expanding the measuring range and so on.

REFERENCES

[1]   Jiang Chongjin, "Correlation Detection Technique and Correlation Algorithm". Journal of Data Acquisition & Processing, vol.10, pp.104-109,1995.

[2]   Mc Donough R N and Whelen A D, " Detection of Signals in Noise".2nd ed. San Diego: Academic Press,1995.

[3]   Trees H L V. Detection, "Estimation and Modulation Theory". New York: John Wiley&Sons,2001.

978-1-4799-1215-5/13 $31.00 © 2013 IEEE

[4] Liu Zhengjun, "Scientific computing and visualization simulation in Matlab". Beijing:Publishing house of electronics industry,2009.

The Intensity Image Mosaic of LADER Based on SIFT

Dejian Meng, Jianfeng Sun, Jian Gao
National Key Laboratory of Science and Technology on Tunable Laser,
Institute of Opto-Electronic of Harbin Institute of Technology
Harbin, China
e-mail: mengdejian2007@163.com

Abstract—**The LADAR(Laser Detection and Ranging) occupies an important position in the laser guidance technology, completing mosaic of LADAR image and achieving LADAR image of big scene is important significance for the laser guidance technology. SIFT algorithm can extract image feature points effectively, and realize feature matching through these feature points precisely, Then using the RANSAC method can obtain homography matrix between two images. Through the homography matrix effect, we can get the image of registration, then we make image to be fusion. The last we obtain the image to be mosaicked. In the experiment, we use LADAR intensity image as input image, and verify good mosaic result of SIFT algorithm through MATLAB software.**

Keywords-SIFT; image mosaic; LADAR intensity image

I. INTRODUCTION

Image mosaic means that some images compose panoramic image of big field, it is worth that these images contain overlap parts. Image mosaic is an important research direction in the digital image processing field, it has application value in many fields, for example medical image processing, aerospace, LADAR guidance, etc. SIFT is initial of scale invariant feature transform, SIFT algorithm is hot spot of image mosaic field, it keeps invariant for image rotation, scale change and brightness change, and keeps stability for the perspective transformation and affine transformation. SIFT feature point is a local feature point of image scale invariant , it has advantages of uniqueness, abundant information, large quantity, high speed, scalability, etc. David Lowe first put forward SIFT in 1999[1~2], and he perfect it in 2004. Brown and Lowe put SIFT algorithm in image mosaic experiment, and the experiment is very successful, SIFT algorithm shows the good and stable matching properties in 2003[3]. Now more and more people research SIFT algorithm, SIFT algorithm has been a useful tool that extracted feature of image.

II. IMAGE MOSAIC PROCESS

The image mosaic process includes image acquisition, feature extraction, feature matching, image registration, image fusion. The flow chat of image mosaic process is shown in Fig.1.

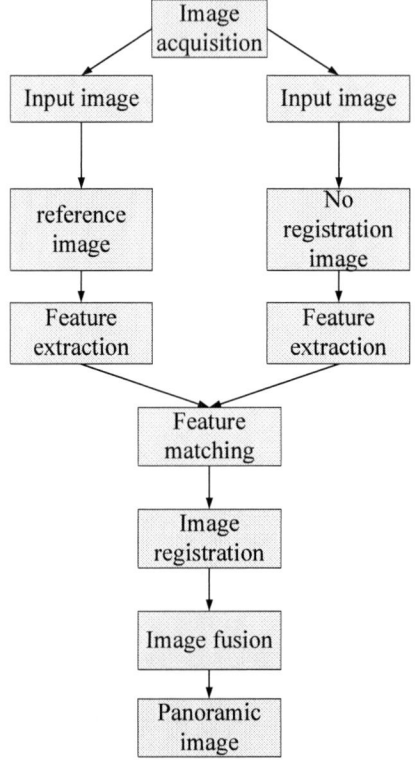

Figure 1. The flow char of image mosaic process

A. Image acquisition

The method of image acquisition is very much, the image is acquired through imaging equipment of LADAR. The two images to be acquired must contain overlap parts. The two images are called "input image", one of the input images is called "reference image", and the other is called "no registration image".

B. Feature extraction

Using SIFT algorithm extract the feature points of input images, there many methods of feature extraction, such as Harris algorithm, SIFT algorithm, etc.

978-1-4799-1215-5/13 $31.00 © 2013 IEEE

C. Image matching

Using the method of Euclidean distance conduct image matching, in the experiment if the ratio of nearest Euclidean distance and litter nearest Euclidean distance more than a threshold, so this two features are considered as matching feature points.

D. Image registration

Before fusion image, the homography matrix was the first. The homography matrix usually contains 9 parameters, as shown in Equation (1~2).

$$\begin{bmatrix} x' \\ y' \\ 1 \end{bmatrix} = \begin{bmatrix} h_1 & h_2 & h_3 \\ h_4 & h_5 & h_6 \\ h_7 & h_8 & h_9 \end{bmatrix} \begin{bmatrix} x \\ y \\ 1 \end{bmatrix} \tag{1}$$

$$H = \begin{bmatrix} h_1 & h_2 & h_3 \\ h_4 & h_5 & h_6 \\ h_7 & h_8 & h_9 \end{bmatrix} \tag{2}$$

The (x, y) and (x', y') are stand for feature point coordinate of reference image and no registration image respectively. Using RANSAC method eliminates error matching feature points, and obtain image homography matrix [4]. The method repeat N times random sampling, through finding the minimum value of matching error, there are the most interior points that consistent with H, and calculation the accurate H value. The homography matrix is applied to the no registration image, so this can obtain the image of registration.

E. Image fusion

After accomplishing the image registration, the image of registration and the image of reference conduct image fusion. There are many image fusion methods, for example, fade out method, weighted average method, etc.

III. PRINCIPLE OF SIFT ALGORITHM

The main idea of SIFT algorithm is that scale-space representation is established firstly, the next, the extreme points are searched in scale-space, the feature description vector is established by extreme points, the last is that similarity matching is realized by using feature description vector. The generation step of SIFT feature-points can be divided into generation of scale-space, detection of extreme points, accurate feature-point localization, eliminating instable feature-point, orientation assignment, generation of feature-point descriptor.

A. Generation of scale-space

Scale-space principle means that scale-space sequences are showed by scale transition of original image, then these sequences are used for scale-space feature extraction. Lindeberg show that Gaussian function is the only possible scale-space kernel [5]. The 2D Gaussian function is definition as equation (3).

$$G(x, y, \sigma) = \frac{1}{2\pi\sigma^2} \exp\left(\frac{-(x^2 + y^2)}{2\sigma^2}\right) \tag{3}$$

Where $G(x, y, \sigma)$ is a normal distribution, and its mean is 0, and its variance is σ^2. In digital image processing, the scale-space of an image is produced from the convolution of a variable-scale Gaussian, $G(x, y, \sigma)$, with an input image, $I(x, y)$. As shown in equation (4).

$$L(x, y, \sigma) = G(x, y, \sigma) * I(x, y) \tag{4}$$

Where σ is standard deviation of Gaussian function, and it can be understood as scale-space factor of scale-space, its size decide the degree of image smoothing. The big size factor is corresponding with generalization of image, and the small size factor is corresponding with the details of image. So it is very important that choosing an appropriate scale-space factor conducts smoothing.

These images that are produced from the convolution of a variable-scale Gaussian with an input image and the difference-of-Gaussian images are shown in Fig.2.

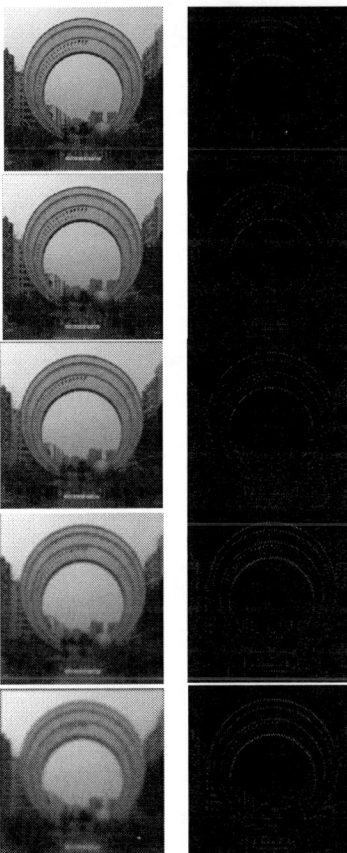

Figure 2. These images are produced from the convolution of a variable-scale Gaussian with an input image, as shown on the left. The DOG (difference-of-Gaussian)images are shown on the right.

978-1-4799-1215-5/13 $31.00 © 2013 IEEE

To obtain feature of image, the difference-of-Gaussian image (DOG) should be obtain firstly. The DOG image can be computed from the difference of two nearby scales separated by a constant multiplicative factor k. As shown in equation (5).

$$D(x,y,\sigma) = \big(G(x,y,k\sigma) - G(x,y,\sigma)\big) * I(x,y)$$
$$= L(x,y,k\sigma) - L(x,y,\sigma) \qquad (5)$$

B. Detection of extreme point

To find the extreme points, in DOG images, each sample is compared to its eight neighbors in the current DOG image and nine neighbors in the scale above and below. It is selected only if it is the largest or the smallest of 26 points. The extreme points selected are called as the candidate feature points. The extreme of the difference-of-Gaussian images are detected by comparing a pixel to its 26 neighbors is shown in Fig.3.

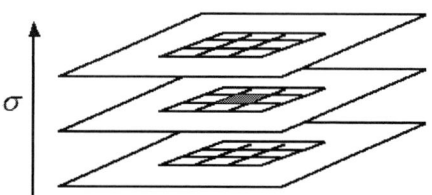

Figure 3. Max value and min value of the difference-of-Gaussian images are detected by comparing a pixel to its 26 neighbors.

C. Accurate featurepoint localization

In the previous step, the extreme points have been obtained, the extreme points are called as accurate feature points, and their feature points are showed by integer form. It is necessary for determining the sub-pixel accuracy, or floating point number coordinate. The accurate extreme points are set up x_m, the distance is $\hat{x}$ between x_m and the candidate feature point. The $D(x_m)$ conduct Taylor expansion of the scale-space function, $D(x,y,\sigma)$. As shown in equation (6).

$$D(x_m) = D(x) + \frac{\partial D^T}{\partial x}(x_m - x) + \frac{1}{2}(x_m - x)^T \frac{\partial^2 D}{\partial x^2}(x_m - x)$$
$$= D(x) + \frac{\partial D^T}{\partial x}\hat{x} + \frac{1}{2}\hat{x}^T \frac{\partial^2 D}{\partial x^2}\hat{x} \qquad (6)$$

Where D is difference-of-Gaussian image, and $x = (x,y,\sigma)^T$. The $\hat{x}$ is determined by taking the derivative of this function with respect to x and setting it to zero, as shown in equation (7).

$$\hat{x} = -\frac{\partial^2 D^{-1}}{\partial x^2}\frac{\partial D}{\partial x} \qquad (7)$$

Though above formula, the offset $\hat{x}$ can be calculated, so the sub-pixel location of x_m can be determined too. If the offset, $\hat{x}$, is larger than 0.5 in any dimension, then it means that the extreme point closer to another point. Then repeating the above calculation, the sub-pixel accuracy location can be obtained. Lastly the candidate feature point location is replaced by the sub-pixel accuracy location.

D. Eliminating instable featurepoint

The extreme points should be screened, and to eliminate the instable points, to enhance stability of feature, to improve the noise resisting ability. The instable points include the points of low contrast and the points of edge.

Gray value of extreme point x_m can be used to judge the contrast high or low, the points of low contrast are easy to be interfered by the noise, so it should be removed. The offset $\hat{x}$ is substituted into the equation (6), then gray value of the extreme points can be obtained. As shown in equation (8).

$$D(x_m) = D(x) + \frac{\partial D^T}{\partial x}\hat{x} + \frac{1}{2}\hat{x}^T \frac{\partial^2 D}{\partial x^2}\left(-\frac{\partial^2 D^{-1}}{\partial x^2}\frac{\partial D}{\partial x}\right)$$
$$= D(x) + \frac{\partial D^T}{\partial x}\hat{x} + \left(-\frac{1}{2}\right)\hat{x}^T \frac{\partial D}{\partial x}$$
$$= D(x) + \frac{1}{2}\frac{\partial D^T}{\partial x}\hat{x} \qquad (8)$$

For the experiments in Lowe's paper, all extreme with a value of $|D(x_m)|$ less than 0.03 were discarded.

The difference-of-Gaussian function will have a strong response along edges, the feature points of edge are easy to be influenced by noise, so it should be removed. The points that are easy to be influenced in the difference-of-Gaussian function will have a large principal curvature across the edge but a small one in the perpendicular direction. The principal curvature can be computed from a 2×2 Hessian matrix, as shown in equation (9).

$$\mathbf{H} = \begin{bmatrix} \dfrac{\partial^2 D}{\partial x^2} & \dfrac{\partial^2 D}{\partial x \partial y} \\ \dfrac{\partial^2 D}{\partial x \partial y} & \dfrac{\partial^2 D}{\partial y^2} \end{bmatrix} \qquad (9)$$

Let α be the eigenvalue with the large magnitude and β be the smaller one, let γ be the ratio between the largest magnitude eigenvalue and smaller one, so that $\alpha = \gamma\beta$. As shown in equation (10).

$$\frac{\mathrm{Tr}(\mathbf{H})^2}{\mathrm{Det}(\mathbf{H})} = \frac{(\alpha + \beta)^2}{\alpha\beta} = \frac{(\gamma\beta + \beta)^2}{\gamma\beta^2} = \frac{(\gamma + 1)^2}{\gamma} \qquad (10)$$

When the two eigenvalue are equal, then the ratio, $\dfrac{\mathrm{Tr}(\mathbf{H})^2}{\mathrm{Det}(\mathbf{H})}$, is the smallest value. However, the difference of the two eigenvalue is the bigger, and the ratio, $\dfrac{\mathrm{Tr}(\mathbf{H})^2}{\mathrm{Det}(\mathbf{H})}$, is the bigger. So to give a threshold, $\tilde{\gamma}$, we only need to check equation (11).

$$\frac{\mathrm{Tr}(\mathbf{H})^2}{\mathrm{Det}(\mathbf{H})} < \frac{(\tilde{\gamma}+1)^2}{\tilde{\gamma}} \qquad (11)$$

This is very efficient to compute. In Lowe's paper, the threshold $\tilde{\gamma}$ is 10.

E. Orientation assignment

In order to achieve invariance to image rotation, the each feature-point should be assigned a consistent orientation though using local image properties. To use gradient and orientation characteristic of feature-point neighborhood pixels, can get gradient magnitude and orientation, as shown in equation (12).

$$m(x,y) = \sqrt{\left[L(x+1,y)-L(x-1,y)\right]^2 + \left[L(x,y+1)-L(x,y-1)\right]^2} \qquad (12)$$

$$\theta(x,y) = \arctan \frac{L(x,y+1)-L(x,y-1)}{L(x+1,y)-L(x-1,y)}$$

Where scale L is separate scale of each feature-point, and $m(x,y)$ is gradient magnitude, $\theta(x,y)$ is orientation. A gradient orientation histogram is formed from the gradient orientations of sample points within a region around the feature-point. Range of gradient orientation histogram is 360 degrees, each 10 degrees represents a gradient orientation, so the gradient orientation histogram has 36 bins. The peak in the histogram is considered as dominant direction of feature-point neighborhood gradient. In order to enhance matching robustness, only more than 80% of the highest peak is used as auxiliary direction of feature-point. Only 15% of points are used multiple orientations, but these contribute importantly to the stability of matching.

Image feature-point detection has finished, each feature-point has 3 factors, include position, scale and orientation.

F. Generation of featurepoint descriptor

The first, in order to achieve orientation invariance, the coordinates of descriptor are rotated to the feature-point orientation. The next, the feature-point is regarded as the center, the 16×16 sample array is selected. Each a small lattice is regarded as a pixel, the direction of the arrow is gradient direction, and the length of the arrow is gradient magnitude. The generation of feature-point descriptor is shown in Fig.4.

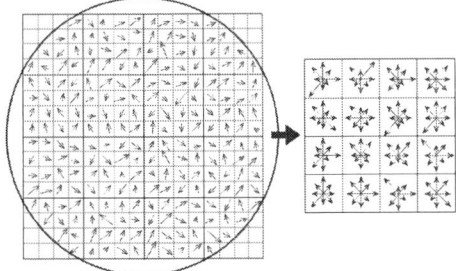

Figure 4.   A feature descriptor is created by first computing the gradient magnitude and orientation at each image sample point, as shown on the left. These samples are then accumulated into orientation histograms summarizing the contents over 4×4 sub-regions, as shown on the right.

Each feature-point generates 16 gradient orientation histogram, the gradient orientation histogram has 8 direction information, so each feature-point has $4\times4\times8=128$ dimension SIFT feature vector.

IV.   THE EXPERIMENTAL RESULT

In this experiment, LADAR intensity image is used as input image, and the mosaic image is obtained by SIFT algorithm, and the effect is very obvious. The result of experiment is shown in Fig.5~7.

Figure 5.   Input image is LADAR intensity image

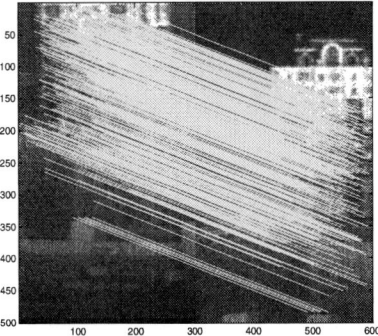

Figure 6.   The result of image matching by SIFT

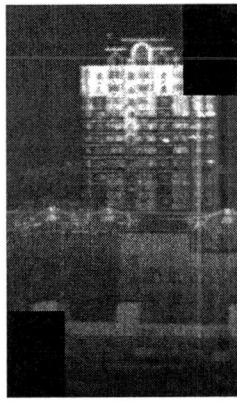

Figure 7.   The mosaic image is obtained by SIFT

Through experiment of the LADAR intensity image mosaic, the SIFT algorithm has good use value.

V. CONCLUSIONS

SIFT algorithm is used for extracting image feature-points, it has excellent matching result for LADAR intensity image. The feature-points extracted by SIFT are relatively stable, and the matching of feature-points is accurate comparatively. The SIFT algorithm can effectively against noise, the change of brightness and the deformation of image, it has good robustness for the translation, rotation, affine transformation and projection transformation of image, so SIFT algorithm has the application value in LADAR image.

ACKNOWLEDGMENT

This work is sponsored by National Nature Science Foundation, contract No. 60901046.

REFERENCES

[1] Lowe D G. Object recognition from local scale-invariant features[C]. Computer vision, 1999. The proceedings of the seventh IEEE international conference on. Ieee, 1999, 2: 1150-1157.

[2] Lowe D G. Distinctive image features from scale-invariant keypoints[J]. International journal of computer vision, 2004, 60(2): 91-110.

[3] Brown M, Lowe D G. Automatic panoramic image stitching using invariant features[J]. International Journal of Computer Vision, 2007, 74(1): 59-73.

[4] Fishier M A, Boles R C, Random Sample Concensus: A Paradigm for Model Fitting with Applications to Image Analysis and Automated Cartography [J]. Comm Assoc Comp Mach, 1981, 24(6): 381-395.

[5] Lindeberg T. Scale-space theory: A basic tool for analyzing structures at different scales[J]. Journal of applied statistics, 1994, 21(1-2): 225-270.

Research on the Key Technology of TOC Detection based on Ultraviolet Optical Absorbable

Shimin Fu, Haitao Chen, Lijie Chen, Jiannan Yu, Likai Sun
The 49th Research Institute of China Electronics Technology Group Corporation
Harbin, China
e-mail: fsmlll@163.com

Abstract—**Total organic carbon (TOC) is the important index, which reflects the water pollution level by organic substance. In order to do a more comprehensive evaluation of the water degree affected by the organic pollution, we have to research the TOC. The traditional TOC analysis technology is based on chemic oxygenation. The analytic process is cockamamie and time-consuming, and brings the expense of chemic reagent. The TOC measure technology is designed based on ultraviolet optical absorbable method in this paper. Lock-in amplifier is used to gain the low harmonic signal for the measure of TOC concentration. The test result indicates that the TOC measure technology possesses the advantages of real-time online, non-contact and small volume, and the resolution could reach 0.5mg/L, can effectively detect TOC in water.**

Keywords-TOC; real time online; non-contact; uv

I. Introduction

Total organic carbon (TOC) represents the sum of the organic matter in the water, it reflects the degree of organic compounds pollution in the water. So the detection of the TOC is a means to comprehensively reflect the water quality[1]. Currently, total organic carbon analysis measurements have been widely used in the monitoring of surface water, drinking water, sea water, industrial water and pharmaceutical water, etc. Actually, TOC has become the main testing methods of water quality control in the world[2-4].

Traditional TOC detection technology is based on chemical oxidation method. The detection process is cumbersome, complicated and time consuming. But it can also cause the loss of chemical reagents, and these monitoring techniques usually indirectly calculate the content of TOC by tested the amount of CO_2, so it is difficult to achieve real-time and continuous TOC measurement[5]. Especially because of containing water oxidation section and gas purification section, lead to volume is relatively large, the structure is relatively complicated[6]. In certain circumstances, need for real-time online monitoring of water quality. The volume of traditional TOC monitor is too big, so it cannot meet these specific needs[7].

This paper proposed an experiment device design based on the ultraviolet optical absorbable TOC measure technology. The ultraviolet radiation at 254nm is used to be the detecting light source, which is the boundary of absorption spectroscopy between organism and mineral. And lock-in amplifier is used to gain the low first harmonic signal for the measure of TOC concentration. Using the ultraviolet optical test method of the total organic carbon content and the corresponding system constitutes technology to break through the ultraviolet absorption TOC testing technology. Achieve technology goals of small size, low power consumption and fast real-time online water quality testing. Developed a water quality monitoring system based on ultraviolet absorption method, to monitor the TOC content of the circulation treating water.

II. Detection Principle of Ultraviolet Optical Absorbable TOC Experimental Apparatus

Ultraviolet optical absorbable TOC measure technology is a method to test the material composition and content by measuring the UV absorption of the sample. A large number of experiments show that all matter could absorb some certain wavelengths of visible of invisible light. But different wavelengths of light are not absorbed by the material with the same degree. Since the energy level transition of various substances need different energy, they show different absorption ability for different wavelengths of light. This property is called selective absorption of light. When a photon interacts with an acceptor, there is collision between the photon and the acceptor. The energy of the photon could be transmitted to the acceptor during a non-continuous process and be absorbed, so produced the absorption spectrum.

Basic laws for spectral absorption are Beer-Lambert law. When one-wavelength laser with the frequency v and the light intensity I0 passes through matter awaiting measurement with a certain concentration, if the laser spectrum and the absorption spectrum of the gas match, light intensity will be significantly reduced due to be absorbed. The output light intensity I_{out} and input light intensity I_{in} meet the Beer-Lambert law.

$$I_{out} = I_{in}e^{-a(v)CL} \approx I_{in}[1-a(v)CL] \qquad (1)$$

where $a(v)$ is the absorption coefficient of material, which is the absorption line of material in a certain frequency v, L is the length of absorption path, C is the concentration of material under test.

From Beer-Lambert law, we can get that:

$$C = \frac{1}{a(v)L} \ln \frac{I_0}{I(v)} \qquad (2)$$

For the monochromatic light in frequency v, the absorption coefficient of the material is certain value. If keep the optical path of absorption L unchanged, there is a certain function relationship between the output light intensity I(v) and the concentration C. Under the ideal condition, the concentration of the material under test could be got from measuring the output light intensity I(v), after the light is absorbed by the material.

III. SYSTEM DESIGN OF ULTRAVIOLET OPTICAL ABSORBABLE TOC

A. Overall Structure Design of Experiment Device

Sensor is formed by the detecting light source, compensatory light source, water sample chamber and photoelectric detection unit. The overall design diagram is shown in Fig. 1:

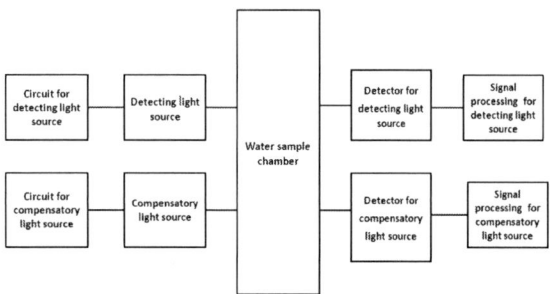

Figure 1. Overall design diagram.

As shown in Fig. 1, the overall structure design of experiment device includes water sample chamber, light source, circuit board and electrical connections, etc. According to the requirements of the appropriate size, carrying on the layout reasonably and selecting appropriate experimental device structure to improve the comprehensive performance of device and avoid the electromagnetic interference between each other. Due to the UV LED power is smaller, the structure design uses the light hood to ensure the exit angle and intensity of the light source. water sample chamber material selects the quartz with ultraviolet transmittance of more than 90%, to reduce the effect on the overall performance of the sensor due to the loss of light intensity.

The water sample under text flows into the water sample chamber through the intake of the quartz tube. LED light passes through the quartz window into the water chamber, which is received by the photodiode. The TOC concentrations of water samples are output after the voltage signal received were processed by the signal processing circuit.

B. Research on the Optical Path Parameter Optimization Technology

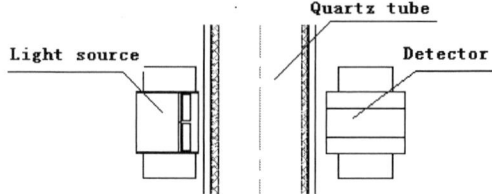

Figure 2. Structure design of double light sources.

From the principle used in this program, we can know that light intensity and optical path are the key factors of affecting the detector output. The longer the optical path, the higher the resolution, and the range will be smaller. The program uses a UV LED light source, whose power is small, so the design of optical path is seriously affected. Therefore, this program uses double light source design method to increase the intensity of the light source, thus the problem of the small light power was solved.

C. Structure Design of Anti Background Light Interference

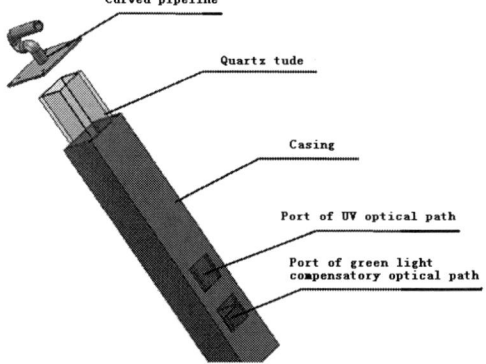

Figure 3. Structure design of anti background light interference.

As background light may affect the testing function of system, quartz tube water sample chamber is sealed in the casing. Leaving the port of UV optical path and the port of green light compensatory optical path on the casing only. In order to avoid the background light which comes from the water inlet and outlet, the inlet and outlet are made into the curved pipeline and the inner wall are treated into blackbody. Thus the background light comes from the water inlet and outlet is shielded. There are light source and detector on both sides of the port of UV optical path and the port of green light compensatory optical path. Light source and detector are mounted at their respective fixed sleeves. By adjusting the screw, the fixed sleeve can contact with the outer wall of casing closely to achieve avoiding light effect. The casing with quartz tube is installed inside the sealed box body to achieve secondary avoiding light effect.

D. Detection Techniques of Light Absorption of Weak Signal

The light signal which is absorbed by the water sample is very weak, so the output photoelectric signal is often submerged in the background noise, and is difficult detected out by the conventional detection methods. Lock-in amplification technique is one of important means which extract the weak signals drowned in the noise, and is widely used in the field of spectral measurements. In this paper, lock-in amplification technique was used to extract the signal of the TOC.

1) Basic Principles of Lock-in Amplifier Technology

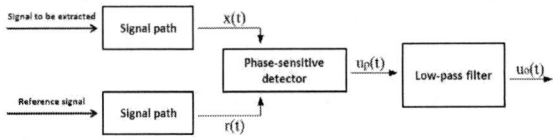

Figure 4. Structure of lock-in amplifier.

Basic structure of lock-in amplifier is shown in the figure 4. The signal path adjusts the signal amplitude and makes it meet the needs of the phase detector. This design selects the square wave with the frequency f ($f = 2\pi\omega$) and amplitude A_r as the reference signal $r(t)$, and the signal to be extracted is $X(t) = I_{out}$. Do Fourier series expansion for $r(t)$, we can get that:

$$r(t) = \frac{4A_r}{\pi} \sum_{n=1}^{\infty} \frac{(-1)^{n+1}}{2n-1} \cos(2n-1)\omega t \quad (3)$$

And the output of phase-sensitive detector is:

$$
\begin{aligned}
u_p(t) &= x(t) \cdot r(t) \\
&= \frac{2A_r A_\omega}{\pi} \sum_{n=1}^{\infty} \frac{(-1)^{n+1}}{2n-1} \cos[(2n-2)\omega t] \\
&\quad + \frac{2A_r A_\omega}{\pi} \sum_{n=1}^{\infty} \frac{(-1)^{n+1}}{2n-1} \cos(2n\omega t) \\
&\quad + N(t) \cdot \frac{4A_r}{\pi} \sum_{n=1}^{\infty} \frac{(-1)^{n+1}}{2n-1} \cos(2n-1)\omega t
\end{aligned}
\quad (4)
$$

Where, the first contains a DC component and $2n\omega$ ($n \in [1, \infty)$) harmonic components. The second contains $2n\omega$ ($n \in [1, \infty)$) harmonic components. The third is very complex due to the N(t). From the previous analysis, the main components of N(t) are 2ω and above high harmonics, so the main components of $N(t) \cdot \frac{4A_r}{\pi} \sum_{n=1}^{\infty} \frac{(-1)^{n+1}}{2n-1} \cos(2n-1)\omega t$ is the every harmonic of ω. So by setting the cutoff frequency of low-pass filter, the high frequency components could be filter out effectively and the DC component of $u_p(t)$ could be kept:

$$u_o(t) = \frac{2A_r A_\omega}{\pi} \quad (5)$$

Reference signal amplitude A_r is a known quantity. First harmonic signal amplitude A_ω of absorption signal could be get through $u_o(t)$, thus the TOC concentration measurement was achieved.

2) Design of Lock-in Amplifier Circuit

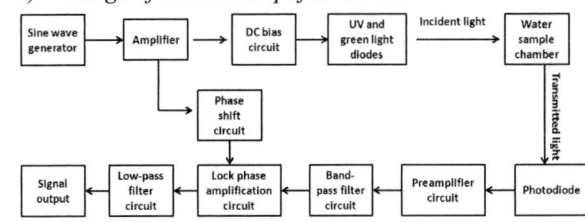

Figure 5. Composition diagram of photoelectric and signal process circuit.

Composition diagram of Lock-in amplifier Circuit is shown in Fig. 5. First use the high impedance low noise preamplifier to amplify the whole signal. Second use the band-pass filter to filter the frequency of the noise except the sine wave frequency. Then use the lock-in amplifier to do the signal phase detection, and do the lock-in amplifying between the received signal and original signal. This process can adjust the frequency of useful signal to the zero frequency. Then the detection signal could be got by the signal passed the low-pass filter to further filter out high frequency signals.

IV. RESULTSTHE TESTING OF THE EXPERIMENTAL DEVICE

TABLE I.   TESTING DATA OF THE EXPERIMENT()

Times	Density (mg/L)			
	5	10	15	20
1	5.6	10.3	15.6	20.6
2	5.8	10.6	15.7	20.7
3	5.2	10.8	15.3	20.9
4	4.7	10.5	15.5	20.7

This experiment chooses the TOC standard liquid of 5, 10, 15 and 20 mg/L. In the conditions of room temperature 25℃ and humidity 80%, the sample solutions were measured by the TOC sensor. The experimental data is shown in Table 1. Experimental results show that: experimental device response time is 40s; the resolution is 0.5mg/L; the accuracy is ±5%FS. So the overall design scheme of Ultraviolet Optical Absorbable TOC experimental facility is feasible.

V. CONCLUSIONS

This technology possesses the advantages of real-time online, non-contact and small volume. It can be proverbially in the TOC measure of drinking water and industry water.

In this design program, the ultraviolet radiation at 254nm is used to be the detecting light source, which is the boundary of absorption spectroscopy between organism and mineral. And lock-in amplifier is used to gain the low first harmonic signal for the measure of TOC concentration. The feasibility of this design is verified by the testing.

The lock-in amplifier technology is applied to the design of ultraviolet absorption TOC experimental device. The problem of that the weak first harmonic signal was extracted difficultly in big background noise has been solved effectively. The whole schematic design is feasible.

Meanwhile, ultraviolet absorption TOC testing technology has the reference value for spectral analysis type sensor designing of other material

REFERENCES

[1] Yamamoto Katuhiro,"Trial of an organic composition examination in natural water by UV spectrua,"Kagaku to Kyoiku,vol.47,pp.338-341,1999.

[2] Chen Li-jie,Fu Shi-min, "Research on ultraviolet optical absorbable TOC detection technology,"Transducer and Microsystem Technologies,vol.32(4),pp.68-71,2013.

[3] Matsche N,Stumwohrer K."Influence of changes of the wastewater composition on the applicability of UV-absorption measurements at combineed sewer ocerflows,"Water Science and Technology, vol.47,pp.73-78,2003.

[4] Kato Yasunobu,Kumagai Tetsu, "Prediction of chemical oxygen demand by UV visible absorption spectrum – parttial lest -squares,"Bunseki Kagaku, vol.48,pp.225-230,1999.

[5] Y.Wang,J.F Wang, "Protective effect of collagen polypeptides from Apostichopus japonicus on the skin of photoaging –model mice induced by UV irradiation,"Journal of China Pharmaceutical University, vol.39,pp.64-67,2008.

[6] Isabella Bisutti, "Determination of total organic carbon-an overview of current methods,"Trends in Analytical Chemistry,pp.716-726,2004.

[7] Langergraber G,"Calibration procedure for UV Spectrometric Quantification of Organic Matter and Nitrate in Wastewater,"Water Research, pp.63-71,2003.

A low-noise MEMS acoustic vector sensor

Jinping Li[1]*, Lijie Chen[1], Zhanjiang Gong[1], Shi Xin[1], Meng Hong[2]

1 The 49th Research Institute of China Electronics Technology Group Corporation, Harbin, China
2 Science and Technology on Sonar Laboratory, Hangzhou, China
blackxeme@163.com

Abstract—A low-noise micro-machined acoustic vector sensor is presented. It is desirable that the application of difference capacitance principle combined with bulk micro-machining silicon process techniques may improve the low frequency sensitivity and dynamic range of the acoustic vector sensor as well as its miniaturization. The microstructure of the hydrophone was fabricated by MEMS technology, and measured by underwater standing wave field. The experiment results show that the acoustic vector sensor has good low-frequency characteristic, the free-field pressure sensitivity is -179.9 dB (dB re 1V/μPa) at 1000 Hz with a about 2 dB one-third octave positive slope over the 20~2000Hz frequency response range, and the dynamic range reaches to 120 dB (100Hz BW).

Keywords-acoustic vector sensor; MEMS; low-frequency; dynamic range

I. INTRODUCTION

Because of the complexity of the marine environment and weakening of underwater target acoustic signal, there are more higher requirements for the underwater acoustic detection system. Different from the traditional scalar acoustic pressure sensor[1] that can only measure acoustic pressure, acoustic vector sensor can detect the scalar of pressure and the vector of particle velocity in the mode of synchronous and concurrent. Meantime in the situation of small scale and low frequency array, acoustic vector sensor can obtain spatial gain certain, and detect the accurate information of underwater target's position[2-3]. From the perspective of energy detection, acoustic vector sensor has the ability of resisting isotropic noise from each direction and realize the far-field target recognition[4-5]. So study on acoustic vector sensor get great attention domestic and international.

The ideas of measuring the acoustic vector sensor, the acoustic particle velocity and the acoustic particle acceleration, which are required for the sensing elements of the acoustic vector sensor, have been around for at least 50 years. The measurement of particle velocity in an underwater sound field was considered as early as 1956[6]. After more than 10 years, a directional pressure gradient hydrophone and a low-frequency pressure gradient sensor for underwater sound fields were developed respectively[7-8]. In 1998, the first piezoelectric, flexural-disk, neutrally buoyant underwater accelerometer[9] was designed, built and tested.

In ten years, China has developed a variety of structural vector hydrophones, e.g. moving-coil vector hydrophone, the velocity hydrophone, piezoelectric co-vibrating type vector hydrophone, fibre-optic hydrophone and MEMS piezoresistive acoustic vector sensor, which achieve a series of the acoustic vector sensor and the diversity of functions [10]. All of those are to meet the different needs of different situation in the field of underwater acoustic measurement.

This paper mainly studies the feasibility of using low noise vector hydrophone basing on the principle of differential capacitance and silicon MEMS sensitive structure, and carry out the related testing. The vector hydrophone system can be used for marine environment monitoring or other acoustic detection of underwater targets.

II. BASIC PRINCIPLE

A. Difference capacitance principle

The difference capacitance principle as shown in Figure1.

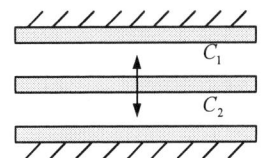

Figure1. principle diagram of difference capacitance

When the sensor senses the external force, the intermediate electrode make C_1、C_2 change the following form:

$$C_1 = C_0(1-\frac{\Delta d}{d_0})^{-1} = C_0\left[1+\frac{\Delta d}{d_0}+(\frac{\Delta d}{d_0})^2+...\right] \qquad (1)$$

$$C_2 = C_0(1+\frac{\Delta d}{d_0})^{-1} = C_0\left[1-\frac{\Delta d}{d_0}+(\frac{\Delta d}{d_0})^2-...\right] \qquad (2)$$

where C_1 is capacitance upper electrode, C_2 is capacitance lower electrode, C_0 is the initial capacitance between electrodes, d_0 is the initial displacement between electrodes.

The equivalent capacitance is:

$$\Delta C = C_1 - C_2 = 2C_0\left[\frac{\Delta d}{d_0}+(\frac{\Delta d}{d_0})^2+...\right] \qquad (3)$$

According to formula above, there is a nonlinear relation between the change of capacitance type and displacement, but the high-order item can be neglected in the micro amount detection, then we get the following formula:

$$\Delta C = 2C_0\frac{\Delta d}{d_0} \qquad (4)$$

The sensitivity is:

978-1-4799-1215-5/13 $31.00 © 2013 IEEE

$$s = \frac{\Delta C}{\Delta d} = \frac{2C_0}{d_0} = \frac{2\varepsilon S}{d_0^2} \qquad (5)$$

B. Acoustics theory of cylinder

Acoustics theory research indicates that for an acoustically small cylinder immersed in fluid when the size of the acoustics cylinder is far smaller than the length of sound wave, under the action of sound wave, the relation of the velocity between the cylinder and the fluid particle is

$$\frac{V}{V_0} = \frac{4}{j(ka)^2 \pi \dfrac{\rho}{\rho_0} \dfrac{dH_1^{(2)}(ka)}{d(ka)} + \pi(ka)H_1^{(2)}(ka)} \qquad (6)$$

Note that the equation is centered using a center tab stop.

where V is the amplitude of the cylinder velocity, V_0 the amplitude of the particle velocity, ρ_0 the density of the fluid, ρ the density of cylinder, k the wave number, and a the crustaceous radius of the cylinder.

When $ka \ll 1$,

$$\frac{V}{V_0} = \frac{4}{2\dfrac{\rho}{\rho_0} + \left[\dfrac{\pi(ka)^2 + 2j}{2}\right]} \qquad (7)$$

So, the relationship between V and V_0 can be expressed as

$$\frac{V}{V_0} = \frac{2\rho_0}{\rho + \rho_0} \qquad (8)$$

This shows that at low frequencies the motion of a cylinder whose density is equal to that of the fluid it displaces is identical to the motion of the fluid particles at this location when the cylinder is removed [11-13]. Consequently, if the cylinder is fixed on an inertial transducer, a signal is produced and can be related to the acoustic particle motion.

III. ACOUSTIC VECTOR SENSOR DESIGN

The MEMS acoustic vector sensor is a difference capacitor design that is operated in a closed-loop configuration with a custom mixed-signal application specific integrated circuit. Wet and dry etching techniques are used to form mechanical structures from bulk silicon which can then be patterned with conductive or dielectric surfaces. The microstructure of MEMS acoustic vector sensor is a wafer stack composed of four silicon wafers bonded together. Within the inner two wafers of the stack, and suspension springs, is a moving structure called the proof-mass. A thin layer of gold is deposited on the top and bottom surfaces of the proof-mass to form a conductive surface. Upper and lower caps with also deposited gold are bonded on either side of the proof-mass support frame to form a differential variable capacitance between the surfaces of the moving proof-mass and the fixed caps. As the acoustic vector sensor is subjected to vibration, the proof-mass moves between the fixed plates which, in turn, causes a change in the differential capacitance. Fig. 2 shows the microstructure cross section of MEMS acoustic vector sensor

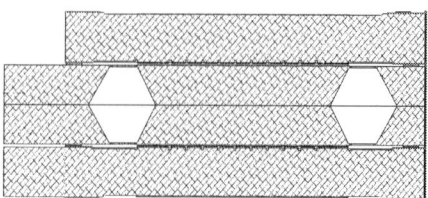

Figure2. the microstructure cross section of MEMS acoustic vector sensor

Detection of small vibration signals with good fidelity requires a sensor with low self-noise. This is an essential requirement for underwater sound field since signal amplitude diminishes rapidly as the distance to a source increases. The two variables in acoustic vector sensor design primarily responsible for sensor self-noise are the mass of the proof-mass and the damping of the resonant structure. The bulk micro-machining process allows relatively large structures to be created when compared with surface micro-machining techniques. The bulk process was deliberately chosen to allow the capability to create an acoustic vector sensor with a larger proof-mass and lower self-noise. The damping of the proof-mass is controlled by packaging the acoustic vector sensor die in a hermetic ceramic package sealed under vacuum to reduce the effects of Brownian movement and damping of the proof-mass. To optimize the performance of the acoustic vector sensor, a custom mixed-signal application specific integrated circuit was designed by analog output modes.

IV. ACOUSTIC VECTOR SENSOR PERFORMANCE

The capacitive bulk microstructure designed was fixed in an acoustic vector sensor with co-oscillating cylinder. The Photo of packaged MEMS acoustic vector sensor is shown in Fig. 3.

Figure3. Photo of packaged MEMS acoustic vector sensor.

The Performance of the acoustic vector sensor which included sensitivity, direction, frequency response and dynamic range was tested in the standing wave field.

A. Sensitivity and direction

The measurement of the hydrophone was processed in a pool of first-class national-defense underwater acoustic calibration station. These were measured in free-field conditions achieved by means of pulse techniques. This method requires a standard hydrophone as a projector and the fabrication hydrophone as the receiver. The reference hydrophone and the acoustic vector sensor are fixed in the calibration house (Fig. 4), and the receiving sensitivity of the hydrophone is -179.9 dB (dB re 1V/μPa) at the frequency of 1000 Hz.

Figure4.   Stage set of testing in standing wave field

As the underwater sound field in the standing wave house is vertically distributed, so the directivity pattern of the hydrophone could be described if it is circumvolved along the horizontal axis and the revolving angle and the open circuit voltage of the sensor are written down at the same time. Fig.5 shows the directivity patterns of the sensor at the frequency of 500 Hz. It is clear that the acoustic vector sensor not only possesses a typical directivity of "8"-shape but also with the depth of pits of the directivity pattern is computed to be about 26.7dB.

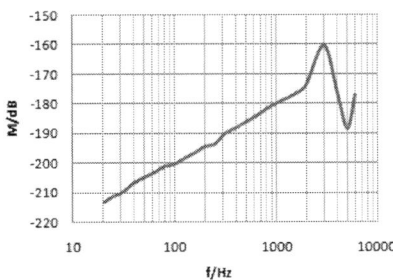

Figure5.   Directivity patterns of MEMS acoustic vector sensor

B.   Frequency response

An ideal acoustic vector sensor will have a good frequency response that allows undistorted measurement of signals at all frequencies. The MEMS acoustic vector sensor frequency response is about 6dB one octave positive slope from DC to 2000Hz and does not require corrections. Measured frequency response data is shown in Fig. 6.

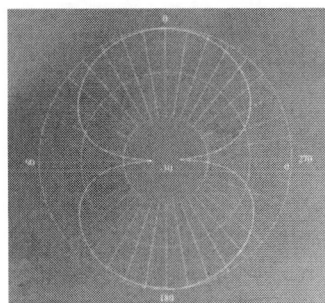

Figure6.   MEMS acoustic vector sensor frequency response measurement

C.   Dynamic range

Since the purpose of underwater sound field is to measure faint signals at great distances, it is important to have a sensor capable of low signal measurements. If the self-noise of an acoustic sensor exceeds the magnitude of the signal, however, the signal will not be detected. The MEMS acoustic vector sensor was carefully designed to minimize the noise floor of the output. The noise floor of the MEMS acoustic vector sensor is shown in Fig. 7.

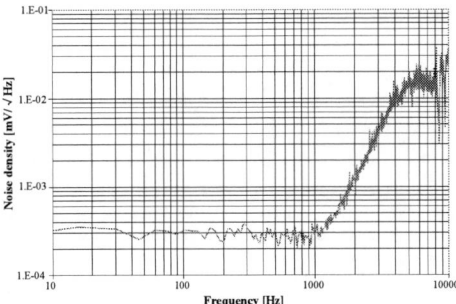

Figure7.   Output noise spectrum for MEMS acoustic vector sensor

The acoustic vector sensor has an equivalence noise spectrum at variable frequency. For example, 60.1 dB at 1000 Hz, 65.4 dB at 500 Hz. This translates into a dynamic range in a 100Hz bandwidth reaches to 120dB. This wide dynamic range provides the capability for underwater sound field to measure large signals as well as to monitor weak signals only one acoustic vector sensor.

V.   CONCLUSION

Based on the theory of difference capacitance and MEMS technology a low noise MEMS acoustic vector sensor was designed and fabricated. This article provides a good solution for solving the problem of miniaturizing vector sensor by the way of using low noise and high sensitivity microstructure. Through the test of standing wave field, the sensor's preliminary performance was verified. The results show that the MEMS acoustic vector sensor has good low-frequency characteristic and can effectively be used as an underwater acoustic sensor within its resonance frequency width. The free-field pressure sensitivity of this sensor is -179.9 dB (dB re 1V/μ Pa) at 1000 Hz with a about 2 dB one-third octave positive slope over the 20~2000Hz frequency response range. The dynamic range reaches to 120 dB (100Hz BW). The directional pattern in the form of " 8 " with the depth of pits of the directivity pattern is about 26.7 dB.

At present, initial design and experiments have been performed. How to optimized the performance of sensor, and to realize the engineer application of sensor will be carried out in the near future.

ACKNOWLEDGMENT

This work was supported by the National 863 Plans Projects of China and Science and Technology on Sonar Laboratory of China.

REFERENCES

[1]   G. Q. Sun, Q. H. Li, "A novel fibre optic hydrophone and vector hydrophone," Physics, 35(8), pp.645-635, 2006,

[2]   A. Nehorai, E. Paldi, "Acoustic vector-sensor array processing,"

[3] IEEE Transactions on Signal Procesing, 42(9), pp. 2481-2491, 1994.

[4] M. Hawkes, A. Nehorai, "Acoustic vector sensor beam-forming and capon direction estimation," IEEE Transactions on Signal Pro.Cessing, 46(9) , pp. 2291-2304, 1998.

[5] L. J. Chen, S. E. Yang, "A design of novel piezoresistive vector hydrophone," Applied Acoustics, 25(5), pp.273-278, 2006.

[6] S. Chen, C. Y. Xue,B. Z. Zheng and B. Xie, "A Novel MEMS Single Vector Hydrophone," Acta Armamentarii, Vol.29, No.6, pp. 673-677, 2008.

[7] C. B. Leslie, J. M. Kendall and J. L. Jones, "Hydrophones for Measuring Particle Velocity," Journal of Acoustical Society of America, Vol.28, No.4, pp. 711-715,1956.

[8] Donald J. Scheiberr "Directional Pressure Gradient Hydrophone," in 21th Navy Symposium on Underwater Acoustics, pp. 1113-1119, October 1969.

[9] Marcia Mongiovi, "Low-Frequency Pressure Gradient Sensor," General Electric Com., Transducer Product Operation, Syracuse NY, Patent #3603921,1974.

[10] M. B. Moffett, D. H. Trivett, P. I. Klippel and P. David. Baird, "A Piezoelectric, Flexural-Disk, Neutrally Buoyant, Underwater Accelerometer," IEEE Trans. On Ultrasonics, Ferroelectrics and Frequency Control, Vol.45, No.5, pp. 1341-1346, September 1998.

[11] L.J. Chen, P. Zheng, X. Y.Xu, eta1., "Overview of vector hydrophone," Transducer and Microsystem Technologies, 25(6), pp.5-8, 2006.

[12] K. Kim, T. B. Gabrielson, G. C. Lauchle, "Development of an accelerometer-based underwater acoustic intensity sensor," Journal of Acoustical Society of America, 116, pp.3384-3392, 1991.

[13] V.A. Shchurov, "The interaction of energy flows of underwater ambient noise and local source," Journal of Acoustical Societyof America, 90(4), pp.1002-1004, 1991.

[14] Binzhen Zhang, Hui Qiao, Shang Chen, et a1., "Modeling and characterization of a micromachined artificial hair cell vector hydrophone," Microsyst Technol, 14, pp:821-828, 2006.

The experimental research on gas magneto-optic properties

Yuanyuan Wang[1]*, Lufei Hong[2], Guoli Song[1]

1 Department of physics, Harbin University, Harbin 150086, China
2 Flight Test Station, Harbin Aircraft Industry Group, Harbin 150066, China
wyy_janet@163.com

Abstract—The rotation angle of gas is too minor to measure, so the experiment has to in harsh terms to measure verdet constant. We designed high-accuracy experiment device to measure minor rotation angle based on frequency multiplication method, which makes the verdet constant of gas can be observed at room temperature. Magneto-optical modulation method for optical polarization measurement based on frequency multiplication signal was investigated by means of computer simulation, this method has well applied value.

Keywords- modulation; magneto-optical rotation; polarized light; frequency multiplication

I. INTRODUCTION

The Faraday's magnetic rotation effect has important application in scientific research [1-2] and practice [3-4], The accurate measurement of polarized light's small angle rotation is an important content. Traditional polarized light detection method mainly has the light extinction method, half shadow method, Magneto-optic modulation method.

light extinction method is based on the transmission light intensity along with the change of partial detector rotation to determine the position of light extinction, Due to the light intensity change rate near extinction position is small, determine the position of light extinction position is difficult, Observed by human eyes to determine, the accuracy is poor, this method accuracy only 1^0[5], Light detector complemented with appropriate detection circuit, the accuracy can be improved, but still can't reach very high accuracy.

Half shadow method can improve the accuracy of measurement, the measurement error is less than 0.1^0, But half shadow method is only suitable for the human eye observation, Precision is difficult to improve, also unable to realize automatic detection.

Magneto-optic modulation method is a kind of important high accuracy of polarized light detection method. Place a magneto-optic modulator in the light path, Combined with sinusoidal ac signal change, using photoelectric detector to detect the change of the output light intensity signal, then can automatically determine the position of light extinction. The measuring accuracy of this method is higher than the light extinction method.

In this experiment, as the rotation Angle of the gas is very small, The above detection method can not achieve accuracy of measurement well, Even it is difficult to detect. So using frequency multiplication method, It is by the ac modulation method is derived. In light extinction location, the output signal become double frequency modulation signal, by observing the signal, we can accurately determine the position of light extinction, realizes the high accuracy measurement, the measuring accuracy is higher than ac modulation method.

II. FREQUENCY MULTIPLICATION METHOD

This experiment adopts the frequency multiplication method, The modulation signal and output signal as X and Y component of oscilloscope, When reach state of frequency multiplication, observe whether the graph on oscilloscope is symmetric, then you can judge whether achieve frequency multiplication, so as to determine the position of light extinction. Because of based on frequency measurement, the instability of light source and the elements in light path generally affects only signal amplitude, not the frequency, so frequency multiplication method also has strong anti-interference [6-7].

In the light path placed a magneto-optic modulator(put magnetic optical medium in the solenoid)In alternating current modulation, $i = i_0 \sin \omega t$ Modulation coil produce alternating magnetic field $B = B_0 \sin \omega t$ So that through modulated medium, the direction of the beam vibration occur cycle oscillation in small increments ,Pendulum Angle is

$$\beta = VLB_0 \sin \omega t = \beta_0 \sin \omega t \tag{1}$$

$$\beta_0 = VLB_0$$

When partial detector light direction and polarization direction of the light beam Angle is φ ,Partial detector output light intensity is

$$I = I_0 \cos^2(\varphi + \beta) \tag{2}$$

Because the Angle β is very small

$$\cos^2(\varphi \pm \beta) = \cos^2 \varphi + \beta^2 \sin^2 \varphi \mp 2\beta \cos \varphi \sin \varphi \tag{3}$$

$$\varphi = \frac{\pi}{2}$$

$$\cos \varphi = 0 \quad, \quad \sin \varphi = 1 \quad, \quad \cos^2(\varphi \mp \beta) = \beta^2 \tag{4}$$

So partial detector output light intensity is

$$I = I_0 \beta_0^2 \sin^2 \omega t = \frac{1}{2} I_0 \beta_0^2 \left(1 - \cos 2\omega t\right)$$

(5)

Show that partial detector in light extinction position, the modulation fundamental frequency signal disappears, appears double frequency signal, which can be used to determine the position of light extinction In experiment and computer simulation, we give introduction of $\pi / 3$ phase delay between the input and output signal, as shown in the graphic, θ_t is Faraday rotation angle(Pendulum Angle) β

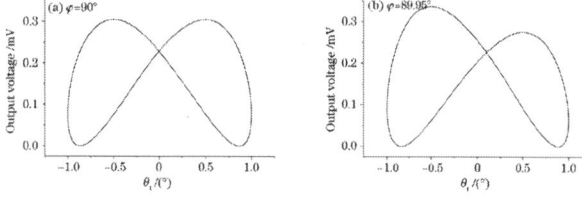

Figure.1   Lissajous figure of Sine wave modulation

（a）State of frequency doubling, similar to ∞ graphics, left and right sides is symmetrical

（b）Deviated the light extinction position 0.05^0, curve has been serious asymmetric

in fact, in our simulation, we found that the measurement accuracy can reach above 0.01^0, Its measurement accuracy is higher than direct observation of a single output waveform on the oscilloscope.

III.   GAS MAGNETO-OPTIC PROPERTIES EXPERIMENT SYSTEM

The experiment system is mainly used for precision measuring magneto-optic small rotation Angle, the experimental light path of the test system and the installation drawing is as follows

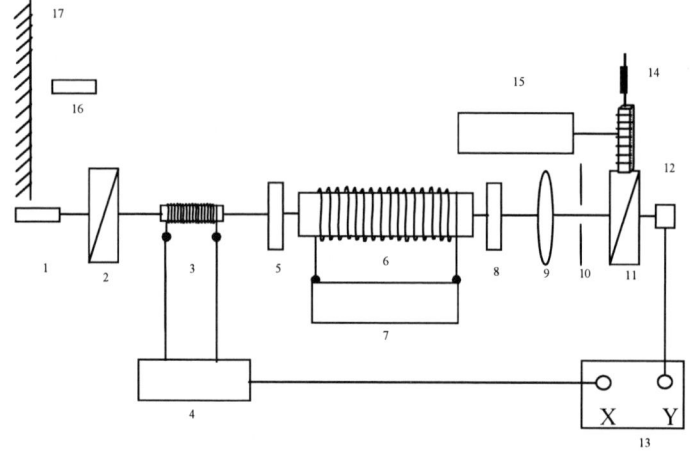

Figure.2   The experimental test device of frequency multiplication method

1: Semiconductor laser   2: The partial device

3: Magneto-optic modulation coil

4: Modulation signal source   5: Plane mirrors:

6: solenoid   7: Solenoid power   8: Plane mirrors

9: Convergent lens   10：Iris   11: Partial detector

12: detector   13: Oscilloscope   14: Rotating mirror

15: Stepper motor   16: light source of Measurement

17: scale

This experiment according to the following steps:

1) collimating light path. All the components are fixed on the

Adjustable bracket, with solid state laser calibration light path, to ensure that the detector receives the light completely, this adjustment can make the basic collimating light path.

2) Adjust the relative position of two plane mirror on both ends of the solenoid, multiple reflections. Through the mirror point of light to calculate the count number of light reflection

3) Adjust the fine-tuning the handle on the location of the plane mirror. Make the center line of planar reflector and fine-tuning the handle shaft, make the light without hindrance to a mirror, the spot arrive on scale through a mirror reflection.

4) Current signal of the modulator as input signal, received signal of detector as the output signals, respectively connected with X axis and Y axis of Oscilloscope. When Reach a state of frequency doubling, adjust the intersection of graphics, with the help of the oscilloscope grid, which is located in the horizontal direction of center position.

IV.   EXPERIMENTAL MEASUREMENTS AND CONCLUSION

Turn off All the light, opened the semiconductor and laser current modulator, open the oscilloscope, first determine the general location of light extinction with rough adjustment, and then open the stepper motor for automatic control, careful observe frequency doubling signal on the oscilloscope and identify the position of light extinction, make sure the starting position of the reflection light spot on scale

Open the solenoid control source, equivalent to air dielectric added magnetic field, then observe frequency doubling signal on the oscilloscope, symmetric center has deviated from the center position, therefore, continue with step motor for automatic control, the frequency doubling signal back to the center position, determine the end of the location of the reflection light on the scale

1) semiconductor laser as light source ,Wavelength of 650 nm, 650 nm, 532 nm, the temperature is 20 degrees which is room temperature. Modulation current is 0.75 A, the light bounces back and forth for 15 times. Get the following different wavelengths of the verdet constant of air medium

TABLE I   THE MAGNETO-OPTIC EFFECT OF GASES WITH SAME
TEMPERATURE . DIFFERENT WAVELENGTH

Wave-length (nm)	solenoid current (A)	field Intensity (GS)	Partial detector Angle(′)	verdet constant (′/GS • cm)
650	3.60	410.760	1.160117683	5.884×10^{-6}
635	3.60	410.760	1.226742765	6.221×10^{-6}
532	3.82	435.862	1.349269148	6.448×10^{-6}

32 cm range of solenoid field intensity is 342.3 GS (current is 3.0 A), Field Intensity B

$$B = \frac{342.3 \times I}{3.0} \qquad (6)$$

verdet constant V

$$V = \frac{\theta}{B \times L} \qquad (7)$$

2) With air as the magneto optical medium, pack a temperature control device in the solenoid for a real-time environment monitoring. semiconductor laser as light source With wavelength of 532 nm, the modulation current is 0.75 A, the temperature ranges from 13 to 26 degrees, the light bounces back and forth for 15 times, Get the results in the table

TABLE II   THE MAGNETO-OPTIC EFFECT OF GASES WITH DIFFERENT
TEMPERATURE . SAME WAVELENGTH

Tempe-rature (^{0})	solenoid current (A)	field Intensity (GS)	Partial detector Angle(′)	verdet constant (′/GS • cm)
13	3.82	435.862	1.387296443	6.631×10^{-6}
17	3.72	424.452	1.335892729	6.556×10^{-6}
20	3.67	418.747	1.296239713	6.448×10^{-6}
23	3.60	410.760	1.248508058	6.332×10^{-6}
26	3.52	401.632	1.112504927	5.770×10^{-6}

From the first table, the verdet constant of the air increase with the decrease of the wavelength, From the second table, with the increase of temperature, verdet constant decreases. the verdet constant of gas is very small, and the higher the temperature, the more difficult to observe, it is also the reason of a lot of literature for gas measurement choose the low temperature freezing. And this experiment system realize the observation of small Angle, which makes the verdet constant of gas can be observed at room temperature

REFERENCES

[1]   Hang chao,Huang Guoxiang "Faraday rotation in a resonant five-level system via electromagnetically induced transparency" [J],chin.opt.lett.2007 5(1):47-50

[2]   Wiegand M. Autonomous satellite navigation via Kalman Filtering of magnetometer data. Acta Astronautica, 1996, 38(4-8), 395~403

[3]   Feng liu,Qing Ye,Jianxin Geng etal "Study of fiber-optic current sensing based on degree of polarization measurement" [J],chin.opt.lett.2007 5(5):267-269

[4]   acobsen B, Jodalen V, Cannon P S, et al. HF radio propagation at high latitudes under quiet and disturbed geomagnetic conditions. IEE Conf Pub, 2000, 187~191

[5]   Liao Yanbiao "Polarization Optics"[M],Beijing:Science press. 2003. 231-233

[6]   Qian Jingren,Liu Fang,SuJue"investigation of frenquency nulldrift in polarimetric fiber laser current sensors [J],.chin.J .Lasers 2006 33(6):791-794

[7]   DiNan,XuXiaoPeng,"Discussion and improvement on MOMDF"[J],Physics Experimentation 2007,27(5):10-16

Research on Capacitance Attitude Self-Correction Vector Hydrophone

Peng Zhang, Xin Shi, Haitao Chen,

The 49th Research Institute of China Electronics Technology Group Corporation

Harbin, China

e-mail: 624132093@qq.com

Abstract—**To solve the underwater target vector detecting problems, the paper introduced the design of a attitude self-correction hydrophone which used its sensitive accelerometer to correct the attitude. The hydrophone's sensitive unit which could response the zero-frequency signal was a MEMS capacitance accelerometer with high sensitivity. Since the vector hydrophone focusing on dynamic signals, we adopted hydrophone's static signal to operate the attitude correction. Therefore the hydrophone's volume and power consumption were both reduced. Another advantage was making the hydrophone and the attitude correction unit concentric. Consequently the system error causing by system assembling was removed. The two-dimension attitude correction was improved. The testing results showed the method was feasible and was well proved. The static correction accuracy was greater than 0.5° which satisfied the hydrophone's attitude correction request.**

Keywords- attitude self-correcting; vector; hydrophone; MEMS

I. Introduction

Acoustic wave is the most effective carriers of long-distance information transmission in the ocean [1-3]. Underwater acoustic is vector information, in addition to the scalar information such as the sound pressure, there are also some vector information such as sound pressure gradient, acceleration, particle vibration velocity, acoustic energy flux, etc. The measuring signal of the vector hydrophone contains the vector information of the sound signal. The vector hydrophone has the advantages of small size, high sensitivity and the directivity is independent with frequency. It is very suitable for the limited volume underwater acoustic detection system such as towed array cable and underwater automatic attacking mine, etc. But due to the complex marine environment and the movement of vectors, the attitude of the hydrophone often changes [4]. There are two types of attitude correction method: the first is correction of particle vibration velocity data; second is correction of DOA estimation [5]. The former has a large amount of calculation, and is suitable for the situation of attitude changing faster. The latter is suitable for the situation of attitude changing slowly [6]. When the vector hydrophone is applied to the linear array detector, attitude change belongs to the slow changes. The attitude correction of vector hydrophone uses the estimation method of correction azimuth in this paper.

The sensitive element of the vector described in this paper is an acceleration sensor based on MEMS technology [7]. Its frequency response can range from zero frequency to 1.5 kHz. Compared with piezoelectric hydrophone, advantage is that it can respond to zero frequency. So we can use this feature to do the attitude correction of hydrophone [8]. The advantages of this design are full use of various properties of element, reducing the number of components, reducing the overall size of hydrophone, meanwhile lowering the installation errors caused by the installation [9].

II. The Structure of the Capacitive Self-Correcting Vector Hydrophone

A. Overall design

Capacitive self-correction vector hydrophone is produced by the using of the acceleration response characteristics of the capacitive acceleration sensor. The core sensitive element is capacitive accelerometer sensor. For verifying the attitude correction capability in the X-axis and Y-axis, the design of the hydrophone uses X and Y two axis design, which uses two capacitive acceleration sensors as sensitive components to be sensitive to X, Y two-axis acceleration signal respectively. Then the pre-processing circuit and the capacitive acceleration sensor will be potting together by the acoustic potting to complete overall packaging of hydrophone. Specific structure is shown in Fig. 1.

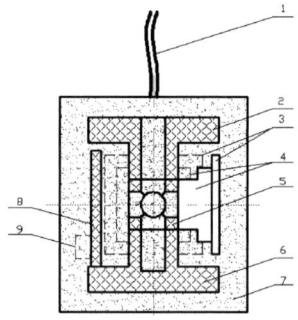

1 cable；2 upper keeper；3 circuit unit；4 accelerometer；5 center cubic；6 under keeper；7 polyurethane；8 side counterweight；9 center counterbalance

Figure 1. Structure of the hydrophone.

B. Structure design of capacitive accelerometer sensor

Capacitive accelerometer sensor is the core sensitive element of hydrophone. The overall sound sensitive

978-1-4799-1215-5/13 $31.00 © 2013 IEEE

performance and attitude correction capability of the hydrophone depend on the distinguish ability of the capacitive acceleration sensor primarily. So the important problem need to be solved is how to improve the sensitivity, broaden the band and Minimize its noise.

The design of the capacitive accelerometer sensor uses four layers of silicon structure to achieve acceleration response. This design includes the upper and bottom electrodes, the upper and lower plates of intermediate electrode, beam membrane structure and mass block, as shown in Fig. 2. This structure has the advantage of using four layer microstructure packaging technology to achieve symmetrical structure package. The top and bottom layers are the symmetric structure, and select the heavily concentration doped conductive silicon material. This will avoid the structure of electrode lead on the upper and bottom, and reduce processing difficulty and asymmetry factor. The electrode leads directly from the upper and lower surfaces to improve the manufacturability and machinability of chip. Meanwhile, the shielding between the chip and outside could reduce outside interference. Middle two layers is symmetrical structure. The two upper and lower mass blocks are supported by four elastic beam structures, whose materials are N-type silicon material. The mass blocks and elastic beam structures are bonded together by bonding way. Electrodes lead through the upper and lower electrode points. Upper and lower structures of the whole chip are symmetric design to reduce the noise caused by interference and achieve high sensitivity and high resolution design. Silicon-silicon bonding and gold-gold bonding technology were used between the intermediate sensitive structures and the upper and lower plates. The four-story structure package together and form an integral chip structure.

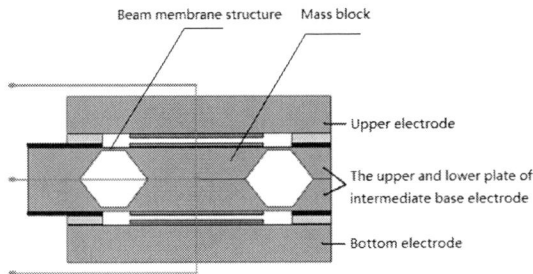

Figure 2.   Structure of the capacitive accelerometer sensor.

III.   EXPERIMENTAL PROCEDURE AND ANALYSIS

When the hydrophone calibrated in the pool, hydrophone attitude can not give accurate positioning and Installation error is larger. Meanwhile, the design of hydrophone uses the same vibration type principle, and its core device is acceleration sensor. So the shaking table test is used instead of acoustic test. This could analog the vibration caused by the sound waves acting on the surface of the hydrophone, and also give the attitude azimuth of the sensor accurately. The hydrophone is a vector hydrophone without digital processing part. The feasibility of the theory is verified by the intercomparison of the angle and the output.

The core experimental basis is that when the sensitive direction of capacitive attitude self-correcting vector hydrophone is consistent with the direction of gravity and there is no outside vibration signal, the static output value of hydrophone in this axis is the output of gravity, that is The output of 1 g. While the dynamic output is 0. When the vibration signal is applied to this state vector hydrophone, the static output of hydrophone unchanged, while the dynamic output is proportional to the vibration signal.

If there is a deflection angle between the sensitive direction of hydrophone and the direction of gravity. When there is no outside vibration signal, the static output of hydrophone is proportional to the deflection angle, while the dynamic output is 0. When the vibration signal is applied to this state vector hydrophone, the static output of hydrophone unchanged (proportional to the deflection angle), while the dynamic output is proportional to the vibration signal. Test plan and test process are shown in Fig. 3 and Fig. 4.

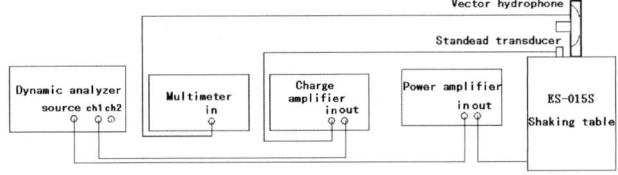

Figure 3.   Test method.

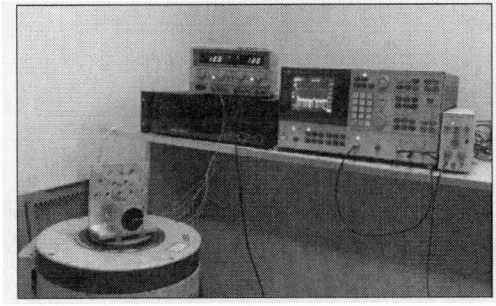

Figure 4.   The photograph of experiment process.

Test one: there is an arbitrary angle a1 between the X-axis of hydrophone and the vibration direction of the vibration table, as shown in 5(a)

Test two: there is an arbitrary angle a2 between the X-axis of hydrophone and the vibration direction of the vibration table, as shown in 5(b)

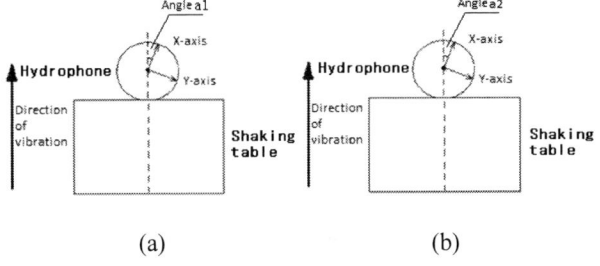

(a)   (b)

Figure 5.   Fitting schematic plan of the hydrophone vibration experiment.

The data of text one are shown in Table 1.

TABLE I.   DATA OF TEST ONE

Test side	Without vibration signal		The angle between the hybrophone and the vibration direction of vibration table	With vibration signal in 60Hz		The angle between the hybrophone and the vibration direction of vibration table
	Dc output (V)	*Ac output (mV)*		*Dc output (V)*	*Ac output (mV)*	
X-axis output	2.5126	8.14	71.26°	2.5134	2.5155	71.32°
Y-axis output	1.2390	9.446	25.55°	1.2380	1.2356	25.52°
Remarks: The sensitivity of X-axis is 2.6532V/g, and the sensitivity of Y-axis is 2.8732V/g.						

The data of text two are shown in Table 2.

TABLE II.   DATA OF TEST TWO

Test side	Without vibration signal		The angle between the hybrophone and the vibration direction of vibration table	With vibration signal in 60Hz		The angle between the hybrophone and the vibration direction of vibration table
	Dc output (V)	*Ac output (mV)*		*Dc output (V)*	*Ac output (mV)*	
X-axis output	2.1846	8.14	55.42°	2.1853	2.1866	55.45°
Y-axis output	1.8758	9.446	40.76°	1.8742	1.8785	40.72°
Remarks: The sensitivity of X-axis is 2.6532V/g, and the sensitivity of Y-axis is 2.8732V/g.						

Through comparing these tow test, we can found that the static dc zero point of hydrophone remains relatively unchanged when there were vibration signals or not. So when the hydrophone is in the case of a fixed posture, the dc zero point of hydrophone is constant and not affected by the outside vibration signals. The static dc zero point is affected only by the angle between hydrophone and the direction of gravity. So the attitude of hydrophone could be judged by measuring the static dc zero point of hydrophone, and then do correcting the attitude of hydrophone.

The results of the calculation are show in Table 3.

TABLE III.   RESULTS OF THE EXPERIMENT

—	The angle sum of the two axis of hybrophone without vibration	The angle sum of the two axis of hybrophone with vibration	Angular deviation
Test one	96.81°	96.84°	0.03°
Test two	96.18°	96.17°	0.01°
System error	0.63°	0.67°	—

The results show that the angle deviation measured in two tests is less than 0.1° and the test system error is less than 0.7°. If remove the systematic errors and fine the test steps, Static accuracy can be controlled within 0.5°. Angle values with and without vibration conditions are basically identical. So doing attitude self-correcting by capacitive vector hydrophone is feasible.

IV. CONCLUSIONS

The capacitive attitude self-correcting vector hydrophone presented in this paper used MEMS chip production technology to produce a vector hydrophone which can respond to zero frequency. The zero-frequency static signal and dynamic response signal are separated by the circuit processing technology. Target signal measurement and attitude self-correcting of hydrophone are achieved by the static and dynamic signal of the same sensitive devices of the hydrophone. The technical feasibility was fully validated by the design, production and equivalent test of hydrophone. This research proposed function multiplexing method of capacitive vector hydrophone, and gave the experimental results. This paper also provides the design solutions basis for the problem of reducing the volume and function optimization which exists in the engineering applications for the hydrophone. Capacitive attitude self-correcting vector hydrophone has a good application prospect in the underwater systems of limitations in size and power consumption limits, such as buoy system, mine system.

REFERENCES

[1] A. Carpenter, M. Silvia and B. A. Cray, "The design of a broadband ocean acoustic laboratory: detailet examination of vector sensor performance," Proc. SPIE. Vol.6231 pp. 62310-62311, 2006.

[2] M. Hawkes, A. Nehorai, "Vector-Sensor Beamforming and Capon Direction Estimation," IEEE Transactions on Signal Processing. 46(9), pp. 2291-2304, 1998.

[3] J. F. McEachern, J. A. McConnell, J. Jamieson, et al. "ARAP-Deep Ocean Vector Sensor Research Array," Oceans 2006, pp. 1-5, 2006.

[4] Han Qiushi, Chen Chen. "Research on Tilt Sensor Technology," IEEE International Symposium on KAM Workshop 2008, pp. 786-789, 2008.

[5] R. Racz, C. Schott, S Huber, "Electronic Compassw Sensor," Proceedings of IEEE Sensors 2004, vol.3, pp. 1446-1449, 2004.

[6] P. Tichavsky, K. T. Wong, M. D. Zoltowski, "Near-field/far-field Azimuth & Elevation Angle Estimation Using a Sigle Vector-hydrophone," IEEE Trans. Signal Processing, 49(11), pp.2498-2510, 2001.

[7] M. Hawkes, A. Nehorai, "Effects of Sensor Placement on Acoustic Vector-Sensor Array Performance," IEEE Journal of Oceanic Engineering, 24(1), pp. 33-40, 1999.

[8] Li Suilao, Jia Jichao, Long Rui,et al. "Calibration of Misalignment Angles of FOG Unit," Control and Decision Conference, pp. 5129-5132, 2009.

[9] Xu Bo, Sun Feng. "A FOG Online Calibration Research Based on High –Precision Three-axis Turntable," International Asia Conference on Digital Object Identifier, pp. 454-458, 2009.

Study of Influence of Pre-pulse Power on Xe Capillary Discharge EUV Source

Qiang Xu[1], Yongpeng Zhao[1*], Yao Xie[2], Qi Li[1], Qi Wang[1]

1 National Key Laboratory of Science and Technology on Tunable laser, Harbin Institute of Technology, Harbin, China
2 State Key Laboratory of Applied Optics, Changchun Institute of Optics, Fine Mechanics and Physics, Chinese Academy of Sciences, Changchun, China
hgdxq@126.com

Abstract—At present, extreme ultraviolet lithograph (EUVL), which used 13.5nm (2% bandwidth) emission as the source, has been researched extensively to achieve 22nm node or even below. In this paper, we use the Al_2O_3 capillary discharge Z pinch technology to generate high density and high temperature Xe plasma and achieve Xe^{10+} 13.5nm emission. By comparing the discharge current and the spectra for main-pulse power discharge and pre-main-pulse power discharge respectively under the different discharge voltage and Xe flow rate, we analysis the influence of the pre-pulse power on the Xe plasma EUV emission and the discharge stability. The results show that the pre-pulse power can increase the discharge stability. Although the spectra are basically the same as that by main-pulse power only. The Xe gas cannot be broken down under the flow rate of 7.0sccm by main-pulse power only. And the gas pressure is too high to generate enough 13.5nm emission. On the other hand, for pre-main-pulse power, the Xe gas can be breakdown even the gas flow rate is decreased to 0.2sccm. The results show that the optimum flow rate of Xe to generate 13.5nm emission is 0.4sccm.

Keywords- EUV source; Capillary discharge; Pre-pulse discharge; 13.5nm emission;

I. INTRODUCTION

The development of an extreme ultraviolet (EUV) source emitting at 13.5nm for high volume photolithography under the 22nm node or beyond has been investigated comprehensive[1]. The choice of the wavelength of 13.5nm for photolithography is mainly for the reason that the molybdenum-silicon (Mo/Si) multilayer mirrors have the maximal reflectivity at this wavelength, which is of the value of over 67% in production optics[2, 3]. A large number of EUV sources, for example the laser produced plasma (LPP)[4], the synchrotron wiggler[5], Z-pinch[6], plasma focus[7] and capillary discharge[8], are able to produce 13.5nm emission and all of them have the potentiality to be used for EUV lithography.

The use of the pre-pulse power in the capillary discharge system is to generate X-ray laser originally[9, 10]. The results show that the pre-pulse is the key parameter to generate the uniform plasma column and to get X-ray laser.[11] For the EUV source, the 13.5nm emission is generated by the spontaneous radiation. In this way, it didn't need to generate very good uniform plasma. And there is no pre-pulse power

for the EUV source originally[12, 13]. However, the exposure of the EUV lithograph shows that the stability of the power for the EUV source has deep influence on the quality of the chip. In this way, ASML, Nikon and Canon propose the requirements for the EUV source, which have the two most important parameters: the power and the power stability of the EUV source[14]. As a result, most of the researchers have introduced the pre-pulse power into the EUV source, which are focused on the engineering application[6, 8]. In this paper, the current stability and the spectra for the Xe are studied to analysis the use of the pre-pulse power in the capillary discharge EUV source.

II. EXPERIMENTAL APPARATUS

Figure 1 shows the cross sectional view of the capillary discharge system used for generating the EUV radiation in the experiment. The main-pulse power voltage between the anode and the cathode is ranged from 20kV to 30kV, and the peak current is ranged from 25kA to 45kA, with the bandwidth is 120ns (FWHM). The voltage of the pre-pulse power is about 20kV, and the current is 20A. The current is measured by a Rogowski coil, whose output is recorded by a digital oscilloscope (Tektronix, P6015A). Xe gas is used as a EUV emitter and is continuously flowing through the discharge capillary. Alumina capillary 7mm in diameter and 12mm in length is used. The discharge device is mounted onto a vacuum pumping device that maintains a stable pressure of Xe gas within the capillary. The capillary pressure is varied from 1Pa to 60Pa, controlled by a mass-flow controller to adjust the gas flow rate from 0.2sccm to 2.0sccm into the capillary, with approximately 0.1Pa of residual air pressure in the capillary prior to the admission of Xe.

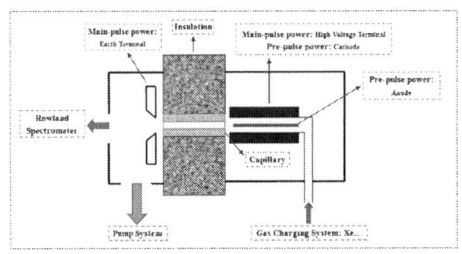

Figure 1. Cross sectional view of the capillary discharge system

978-1-4799-1215-5/13 $31.00 © 2013 IEEE

The EUV spectra are obtained by using a 1m grazing-incidence Rowland-circle spectrometer (Mcpherson 248/310G) connected to the discharge chamber, which is located along the axis of the capillary. The instrument has a 1200 l/mm Au-coated grating and an X-ray CCD (Andor, DO920P-BN). The entrance slit is reduced to be 10μm, resulting in a wavelength resolution of $\Delta\lambda_g=0.04$nm.

III. RESULTS AND ANALYSIS

The comparisons of main-pulse currents for the main-pulse power discharge only and main-pre-pulse power discharge at the flow rate of Xe 10.0sccm are shown in Fig.2 and Fig.3.

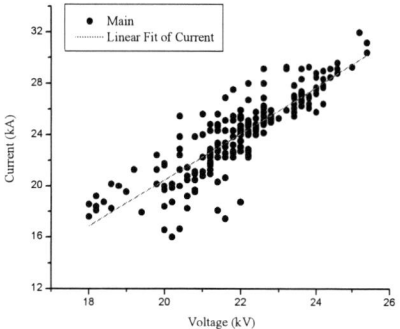

Figure 2. Relationship between the main-voltage and the current when discharging by the main-pulse only

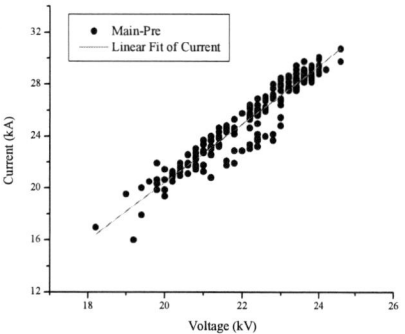

Figure 3. Relationship between the main-voltage and the current when discharging by the main-pre-pulse

As we all know, when the gas is broken down, it is pinched to the center and generated high temperature and high density plasma. And then it can be treated as the electrical conductor. The impedance and the inductance of the plasma should be very small[15] by comparing with the total value of the circuit. In this way, when the gas pressure is the same, the relationship between the voltage and the current should be linear. According to Fig.2 and Fig.3 it can be found that the relationship between the voltage and the current are not suiting the liner fitting very well. The slopes for the two conditions are all 0.92, which are the reflexes of the impedance and the inductance for the circuit under ideal condition. However, the deviation of the current for Fig.2 and Fig.3 is 8.5% and 3.0% respectively. The same main current means the same EUV power. In this way, the

stability of the current responses the stability of the EUV power. As a result, by leading the pre-pulse power into the power system, the source power stability can increase about 70%, which is very useful to the EUV source.

In order to analysis the influence of the pre-pulse power on the plasma and the EUV emission, the spectra are measured for the main-pulse power discharge only and main-pre-pulse power discharge at the flow rate of Xe 10.0sccm. Fig.4 shows the experimental results.

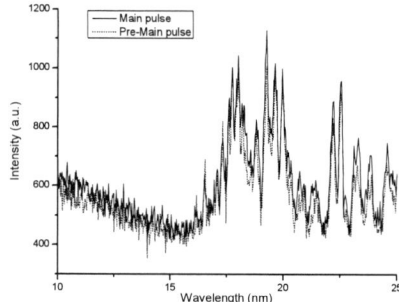

Figure 4. Spectra comparisons generated by main pulse and pre-main-pulse, current 32.0kA, Xe flow rate 10.0sccm

From Fig.4 it can be found that the spectra are almost generated by the Xe^{7+} and Xe^{8+} ions. The gas pressure is too high to be pinched efficiently and to generated high enough degree ions. In this way, the Xe^{10+} ions and 13.5nm emission cannot be got. In the perspective of energy transformation, the energy injected by the pre-pulse power is much smaller than the main-pulse power. In this way, when the discharge voltage and discharge current is the same, the injected is almost the same, which means that the energy converted to the internal energy of the Xe plasma is the same. As a result, the spectra should be almost the same for them. In fact, the spectra under different flow rate of Xe and different discharge current are almost the same.

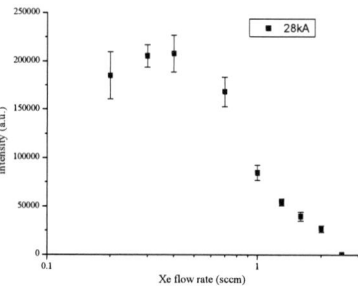

Figure 5. Relationship between the 13.5nm (2% bandwidth) emission and the flow rate of Xe

On the other hand, for the main-pulse power discharge only, the gas cannot be broken down when the Xe flow rate is lower than 7.0scmm. In this way, it cannot generated 13.5nm emission as well. For the pre-main-pulse power discharge, the gas can be broken down even the gas flow as low as 0.2sccm. And the temperature is high enough to generate Xe^{10+} ions and get 13.5nm emission. The spectra can be seen in the prevenient paper[16]. Fig.5 shows the relationship between the 13.5nm (2% bandwidth) emission and the flow rate of Xe under the pre-main-pulse power discharge and the current of 28kA. The results show that the 13.5nm (2% bandwidth) emission get the maximum value when the flow rate of Xe is 0.4sccm. And the 13.5nm (2% bandwidth) emission will disappear when the flow rate of Xe is 2.5sccm. Too lower gas pressure will lead to overmuch ionization, and the density of Xe^{10+} ions will be too low. On the other hand, too higher gas pressure will lead to too lower electron temperature. In this way, it cannot be got enough density of Xe^{10+} ions.

As a result, when the power is worked with the main-pulse only, it cannot be got 13.5nm emission. However, by pre-main-pulse discharge, the gas can be broken down as low as the flow rate of 0.2sccm. And the 13.5nm (2% bandwidth) emission gets the maximum value when the flow rate of Xe is 0.2sccm. In a word, it can get high enough emission by the pre-main-pulse discharge.

IV. CONCLUSION

The effect of the pre-pulse power on the power stability and the 13.5nm (2% bandwidth) emission for Xe gas capillary discharge EUV source are studied. The results show that the power stability can increase about 70% by drawing the pre-pulse power into EUV source. The pre-pulse power has little influence on the discharge spectra. However, when the source is worked the main-pulse power only, the gas cannot be broken down when the gas flow rate is lower than 7.0sccm. And it cannot achieve 13.5nm emission. On the hand, with the pre-pulse power, it can make the gas be broken at the flow rate as low as 0.2sccm. And the 13.5nm (2% bandwidth) emission gets the maximum value when the Xe gas flow rate of 0.4sccm.

ACKNOWLEDGMENT

This work is supported by the Project supported by the Key Program of the National Natural Science Foundation of China((No: 60838005) and the Project supported by the National Science and Technology Major Project of the Ministry of Science and Technology of China (2008ZX02501).

REFERENCES

[1] J. V. Hermans, D. Laidler, C. Pigneret, A. V. Dijk, O. Voznyi, M. Dusa, and E. Hendrickxa, "Overlay progress in EUV lithography towards adoption for manufacturing", Extreme Ultraviolet (EUV) Lithography II, 2011, vol. 7969, pp. 79691M.

[2] C. Zaczek, S. Müllender, H. Enkisch, and F. Bijkerk, "Coatings for next generation lithography", Advances in Optical Thin Films III, 2008, vol. 7101, pp. 71010X.

[3] S. Bajt, and D. Stearns, "High-temperature stability multilayers for extreme-ultraviolet condenser optics", APPL OPTICS, 2005, vol. 44, pp. 7735-4116.

[4] G. O'sullivan, and B. Li, "Development of laser-produced plasma sources for extreme ultraviolet lithography", Journal of Micro/Nanolithography, MEMS, and MOEMS, 2012, vol. 11, pp. 021108.

[5] P. P. Naulleau, P. E. Denham, B. Hoef, and S. Rekawa, "A design study for synchrotron-based high-numerical-aperture scanning illuminators", Optics Communications, 2004, vol. 234, pp. 53-62.

[6] I. Song, K. Iwata, Y. Homma, S. R. Mohanty, M. Watanabe, T. Kawamura, A. Okino, K. Yasuoka, K. Horioka, and E. Hotta, "A comparative study on the performance of a xenon capillary Z-pinch EUV lithography light source using a pinhole camera", PLASMA SOURCES SCI T, 2006, vol. 15, pp. 322-327.

[7] S. Mohanty, T. Sakamoto, Y. Kobayashi, I. Song, M. Watanabe, T. Kawamura, A. Okino, K. Horioka, and E. Hotta, "Miniature hybrid plasma focus extreme ultraviolet source driven by 10 kA fast current pulse", Review of Scientific Instruments, 2006, vol. 77, pp. 043506.

[8] P. Zuppella, A. Reale, A. Ritucci, P. Tucceri, S. Prezioso, F. Flora, L. Mezi, and P. Dunne, "Spectral enhancement of a Xe-based EUV discharge plasma source", PLASMA SOURCES SCI T, 2009, vol. 18, pp. 025014.

[9] W. Qi, C. Yuanli, Z. Xinlu, L. Peng, Z. Wudi, and P. Huiming, "Study on Capillary Discharge-pumped Soft X-ray Laser in Ne-Like Ar", Chinese Journal of Laers, 2002, vol. 29, pp. 97-100.

[10] W. Qi, X. Yao, Z. Yongpeng, L. Bohan, and Z. Qiushi, "Capillay Discharged Soft-X-Ray Laser and Application", Chinese Journal of Laers, 2010, vol. 37, pp. 5-17.

[11] C. Yuan-Li, L. Bo-Han, W. Yin-Chu, Z. Yong-Peng, W. Qi, Z. Wu-Di, P. Hui-Min, and Y. Da-Wei, "Effect of pre-pulses on capillary discharge soft x-ray laser", ACTA PHYSICA SINICA, 2005, vol. 54, pp. 4979-4984.

[12] M. A. Klosner, and W. T. Silfvast, "Intense xenon capillary discharge extreme-ultraviolet source in the 10-16-nm-wavelength region", optics letters, 1998, vol. 23, pp. 1609-1611.

[13] L. Juschkin, A. Chuvatin, S. Zakharov, S. Ellwi, and H. Kunze, "EUV emission from Kr and Xe capillary discharge plasmas", Journal of Physics D: Applied Physics, 2002, vol. 35, pp. 219-227.

[14] U. Stamm, J. Kleinschmidt, K. Gäbel, H. Birner, and I. Ahmad, "EUV Source Power and Lifetime: the Most Critical Issues for EUV Lithography", Emerging Lithographic Technologies VIII, 2004, vol. 5374, pp. 133.

[15] W. Qi, L. Peng, and Z. Yong, "Study of Capil lary Impedance Characteristic", Chinese Journal of Laers, 2003, vol. 30, pp. 497-500.

[16] Q. Xu, Y. Zhao, Y. Liu, Q. Li, and Q. Wang, "Effect of He/Ne/Ar on EUV emission and Xe plasma pumped by capillary discharge", The European Physical Journal D, 2013, vol. 67, pp. 1-5.

Research of high accuracy straight line parameter estimation algorithm for ultralow-pixel CCD

Jie Tang[1]*, MaoJun Fan[2], Jun Hu[2]

1 school of mechanical engineering, Nanjing University of Science and Technology, Nan Jing, CHINA
2 The third research institute of china electronics technology group corporation, Bei Jing, CHINA
tangjiemail@126.com

Abstract—In high-speed imaging systems, the number of CCD's pixels is limited. To compensate for this deficiency, the research is carried out for high accuracy linear parameter estimation algorithm for ultralow-pixel CCD. In order to estimate line segment parameter, introduce the TOEPLITZ matrix reconstruction algorithm from array signal processing method. Thus the data covariance matrix can be constructed from one image unit. The merit is the avoiding the needing of many image units, in which the line segment parameters are required to be nearly the same. At the last some experiments on true images demonstrate the efficiency of the algorithm.

Keywords-Line segments extraction; flexible constraints; image edges processing

I. INTRODUCTION

In the optoelectronics (laser and infrared) monitoring field, image processing increasingly plays a more important role. For example, independent real-time location estimation of automation equipment by photoelectron image processing, plays an important role in industrial control.

In practical applications, the relative displacement of feature is often calculated by calculating the straight portion parameters change of the edges. J.M.M. Montiel discussed[1] the method which extracted a straight line from complex images, and achieved the camera calibration and the object structure motion estimation. David Ribas[2] explored the method extracting straight lines from the mechanical scanning sonar image to create the underwater topographic maps. Derek C. W. Pao[3] extracted the lines from the complex image to achieve the object shape recognition more effectively.

Now there are many images multi-linear parameter estimation algorithms, and primarily based on the HOUGH transform[4] and its improved algorithm. The basic characteristics of the HOUGH are analyzed in the literature[4]. In HOUGH transform, each point on the edge is generated, the existence and parameters of the line can be determined by calculating the extreme points of the density distribution in the parameter space. In order to solve the problem of large operations in the HOUGH transform, the improved algorithms have been proposed, such as fast HOUGH transform[5], adaptive HOUGH transform[6], some other improved algorithm[7], and so on.

So far these algorithms still exist the following problems: (1) there need a large amount of computation not only in the conversion between the image space and the mapping parameter space ,but also in the voting accumulation process;

(2) discrete parameter space make inevitably estimation errors of the straight line parameter. If the quantization interval of the parameter space refined, the peak in the parameter space certainly will broadened, thus affecting the parameter detection; (3) in the most practical applications, the number of the linear image is unknown, occurring an error detection because the local parameter space maxima caused by of the noise.

To overcome the above the drawback of the linear parameter estimation algorithm, especially effectively detect the linear number, Hamid K.Aghajan[8] use high-resolution spatial spectrum estimation algorithm to the image multi-linear parameter estimation, converting the line number determination to the number of signal sources estimation in the array signal processing, which have been extensively studied before. He also used high-resolution spectral estimation ESPRIT algorithm to directly obtain the linear parameters, avoiding the discrete computation.

Hamid K.Aghajan's algorithms has a major problem limiting the application in practice, which need a dozen or more of the image matrix to constituting the data covariance matrix in is the process convert the image matrix to the data covariance matrix. Thus the image matrix linear parameters must be kept constant. In other words, the number of the line and parameters in the image matrix cannot be changed, thus improving the knowledge requirements controlling the image.

In order to solve the practical application of the above problems of Hamid K.Aghajan algorithm ,TOEPLITZ reconstruction method is used to transform the image matrix to data matrix, and allows a single image matrix to generate the valid data covariance matrix , thus avoiding the requirement that image lines in the matrix must have the same parameters and improving the practicality of linear parameter estimation algorithm.

II. DESCRIPTION OF THE ALGORITHM

A. Construction of data covariance matrix

As most of the other linear parameter estimation extraction algorithm, this algorithm is also the processing on the complex image edge contour , and estimation of the straight edge contour parameters. Canny[9] edge detection algorithm is used to get the edge image. The algorithm divides edge image into many processing unit, without loss of generality. Assume that the edges unit M is N rows × N columns, in the image data matrix, the elements corresponding with edges are assigned 1, others are assigned 0. This algorithm also generates one-

---

*Fund: Funded by major national science and technology projects (2011ZX05026-005)

978-1-4799-1215-5/13 $31.00 © 2013 IEEE

dimensional vector K, $\mathbf{K} = (e^{\mu \times i}, e^{2\mu \times i}, \cdots, e^{N\mu \times i})^T$, where μ is a number smaller than 2π, which can be set up by yourself, i is a plural, $\mathbf{T} = \mathbf{M} \bullet \mathbf{K}$. Assuming R as the cross-covariance matrix of vector T, then $\mathbf{R} = \mathbf{T} \bullet \mathbf{T}'$, where ' denotes the conjugate transpose. The matrix R is covariance matrix used to estimate the line parameter.

Assuming image matrix including w straight lines and defining the origin point of the image matrix as the first point in the upper-left corner，downwards form the origin of the Y-axis and right from the origin of the X-axis is positive direction. So the equations of w straight lines can be written in the form, s = 1,2, ..., w , s is the line number. Then T can be expressed as:

$$\mathbf{T} = \mathbf{M} \cdot \mathbf{K} = \begin{pmatrix} t_1 \\ t_2 \\ \vdots \\ t_N \end{pmatrix} = \begin{pmatrix} \sum_{s=1}^{w} e^{i\mu \times (a_s \times 1 + b_s)} \\ \sum_{s=1}^{w} e^{i\mu \times (a_s \times 2 + b_s)} \\ \vdots \\ \sum_{s=1}^{w} e^{i\mu \times (a_s \times N + b_s)} \end{pmatrix} =$$

$$\begin{pmatrix} e^{i\mu \times (a_1 \times 1)} & e^{i\mu \times (a_2 \times 1)} & \cdots & e^{i\mu \times (a_w \times 1)} \\ e^{i\mu \times (a_1 \times 2)} & e^{i\mu \times (a_2 \times 2)} & \cdots & e^{i\mu \times (a_w \times 2)} \\ \vdots & \vdots & \vdots & \vdots \\ e^{i\mu \times (a_1 \times N)} & e^{i\mu \times (a_2 \times N)} & \cdots & e^{i\mu \times (a_w \times N)} \end{pmatrix} \cdot \begin{pmatrix} e^{i\mu \times b_1} \\ e^{i\mu \times b_2} \\ \vdots \\ e^{i\mu \times b_w} \end{pmatrix}$$

(1.1.1)

Assuming:

$$\mathbf{A} = \begin{pmatrix} e^{i\mu \times (a_1 \times 1)} & e^{i\mu \times (a_2 \times 1)} & \cdots & e^{i\mu \times (a_w \times 1)} \\ e^{i\mu \times (a_1 \times 2)} & e^{i\mu \times (a_2 \times 2)} & \cdots & e^{i\mu \times (a_w \times 2)} \\ \vdots & \vdots & \vdots & \vdots \\ e^{i\mu \times (a_1 \times N)} & e^{i\mu \times (a_2 \times N)} & \cdots & e^{i\mu \times (a_w \times N)} \end{pmatrix}$$

So we obtain that

$$\mathbf{T} = \mathbf{A} \cdot \begin{pmatrix} e^{i\mu \times b_1} \\ e^{i\mu \times b_2} \\ \vdots \\ e^{i\mu \times b_w} \end{pmatrix} = \mathbf{A} \cdot \mathbf{U} \qquad (1.1.2)$$

Then T's cross-covariance matrix can be written as

$$\mathbf{R} = \mathbf{T} \cdot \mathbf{T}' = (\mathbf{A} \cdot \mathbf{U}) \cdot (\mathbf{A} \cdot \mathbf{U})' = \mathbf{A} \cdot \mathbf{U} \cdot \mathbf{U}' \cdot \mathbf{A}' = \mathbf{A} \cdot \mathbf{R}_U \cdot \mathbf{A}'$$

(1.1.3)

$\mathbf{R}_U = \mathbf{U} \cdot \mathbf{U}'$ is ranked 1, which is the same with the form of related information source's covariance matrix. So the spatial spectrum estimation MUSIC or ESPRIT algorithm can be used to calculate A, meanwhile get the lines' slope of image matrix. To calculate linear parameters correctly, N must be greater than the maximum w, then $\mathbf{U} = (\mathbf{A}' \cdot \mathbf{A})^{-1} \cdot \mathbf{A}' \cdot \mathbf{T}$ can be calculated by the method of least squares from (1.1.2), and the corresponding intercept $-\frac{b_1}{a_1}, -\frac{b_2}{a_2}, \cdots, -\frac{b_w}{a_w}$ are obtained. This is the processes of calculating image matrix line parameters by the spatial spectrum estimation theory. The paper only has a discussion on the slope of straight lines. The estimation of the intercept is not ideal in the practical application, thus needs further study in another paper.

In the spatial spectrum estimation algorithm, there are usually two algorithms to estimate the A: one is the spatial smoothing method[8], which used in the image matrix with multiple mean covariance matrix seeking obtained after decorrelation covariance matrix, but this need line parameters consistent among different images, which often requires manual intervention in practice; another method obtains de-correlative data covariance matrix after remodeling the relative covariance matrix by the matrix TOEPLITZ. The image matrix line parameters can be estimated by an image matrix.

B. TOEPLITZ remodeling of data covariance matrix

In the classical theory of spatial spectrum estimation, the matrix TOEPLITZ reconstruction and related algorithm uses data covariance matrix of the related information source to fit covariance matrix of the independent information source. In the formula(1.1.3), if the rank of $\mathbf{R}_U$ is w, though the covariance matrix R satisfies TOEPLITZ matrix characteristics of the independent source covariance matrix, due to the rank of R is 1, R is not TOEPLITZ matrix. There are many ways to make R fit to the nearest TOEPLITZ, the most commonly used is matrix reconstruction method, which substitute the average value for each element in the axis paralleling to the matrix main diagonal, and generate the matrix with characters of TOEPLITZ matrix. The essence of TOEPLITZ de-correlation matrix reconstruction algorithm finds a way to make the data covariance matrix fit the TOEPLITZ matrix.

C. the application of high-resolution spatial spectrum estimation algorithm

The covariance matrix reconstructed by TOEPLITZ matrix can be applied to estimate the lines parameters of image matrix by high-resolution spatial spectrum estimation algorithm. The most classic cases are the MUSIC algorithm and ESPRIT algorithm. MUSIC algorithm can be applied to low SNR (Signal to Noise Ratio) case with low evaluated error , but it still needs search for the peak point of the curve ,resulting with a relatively large amount of computation. The ESPRIT algorithm can be used to directly calculate A and the slope of the line with a small amount of computation. If the total least squares ESPRIT algorithm used, it is not only suitable for low signal to noise ratio case, but also can reduce the estimation error. For the case of non-continuous curve, total least squares ESPRIT algorithm actually calculates the lines parameters after straight-line approximation method is used to the edge curve by least squares. About the content of total least squares ESPRIT algorithm, the reader can refer to the literature [10].

High-resolution spatial spectrum estimation algorithm prevents HOUGH parameter space quantization interval from computational constraints, thereby reducing the calculation error, saving a great amount of computation and meeting real-time operation requirements because of the avoiding of the search operation in algorithm. More details can be seen in the literature [8].

III. SIMULATION ANALYSIS

To verify the technical feasibility of High-resolution spatial spectrum estimation, the estimation of some straight lines slope is examined. The image area with 40 × 40 pixels are shown in Figure 1, the slopes of the two straight lines are 1 and -1 respectively. Vector T is constructed according to (1.1.2) , and covariance matrix R is obtained according to (1.1.3). R is reconstructed by TOEPLITZ ,and the direction vector matrix A is obtained by the processing of MUSIC or the total ESPRIT least squares algorithm, thus the intermediate variables $a_1, a_2, \cdots, a_w$ and the slope of the straight lines can be calculated. MUSIC algorithm waveform is shown in Figure 2, where the Abscissa values of the peak of the waveform represents the estimated value of $a_1, a_2, \cdots, a_w$, so the estimated slopes by MUSIC are 1 and -1 respectively. The values of $a_1, a_2, \cdots, a_w$ calculated by Total Least Squares ESPRIT algorithm are 0.9894 and -1.0028 respectively, so the slopes estimated are 0.9894 and -1.0028.

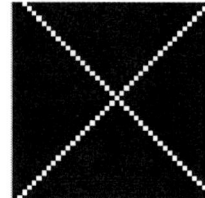

Fig.1 A line pattern in image unit

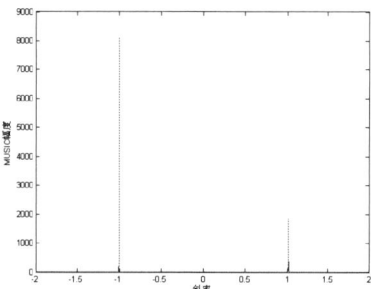

Fig.2 MUSIC spectrum estimation

Figure 1 is rotated 15 degrees right using the image processing software to and get Figure 3, the slopes of the two straight lines become about -1.732 and 0.577,and the corresponding intermediate variables $a_1, a_2, \cdots, a_w$ are-0.5774 and 1.7331. In the same way, MUSIC spectrum waveform can be obtained and shown in Figure 4, where the Abscissa values of the peak of the waveform represents the estimated value of $a_1, a_2, \cdots, a_w$. Using Total Least Squares ESPRIT algorithm $a_1, a_2, \cdots, a_w$ can be calculated out, -0.5282 and 1.6701 respectively. Clearly, both the image noise introduced by the rotation of image and the discontinuity of the lines pixel can cause the estimation error.

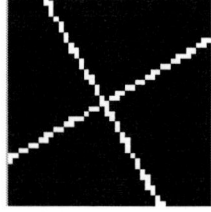

Fig.3 Image unit after rotated

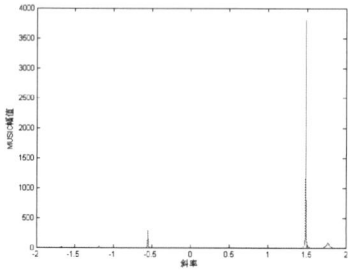

Fig.4 MUSIC spectrum estimation in rotated image unit

IV. THE DISCUSSION ON THE INDOOR POSITIONING APPLICATIONS OF ROBOTICS

In the Indoor robot localization, the computer vision method is used to deal with a sequence of real-time images. The motion parameters are estimated according to the parameter variations of straight-line edge, including the part with straight edges or straight texture. Since the practical applications of spatial spectrum estimation algorithm into estimate the intercept need further research, now emphasis is on estimating changes in the slope of the straight-line contour to estimate the orientation of the robot's walk. The movement distance can be measured by measuring the number of revolutions of the traveling wheels.

In the indoor positioning algorithm, a picture is first extracted from an image sequence, and the edge of the scene is extracted using canny edge detection algorithm[9]. Then we set the image data matrix in a straight-line edge, and use the high-resolution spatial spectrum estimation algorithm to estimate the parameters of straight-line edges distribution. In the same way the edge line parameters of the adjacent images can be estimated, and the direction changes of the robot according to the slope changes of the adjacent images are determined.

V. EXPERIMENTAL RESULTS ANALYSIS

To verify the effect of the algorithm, the object image is selected as the target image, shown in Figure 5, and the edge detection image can be obtained, shown in Figure 6 ,after the processing of the canny edge detection algorithm. 40×40 pixels of the left bottom corner are selected as the image data matrix, shown in Figure 7, and substituted into linear parameter estimation algorithm. The image data covariance matrix can be constructed, and reconstructed by TOEPLITZ method. Using the least squares of the ESPRIT algorithm to process the reconstructed covariance matrix, the intermediate parameters $a_1, a_2, \cdots, a_w$ can be estimated 1.0771, -0.0382 and -2.1020 .from the above the slope of the line $\frac{1}{a_1}, \frac{1}{a_2}, \cdots, \frac{1}{a_w}$ are 0.9285, -26.1731 and -0.4757. This result is consistent with that of visual estimation.

978-1-4799-1215-5/13 $31.00 © 2013 IEEE

Fig.5 Object picture

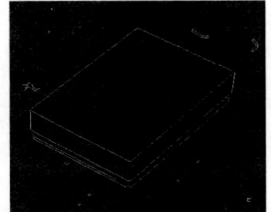

Fig.6 Edge picture after edge detecting

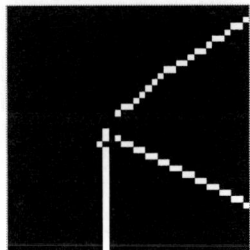

Fig 7 A corner edge picture

ACKNOWLEDGMENT

This article substitute TOEPLITZ matrix reconstruction method in the array signal processing for the spatial smoothing algorithm [8], thus only an image unit can form the data covariance matrix with the method using high-resolution spatial spectrum estimation algorithm to estimate linear parameters, and this method needs not many image units to form the data covariance matrix. This algorithm is proved valid through simulation experiments and actual verification. It must be pointed out that high-resolution spatial spectrum estimation algorithm is a sub-pixel estimation algorithm for the straight-line parameters estimation.

REFERENCES

[1] J.M.M. Montiel, J.D. TardoHs, L. Montano. Structure and motion from straight line segments [J]. Pattern Recognition. 2000, 33, 1295-1307

[2] David Ribas, Pere Ridao, Jos´e Neira, et al. Line Extraction from Mechanically Scanned Imaging Sonar [J]. IbPRIA 2007, Part I, LNCS 4477, 2007, 322–329

[3] Derek C. W. Pao, Hon F. Li, and R. Jayakumar, Shapes Recognition Using the Straight Line Hough Transform: Theory and Generalization[J], IEEE TRANSACTIONS ON PATTERN ANALYSIS AND MACHINE INTELLIGENCE, VOL. 14, NO. 11, NOVEMBER 1992,1076-1089

[4] Duda, R.O., Hart, P.E., Use of Hough transformation to detect lines and curves in pictures[J]. Commun. ACM ,1972,15 (1), 11–15

[5] Li, H., Lavin, M.A., Le Master, R.J., Fast Hough transform: A hierarchical approach[J]. Computer Vision Graphics Image Process, 1986, 139–161

[6] Illingworth, J., Kittler, J., The adaptive Hough transform[J]. IEEE Trans. Pattern Anal. Mach. Intell. 1987,9 (5), 690–698

[7] Xu Sheng-hua, Zhu Qing, Liu Ji-Ping, Straight Line Extraction via Multi-scale Hough Transform Based on Pre-storage Weight Matrix [J], ACTA GEODAETICA et CARTOGRAPHICA SINICA, 2008, 1(37), 83-88.

[8] Hamid K.Aghajan, and Thomas Kailath, Sensor Array Processing Techniques for Super Resolution Multi-Line-Fitting and Straight Edge Detection[J], IEEE TRANSACTIONS ON IMAGE PROCESSING, Vol. 2, No. 4, OCTOBER 1992, 454-465.

[9] Canny, J.F. A computational approach to edge detection[J]. IEEE Trans. Pattern Anal. Mach. Intell. 1986, 8 (6), 679–698

[10] Tripathy. P., Srivastava. S.C., Singh.S.N., A Modified TLS-ESPRIT-Based Method for Low-Frequency Mode Identification in Power Systems[J], IEEE Transactions on Power Systems, 2011, 2(26) , 719 – 727.

CA optimization based on simulation annealing in BIST

Bei Cao[1], Yongsheng Wang[2], Yanwei Dou[1], Dan Bu[1], Bin Zhou[2]

1 Electronic Science and technology Post-Doctoral Research Center, Heilongjiang University, Harbin, China
2 Microelectronic Center, Harbin Institute of Technology, Harbin, China
cao_bei@163.com

Abstract—**The deterministic test patterns generator in BIST often suffer from the problems that it requires extra test power consumption, area overhead and the idle test cycles between the test patterns. This paper proposes an efficient strategy for synthesizing a built-in test pattern generator that can generate a given set of predetermined low power test patterns for reducing the test power of a circuit under test (CUT) without modifying the initial fault coverage. The technique is based on the cellular automata (CA) model for testing combinational circuits. The algorithm we present based on simulation annealing (SA) that can optimize a CA structure to generate given low power test sequence by adjusting dynamically cell neighborhood range of CA. The results of simulation using benchmark combinational circuits showed that the designing generator is efficient to generate the deterministic test sequences in terms of power consumption, fault coverage, test time and area overhead compared to alternative solution.**

Keywords-simulation annealing; BIST; CA; test pattern generatio

I. INTRODUCTION

Build-In Self-Test (BIST) is an alternative and effective testing technique because of its advantages such as improved testability, at-speed testing and reduced dependence on automatic test equipment. Using BIST scheme can achieve the combination of higher quality and low cost test requirements and at-speed testing. A typical BIST structure is consisted of test pattern generator (TPG), the circuit under test (CUT) and output response analyzer. The main function of the TPG is to provide the test patterns to the CUT as input. The design of TPG is an important work to BIST technique. Normally, the TPGs often use linear feedback shift register (LFSR), cellular automata (CA) or counter. The efficient TPG in BIST scheme must satisfy some main objectives, such as achieving high fault coverage, short test lengths and minimal area overhead. While in recent years, this objective still remains important, reducing power dissipation during test application is also becoming an important objective. Since power consumption in test mode could be higher than the power consumption during normal operation [1], special care must be taken to ensure that the power consumption of the CUT is not exceeded during test mode.

In this paper, we first reorder the deterministic test patterns obtained by an ATPG tool. A pre-computed set of test patterns with the low power is generated and used in the synthesis of CA structures. The deterministic test patterns generator based LFSR reseeding often suffer from the problems that it requires the idle cycles between the test patterns, extra hardware size and test power consumption. This paper is devoted to solve the above problem by proposing the algorithm that can generate given low power test patterns using a scheme based on CA structure without increasing the idle test time. For the purpose of generating the low power and deterministic test patterns, CA structures is a good selection. CA structure is optimized based on simulation annealing algorithm. CA has been deeply studied and applied as effective test pattern generators [2-6].

The rest of paper will be organized as follows. In section II, the main definitions of CA and a review of previous work are given. Section III includes the pre-compute of test patterns and the details of proposing CA synthesis algorithm based on simulation annealing. The experimental results were given in section IV. The last is conclusions.

II. CELLULAR AUTOMATA

CAs were originally presented by Von Neumann. A CA is a system composed of cells interconnected in a regular manner, whose behavior advances in time in discrete steps. Wolfram proposed the 2-state, 3-neighborhood CA with cells arranged linearly in one dimension [7]. Each cell essentially comprises of a memory element and a combinational logic that generates the next-state of the cell from the present state of its neighboring cells-left, right, self. A cell structure is shown in figure 1. In each clock cycle, the states of the cells are discrete and are updated synchronously according to a local rule operating on a given neighborhood. Rule is a function that computes the cell's successor state based on the value received from its neighbor cells. The rule is implemented by the combinational logic. There are two types of the boundary, null or cyclic. Null boundary conditions behave as if the boundary always supplies a "0" from beyond the boundary into the sum that determines the state of the end cell. Cyclic boundary conditions connect the ends so that the cellular array forms a circle. The boundary conditions have little or no effect on the bit-stream generated [8]. In this paper the null boundary conditions are supposed.

CAs have been deeply researched as pseudorandom and deterministic test pattern generators to be applied to combinational circuits. Our focus of attention is synthesis of low power deterministic and optimized structure based on CAs as test pattern generators in BIST. A deterministic hardware pattern generator is a circuit that autonomously generates pre-determined test patterns with high fault coverage, minimal test time. In recent years, CAs have been proved the existence of the isomorphism with LFSR and proposed as an alternative to

978-1-4799-1215-5/13 $31.00 © 2013 IEEE

LFSR, in applications such as TPG, because of some unique advantages of CA. Regular, modular and cascadable structure of CA with local interconnections makes it ideally suited for TPG implementation in BIST [9,10]. Another significant advantage is higher speed due to their local interconnection structures. CA is an attractive option for TPG in BIST.

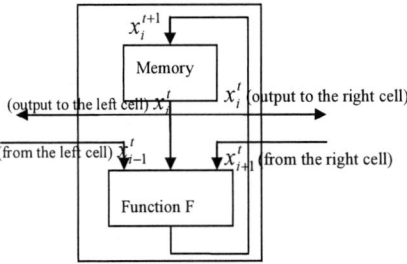

Figure 1. A CA cell structure

In previous work, the use of CA for the generation of deterministic test pattern for combinational circuits has been investigated. Boubezari presented two techniques for generating pre-computed test patterns based on CA [4]. The first technique used the reordering the column and adding the link column in given test pattern set for satisfying the property of evolution to synthesize the CA structures. The second technique considered the vector of rows in test patterns. The schemes are efficient as the synthesis of deterministic TPG based on CA, but some important problems are omitted for BIST. These techniques, such as additional link column and link row vectors that extend the original test pattern set, require extra CA cells or idle clock cycles and increase the test time. Test power consumption and hardware consumption is extremely increased due to additional link values and without considering the switch activity between the test patterns. Power consumption is also a major concern during BIST recently.

This paper proposes a new solution to the problem of high power consumption in deterministic testing, able to synthesize a TPG based on CA structure and generate the given test patterns with minimal switching activity. In other words, the power consumption of the test pattern set generated by TPG presented in this paper is the minimum. The key idea of algorithm is to extend dynamically the neighborhood of CA, not only 3-neighborhood, but also 5-neighbors defined by the range of the CA rule. Our scheme avoids using link columns or link vectors and makes the original test pattern set keep the minimum. The test power consumption and test time keep also the minimum for test patterns generated by TPG based on CA structure. The numbers of cells in TPG equal to the numbers of input pin of the benchmark circuit. It means that the area overhead is low due to without any link elements. But for achieving the optimal the CA structure, we use the simulation annealing algorithm in the synthesis scheme.

III. THE ALGORITHM

Two types of problems are defined to CA research. They are the forward problem and the inverse problem. The forward problem is mainly to find CA's properties to a given CA structure through a finite clock cycles. The inverse problem is to find a rule or a set of rules for a given evolution property or other properties. The CA structure can be synthesized by the set of rules.

Given a set of low power deterministic test patterns, our focus of attention is to find a cascading CA cell structures that can generate the set by the same evolution behavior. It is obvious that our research work belongs to the CA inverse problem. A theorem is as follows to the inverse problem [7].

Theorem: A CA structure will exist to generate a given binary sequence if this binary sequence satisfies the property of a CA evolution.

The most important problem for synthesis a CA structure is that a CA evolution property must be satisfied in the inverse problem. We define the CA evolution property of a given set of test patterns, here only assuming the 3-neighborhood CA for simple description. It is similarity to extension of neighborhood. Suppose the test set is $V = \{c_1, c_2, \cdots c_{n-1}, c_n\}$, where n is the number of bits in the test patterns which is equal to the number of the prime inputs in the CUT, and $c_i = \{c_i^1, c_i^2, \cdots c_i^t\}^T$.

Define: Let $V = \{c_1, c_2, \cdots c_{n-1}, c_n\}$ be the given deterministic test patterns with n columns, and m bits each column. The evolution of an ith CA cell "c_i" need verify a sequence with three neighbor, such as $\{c_{i-1}, c_i, c_{i+1}\}$. In the t time step, its binary state value is $\{c_{i-1}^t, c_i^t, c_{i+1}^t\}$, next time step, they produce c_i^{t+1} in the ith CA cell bit. For $\{c_{i-1}^j, c_i^j, c_{i+1}^j\} = \{c_{i-1}^k, c_i^k, c_{i+1}^k\}$, here, j and k being any time step, and $j \neq k$, if $c_i^{j+1} = c_i^{k+1}$ exists, we call the test pattern for the ith CA cell satisfies the evolution property. If $c_i^{j+1} \neq c_i^{k+1}$ exists, the evolution is violated. The instance of benchmark c17 about evolution violating is showed in figure 2. Based on above theorem and define, we can conclude that the CA structure can't be synthesized if evolution property is violated in a given test pattern set.

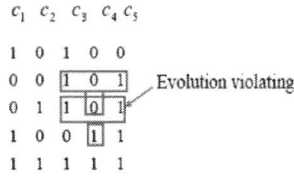

Figure 2. The example of evolution violating

We first order the deterministic test pattern set generated by ATPG tool based on the Hamilton path. The objective is to find an optimal test pattern ordering with the minimum Hamilton path between test patterns such that the switching activities are minimal in the CUT. The algorithm can guarantee a decrease in average power consumption and peak power consumption without modifying the initial fault coverage. The pre-

978-1-4799-1215-5/13 $31.00 © 2013 IEEE

computing process isn't main focus in this paper so that we don't pay more attention to discuss the pre-computing step.

The basic idea behind low power BIST generator is to generate test patterns with the minimal switching activities. In the next step of the algorithm, the evolution property of pre-computing set is checked. The characteristic of the deterministic test patterns with low power consumption is that the order of test vectors can't be changed. How to deal with the violation of evolution property is the key when the violation exists. The proposed algorithm based on adjusting and optimizing dynamically cell neighborhood range of CA is efficient on find a CA structure to generate low power test patterns. The key of the proposed techniques for synthesizing a TPG based on CA to generate a given low power deterministic test patterns is to rectify evolution violation. The algorithm adjusts dynamically cell neighborhood range of CA and uses the extension of neighborhood based on the simulation annealing algorithm. The details are described here.

For the algorithm, the exchange of column is adopted when violation occurs. If the ordering can't find the appropriate column to satisfy the evolution, the range of neighborhood specification is extended to 5-neighborhood. In some worse cases where the 5-neighborhood can't still verify the evolution property and two schemes can be used, the first, 7-enighborhood, the second, using link cells. But in this paper, we only extend to the 5-neighborhood and don't use the link cell for low hardware cost. The first and the last cells are difficulties to satisfy the evolution property and they must keep 3- neighborhood for the null boundary.

The algorithm is implemented with a large of 5-neighborhood for satisfying the evolution property. The numbers of 5-neighborhood means the hardware cost is bigger. For reducing the area overhead consumption and simple the connection, we combine the simulation annealing algorithm and extension of neighborhood and try to find the 3-neighborhood CA structure. In this algorithm, the objective function is the number of the CA neighborhood for simulation annealing algorithm, and solution space is the set of cell ordering with evolution. The every result in solution space must be efficient answer, in other words, must satisfy the evolution property for every CA cell. The minimal values of objective function mean the optimization of area overhead and simple cell connection. The main process is introduced as follows for CA synthesis algorithm based on simulation annealing.

Step1: set the initial simulation annealing parameters, such as initial and final temperature, t0 and tf; the number of the iterative, L.

Step2: generate the initial answer and objective function values.

Step3: if the current temperature is bigger than the final temperature "tf", executes the step4; or else step6.

Step4: circulate iterative L; in every iterative step the "generate" function is used to generate new answer, the return value is the difference of objective function and new cell orders. If difference of objective function is smaller than "0",

new answer is accepted as the current value. Otherwise accept the new answer according to probability "P".

Step5: attenuation of temperature. Return the step3.

Step6: compute rule function of the final answer. End program.

IV. EXPERIMENTS AND DISCUSSION

We wrote a prototypical tool that implements the TPG synthesis algorithm and performed the experiments with the ISCAS'85 combinational benchmark. We come back the classical method that inserts the link cell for evaluating the efficient of the proposed algorithm. The algorithm of inserting the link vectors can't implement low power test patterns, so we don't discuss and compare this method.

The results of the pre-compute are shown in the table I. Total hamming distance is calculated for the test set generated from ATPG tool TetraMAX and is represented as Original HD in column 2. The next column is the results with the minimal hamming distance using low power pre-computed technique. We can compute the saving ratio of the average power dissipation and peak power dissipation from power simulation tool. The reordered test set can keep the lower average power consumption than the original set. We use the pre-computed test set to synthesis the low power generator based on the CA structure.

TABLE I.   POWER PRE-COMPUTER RESULTS

Circuit	Ori. HD	Opt. HD	Power saving ratio (%)		
			HD	*Average power*	*Peak power*
C17	9	8	11.11%	6.00%	11.50%
C432	270	214	20.74%	6.32%	15.26%
C499	181	114	37.02%	19.48%	32.25%
C880	275	222	19.27%	11.66%	-65.67%
C1355	214	174	18.69%	5.10%	13.19%
C1908	124	96	22.58%	16.84%	-45.00%
C3540	382	308	19.37%	1.59%	15.17%
C6288	93	79	15.05%	7.41%	15.47%

Table II shows the results obtained for some ISCAS'85 circuits by applying the presented algorithm in section III. The number of original cell is represented as #O.C and #L.C means the number of the link cell. T.C is the total number of the CA cell for generating pre-computed low power test patterns using the classical method in reference [4].

TABLE II.   THE COMPARISON OF SYNTHESIS RESULTS

Circuit	Algorithm based on link cell [4]			Power saving ratio (%)		
	#O.C	#L.C	#T.C	*#3-N*	*#5-N*	*#T.C*
C17	5	1	6	4	1	5
C880	60	19	79	52	8	60
C1355	41	16	57	34	7	41
C1908	33	6	39	28	5	33
C6288	32	2	34	30	2	32

In proposed algorithm, it doesn't use the link cell. The total number of CA cell is equal to the number of the circuit's input port. We use the simulated annealing to reduce the number of 5-neighborhood. In the fact, some circuits can't find the appropriate CA structure if only using 3-neighborhood and 5-neighborhood because a large number of violations can't avoid only using neighborhood extension. But neighborhood extension of CA is a feasible technique for synthesis CA generator under power aware constraints in deterministic BIST.

V. SUMMARY AND CONCLUSIONGS

Power consumption during test motivates the search for effective solutions able to reduce power dissipation without affecting other factor, such as test length, hardware cost. In this paper, a novel CA synthsis algorithm is presented for selecting an optimal CA structure that can generate given low power test sets. The deterministic test patterns generator based on LFSR in BIST often suffer from the problems that it requires extra test power consumption, area overhead and the idle test cycles between the test patterns. The generator based on the CA can resolve these problems. The technique of neighborhood extension is an efficient strategy for synthesizing CA generator that can generate a given low power test set and low test time. The results of simulation using benchmark combinational circuits showed that the designing generator is efficient to generate the deterministic test sequences in terms of power consumption, fault coverage, test time and area overhead compared to alternative solution.

ACKNOWLEDGMENT

This work is supported by the research foundation of education bureau of Heilongjiang province under grant No.12531479.

REFERENCES

[1] Y. Zorian, "A distribued BIST control scheme for complex VLSI devises," Proc. 11th IEEE VLSI Test Symp. Los Alamitos Calif., pp. 4–9, 1993.

[2] H.Jabbari, J.C.Muzio, L.Sun, "A new class of cellular automata," In 10th Euromicro Confrence Digital System Design Architectures, Methods and Tools. pp.331-338, 2007.

[3] R.Santoro, S.Roy and O.Sentieys, "Search for optimal five-neighbor FPGA based cellular automata random number generators," IEEE International Sympsium Singles, Systems and Electronics. pp. 343-346, 2007.

[4] S.Boubezari. "A deterministic BIST generator based on CA structures," IEEE Transaction on Cmputer. vol44. pp. 805-816. 1995.

[5] Bei Cao, Liyi Xiao and Yongsheng Wang. "A low power deterministic test pattern generator for based on cellular automata," IEEE International Sympsium on Electronic Design, Test & Applications. pp.266-269. 2008.

[6] M.M.Arjmand, M.Soryani, K.Navi and M.A.Tehrani. "A novel ternary-to-binary converter in quantum-dot cellular automata," IEEE Computer society Annual Symposium on VLSI. pp.147-152. 2012.

[7] P.D Hortesius. "Cellulare atuomata based pseudorandom number generators for built-in self-test," IEEE Transactions on Computer-Aided Design. vol 18. pp. 842-859. 1989.

[8] D.Martinez and A.F.Sabater. "Cryptographic design based on cellular automata." In proc. of IEEE Intermational Symposium on Information Theory. pp. 180. 1997.

[9] P.H.Bardell. "Ananlysis of cellular automata used as pseudorandom pattern generators," In Proc. of International Test Conference. pp. 762-768. 1990.

[10] A.Piwonska, F.Seredynski and M.Szaban. "Discovering cellular automata rules for binary classification problem with use of genrtic algorithm," IEEE 26th International Parallel and Distributed processing Symposium Workshop & PhD Forum, pp. 649-655. 2012.

Fabrication and Optical Properties of Linear-core-array Multicore Fiber

Qiang Dai[1], Hong Bo Bai[1], Xiao Liang Zhu[12], Li Jia Ma[1], Tao Zhang[1]*

1 Key Lab of In-fiber Integrated Optics, Ministry Education of China, Harbin Engineering University, Harbin, China
2 Zhejiang Gongshang University, Hangzhou, China
zhangto2@163.com

Abstract—This paper presents a linear array core optical fiber that can be directly connected and coupled with a single core fiber. We propose a convenient preparation technique. With this production technology, we have successfully produced a 5-core and 23-core optical fiber linear array core optical fiber, and we tested its refractive index distribution and transmission properties of light. Further, we calculated optical transmission coupling characteristics of each core and supermodel properties of a linear array 23 core optical fiber as a representative. This study can provide the basis for special optical fiber design, production, and has a reference for other similar microstructure fiber analysis and application.

Keywords- linear array core; multi-core optical fiber; coupling; supermodel; mode field converter

I. INTRODUCTION

The usual meaning of multicore fiber refers to a fiber formed by a group of mutual parallel multiple cores along the fiber axis placed in fiber cladding. Optical properties of such multicore fiber are implemented through leading the light into multiple fiber cores. In general case, a single optical fiber core in diameter ranges a few micron to tens micron. The diameter of the whole fiber is 125 micrometer. In order to obtain the desired optical properties, fabrication of an optical fiber perform is deposited on common quartz material usually used the elements Ge and P etc. So that achieves its optical fiber core refractive index higher than the refractive index of the cladding. In recent years, with the improving of theoretical study of multicore microstructure fiber and fabrication technology, the special optical properties of multicore microstructure fiber attracted prevalent attention. It provides a new type of technical way of optical fiber sensor and optical fiber communication devices [1-3]. The key to fabricate multicore fiber is that each individual fiber core have the same diameter and precise positional relationship, so that it can ensure that the transmission performance of optical signals in optical fiber.

This paper presents a linear-core-array multicore fiber can directly connected with the standard single-mode fiber. This fiber manufacturing technology could fabricate linear-core-array multicore fiber. It can also reduce an optical fiber perform fabrication difficulty, to solve economical problem of perform fabrication. With these preparation techniques, we were making a 5 core linear-core-array multicore fiber and 23 core linear-core-array multicore fibers and discussed on optical properties of such multicore fiber.

II. MULTICORE FIBER FABRICATION METHOD

Study on uniform periodic linear-core-array multicore fiber is presented in Figure.1 (a). There are multiple doped fiber cores arranged in a linear array in the inner cladding. The diameter and doping concentration of each core are equal. Each core is the single-mode, and the distance d among fiber cores are near. Fiber core radius is r. The fiber core and cladding refractive indices are n_0 and n_1, respectively, showed as schematically in Figure.1 (b).

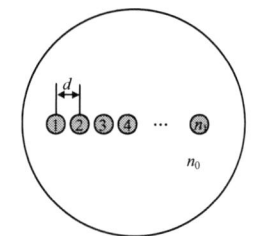

(a) Cross-section of the linear-core-array fiber

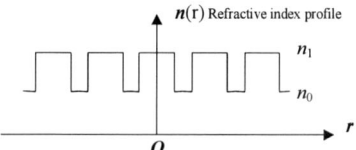

(b) Refractive index profile of the linear-core-array

Figure 1. Structure diagram of linear-core-array multicore fiber

Firstly, an optical fiber perform containing the desired refractive index profile is fabricated, taken by MCVD process. After fiber perform preparation, the fiber core rods are drawn by fiber drawing tower. Secondly, the other appropriate size quartz glass tube and high-purity quartz glass rod are cleaned and reserved. The quartz glass tube is drawn out appropriate size capillary. The high-purity quartz glass rod is cut in half along the diameter direction by glass cutter, getting two semi-cylindrical quartz glass rods. Then grind both the cut quartz glass rod surface, formation of two D-type quartz glass rods, reserved. Two D-type glass rods are both milled out of should be the same size, which size should be equal to the size of the inserted fiber core rods diameter. Thirdly, the reserved fiber core rod is inserted into the prepared thin quartz capillary tube. Then, the reserved fiber core rod with the quartz capillary tube is fused on the fiber drawing tower. At last, the quartz glass

978-1-4799-1215-5/13 $31.00 © 2013 IEEE 142

tube and the two D-type glass rods and the fiber core rod thin quartz capillary tube are combined to fabricate a linear-core-array multicore fibers perform, as show in Fig. 2.

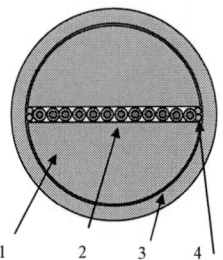

Figure 2. Cross-section of the linear-core-array fiber perform

In above, two D-type glass rod (1) and the fiber core rod with the quartz capillary tube (2) dimensions should be exactly equal to the inner diameter quartz glass tube (3), tightly. The distribution of several fiber core rods with the quartz capillary tube arrange in linear-core-array structure. If the dimension of linear-core-array rods inside the quartz tube is slightly smaller than the inner diameter of the quartz glass tube, then two rod ends (4) are filled the left gaps. Since the closely combination of (1) (2) (4), it can ensure the fiber core mutual position accuracy. The prepared linear-core-array multicore fiber perform will be drawn in fiber drawing tower in one end. Meanwhile, the other end of the fiber performs vacuums in the drawing process. The high purity quartz materials become the cladding and the fiber core rods become linear-core-array multicore of fiber.

We use this technique drawing 5 core linear-core-array fibers, as shown in Fig.3 (a). When implementing the same process, different kind of cores structures and cores quantity linear-core-array multicore fiber can be prepared. A fiber of 23 core linear-core-array fiber is drawn with this technique, as shown in Fig.3 (b).

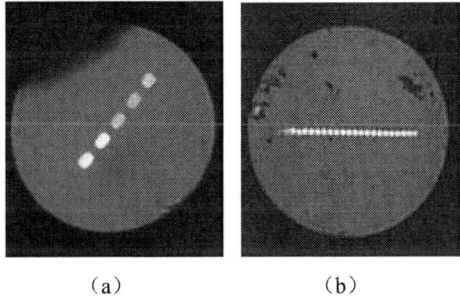

（a）   （b）

（a）5 core linear-core-array fiber（b）23 core linear-core-array fiber

Figure 3. Linear-core-array multicore fiber

III. OPTICAL PROPERTIES OF LINERR-CORE-ARRAY MULTICORE FIBER

The produced linear array core fiber can be used on refractive index analyzer for testing. Testing principle is near-field method. For testing 5 core linear-core-array

fibers, the three-dimensional refractive index profile result is shown in Fig.4. Fiber core and cladding refractive indices is $n_0 = 1.466$ and $n_1 = 1.463$, respectively. Similarly, fiber core and cladding refractive indices of 23 core linear-core-array fiber is $n_0 = 1.463$ and $n_1 = 1.452$, respectively.

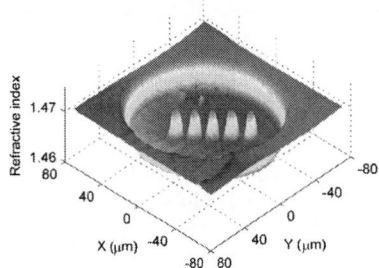

Figure 4. 5-core-array optical fiber 3-D distribution of RIP

The produced 5 core and 23 core array-core fibers having core common features can be used directly with a standard single-mode Single Fiber-coupled. So it can be easily used for optical fiber sensing or integrated devices. After a 5 array-core fibers splice with standard single mode optical fiber, the single mode fiber core aligned with central core a 5-pin optical fiber array. Optical excitation can be directly implemented, as shown in Figure. 5. The incident light Gauss field directly converted to 5-pin linear array core fiber mode field output. Seen from Figure.5, the conversion performance is good, so the production of a linear array of optical fiber can be used as the mode field converter.

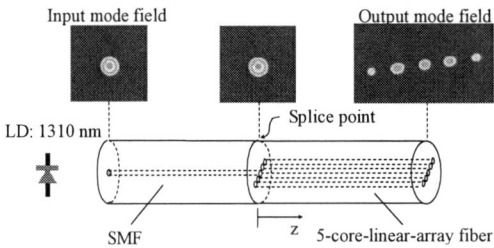

Figure 5. Inspirit 5-core-linear-array fiber by fusion splice

After single mode optical fiber and 23 core linear array fiber optical fields welding, a 23 core linear array fiber optical field output shown in Figure. 6. Figure.6 (a) is two-dimensional light field; Fig. 6 (b) is three-dimensional optical field map. As shown in Figure. 6, the optical power of fiber core is not uniformly distributed, which is due to the transmission distance different of optical fiber. It shows cyclical coupling curve. Here we use the coupled mode theory to explain this phenomenon.

978-1-4799-1215-5/13 $31.00 © 2013 IEEE   143

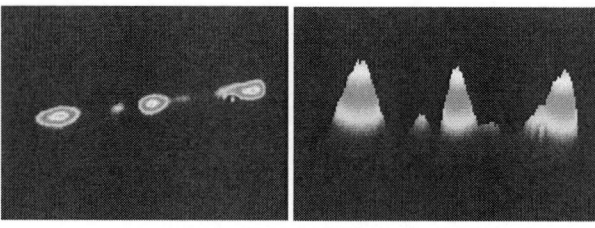

(a) two-dimensional map (b) three-dimensional map

Figure 6.   23 core-core fiber optical field linear array

The analysis assumes the 23 fiber cores with the same geometry and doping concentration. Each core is single mode waveguides, no loss, and the same distance between the neighboring cores, the core as a circular approximation. When the core interval is infinite in each isolated fiber core fundamental mode propagation constant $\beta_1 = \beta_2 = \beta_3 = \cdots = \beta_{23} = \beta_0$, if only to consider each of the core of the model is only coupled to the core, the adjacent core of coupling coefficient $\kappa_{12} = \kappa_{21} = \kappa_{n,n+1} = \kappa_{n+1,n} = \kappa$, according to the parallel waveguides coupled mode theory, the coupling between different cores can use the following mode coupling equations describe:

$$\frac{d}{dz}\begin{bmatrix} a_1(z) \\ a_2(z) \\ a_3(z) \\ a_4(z) \\ \vdots \\ a_{23}(z) \end{bmatrix} = -j\,\beta_m \begin{bmatrix} a_1(z) \\ a_2(z) \\ a_3(z) \\ a_4(z) \\ \vdots \\ a_{23}(z) \end{bmatrix} = -j \begin{bmatrix} \beta_0 & \kappa & & & & \\ \kappa & \beta_0 & \kappa & & 0 & \\ & \kappa & \beta_0 & \kappa & & \\ & & \kappa & \ddots & \ddots & \\ & 0 & & \ddots & \ddots & \kappa \\ & & & & \kappa & \beta_0 \end{bmatrix} \begin{bmatrix} a_1(z) \\ a_2(z) \\ a_3(z) \\ a_4(z) \\ \vdots \\ a_{23}(z) \end{bmatrix} \tag{1}$$

Where β_m ($m = 1, 2, \cdots, v$) is the new multi-core optical fiber propagation constant presence supermodel, v is the supermodel motif number, $a_i(z)$ ($i = 1,2,3....,23$) is the field amplitude at the inner fiber [4]. We analyze the coupling properties of the modes in multicore fiber to see how the power incident into one core is transferred to other cores in this section. The amplitudes of all eigenmodes are determined by the excitation field at $z = 0$. When lth core is excited at $z = 0$ by a unit electric field, the complex amplitude ($e_i(z)$) of the electric field of the ith core at a distance z is derived using Eq. (1) as follows

$$e_i(z) = \frac{2}{n+1} e^{-j\beta_0 z} \sum_{m=1}^{n} \sin\frac{ml\pi}{n+1} \cdot \sin\frac{mi\pi}{n+1} \cdot e^{-j\kappa z \cos(m\pi/(n+1))} \tag{2}$$

The normalized power in the individual core is obtained by

$$P_i(z) = e_i(z) e_i^*(z) \tag{3}$$

Figure. 7 show the normalized power in each core versus the propagation length when 12th core i.e. middle core is excited. The normalized power of the middle core in the proximity of 16.6mm is enhanced to 56% again when the 12th middle core is excited (see the inset of Fig.7). Though the coupling characteristic between the multicore is quite complex,

the intriguing properties above is instructive to novel fiber-based device, such as optical switch and coupler.

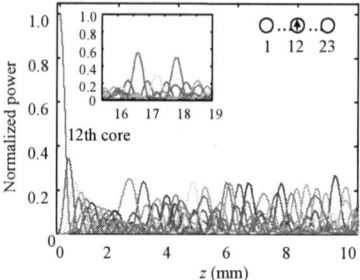

Figure 7.   Normalized power in each core versus the propagation length when 12th core is excited.

In addition, for example of the linear array of 23 cores optical fiber, we also studied the linear array core fiber supermodel coupling characteristics. As a first order approximation of the conditions, the 23 cores linear array fiber coupling between non-adjacent cores is much smaller than the coupling between the adjacent cores. So as an approximation, we consider only the coupling between two adjacent cores. The linear array of optical fiber core radius, $r_0 = 1.5$ μm, light wavelength 0.98 μm, the core distance d is 3.7 μm, set the initial power of the excitation light 1. According to the coupled mode theory and mode superposition theory [5], 23 core linear array of optical fiber core from left to right (see Fig 3 (b)) the incident power of the central core 1 and the other is 0. In this initial condition, solve the coupled mode equations eigenvalues and eigenvectors (according to the literature [6]). Based on the theory of linear superposition supermodel amplitude distribution [7]

$$E^v(x,y,z) = \left[\sum_m E_m^v E_m(x,y)\right]\exp(\lambda_v z) \tag{4}$$

Where $E^v(x,y,z)$ is the amplitude of the v supermodel, which E_m^v means that the eigenvectors m-th component of E^v, λ_v is the coupled mode equations for the first v eigen values. $E_m(x,y)$ is the core of the m-th transverse field distribution, and its expression is

$$E(x,y) = \sqrt{\frac{2}{\pi}} \frac{1}{\omega_0} \exp(-\frac{x^2 + y^2}{\omega_0^2}) \tag{5}$$

Where ω_0 is the mode field radius.

By Equation (1) shows, when the 23-core linear array optical fiber core refractive index is determined, the supermodel intensity distribution is only relate to the core interval, whereas independent of the length of the transmission z. Substitution equation (5) into equation (4), it can be calculated as a 23 core linear array fiber supermodel field which is shown in Figure 8. Each mode labeled with integer v, where v = 1 for the same phase supermodel (all core in same phase). show the near field plot of supermodel field of linear-

978-1-4799-1215-5/13 $31.00 © 2013 IEEE   144

core-array multicore fiber. As Fig.8 shown we can see, model number v sequencing increase with the propagation constant descending. Each co-phase mode of common core（v=1）has the same phase. It has the best beam quality. The working mode can improve the performance of fiber laser beam, which field outline is Gaussian-like distribution. Contrast, （when v=23）, the phase of adjacent core opposite to each other, the field outline is approximately sinusoidal.

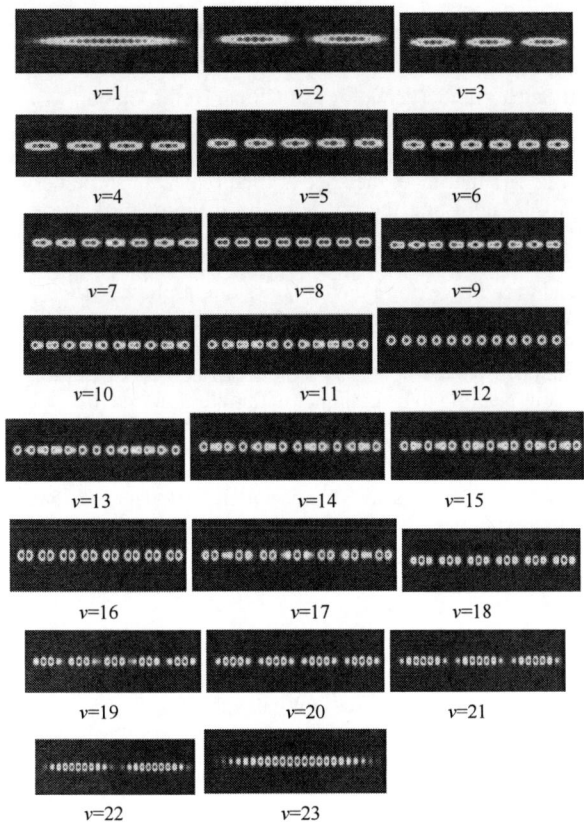

Figure 8.   The near-field emission patterns of linear-core-array fiber.

IV. CONCLUSION

This paper presents a convenient linear array multicore optical fiber preparation technology, which can greatly save multicore fiber preparation costs. The fabrication method results show that this kind of linear array multicore optical fiber can be used for optical mode field converter or optic coupler furthermore it can used in fiber integrated sensors. This paper calculates and gives results of a 23-core linear array optical fiber coupling characteristics and supermodel characteristics. This study can provide theoretical basis for a multi-core fiber or similar microstructure fiber design. It provides a reference for other similar microstructure fiber in the mode conversion and optical field coupling.

ACKNOWLEDGMENT

This work was supported by the Key Laboratory Program for in-Fiber Intergrated Optics of the Education Ministry of China, partially supported by the National Nature Science Foundation of China(No. 41174161, No. 61307005), and supported by the Fundamental Research Funds for the Central Universities of China to Harbin Engineering University and by China Postdoctoral Science Foundation Grant(2013M531013).

REFERENCES

[1]   Method of Manufacturing a Multicore Optical Fiber, United States Patent, Patent Number 5,792 ,233 , 1998

[2]   Multicore Glass Optical Fiber and Methods of Manufacturing such Fibers, United States Patent, Patent Number6,154 ,594 ,2000.

[3]   Method for Producing Parallel Arrays of Fibers, United States Patent, Patent Number7,209 ,616 B2, 2007

[4]   Peterka. P, Kasik. I., Kanka. J., Honzatko. P., et. al. Twin-Core Fiber Design and Preparation for Easy Splicing. IEEE Photonics Technology Letters, 2000, 12(12): 1656-1658P

[5]   Xiaoliang Zhu, Libo Yuan, Jun Yang, and Shouxiu Cao, 'Coupling model of standard single-mode fiber and capillary fiber," Appl. Opt., vol. 48, pp. 5624-5628, 2009.

[6]   Chunying Guan, Libo Yuan, Qiang Dai, and Fengjun Tian, "Supermodes Analysis for Linear-Core-Array Microstructured Fiber," J. Lightwave Technol. vol. 27, pp. 1741-1745, 2009.

[7]   J. A. Besley, J. D. Love, "Supermode analysis of fibre transmission," IEE Proc. –Optoelectron., vol. 144, pp. 411-419, 1997.

Fabricate phase-shifted fiber Bragg grating based on a piezoelectric ceramics

Lin Jiping[1,3]   Gui Yonglei[1,2]   Cao Zhigang[1]

1 Key Laboratory of Optic-electronic Information Acquisition and Manipulation Minister of Education, Anhui University, Hefei, China

2 Physical Sensors Technology Research Center of NO. 49th Research Institute of China Electronics Technology Group Corporation, Harbin, China

3 National Synchrotron Radiation Laboratory, University of Science and Technology of China, Hefei, China

Abstract—**A novel controlling means of introducing a given phase-shift in the fiber Bragg grating (FBG) with a uniform period phase mask was introduced, and at the same time a three-level mechanical displacement platform based on piezoelectric ceramic was developed in order to precisely control the motivation of the phase mask, which will ensure the given phase-shift to be lead into the exposure section of the fiber. By the experimental verification, the transmittance of the transmission peak in the stop band is enhanced markedly .In a conclusion, this is a simple process realizing phase shift fiber gratings.**

Keywords-component; phase-shifted fiber Bragg gratings; mechanical platform; phase mask;

I.   INTRODUCTION

The reflective and the transmissive properties of phase-shifted fiber Bragg grating have been well studied and used in many applications, such as dispersion compensators for optical fiber communication links, discrete sensors for strain and temperature measurements and narrowband wavelength selective elements for doped fiber lasers. There are two most commonly used fabrication methods of phase-shifted fiber Bragg grating. The one is fabrication thermal processing and another is stepped exposure method. The former is hard to accurately control the length of the thermal processing segments and core refractive index modulation depth, the latter mainly depends on the expensive precision stepper motor for most [1]. This article proposed a simple and practical method which is based on a piezoelectric ceramics (PZT). The simple method achieved phase-shifted fiber Bragg grating with the expectation of good performance and lower complexity.

II.   PHASE-SHIFTED FIBER BRAGG GRATINGS

A.   The phase-shifted fiber Bragg grating

The phase-shifted grating is a non-uniform period grating with discontinuous refractive index distribution. Its formation mechanism is that a phase shift is introduced at a specific part in a conventional optical fiber Bragg grating so as to generate two segments of gratings with different phases. The two-segment grating is similar to a wavelength selection resonator, which allows resonant wavelength to inject into the stop-band and open a narrow transmission window in the stop-band. Compared with a common FBG, a phase-shifted Bragg grating has many advantages, such as higher wavelength selectivity,

lower insertion loss and is independent of the polarization state etc. Filters that meet the special requirements can be designed easily by adjusting the position and magnitude of phase shift [2], which significantly promotes the applications in all-optical communication [3]. Especially in the dense wavelength division multiplexing (DWDM) system, a phase-shifted Bragg grating is used as a demultiplexer[4,5]. A phase shift FBG operates under the transmission mode (it acts as a band-pass filer not a band-stop filter) which can act as a demultiplexer without needing to be conjunct with other devices [6].

In short, the phase-shifted fiber gratings are superior to common FBGs. Therefore, they have obtained a wide range of applications in narrow-band filter, wavelength division multiplexing/demultiplexing, erbium-doped fiber gain flatness and single frequency fiber laser and other fields [7].

B.   Fabrication of phase shifted fiber gratings

FBG fabrication principles: Firstly, enhance the photo-sensitivity of the fiber core, for example, by way of germanium doped or hydrogen loaded, (Whatever technology to produce fiber Bragg gratings, the fiber core must be sensitive to the writing light). Then use ultraviolet light (usually 248nm) through the phase mask or amplitude mask to exposure the fiber, causing periodic modulation of the refractive index in the core, thereby forming an optical fiber gratings [8].

The phase shift gratings can be either directly produced or written on ordinary FBGs by post-processing methods. Direct production method refers the use of a phase mask which possess both phase shift and waveform structure [9]. This direct writing method is more easily controlled and more accurate compare with other writing methods. Another method is the optical fiber micro-displacement method, which is to write partial grating, then move the fiber to achieve the required optical phase shift and write the rest part of grating. This technique requires the interferometer to control the optical bench in order to precisely control the micro-displacement. The primary post-processing methods are: UV irradiation method [10], local heating method [11], ect. But the common weakness is that they are difficult to control [8].

III. PHASE MASK

A. Analysis of phase mask diffraction field

A phase mask is fabricated by electron beam lithography or holographic exposure etching, which is the one-dimensional periodic (the period is Λ_{PM}) transmission phase grating. Essentially, it is a specially designed optical diffraction component (see Fig.1).

Let the plane wave with unit amplitude and zero initial phase income into the phase mask at an angle of θ_i (the absorption of mask is negligible).When the width of the teeth is equal to that of the groove, the transmittance function of the phase mask can be given as follows:

$$T(z) = \begin{cases} \exp(i\varphi_1), \left(\left|z-(2J+1)\dfrac{\Lambda_{PM}}{2}\right| < \dfrac{\Lambda_{PM}}{4}\right) \\ \exp(i\varphi_2), \left(\left|z-J\Lambda_{PM}\right| < \dfrac{\Lambda_{PM}}{4}\right) \end{cases} \quad (1)$$

Where $J = 0, \pm 1, \pm 2, \cdots$ is the period number. φ_1 and φ_2 are the phase delays when the light pass through the tooth and the groove of the phase mask

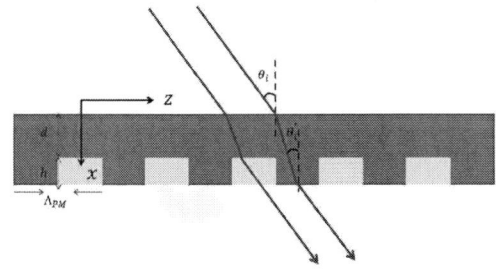

Figure 1. The phase delay generated when light passes through the phase mask

So the relative intensity of 0-order diffraction light and the m-order diffraction light are:

$$\begin{cases} I_0 = |c_0|^2 = \dfrac{1}{2}\left[1+\cos(\varphi_1-\varphi_2)\right] \\ I_m = |c_m|^2 = 2\dfrac{\sin^2(\frac{m\pi}{2})}{m^2\pi^2}\left[1-\cos(\varphi_1-\varphi_2)\right] \\ (m = \pm 1, \pm 2, \cdots\cdots) \end{cases} \quad (2)$$

The (2) shows that the high order diffraction light intensity is quite weak, usually only 0-order and ±1 order diffraction light are considered.

B. Normal incidence writing light

Grating equation at normal incidence is (3):

$$\sin\theta_m = m\frac{\lambda_W}{\Lambda_{PM}}. \quad (3)$$

λ_W is the writing wavelength ,obviously, when $\Lambda_{PM} < \lambda_W$, only 0-order diffraction light exits and does not produce interference fringes. Only when $\Lambda_{PM} > \lambda_W$, the normal incidence will produce high-order diffraction light and ±1-order diffraction light are symmetrical distributed with respect to incident light. i.e. $\theta_{+1} = \theta_{-1} = \theta$. (see Fig.2)

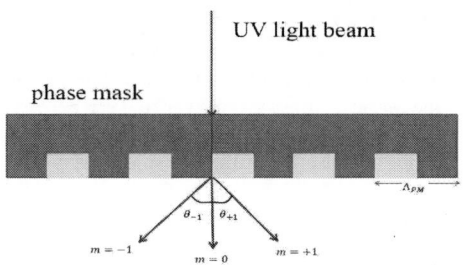

Figure 2. The diffraction light behind a phase mask at normal incidence

There exits 0-order extinction when h meets (4)

$$h = \frac{\lambda_W}{2(n_g-1)}. \quad (4)$$

h is the tooth height of a phase mask(see Fig.1), n_g is the refraction index of a phase mask[12]. At this point the intensity distribution of interference fringes is:

$$I(z) = 2A_1^2\left[1+\cos(2K\sin\theta_z)\right] \quad (5)$$

K is the wavenumber and θ_z is the diffraction angle behind the phase mask at z .Therefore, the period of the interference fringes can be expressed as:

$$\Lambda = \frac{\lambda_W}{2\sin\theta} = \frac{\Lambda_{PM}}{2} \quad (6)$$

Equation (6) shows that the period of the interference fringes is half of the period of phase mask at normal incidence.

C. The key parameters of the phase mask method

The distance between fiber and phase mask is a key parameter in experiments. The ±1-order diffraction light with a distance of y form interference fringes in the fiber. When the distance from the fiber to +1-order diffraction is equal to that to -1-order diffraction, the temporal coherence is not the key factor to decide a high contrast ratio. However, the two interfering beams increases with the increasing distance from fiber to the phase mask the distance between. Therefore, a good spatial coherence is a key factor to form the high contrast ratio of the interference fringes. When the increasing distance from fiber to the phase mask beyond the spatial coherence of the incident beam, the contrast of the interference fringes will deteriorate and eventually completely unable to form interference fringes. However, it will jeopardize the fine grating corrugated structure to place fiber close to the phase mask. Thus, the distance between fiber and phase mask is a key parameter in writing process, the distance from fiber to the phase mask is generally greater than tens of microns [1].

IV. DESIGN OF THE PHASE SHIFTED GRATING PREPARATION SYSTEM

A. Overall designs

The design is based on a phase mask method, by moving phase mask during the engraving to introduce a given phase shift in the optical fiber grating. The system consists of a mechanical displacement module and a PZT piezoelectric ceramic module, ensure accurate displacement of the phase mask.

B. Design of the system workbench

A workbench is designed to achieve one-dimensional displacement precision positioning, the platform consists of a coarse level & a fine level positioning mechanism and a PZT precision displacement bodies (see Fig.3). This platform can achieve 0 ~ 20mm range, 100nm theoretical resolution positioning. The fiber Bragg grating fixed on PZT realizes high-precision closed-loop test of bench displacement. XP 6 * 6/18 type piezoelectric ceramic was employed as the actuator in experiments and matched a dedicated driver power to ensure accuracy displacement. Graphite powder was used as a lubricant to reduce the sliding friction between the phase mask and the test platform.

It's the core idea of the experimental program that we translate the phase mask along the fiber axis to introduce a specific phase shift. And the three levels precision displacement platform just meets this requirement.

There exists an optimum working distance between the fiber and the phase mask (see paragraph C of part III). In actual operation, the fiber should be placed as close as possible to the phase mask as long as the phase mask will not be damaged. We use two three-dimensional adjustment brackets to replace the original fixed brackets as fiber fixtures. By adjusting the three-dimensional adjustment brackets we can make the fiber and the phase mask in the best working distance nearby. (see Fig.3).

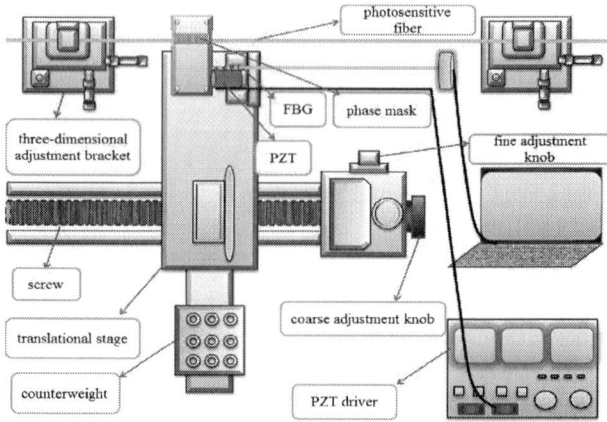

Figure 3. Complete phase shift grating production system diagram

C. Selection of a suitable phase mask

Table 1 shows the periods of three phase masks used in the laboratory. Since the adjustment precision of the mechanical displacement module is only 0.1um, we have to choose appropriate displacements to reduce the deviation. As can be seen from Table 1, the magnitude of the phase shift in each column increase from π/2 at an increment of π/2(the cases of 2nπ can be directly omitted). Phase Mask 1 has eight data available (see the data with bold font), so it should be selected.

After the second selection, the available data is as shown in the color parts of Table 1.The red zone (9π/2 area) shows the corresponding displacement of the introduction of π/2 phase shift; green zone (3π, 9π area) shows that of π phase shift; blue zone (3π/2,15π/2 area) shows that of 3π/2.

TABLE I. DISPLACEMENTS CORRESPOND TO PHASE SHIFTS(unit: um)

Mask Number	Mask Period	π/2	π	3π/2	5π/2	3π	7π/2	9π/2
1	1.0651	0.26628	0.53255	0.79883	1.331375	1.59765	1.86393	2.39648
2	1.0607	0.265175	0.53035	0.795525	1.325875	1.59105	1.856225	2.386575
3	1.0574	0.26435	0.5287	0.79305	1.32175	1.5861	1.85045	2.37915

Mask Number	Mask Period	5π	11π/2	13π/2	7π	15π/2	17π/2	9π
1	1.0651	2.66275	2.929025	3.461575	3.72785	3.99413	4.526675	4.79295
2	1.0607	2.65175	2.916925	3.447125	3.71245	3.977625	4.507975	4.77315
3	1.0574	2.6435	2.90785	3.43655	3.7009	3.96525	4.49395	4.7583

D. Experimental procedures

In experiment, an argon ion frequency doubling ultra-violet continuous laser with a wavelength of 248nm is employed to expose the germanium-doped optical fiber.

We employ the dedicated drive power to drive the PZT actuator, the driving voltage increases from 0 to 10V with a 10mv increment, while a fiber Bragg grating is stuck on PZT to measure micro-strain. Then obtain the magnitude of displacement via the terminal processing software.

E. Experimental Results

Fig.4 shows the transmission spectrum of a uniform fiber Bragg grating without phase shift. Its stop-band center is located in 1549.63nm and determined by (7), and the full width at half maximum (FWHM) is 0.1nm.

978-1-4799-1215-5/13 $31.00 © 2013 IEEE

$$\lambda_B(m) = \frac{2}{m} n_{eff} \Lambda \qquad (7)$$

Where n_{eff} is effective refraction index. Λ is the fringes period and is half of Λ_{PM} .[1,8].

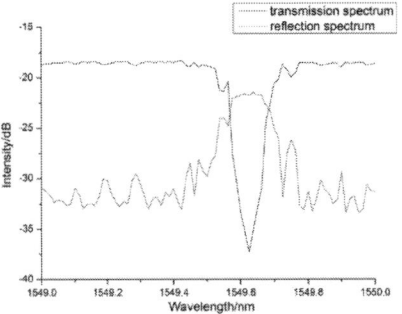

Figure 4.   The spectrum response of FBG

According to the coupled mode theory, when the light-induced refractive index modulation reached the amplitude of 10^{-4} the fiber grating written is Type I grating [13,14]. A notable feature of Type I grating spectrum is that its reflection spectrum and transmission spectrum are complementary and reflection bandwidth is at a magnitude of 10^{-1} nm(see Fig.4). Obviously, the fiber grating we obtained is just the Type I grating. The center wavelength can be obtained from (7).

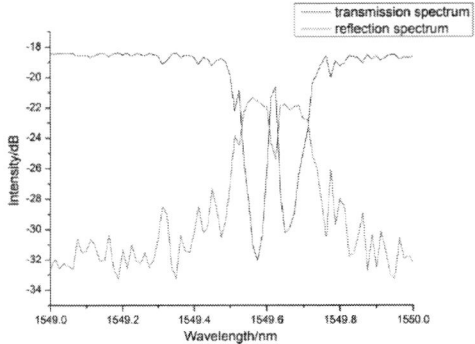

Figure 5.   The spectruml response after introducing a π phase shift

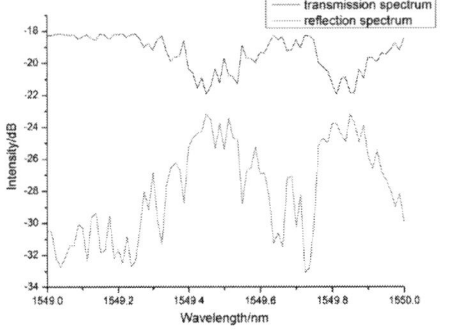

Figure 6.   The spectrum response with a tilted phase shift FBG

When introduced a π phase shift, as is shown in Fig.5, an extremely narrow transmission peak appears in the stop-band, which has a central wavelength of 1549.63nm and a FWHM of 0.04nm.

F.   Experimental advantageous

We introduce three dimensions regulation brackets into the system. Therefore, it's easy to adjust the distance and the parallelism between the phase mask and the fiber. When the parallel between the phase mask and the fiber is not controlled well, the interference field on the fiber will become vague. Fig.6 shows the transmission and reflection spectrum of the fiber grating fabricated in this interference field, which is called titled fiber grating or blazed fiber grating [15,16] .The system can avoid that case in fabricating phase-shifted fiber gratings.

V.   CONCLUSION

The method proposed can achieve the phase-shifted grating. The experimental platform designed is on the basis of the existing equipment and omits the complexity of the underlying platform, thus achieves phase-shifted grating with a relatively simple way. Experiments have proved that the system has good stability, with relatively simple operation and is ideal to fabricate phase-shifted fiber Bragg gratings.

REFERENCES

[1]   Turan Erdogan,Fiber Grating Spectra [J].JOURNAL OF LIGHTWAVE TECHNOLOGY, 1997,15(8).

[2]   Li Wei, JOHN W. Phase shifted Bragg grating filter with symmetrical structures [J]. Light-wave Technol,1997,15(8):1405-1410.

[3]   GILES C R. Lightwave applications of fiber Bragg gratings[J]. Joumalof Lightwave Technology,1997, 15(8):1391-1404.

[4]   KIM S Y, LEE S B. Channel switching active add/dropMultiplexer with tunable grating[J]. Electron Lett, 1998,34: 104-105.

[5]   AGRAWAL G P, DUTTA N K. Semiconductor lasers[M]. NewYork: Van Nostrand Reinhold, 1993.

[6]   Zhang fang,Li Qianghua Analysis for reflected spectrum of cascaded phase-shifted fiber grantings [J]., Natural Science Journal of Harbin Normal University 2007,23(2).

[7]   Lu Shaohua,Xu Ou,Jian Shuisheng.Analysis for the Reflective Spectrum Characteristics of Phase-Shifted Fiber Gratings [J]. CHINISE JOURNAL OF LASER,2008,35(4).

[8]   Rao Yunjiang,Wang Yiping,Zhu Tao. **The principles and applications of fiber grating** [M].BeiJing:Science Press 2006: 83-84.

[9]   kashyap, Mckee P F, Armes D.UV written reflection grating structures in optical fibers using phase-shifted phase-masks[J].ElectronLetters,1994,30:1977~1978.

[10]   CanningJ, Sceats M G. Pi-phase-shifted periodic distributed structures in optical fibers by UV post-processing[J]. ElectronLetters. 1994, 30:1344~1345.

[11]   Uttamchandani D, Othonos A. Phase shifted Bragg gratings formed in optical fibers by post-fabrication thermal processing[J]. Optics Communications. 1996,127:200-204.

[12]   Li Chuang,Zhang Yimou,Zhao Yonggui.Fiber graing principles techniques and sensing applications[M]. BeiJing:Science Press, 2005: 53-57.

[13]   Dyer P E, Farley R J, Giedl R. Analysis of Grating Formation With Excimer Laser Irradiated Phase Masks[J].Optics Communications,1995,115:327.

978-1-4799-1215-5/13 $31.00 © 2013 IEEE   149

[14] S.P.Yam1,Z.Brodzeli1,B. P.Kouskousis1,C. M. Rollinson1. A -phase-shifted Fiber Bragg Grating Fabricated using aSingle Phase Mask [J]. IEEE , 2009 ,978-1-4244-4103-7/09.

[15] T.Erdogan,J.E.Sipe   Tilted   fiber   phase   gratings [J].Opt.Soc.Am.A/Vol.13,No.2/February 1996

[16] Seungin Baek,Yoonchan Jeong,and Byoungho Lee Characteristics of short-period blazed fiber Bragg gratings for use as macro-bending sensors[J].Applied Optics 1Februray 2002/Vol.41.No.4

Research of Effect of Underwater Backscattering On the Detection of Sea Wave Using a Slit Streak Tube Imaging Lidar

Jian.Gao, J.F.Sun, Qi.Wang

National Key Institute of Tunable Laser Technology, Harbin Institute of Technology,
Harbin 150001, China
Email:59122892@163.com

Abstract—**Slit Streak Tube Imaging Lidar (STIL) is a promising imaging system as its high frame rate and good image quality. It can output the 4-D image (3-D range image + 1-D intensity image). And its functionality demonstrated for imaging of short scale ocean surface waves. Accuracy of sea-wave height measurement requires high resolution range supplied by this system. But the backscattered of sea water has an effect on the extraction of range and intensity information. In this paper, according to STIL, we deduce the signals of energy of reflected and backscattered light and analysis the effect on the extraction of range and intensity information.**

Keywords- Backscattering; Slit Streak Tube; Lidar; Sea Wave Imaging

I. INTRODUCTION

The ocean surface wave field plays an important role in the oceanographic and coastal measurements[1]. Especially, the short scale waves whose wavelength and height are less than 10cm contain more information (currents, underwater vehicle, shoals of fish, sea wind) [2].

Flash imaging lidar is a promising imaging system as its high frame and spatial resolution. It can image the whole area of sea fast [3]. Especially, the flash imaging lidar using a streak tube camera as a receiver can directly provide the range and intensity information with high resolution [4]. It can get time of different photons reflected from sea surface through deflecting voltage below highly sensitive photocathode. According to the time and the quantity of photons, it can directly supply range and intensity information so as to provide 3D imaging of short scale surface wave field. According to the intensity information, we can recognize the surface blistering and films of oil. Its time resolution is in the region of a picosecond and its range accuracy is about centimeter-level. According to the wave height, it is feasible for Streak Tub Imaging Lidar(STIL) to detect short scale ocean surface waves.

But when the light reaches the sea surface, it will continue to travel in the sea water. The backscattering of sea water causes errors in the height measurement of short scale wave.

In this paper, we analyze the streak imaging of the short scale ocean waves. Through the simulation of the echo light signal, we get the expression of energy of received light. And we analysis the effect on the extraction of range and intensity information.

II. IMAGING THEORY OF STIL

A. Streak tub imaging lidar(STIL)

STIL can identify photons of different time with high sensitivity and record the time and photon energy. It is used as a receiver that the reflected light from the ocean surface is imaged onto a slit in front of the streak tube photocathode by a conventional lens, and the time (range) is resolved by electrostatic sweep within the streak tube, that generating a 2-D range-azimuth image on each laser pulse [5,6], as is shown in Fig. 1. By orienting the fan beam perpendicular to the vehicle track, the along-track dimension is sampled by adjusting the Pulse Repetition Frequency (PRF) of the laser to the forward speed of the vehicle, thus sweeping out the three-dimensional ocean volume.

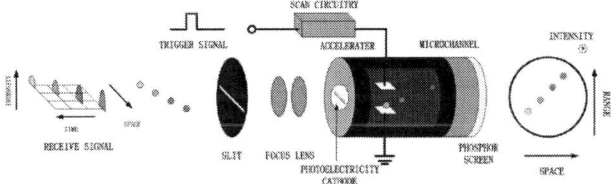

Figure 1. The imaging principle of streak tub detector

Through receiving the reflected light from the ocean surface, we get the streak imaging of sea waves. According to extraction algorithm of range and intensity information, we can also get the 3D-imaging of ocean surface waves and 2D intensity imaging of the sea surface waves field [7]. As is shown in the Fig. 2, picture (A) is the map of experimental site. Picture (B) is the real streak imaging of sea waves that our workgroup get near Liugong Island in the Yellow Sea, China, 2011. Picture (C) is a column of the streak imaging. But we can see the echo does not accord with Gauss distribution. So it is not accurate that getting the range and intensity information of sea surface through the peak of the echo.

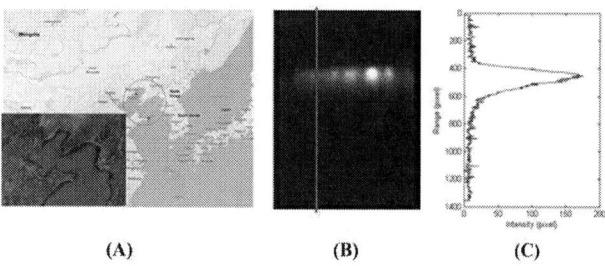

(A) **(B)** **(C)**

Figure 2.   Streak imaging of ocean waves using STIL. (A) The map of experimental site. (B) A piece streak imaging of the building. (C) A column of the streak imaging.

B. Imaging model of ocean waves

According to the principle of Streak Tube Imaging Lidar, we build our experiment platform using Slit Streak Tube Imaging Lidar. As is shown in Fig 3, it is the photograph of Slit Streak Tube Imaging Lidar system.

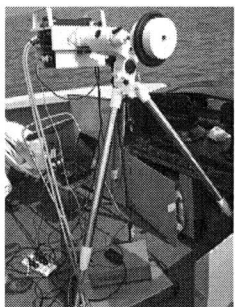

Figure 3.   The Slit Streak Tube Imaging Lidar system

We suppose that the laser pulse is Gaussian beam, the range between STIL and reflection surface is S, and the time when the light reaches the water surface is t_0, as is shown in Fig. 4 (A).

$$t_0 = \frac{S}{c} \qquad (1)$$

At this moment, the laser pulse is:

$$E(t) = E_0 e^{-\frac{(t-t_0)^2}{\sigma^2}} \qquad (2)$$

where σ is the half of pulse width.

When the target is a building, the light is reflected. If the reflectivity is ρ and the atmospheric attenuation is not considered, the received echo signal is:

$$E_1(t) = \rho E_0 e^{-\frac{(t-2t_0)^2}{\sigma^2}} \qquad (3)$$

As is shown in Fig.4(B), in the peak of the echo, we can get the range S_1, and the energy E_1 .which can be written as

$$S_1 = \frac{c \cdot 2t_0}{2} = c \cdot t_0 = S \qquad (4)$$

$$E_1 = \rho E_0 \qquad (5)$$

The intensity is in proportion to the reflected energy E_1. Here is the range and intensity imaging extraction method of reflection.

When the imaging target is sea surface, the light is partial reflected. As is shown in Fig.4(C), if the reflectivity is ρ_1 and the atmospheric attenuation is not considered, the reflected light signal is:

$$E_2(t) = \rho_1 E_0 e^{-\frac{(t-2t_0)^2}{\sigma^2}} \qquad (6)$$

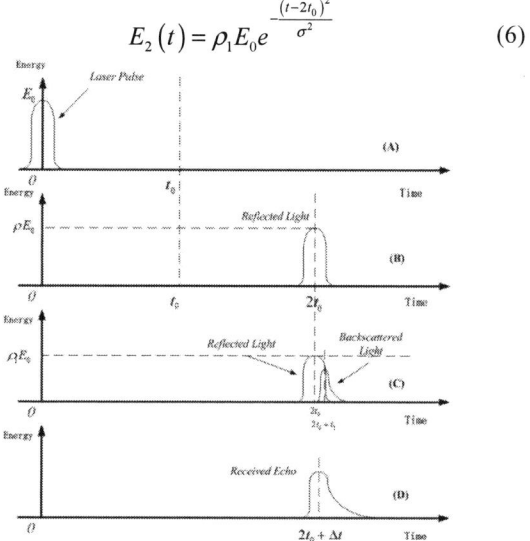

Figure 4.   Relationship between echo energy and time. (A) Energy of laser. (B) Echo energy reflected by building. (C) Echo energy reflected and backscattered by seawater. (D) Echo energy of sea water received by STIL

C. Effect of back scattering

We assume that the energy reaching the sea surface is $E_{es}(t)$, and the energy of reflected light is $E_{er}(t)$. Then they can be written as

$$E_{es}(t) = E_i(t)\exp(-kH) = E_0 \exp\left(-\frac{(t-t_{S1})^2}{\sigma^2}\right)\exp(-kH) \qquad (7)$$

$$E_{er}(t) = rE_{es}(t) = rE_0 \exp\left(-\frac{(t-t_{S1})^2}{\sigma^2}\right)\exp(-kH) \qquad (8)$$

$$t_{S1} = \frac{H}{C} \qquad (9)$$

where H is the range between STIL and sea surface; r is the reflectivity of sea surface, t_{S1} is the time when the light travels from the STIL to sea surface.

978-1-4799-1215-5/13 $31.00 © 2013 IEEE   152

Laser penetrates the sea surface, and it travels in the sea water. We assume the velocity of light in the sea water is C_w. Then the energy of laser ($E_w(t)$) when it reaches the area with the deep of L in the sea water can be written as

$$E_w(t) = (1-r)E_0 \exp(-kH)\exp(-c_1 L)\exp\left(-\frac{(t-t_{S2})^2}{(\sigma n)^2}\right)$$

(10)

$$t_{S2} = \frac{H}{C} + \frac{L}{C/n} = \frac{H+nL}{C}$$

(11)

Where c_1 is attenuation coefficient of seawater, n is refraction coefficient.

When the laser travels in the sea water, it will be scattered in any directions. The receiver can catch the scattered light only along the opposite direction of the incident light transmission. So we care about the backscattered light only.

When the laser reaches the area with the deep of L in the sea water, the energy of backscattered light ($E_{ws}(t)$) can be written as

$$E_{ws}(t) = (1-r)E_0 \exp(-kH)\exp(-c_1 L)\beta(\pi)\exp\left(-\frac{(t-t_{S2})^2}{(\sigma n)^2}\right)$$

(12)

Where $\beta(\pi)$ is the backscattering coefficient.

So the energy of light reflected by sea surface in the receiver ($E_{rr}(t)$) can be written as

$$E_{rr}(t) = rT_1 T_2 E_0 \exp\left(-\frac{(t-2t_{S1})^2}{\sigma^2}\right)\exp(-2kH)$$

(13)

Where T_1 is the transmission when the light travels from seawater to air; T_2 is transmission of optical system.

The energy of backscattered light in the receiver ($E_{rb}(t)$) can be written as

$$E_{rb}(t) = \int_0^\infty (1-r)T_1 T_2 E_0 \exp(-2kH)\exp(-2c_1 L)\beta(\pi)\exp\left(-\frac{(t-2t_{S2})^2}{(\sigma n)^2}\right)dL$$

(14)

So the total energy of light captured by receiver is:

$$E_{rt}(t) = E_{rr}(t) + E_{rb}(t)$$

(15)

$$E_{rt}(t) = rT_1 T_2 E_0 \exp\left(-\frac{(t-2t_{S1})^2}{\sigma^2}\right)\exp(-2kH)$$
$$+\int_0^\infty (1-r)T_1 T_2 E_0 \exp(-2kH)\exp(-2c_1 L)\beta(\pi)\exp\left(-\frac{(t-2t_{S2})^2}{(\sigma n)^2}\right)dL$$

(16)

TABLE I. PARAMETER OF SIMULATION OF UNDERWATER BACKSCATTERING

$E_0 = 20 mJ$	$c = 0.3/m$
$k = 0.17/km$	$H = 10m$
$n = 1.33$	$\sigma = 10ns$
$r = 0.02$	$T_1 = 0.95$
$\beta(\pi) = 0.026$	$T_2 = 0.8$

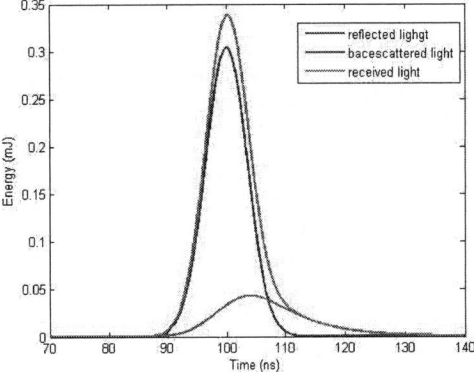

Figure 5. Simulation of the reflected and backscattered light

As is shown in the Fig. 5, the blue line shows the reflected light from sea surface. The green line shows backscattered light underwater. And the red line shows the light energy catch by the receiver. The peak time of blue line (reflected light) is 100ns. The peak time of red line (received light) is 100.17ns.

The distance between STIL and the sea surface is H_0.

$$H_0 = \frac{100ns \times 3\times 10^8 m/s}{2} = 15m = H$$

(18)

But the range received is H_1

$$H_1 = \frac{100.17ns \times 3\times 10^8 m/s}{2} = 15.26m > H$$

(19)

The range error caused by the sea water backscattering is 1.7%. We consider that choosing the multiple Gauss fitting method to simulate the reflected and backscattered light, so as to extract the range and intensity information from the reflected light.

As is shown in Fig. 5, the blue line shows the backscattered light energy in the sea water. The Equation (14) can be written as

$$E_{rb}(t) = \int_0^\infty (1-r)T_1 T_2 E_0 \exp(-2kH)\exp(-2cL)\beta(\pi)\exp\left(\frac{(t-2t_{S2})^2}{(\sigma n)^2}\right)dL$$
$$= \frac{4(1-r)T_1 T_2 E_0 \exp(-2kH)\beta(\pi)}{\sigma^2 C^2}\exp\left(-\frac{(\sigma^2 C^2 c_1 - 4Ct)(\sigma^2 C^2 c_1 + 8H)}{16n^2}\right)$$
$$\cdot \int_0^\infty \exp\left(-\left(L+\frac{\sigma^2 C^2 c_1 - 2Ct + 4H}{4n}\right)^2\right)dL$$

(20)

We assume that

$$A = \int_0^\infty \exp\left(-\left(L + \frac{\sigma^2 C^2 c_1 - 2Ct + 4H}{4n}\right)^2\right) dL \qquad (21)$$

With the deep (L) increased, the value of A approaches to a constant. So the backscattering energy nearly approaches to a negative exponential function. So the first part of total energy of light captured by receiver is a Gauss function, and the second part is a negative exponential function.

III. EXPERIMENT OF SEA SURFACE IMAGING

Our work team do the experiment of sea waves imaging in the Bohai Sea, where is near Huludao City, Liaoning Province from August, 2012. We do the experiment using STIL carried by the ship.

Using STIL, we can get short scale sea waves field imaging. Through sea wave forms, we can retrieval wind field on the sea surface. Through reflected intensity by sea surface, we can identify the floater on the sea surface. It has great advantage on remote sensing, sea surface imaging, target detecting…and other applications.

As is shown in Fig. 6, there is the imaging of ocean surface, we get in Bohai sea through Gauss and exponential fitting extraction method to eliminate the effect of water backscattering and improve the accuracy. Picture (A) is the 3D-imaging of ocean surface waves. Picture (B) is the 2D intensity imaging of the ocean surface waves. The area is about 300m×1500m. Through the color in the 3D-imaging, we can get the wave height (range). And through the 2D intensity imaging, we can get the reflection of ocean surface. In the Fig. 6 (B), we can see the foam with high reflection.

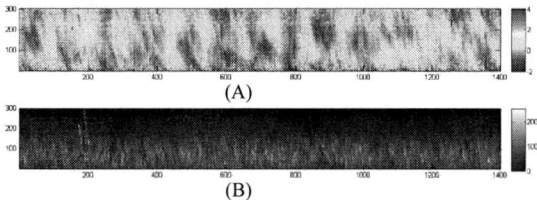

Figure 6.   3D-imaging and 2D intensity imaging of real ocean surface waves.

IV. CONCLUSION

According to the wave height of short scale waves, STIL has great advantage in imaging the ocean surface waves field because of its high range resolution. It can image the ocean surface fast, so as to improve efficiency for detection. We get echo energy reflected by ocean surface and underwater back scattered. According to the expression of the energy of underwater backscattering, we choose the extraction method of range and intensity information that based on Gauss and exponential fitting to eliminate the effect of water backscattering and improve the accuracy. Through this method, we get the 3D-imaging and 2D intensity imaging of ocean wave field from the real streak imaging in Bohai sea, China.

ACKNOWLEDGMENT

This research is funded by National Natural Science Foundation for Young Scholars (60901046).

REFERENCES

[1] Sun J, Wang Q. 4-D image reconstruction for streak tube imaging lidar[J]. Laser physics, 2009, 19(3): 502-504.

[2] Liu J, Wang Q, Li S, et al. Research on a flash imaging lidar based on a multiple-streak tube[J]. Laser physics, 2009, 19(1): 115-120.

[3] Wei Jinsong, Chen Yuanli, Xu Qiang, "Imaging by single-slit streak tube laser lidar," Chinese J. Lasers, 35(4), 496-500 (2008).

[4] Hongru Yang, Lei Wu,* Xiaopeng Wang, et al. Signal-to-noise performance analysis of streak tube imaging lidar systems. I. Cascaded model [J]. Appl. Opt. 2012, 51(36):8825-8835.

[5] Sining Li, Qi Wang, Jinbo Liu, "Research of range resolution of streak tube imaging system," SPIE, 2007, 6279-6280.

[6] Jian Gao, Jianfeng Sun, Jingsong Wei, Qi Wang, "Research of underwater target detection using a Slit Streak Tube Imaging Lidar", Optoelectronics and Microelectronics Technology (AISOMT), 2011 Academic International Symposium, Digital Object 2011 , 240-243.

[7] Qi Wang, Jian Gao, Sun J.F.,Wei J.S., "A new method of detection of short scale ocean waves using a slit Streak Tube Imaging Lidar", Optoelectronics and Microelectronics (ICOM), 2012 International Conference,182-184.

978-1-4799-1215-5/13 $31.00 © 2013 IEEE

Fitting Operate Mode for Incoherent Mie Doppler Wind Lidar

Jun Du, Xiang Yang Cheng, Yan Chen Qu*, Wei Jiang Zhao, De Ming Ren, Zhen Lei Chen, Li Jie Geng

National Key Laboratory of Tunable Laser Technology, Harbin Institute of Technology, Harbin, China
quyanchen@hit.edu.cn

Abstract—**For direct detection Mie Doppler wind lidar, the method of use Fabry-Perot interferometer as a dynamic Doppler shift discriminator is proposed and investigated theoretically. For the highest measuring sensitivity part of the FPI transmission curves is utilized to measure any possible laser Doppler shift, its measuring accuracy is very high. With no loss of this significant superiority, any actual wind speed range requirement can be met by repeating this way in a measurement. The simulations for different atmosphere conditions prove that this method is more suitable for wind measurement of higher part troposphere than previous.**

Keywords-wind Lidar; Doppler shift, measuring accuracy; double-edge technique; Fabry-Perot interferometer

I. INTRODUCTION

The high accuracy atmosphere wind data are the most important parameters for improving forecasting, climate studies, hurricane tracing, and so on. The Doppler wind lidar (DWL) has demonstrated its capability for these parameters measurement [1-3]. Two kinds of detection mechanisms have been investigated and applied for it, the coherent (or heterodyne) detect [4, 5] and incoherent (direct) detect [6]. For the aerosols backscatter dependence of the coherent DWL systems, it is limited to the lower troposphere. For incoherent detection technique can be extended to the Raleigh backscatter from air molecules for the whole atmosphere wind observation [7-9], more advances and discussions have been concentrated on it over the past decade [10-12]. The main kind of incoherent DWL system is utilizing the double-edge technique based on the Fabry-Perot interferometer (FPI) [13-16]. This kind DWL system with FPI can be operated at the wavelength of 1064 nm or 355 nm for lower or higher part atmosphere wind measurement with Mie or Raleigh backscatter respectively. The FPI system with design flexibility allows the desensitization of either molecular or aerosol for wind measurement of different heights respectively [17]. But the edge technique is not perfect, for example, in the Mie DWL system, the measuring accuracy decreases evidently with the measured value increasing. In addition, the measuring range is fixed and narrow for the measuring accuracy limit. The purpose of this paper is to propose a different working way for the FPI double-edge structure in incoherent DWL to improve its measuring accuracy and range for the wind.

II. BACKGROUND OF THE DOUBLE-EDGE MIE DWL

In the incoherent Mie DWL, the backscatter light Doppler shifted due to the atmosphere bulk movement is split into two beams. One beam with small percentage of energy goes into the energy detector. The other as the signal light is incident on the double-channel FPI structure in which different cavity spaces are used to construct the double-edge Doppler frequency discriminator. The frequency of outgoing laser is locked at the crossing point of two opposite transmission curves of this double-channel FPI. The differential measurement of the transmittances for outgoing laser and atmospheric backscatter signal can be used to retrieve the Doppler shift, which can be used to determine the wind speed in the line of sight (LOS) of the outgoing laser [1, 2].

Assume the surface defect of the FPI to follow the Gaussian probability distribution simply, the transmission function of FPI can be shown as fallow [18]

$$h_1 = \frac{(1-R-A)^2}{1-R^2}\left\{1+2\sum_{n=1}^{\infty}R^n\cos(n\varphi)\exp\left[-\frac{1}{4}n^2\varphi^2\left(\frac{\Delta d_D}{d_0}\right)^2\right]\right\} \quad (1)$$

In Eq. (1), $\varphi = 4\pi\mu d_0\nu_0\cos\theta_0/c$, where μ is the refractive index between the FPI plates, d_0 is the plate spacing, ν_0 is the frequency of the incident light, θ_0 is the incident angle of light between plates, A denotes any scattering and absorptive losses in the FPI plates, R is the plate reflectivity, c is the light velocity in air, and Δd_D denotes the FPI surface defect.

The FPI is usually illuminated by light collimated with certain angular field of view θ_M (semi-angle). Therefore, the transmission function can be written as

$$h_2 = (2/\theta_M^2)\int_0^{\theta_M}\sin(\theta)h_1(\nu_0,\theta)\,d\theta$$

(2)

Both the aerosol and molecule backscatters spectra, $f_M(\nu)$ and $f_R(\nu)$, can be expressed with Gaussian profiles for simplicity. Because the Brownian motion of aerosol particle does not broaden the aerosol spectrum significantly, it can be replaced by the line width of outgoing laser $\Delta\nu_L$. Hence, $f_M(\nu)$ is replaced by the outgoing laser spectrum $f_L(\nu)$ in this paper. The line width of the atmosphere Raleigh spectrum is $\Delta\nu_R = (32kT\ln 2/\lambda^2 M)^{1/2}$, where k is the Boltzmann constant, T is the atmosphere temperature, M is the averaging mass of the atmospheric molecules, and λ is the laser wavelength.

978-1-4799-1215-5/13 $31.00 © 2013 IEEE

$$
\begin{cases}
T_{\mathrm{Mi}} = \displaystyle\int_{-\infty}^{\infty} h_2(v_0 - v') f_{\mathrm{L}}(v')\,\mathrm{d}v' \\[2mm]
T_{\mathrm{Ri}} = \displaystyle\int_{-\infty}^{\infty} T_{\mathrm{Mi}}(v_0 - v') f_{\mathrm{R}}(v')\,\mathrm{d}v'
\end{cases}
\tag{3}
$$

The transmission function of the FPI two channels (noted as $i = 1, 2$) for Mie and Raleigh backscatter are written as convolution functions [2]

The transmitted photon number measured on two FPI channels and energy channel are

$$
\begin{cases}
N_i = a_i\left[N_{\mathrm{M}} T_{\mathrm{Mi}}(v) + N_{\mathrm{R}} T_{\mathrm{Ri}}(v) \right] \\[2mm]
N_{\mathrm{E}} = a_3\left(N_{\mathrm{M}} + N_{\mathrm{R}} \right)
\end{cases}
\tag{4}
$$

where a_i and a_3 are calibration constants, N_{M} and N_{R} are total Mie and Raleigh backscatter photon number respectively. So Mie complement $N_{ic} = a_i N_{\mathrm{M}} T_{\mathrm{Mi}}(v)$ measured by signal channel i can be obtained from the basic measured signals N_1, N_2 and N_{E}, its variance is given as

$$
\sigma_{N_{ic}} = c_1^2 N_1 + c_2^2 N_2 + c_3^2 N_{\mathrm{E}}
\tag{5}
$$

The expressions of parameter in Eq. (5) see reference [19]. For the ratio of two Mie complements $N_{1c}/N_{2c} = T_{\mathrm{M1}}(v)/T_{\mathrm{M2}}(v)$ when $a_1 = a_2$, it can be used to retrieve the Doppler shift.

The measuring uncertainty in the LOS wind speed is given as [19]

$$
\varepsilon = \frac{1}{(S/N)\Theta}
\tag{6}
$$

where S/N is the signal-to-noise ratio (SNR) for the double-channel system

$$
S/N = \left[\left(\sigma_{N1c}/N_{1c} \right)^2 + \left(\sigma_{N2c}/N_{2c} \right)^2 \right]^{-1/2}
\tag{7}
$$

and measurement sensitivity Θ is the fractional change of the measuring metric $R(v) = T_{\mathrm{M1}}(v)/T_{\mathrm{M2}}(v)$ for a unit wind velocity v in the LOS

$$
\Theta = (2/\lambda)\left[1/R(v)\right]\left[\mathrm{d}R(v)/\mathrm{d}v\right]
\tag{8}
$$

III. FITTING WORKING WAY FOR THE FPI DOUBLE-EDGE TECHNIQUE

In the double-edge system of incoherent Mie DWL, the transmission curves of FPI for Mie and Raleigh backscatters are calculated with Eq. (3), and shown in Fig.1(a), with the $\lambda = 1064\text{nm}$, $R = 0.866$, $A = 0.0025$, $\theta_{\mathrm{M}} = 0.5\text{mrad}$, $\Delta v_{\mathrm{L}} = 90\text{MHz}$, $d_1 \approx 42.86\text{mm}$, $d_2 \approx 42.86 - 3.4 \times 10^{-5}\text{mm}$, $T = 273\text{ K}$. The design has taken into consideration of the possible wind velocity range in the LOS. When the frequency of outgoing laser is fixed at the

relative frequency location 0MHz , the frequency shift measuring dynamic range of backscatter light is $-110 \sim 110$ MHz (about $-58.5 \sim 58.5\,\text{m/s}$ wind velocity in the LOS). For the shot noise limited case, its measuring uncertainty curve is calculated by Eq. (6) and shown in Fig.1 (b), under the assumptions of Mie and Raleigh backscatter ratio $Ra = 2$ and total backscatter photon number $N_0 = 5000$. In Fig.1 (b), it can be found that the measuring uncertainties ($2.8 \sim 9.7\,\text{m/s}$) are not the same in the measuring range. Its minimum is at the zero Doppler shift and its value increases with the Doppler shift increasing.

With the frequency of outgoing laser locked at the cross-point on the FPI transmission curves, the FPI is used as a static filter to retrieve the Doppler shift. As a result, wide enough FWHM and interval of FPI transmission peaks are necessary to increase the measured dynamic range for possible wind measurement requirement, while the wider measuring range and the lower measuring sensitivity. In most cases, the LOS wind velocity is small and may not reach this wide range, so this is a kind of waste for the measuring accuracy in a certain degree. Moreover, at some extremely heavy weather condition, for example the Category five hurricanes ($\geq 70\text{m/s}$), this fixed measuring range will not be enough at all. And the characteristic of measuring accuracy decreasing with the wind velocity increasing seriously, as shown in Fig.1 (b), may make it difficulty for the high wind velocity measurement to reach the accuracy requirement.

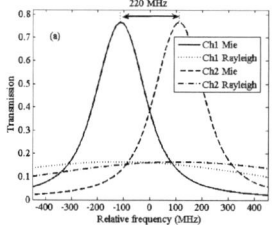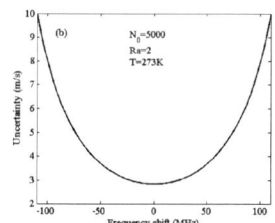

Figure 1. (a)Transmission spectra of the double-edge FPI system for Mie and Rayleigh components in backscatter signals (b) Measuring uncertainties of the double-edge FPI system

Based on the discussion above, we suggest dynamic adjusting outgoing laser frequency relative location on the transmission curves of the FPI for always using the highest measuring accuracy point (the cross-point of curves) or small range around it to measure any wind velocity value, the defect of the edge technique will be made up. We name this method as the fitting double-edge technique. For comparison simplicity, we call the previous way as the fixed double-edge technique.

A. One-point fitting double-edge technique

Because the fitting double-edge technique does not need such wide FWHM and interval of the FPI transmission peaks as the fixed type for measuring dynamic range requirement, these two parameters of the double-channel FPI can be designed smaller for higher measuring sensitivity by increasing its thickness. The thicknesses of the FPI two channels are designed to be about $d_1 \approx 82.86\text{mm}$,

$d_2 \approx 82.86 - 4.2 \times 10^{-5}$ mm respectively (the parameters may not be optimum, because the optimization of the FPI parameters for this type is not the purpose of this paper). The transmission curves of these two channels for Mie and Raleigh backscatter are calculated and shown in Fig.2 (a), assuming that the other parameters of the FPI and atmosphere do not change.

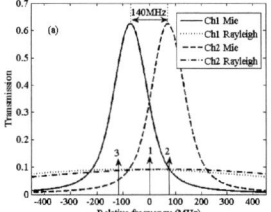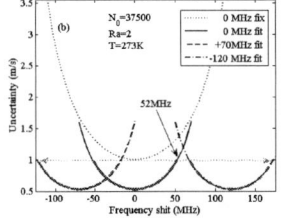

Figure 2. (a) Transmission spectra of the fitting double-edge FPI system for Mie and Rayleigh components in backscatter signals (b) Measuring uncertainty curves of the double-edge FPI system for fixed and one-point fitting types

Under the assumptions of total return photon number $N_0 = 37500$ and outgoing laser frequency at relative locations 0MHz, 70MHz and −120MHz on the coordinate axis respectively (corresponding to1, 2and 3 points in Fig.2 (a)), the measuring uncertainty curves of the fitting double-edge technique are calculated and shown in Fig.2 (b) with different line types. For the comparison, the uncertainty curve of the fixed type is also calculated with the same photon number and expressed with dot line in the same figure. It can be found, from Fig.2 (b), that only the 0MHz Doppler shift measured value of the fixed type reaches 1m/s accuracy requirement, while that of the fitting type (with outgoing laser at 0MHz location) in the range of $-52 \sim 52$MHz (corresponding to $-27.7 \sim +27.7$ m/s wind in LOS). In many cases, the wind velocity of LOS is around 0m/s with a small range so it is enough to use the fitting type with the outgoing laser frequency at 0MHz location. If the LOS wind velocity is out of this measuring range, for example about $-27.7 \sim 27.7$m/s around -37.2m/s (corresponding to −70MHz Doppler shift). We can tune the outgoing laser frequency to be at 70MHz, and the frequency of backscatter laser will locate around the cross-point of the FPI transmission curves. The measuring uncertainties of it are expressed with dash line in Fig.2 (b), with no more differences than that of outgoing laser at 0MHz except of measuring range center. For some extreme heavy weather condition, the LOS wind velocity may exceed the ±110MHz Doppler shift range, such as hurricane, the fixed type can not fulfill its measurement. But this extremely wind measurement can be finished with the fitting type. For example, we can make the outgoing laser frequency of the fitting type be at −120MHz, the120MHz Doppler shift (corresponding to the 63.8m/s wind velocity in the LOS) will also make its frequency locate at the cross-point of the transmission curves. The uncertainty curve of it is expressed with dot-dash curve in Fig. 2 (b) with the same pattern except of measuring range center too.

B. Two and three-point fitting double-edge technique

While the wind velocity range increases, the excellent part around the cross-point of FPI transmission curves may be smaller than it. The simple usage of the fitting type as above will not work. But we can divide the wide wind velocity range into two or three small parts and twice or three times use this curve part to measure them respectively by choosing two or three different outgoing laser frequency locations. This is called two or three-point fitting type. For two-point type, first we tune the location of outgoing laser to be at 35MHz and shot laser to atmosphere, then tune the location to be−35MHz and shot. The same wind of atmosphere is measured twice in this way, and the measuring uncertainty curves of them are shown with dash and solid lines in Fig.3 (a). For three-point type, three outgoing laser frequency locations are chose, 73MHz, 0MHz and −73MHz. The same wind is measured three times at these points, and these three times measured values, which are corresponding to $-110 \sim -36.7$MHz, $-36.7 \sim 36.7$MHz and $36.7 \sim 110$MHz respectively, will locate around the cross-point and be picked out. The measuring uncertainty curves of them are shown in Fig.3 (b).

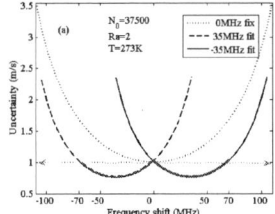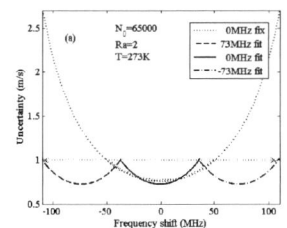

Figure 3. Measuring uncertainties of the double-edge FPI system for fixed and fitting type(a)two-point fitting types (b)three-point fitting types

If total backscatter photons need the outgoing laser pulse number to be n, then $n/2$ or $n/3$ pulses need shoot to measure winds at each frequency location of two or three-point fitting type for demanding not more time than the fixed type in real measurement. So the edge channels get less backscatter photons for fitting type at each location, but the energy channel do not obtain less photon because at each laser location the energy channel all receive photons. Therefore the superiority of two or three-point fitting type is not as evidently as that of the one-point type.

C. Measuring uncertainties of fitting double-edge technique under different condition

With the detective height increasing, temperature T and Mie to Raleigh backscatter ratio Ra of the atmosphere will decrease. So the Raleigh component is higher than Mie component in the measured signal. Though the retrieval method of the double-edge technique can remove the Raleigh component from the signal, the shot noise of it, which is proportional to its intensity, can not be deal with. So the measuring uncertainty will increase with the detective height increasing. And the more backscatter photons will be needed to reach the same measuring accuracy requirement. The usual solution for this is to increase the resolving distance and add pulse cumulative

978-1-4799-1215-5/13 $31.00 © 2013 IEEE

time. The simulations are made for comparing the fixing and fitting type double-edge techniques under the assumptions of $T = 233K$ and $Ra = 0.5$, as shown in Fig. 6. It can be found that the fitting type still can work for this kind atmosphere condition. When the measuring ranges of $-110 \sim 110MHz$ for three- point fitting type can fulfill the accuracy requirement of $1m/s$, corresponding measuring ranges of the fixed type ($-43 \sim 43MHz$) under this condition is narrower than that ($-50 \sim 50MHz$) under the previous condition. So it can be conclude that for higher detection height, the advantage of fitting type is evidently.

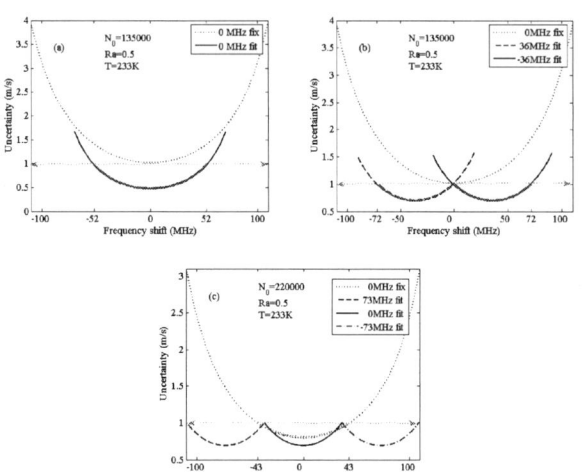

Figure 4. Measuring uncertainties of the double-edge FPI system for fixing and fitting types under different atmosphere condition (a) one-point fitting type and fixing type (b) two-point fitting type and fixing type (c) three-point fitting type and fixing with

IV. CONCLUSION

In summary, the fitting method for the direct detection DWL based on double-channel FPI is proposed and investigated theoretically. With this method, its measuring rang will not be restricted by the FWHM and interval of the FPI transmission peaks as before and can be chosen for high measuring accuracy based on the demand of actual wind speed. Under the same parameter, this method has better measuring accuracy. For the shot noise limited case, while only the 0MHz Doppler shift measured valve can reach 1m/s measuring accuracy for the fixed type, the measured values in the range of $-52 \sim 52MHz$ or $-70 \sim 70MHz$ for one-or two-point fitting types scan beyond this accuracy. While the measured values in the range of $-110 \sim 110MHz$ are beyond this accuracy for three-point fitting types, only $-50 \sim 50MHz$ measured values for fixed type can reach this accuracy. The simulation of this method for different atmosphere condition proves that it superiority is more evident for higher part troposphere wind measurement.

ACKNOWLEDGMENT

This work is supported by "the Fundamental Research Founds for the Central Universities" (Grant

No.HIT.NSRIF.2014043); "the Funds for National Key Laboratory of Science and Technology on Tunable laser".

This paper is dedicated to our colleague and friend Professor De Ming Ren who lost his life shortly after this work was completed for longtime overwork. He brightened the life of all who knew him.

REFERENCES

[1] F. Shen, H. Cha, D. Sun, D. Kim and S. O. Kwon, "Low tropospheric wind measurement with Mie Doppler lidar," Opt. Rev. vol.15, pp. 204-209, 2008 .

[2] H. Xia, D. Sun, Y. Yang, F. Shen, J. Dong and T. Kobayashi, "Fabry-Perot interferometer based Mie Doppler lidar for low tropospheric wind observation," Appl. Opt. vol.46, pp. 7120-7131, 2007.

[3] B. M. Gentry, H. L. Chen, S. X. Li, "Wind measurement with 355-nm molecular doppler lidar," Opt. Letters, vol. 25, pp. 1231-1233, 2000.

[4] Y. Bai, D. M. Ren, W. J. Zhao, Y. C. Qu, L. M. Qian, and Z. L. Chen, "Heterodyne Doppler velocity measurement of moving targets by mode-locked pulse laser," Opt. Express , vol.20, pp. 764-768 , 2012.

[5] Y. Bai, D. M. Ren, W. J. Zhao, Y. C. Qu, L. M. Qian, Z. L. Chen and Y. Liu, "Research on heterodyne detection of a mode-locked pulse laser based on an acousto-optic frequency shift," Appl. Opt. vol.49, pp. 4018-4023, 2010.

[6] A. Belmonte and A. Lázaro, "Measurement uncertainty analysis in incoherent Doppler lidars by a new scattering approach," Opt. Express , vol.17, pp. 7699-7708, 2006.

[7] C. Flesia and C. L. Korb, "Theory of the double-edge molecular technique for Doppler lidar wind measurement," Appl. Opt. vol.38, pp. 432-440, 1999 .

[8] C. Souprayen, A. Garnier and A. Hertzog," Rayleigh-Mie Doppler wind lidar for atmospheric measurements. □. Mie scattering effect, theory, and calibration," Appl. Opt. vol.38, pp. 2422-2431,1999.

[9] H. Xia, X. Dou, D. Sun, Z. Shu, X. Xue, Y. Han, D. Hu, Y. Han and T. Cheng, "Mid-altitude wind measurements with mobile Rayleigh Doppler lidar incorporating system-level optical frequency control method," Opt. Express 20, pp. 15286-15300, 2012 .

[10] M. J. Foster, R. Bond, J. Storey, C. Thwaite, J. Y. Labandibar, I. Bakalski, A. Hélière, A. Delev, D. Rees and M. Slimm, "Fabry-Pérot optical filter assembly: a candidate for the Mie/ Rayleigh separator in EarthCARE," Opt. Express, vol. 17, pp. 3476-3489, 2009 .

[11] D. Kim, S. Kwon, H. Cha, Y. Kim and J. Sunwoo, "A newly designed single etalon double edge Doppler wind lidar receiving optical system," Rev. Sci. Instrum. vol. 79, pp. 123111, 2008 .

[12] L. Tang, Z. Shu, J. Dong, G. Wang, Y. Wang, W. Xu, D. Hu, T. Chen, X. Dou, D. Sun and H. Cha, "Mobile Rayleigh Doppler wind lidar based on double-edge technique," Chin. Opt. Lett. vol. 8, pp. 726-731, 2010 .

[13] B. M. Gentry and C. L.Korb, "Edge technique for high-accuracy Doppler velocimetry," Appl. Opt. vol. 33, pp. 5770-5777, 1994 .

[14] C. L. Korb, B. M. Gentry and S. X. Li, "Edge technique Doppler lidar wind measurements with high vertical resolution," Appl. Opt. vol.36, pp. 5976-5983, 1997.

[15] J. A. Mckay, "Fabry-Perot etalon aperture requirements for direct detection Doppler wind lidar from Earth orbit," Appl. Opt. vol.27, pp. 5859-5866, 1999 .

[16] M. J. McGill and J. D. Spinhirne, "Comparison of two direct-detection Doppler lidar techniques," Opt. Eng. vol.37, pp. 2675-2686, 1998 .

[17] C. Y. She, J. Yue, Z. A. Yan, J. W. Hair, J. J. Guo, S. H. Wu and Z. S. Liu, "Direct-detection Doppler wind measurement with a Canbannes-Mie lidar: A. Comparison between iodine vapor filter and Fabry-Perot interferometer methods, " Appl. Opt. vol.46, pp. 4434-4443, 2007 .

[18] C. Y. She and J. R. Yu, "Simultaneous three-frequency Na lidar measurements of radial wind and temperature in the mesopause region," Geophys. Res. Lett. vol.21, pp. 1771-1774, 1994 .

[19] C. L. Korb, B. M. Gentry, S. X. Li and C. Flesia, "Theory of the double-edge technique for Doppler lidar wind measurement," Appl. Opt. vol.37, pp. 3097-3104, 1998 .

High-efficiency Focusing Grating Coupler for Large-Scale Silicon Photonic Integration

Biao Yang, Zhi-Yong Li, Xi Xiao, Jin-Zhong Yu, Yu-De Yu*

State Key Laboratory for Integrated Optoelectronics, Institute of Semiconductors,
Chinese Academy of Sciences, Beijing 100083, China
* Email: yudeyu@semi.ac.cn

Abstract—A novel high-efficiency compact optical coupler employing curved gratings with non-uniform pitches is proposed to coupling light off or into silicon photonic chips for large-scale integrated functions. The optimized non-uniform pitches improve the coupling efficiency of single-mode fiber and waveguide by reducing the mode matching loss between the fiber and grating coupler. This kind of focusing grating coupler has a decreased transition length between the grating and single-mode waveguide.

Keywords –grating coupler, SOI, photonic integrated circuits, silicon photonics

I. INTRODUCTION

As out-of-plane coupling structures, grating couplers have attracted a lot of attentions for advantages of on-chip testing, no need of chip preprocessing, near vertical propagation, low space limitation, etc. However, grating couplers still perform unsatisfactory to actual applications. In recent years, grating couplers are developed rapidly profitable from technological level's improvement. In order to improve coupling efficiency, grating couplers with bottom reflective film, multilayer antireflective dielectric films[1], large filling factor structure[2], non-uniform structure[3] and double-etched apodized structure[4] were investigated. Among of them, non-uniform grating couplers are considered to be the most possible structure to match the mode of the fiber to achieve the highest coupling efficiency.

At the same time, the large-scale integration needs device miniaturization. Defined on 12μm wide multi-mode waveguides, carves of the gratings are perpendicular to the axis of the waveguide and the whole grating section is generally measured 20μm long. The grating section is then adiabatically linked to a single mode waveguide by a long taper. The adiabatic transition determines the length of the taper (~150μm) between the grating structure and the single-mode waveguide. Focusing grating couplers use curving lines' focusing character to obviate the need for a long adiabatic transition. This will decrease the taper's length and result in a substantial length decrease, resulting in a higher degree of integration.

In this paper, we combine the advantages of non-uniform gratings and focusing structure to improve the design of grating couplers with high coupling efficiency and small footprint on SOI substrates.

II. COUPLING MECHANISM

For a non-uniform grating coupler, the structure is determined by transmission mode of device the grating couples to, usually single-mode fiber. Assuming that the grating coupler's output mode is Gaussian mode in single-mode fiber, leakage factor $\alpha(z)$ along the optical transmission direction on the surface of the chip can be calculated by using the formula[5]:

$$\alpha(z) = \frac{B^2(z)}{2\left[A_0^2 - \int_{z_0}^{z} B^2(t)dt\right]} \quad (1).$$

Here A_0 is the electric field of the guided wave at the beginning z_0 of the grating. The electric field distribution of out-coupled light $B(z)$ represents a Gaussian beam profile:

$$B(z) = \frac{C_g}{\sqrt{2\pi}\delta_g} \exp\left[-\frac{(z-z_m)^2}{2\delta_g^2}\right] \quad (2).$$

The leakage factor $\alpha(z)$ is determined by the depth, period and filling factor of grating cells. According to the experience of the predecessors and simulation, correspondence of these parameters and leakage factor can be obtained. To lower technical difficulty, etch depth is set to a constant value. Every cell along optical transmission direction would be obtained and non-uniform grating couplers would be the combination of these cells after further simulation.

In the case of focusing grating coupler, efficient coupling between single-mode fiber and TE-mode single-mode waveguide on SOI material is studied. Plain wavefront from fiber is curved by focusing grating and focuses on common focus of grating lines. When the top surface of the waveguide is chosen to be (Y, Z) plane of a right handed Cartesian coordinate system, with z along the waveguide axis and the origin is chosen to be in desired focal point. The grating lines must follow the condition of constructive interference[6]:

$$q\lambda_0 = n_{eff}\sqrt{y^2 + z^2} - zn_t\cos\theta_c \quad (3).$$

Here q is an integer number for each line, θ_c is the angle between fiber and chip surface, n_{eff} is the refractive index of the grating, n_t is the refractive index of the surrounding. This condition defines the grating lines' degree of curve.

In line-shape grating coupler, the taper is usually quite long for mode transition, which turns the light from the waveguide to nearly parallel one with low mode transmission loss between the single-mode waveguide and grating. But in curved structure, the grating needs elliptical wavefront to match the curved pitches. In this case, the taper is designed to transfer the plane light from the waveguide to elliptical one.

III. SIMULATION AND DESIGN

We design grating couplers as shown in fig.1 and grating lines are defined by the construction condition, in which n_{eff} is approximate by the well-known grating equation:

$$n_{eff} = n_c \cos\theta_c + m\frac{\lambda}{\Lambda} \qquad (4).$$

Here, n_c is the index of the surroundings(~1.0), Λ is the grating period (610nm), θ_c is 80°. The elliptical wavefront inconsistency with curved grating would result in reflection loss. What's more, the formula(3) is valid with the assumption that the effective refractive index in the grating and the taper region is equal :

$$n_{eff}^{grating} = n_{eff}^{taper} \qquad (5).$$

So the taper need to be designed to realized specific elliptical wavefront and to approximate formular(5). We simulate the effect of the taper's length and angle respectively by FDTD. By simulation, the taper's angle and length are chosen to 30° and 11.5μm for the high directionality. The whole size of the grating coupler is designed to be 12μm × 30μm. Power diffracted up from the grating is 88% at 1571.15nm as shown in fig.2 and the mode field distribution is quite approximate to the Gaussian mode as shown in Fig. 3.

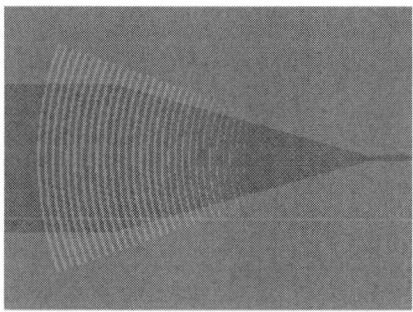

Figure 1.   Non-uniform focusing grating

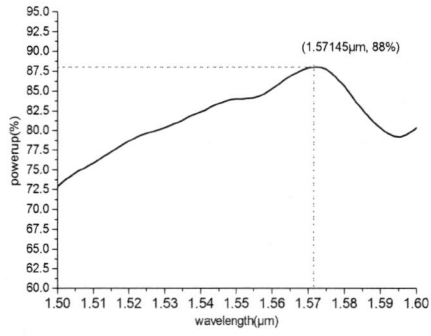

Figure 2.   Power diffracted up of the non-uniform focusing grating coupler

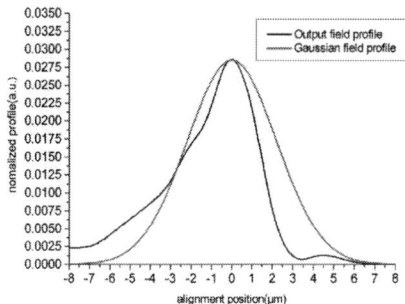

Figure 3.   Distribution of the output field

IV. CONCLUSION

A non-uniform focusing grating coupler on SOI is designed and simulated. Results of calculation on light coupling of a silicon optical waveguide with a single-mode fiber shows that the power diffracted from the grating is 88% at the wavelength of 1571.5nm and the distribution of the output field is approximate to the field profile of the single-mode fiber. The whole size of the grating coupler, including the taper, is designed to be only 12μm × 30μm, which has a small footprint used for photonic integration.

ACKNOWLEDGMENT

The project is supported by the State Key Development Program for Basic Research of China (Grant No. 2011CB301701), the Main Direction Program of Knowledge Innovation of Chinese Academy of Sciences (Grant No. KGCX2-EW-102) and the National Natural Science Foundation of China (Grant No. 61107048, 61275065).

REFERENCES

[1] A Nemkova, et al., "Design and Characterization of a Top Cladding for Silicon-on-Insulator Grating Coupler," Chinese Physics Letters, vol.29, no. 11, pp.114213, November 2012.

[2] L Zhou, et al., "A novel highly efficient grating coupler with large filling factor used for optoelectronic integration." Chinese Physics B, vol.19, no.12, pp. 124214. 2010.

[3] C Zhang, et al., "High efficiency grating coupler for coupling between single-mode fiber and SOI waveguides," Chinese Physics Letters, vol. 30, no. 1, pp. 014207, January 2013.

[4] C Li, H J Zhang, M B Yu, G Q Lo, "CMOS-Compatible Silicon Double-etched Apodized Waveguide Grating Couplers for High Efficient Coupling," Optical Fiber Communication Conference, March 2013.

[5] R Waldhäusl, B Schnabel, P Dannberg, E BKley, A Bräuer, and W Karthe, "Efficient coupling into polymer waveguides by gratings," Applied Optics, vol. 36, no.36, pp. 9383-9390, July 1997.

[6] F Van Laere, T Claes, et al. "Compact Focusing Grating Couplers for Silicon-on-Insulator Integrated Circuits," *IEEE* Photonics Technology Letters, *vol.* 19, no. 23, pp. 1919-1921, December 2007.

The Study and Realization of Real-Time Supervision and Control of Medical Image System

Chunqiu You

The Physics and Electronic Engineering College,Harbin Normal University , Harbin, China

Abstract—**These problems about reducing the noise of medical image, increasing the ratio and clarity of image, and decreasing X-ray dose of medical perspective were studied and analyzed in this article. The digital processing and control system of medical perspective image was designed. The whole structure and working principle of the system are introduced. The design idea , implementation step and function result analyzing of the technique of real–time supervision and control are introduced in detail.**

Keywords-capturing image; moving noise by adding-averaging; real-time supervision and control; true color image

I. Introduction

Digitized real-time video surveillance is important application in the field of intelligent information processing technology [1][2], no matter using which kinds of X-ray equipment (such as X-ray film, CT film), there is a certain dose of radiation while diagnosing and treating to patients, and a long time accumulated dose could cause the other lesions. In order to try to reduce the radiation to the patient, shorten diagnosing radiation time, set out to design the medical perspective digitized image processing and control system, hereinafter referred to as is the medical image system. The real-time monitoring technique is an important part of medical image system, which uses pulse X-ray as the irradiation source[3]. The monitoring image is multiframe superposition average denoising image and the monitor displays real-time patient's physiological movement process. The system can set up diagnosing time, and maximally decrease X-ray irradiation during the process of diagnosis and surgical, minimal damage to patients. Theoretical analysis and preliminary experimental results verify the feasibility of real-time monitoring technique.

II. Medical Image System

A. The structure of medical image system

The whole system includes six parts: CCD camera, video capture card, the GPIO card, computer, the control circuit, light source[9], as shown in figure 1.

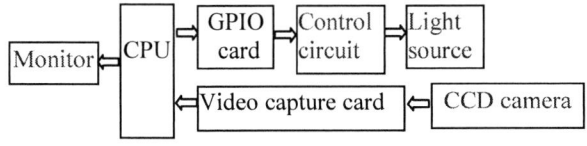

Figure 1. The structure diagram of the medical image system

B. Working principle of the system

In order to reduce the radiation dose to patients, try to shorten the inspection irradiation time, the system adopts pulse X-ray source[4][5]. Considering the safety of the experiment, the system adopts visible light source to replace X-ray source.

The computer starts GPIO card to output a high voltage, then the control circuit drives the light source to work. The same time the computer starts the video capture card to capture image from the camera and send image acquisition to the display window. After capturing some frames, GPIO card outputs a low voltage to close the light source, image capture card stop working at the same time.

III. The Technique of Real-Time Supervision and Control

A. Design idea

According to the visual inertia of human eye, the human eye's visual persistence time is about 0.1 seconds[6], the characteristics of real-time monitoring technology adopting the pulse X-ray sources to reduce the X-ray radiation dose and real-time displaying physiological movement process[11]. The control system not only is used for diagnosis, but also used for the surgical treatment, causing minimal damage to the patients in the process of observation. At the same time, image with a lot of random noise during the process of image acquisition and transmission, so we adopt multiframe superposition average to reduce random noise of real-time image acquisition, then send the denoising image to video window for the doctor to better diagnosis and treatment.

B. Implementation step

Real-time video surveillance system can set up video surveillance parameters[12], such as monitor time, capture image frequency, the number of capture image frames, the delay time, as shown in fig.2[7]. Real-time monitoring technology generally include capture image, multiframe images superposition average, X-ray control and so on, it is composed of the following eight steps .

978-1-4799-1215-5/13 $31.00 © 2013 IEEE 163

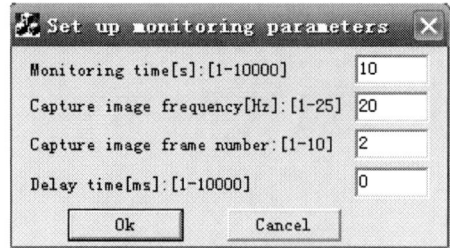

Figure 2. The dialog box of setting up

1) Starting video capture card and GPIO card;

2) GPIO card outputs a high voltage;

3) Capture the image;

4) GPIO card outputs a low voltage;

5) Multiframe superposition average creates a frame image;

6) Send denoising image to the monitor;

7) Delay time T;

8) Return to 2).

Delay time T set in low voltage, indirectly control monitoring image frequency.

If the monitoring time is t_1, cycle time is t_2, the number of cycle times is n(the frames number of monitor),then the relationship among them is: $n=t_1/t_2$.

Capture image and multiframe superposition average modules are not introduced here, only light source control module is analyzed in detail.

C. Light source Control

Medical image system set the TTL voltage of the GPIO card as output, the high voltage is +5V, the low voltage is 0V. The GPIO card outputs a signal to an external control circuit. When the GPIO card outputs high voltage signal , the signal is greatly enhanced by two levels NOT gates(TC4096)of driving ability, and then added to the input end of the solid state relay(SSR), SSR conduction, connected to an external circuit, turning on the light source. When the GPIO card output low voltage, there was no signal output, SSR disconnected, turning off the light source. The principle diagram of the circuit is shown in Figure 3.

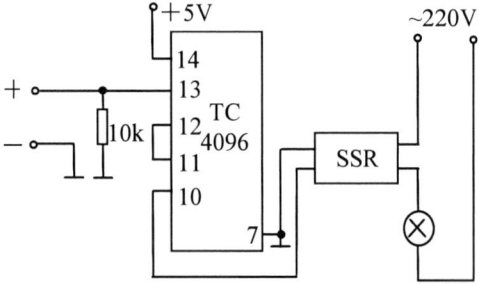

Figure 3. The exterior control circuit

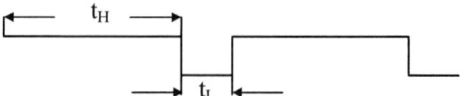

Figure 4. The control pulse

The time of high voltage is X-ray irradiation time, the time of low voltage is no X-ray irradiation time, as shown in Fig. 4.

Set the time of high voltage:

$$t_H = \frac{1}{f} \times m + (5m+1)t \qquad (1)$$

Set the time of low voltage:

$$t_L = \{75 + biHeight[1 + 2m + biWidth(6m+6)]\}t + T \qquad (2)$$

The two formulas are based on the source program of the medical image system[7], and various parameters depend on the specific situation. Because the visual persistence time of human eye is 0.1 seconds, so make sure:

$$t_H + t_L \le 0.1 \qquad (3)$$

Where m is capture image number; f is capture image frequency; t is the computer time of implementing a program statement; BiHeight is horizontal rows of a frame image; BiWidth is vertical rows of a frame image ; 1,6,75 refers to the number of statements in a source program ; T is the delay time.

When GPIO card outputs a high voltage signal, image capture card captures sequential images to display window and memory, after capturing some frames, GPIO card outputs a low voltage signal, at the same time the images of memory was added and averaged, and then denoising images sent to monitor window, as shown in figure 5.

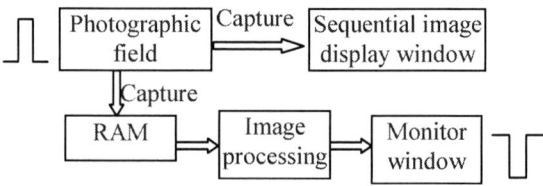

Figure 5. The structure diagram of real-time monitoring control

D. Result and Analysis

1) The system parameters were modified many times , debugged repeatedly, so that not only maximum reduce the X-ray irradiation, but also make the image clear . But found by experiment, reduce the X-ray irradiation quantity is inversely proportional to the image clear, more superposition frames, better denoising effect, but the time of X-ray irradiation is more longer, so to choose parameters depending on the specific situation. Fig.6 is adding-averaging denoising simulation results. Fig.(a) is a 32 bits ture color

bitmap,Fig.6(b) is a denoising bitmap by two frames adding-averaging ,and Fig.6(c) is a denoising bitmap by ten frames adding-averaging.

(a)　　　　　　(b)　　　　　　(c)

Fig.6 The simulation results

2) Except the image enhancement method of adding-averaging denoising, some image enhancement methods are used ,such as gray linear transformation, histogram equalization, median filtering, smoothing etc, according to the specific circumstances of a single image[6], to achieve the result that we want.

3) The ratio of no decreasing X-ray irradiation time and decreasing X-ray irradiation time is:

$$\frac{t_H + t_L}{t_H} = 1 + \frac{t_L}{t_H} \quad (4)$$

4) If the parameter value is: m = 2, f = 20 hz, biHeight = 227, biWidth = 236, T = 0, there is:

$$t_H = \frac{1}{f} \times m + (5m+1)t = 0.1 + 11t \quad (5)$$

$$t_L = \{75 + biHeight\,[1 + 2m + biWidth\,(6m+6)]\}t = 58868\,t$$

If the CPU frequency is 1.1Ghz and the CPU calculates each program statement nearly need 50 times[8] , then t = 0.05us.

$$\frac{t_L}{t_H} = \frac{58868\,t}{0.1 + 11t} = \frac{58868}{\frac{0.1}{t} + 11} = 0.12 \quad (6)$$

$$\frac{t_H + t_L}{t_H} = 1 + 0.12 = \frac{1.12}{1} \quad (7)$$

Therefore we conclude that if the amount of no decreasing X-ray irradiation time is 1.12 seconds and decreasing X-ray irradiation time is 1 second, then no X-ray irradiation time is 0.12 seconds, so decreasing irradiation is very considerable.

If the parameter is selected as Fig.2:

$$t_H + t_L = 58868\,t + 0.1 + 11t = 58879\,t + 0.1 \approx 0.1 \quad (8)$$

The image frequency of the monitor:

$$F = \frac{1}{t_H + t_L} \approx 10Hz \quad (9)$$

5) Monitor image don't need to be continuous in the interventional surgery, so clinical T value be set up according to particular case, to greatly reduce the X-ray irradiation.

6) The advantages of the video surveillance system.

Use recycle memory to reduce memory occupancy;

Can set up monitor time, capture frequency , capture frame;

Image is clear;

Minimize the X-ray irradiation.

7) The disadvantages of the video surveillance system.

Objects not move too fast, otherwise the image appears fuzzy.

IV. CONCLUSION

The medical perspective digitized image processing and control system runs under Windows XP environment, by Visual c++ to implement the software. The image format is BMP format. User interface is beautiful and convenient. The medical system not only has strong gray image processing functions, but also provides a distinctive true color image processing functions, is a kind of very practical medical image digitized processing system. The image system realized the digital image processing and improved the image quality, better serviced for the clinical diagnosis, and reduced the X-ray dose, specially designed for medical radiation inspection and intervention treatment. At present the medical image system is essential digitized interface device of X-ray machine at home and abroad.

ACKNOWLEDGMENT

This work is supported by Scientific Research Fund of Harbin Normal University(KM2007-01),the National Natural Science Foundation of China (11247252) and Scientific Research Fund of Heilongjiang Provincial Education Department (12511163).

REFERENCES

[1] B. Georis,F. Br' emond, M. Thonnat, "Real-Time Control of Video Surveillance Systems with Program Supervision Techniques," Machine Vision and Applications ,2007,(18): pp.189–205.

[2] Liu Fuqiang, Digital video monitoring system development and application,Beijing: mechanical industry press, 2003: pp.278-279.

[3] E.I. Pal'chikov, V.I. Kondrat'ev, E.V. Golikov, A.N. Cheremisin, "Experimental study of a BaFBr:Eu image plate detector depending on dose, spectrum of pulse X-ray source, and scan number," Journal of Surface Investigation. X-ray, Synchrotron and Neutron Techniques , Volume 4, Issue 4, pp 622-629,August 2010.

[4] I. V. Lavrinovich, N. V. Zharova, V. K. Petin, N. A. Ratakhin, V. F. Fedushchak, S. V. Shlyakhtun, A. A. Erfort, "A compact pulsed X-ray source for high-speed radiography," Instruments and Experimental Techniques , Volume 56, Issue 3, pp 329-334,May 2013 .

[5] Rad-icon Imaging Corp,"Imaging with Pulsed X-Ray Sources ,"3193 Belick Street, Unit 1 Santa Clara, CA 95054-2404.

[6] Zhang Yujin, Image processing and analysis, Beijing: tsinghua university press, 1999.

[7] Yang Zhiling, Visual c++ digital image acquisition processing and practical applications, Beijing: people's posts and telecommunications press, 2003.

[8] Rong Guan'ao, Computer image processing,Beijing: tsinghua university press, 2000.

[9] Avanzi, A., Brémond, F., Tornieri, C., Thonnat, M."Design and assessment of an intelligent activity monitoring platform," EURASIP

Journal on Applied Signal Processing,special Issue on Advances in Intelligent Vision Systems:Methods and Applications,2005, (14), 2359–2374.

[10] Huang Xiaoling, "The research of X-ray photo enhancement based on wavelet analysis", Journal of wuhan university, 1998, 44 (1) : 121

[11] List, T., Bins, J., Fisher, R., Tweed, D,"A plug-and-play architecture for cognitive video stream analysis," In: Proceedings of the IEEE International Workshop on Computer Architecture for Machine Perception (CAMP05),Palermo,Italy,2005,pp. 67–72.

[12] Sung Chun LeeRam Nevatia, "Hierarchical abnormal event detection by real time and semi-real time multi-tasking video surveillance system," Machine Vision and Applications,May 2013.

A Micro-machined Optical Fiber Acoustic Sensor based on Fabry-Pérot Interferometer

Shi-Ning Wang, Mei-Yu Zhang, Yong-Hai Cao, Li-Jie Chen

The 49th Research Institute
China Electronics Technology Group Corporation
Harbin 150001, China
edmundhlj@sina.com

Abstract—**A optical fiber acoustic sensor based on Fabry-Pérot (F-P) interferometer was introduced. The ratio of two outputs of the CWDM was theoretically analyzed and experimentally tested. A prototype of the acoustic sensor was fabricated, assembled, and tested.**

Keywords—fiber-optic sensor; Fabry-Perot interferometer; acoustic sensitivity

I. INTRODUCTION

The development of Fiber-optic acoustic sensors can be traced back to the late 1970s [1-2]. These widely utilized interferometric techniques included Mach–Zehnder, Michelson, Sagnac, and Fabry–Pérot interferometers [3]. Optical fiber sensors based on F–P interferometer are widely used in acoustic and pressure detection due to their high sensitivity and wide frequency spectrum response, which are determined by the acoustic diaphragm fabricated with the MEMS technology [4]. Researchers employed many different techniques to sense a wide variety of measurands which include pressure [5-7], sound [4], strain [8-10], temperature [10], and displacement [11], while employed different diaphragm structures which include flat diaphragm [4-11], corrugated diaphragm [12], mesa-diaphragm [6], and photonic-crystal diaphragm [13-15]. The diaphragm of the F-P sensor sensing acoustic signals is a separate extrinsic sensing element. This kind of F-P sensor compared with existing underwater fiber acoustic sensors, shows unique advantages, including small size and weight, low large manufacturing cost, and high frequency bandwidth [15]. Besides, this of F-P sensor is suitable for planar array application.

In this paper, we propose a diaphragm-based F-P acoustic sensor, and test the acoustic sensitivity. Then the compensation technique for fluctuation of static pressure is estimated briefly.

II. STRUCTURE OF SENSOR

The proposed acoustic sensor head consists of 5 parts: a acoustic corrugated diaphragm, a cavity chip, a single mode fiber, a silica glass capillary and organic glass, as shown in figure 1. The F-P cavity, composed with a silicon corrugated diaphragm and a flat tip of single mode fiber, is an air-backed chamber. To gain the reflectivity of diaphragm, there is a thin film of gold on the diaphragm, which slightly affects the sensitivity, but the interference of two faces of the diaphragm is completely eliminated.

The proposed acoustic sensor consists of four parts: a broadband light-emitting diode (LED) light source, a photo-electricity conversion component, amplifying circuit and signal processing component, as shown in figure 2. In figure 2, these symbols of (a), (b), (c) and (d) separately have a relationship with the symbols in figure 3.

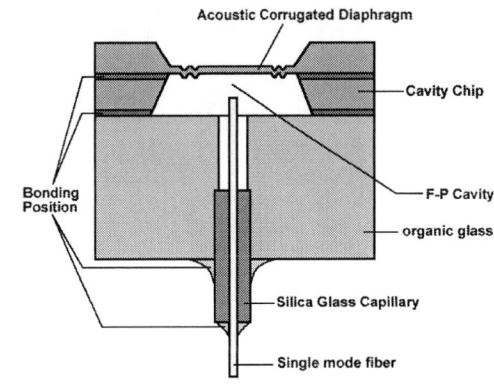

Figure 1. Cross section of the acoustic sensor head with F-P cavity. The sensor head drawing is not to scale.

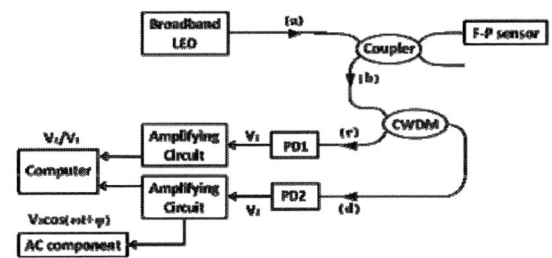

Figure 2. Schematic drawing of the optical detection of the acoustic sensor based on a F-P interferometer.

The output light of broadband LED source is coupled into a single mode fiber coupler and propagates to the F–P sensor head. The light reflected from the F–P sensor is splitted by the Coarse Wavelength Division Multiplexer (CWDM). The CWDM transmits narrowband spectrum with the range of 1303.5–1316.5 nm, and reflects a broadband spectrum without the narrowband spectrum. The narrowband spectrum transmits to the photodetector PD2, while the broadband spectrum to the photodetector PD1. These two optical signals are the functions of the F-P cavity length. The photo-electricity conversion signal from the narrowband eliminates DC component, and the

978-1-4799-1215-5/13 $31.00 © 2013 IEEE

AC component ($V_2\cos(2\pi ft+\varphi)$, f is the frequency of the sound) is reserved to demodulate the sound signal applied to the diaphragm of the F-P sensor head. The computer is used to calculate the division between V_2 and V_1 for pressure detection.

Figure 3 shows the intensity spectrum distribution in experiments. The center wavelength of the LED source is about 1290 nm and the FWHM is about 50 nm. The center wavelength of the narrowband output of the CWDM is about 1310 nm, and the bandwidth of narrowband is about 13 nm.

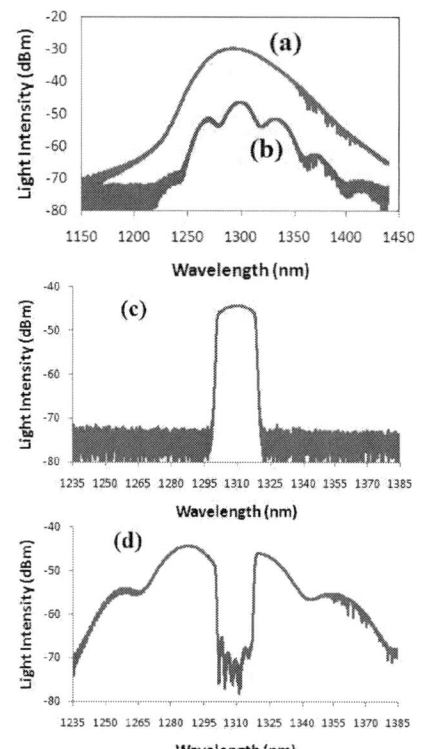

Figure 3. The intensity spectrum distribution in experiments: (a) the intensity spectrum distribution of LED; (b) the reflectance spectrum distribution of F-P sensor; (c) the transmission spectrum distribution of CWDM; (d) the reflectance spectrum distribution of CWDM.

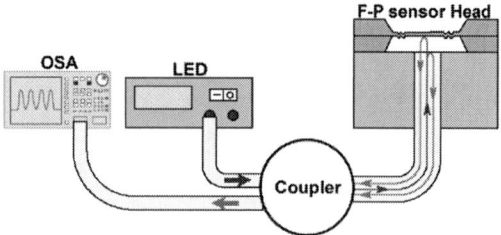

Figure 4. Schematic drawing of the test system for F-P sensor's measurement

III. PRINCIPLE

The spectrum distribution of LED, the reflectance spectrum distribution of F-P sensor, and these two splitted spectrum are separately shown in figure 3, which are measured with Optical Spectrum Analyzer (OSA) AQ6370C. The interferometric spectrum of F-P sensor was obtained with the LED source and

OSA as shown in figure 4, and then we used a spectrum analysis algorithm to calculate F-P cavity length L:

$$L = \frac{\lambda_1\lambda_2}{2(\lambda_2 - \lambda_1)} \quad (1)$$

where, λ_1 and λ_2 are the wavelengths of two valley points of the interferometric spectrum. Hence, the F-P cavity length can be obtained from figure 3(b), L=22.908 μm.

Based on the properties of silicon, the center flexibility of the diaphragm can be theoretically calculated as follows, and the curves are shown in figure 5:

$$w_0 = 0.0151\left(1 - \mu^2\right)\frac{a^4}{Eh^3}p \quad (2)$$

where, μ is the Poisson's ratio, E is the Young's Modulus, a is the side length of the diaphragm, h is the thickness of the diaphragm, p is the outside pressure.

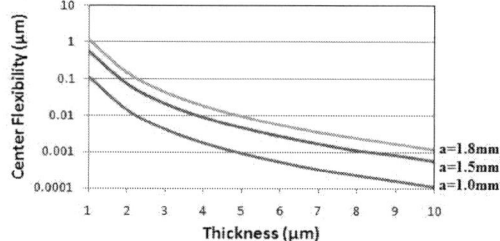

Figure 5. Theoretical calculated center flexibility of the diaphragm for different side length and thickness.

The ratio Ra ($Ra = V_2/V_1$) [16] can be theoretically expressed as a function in a cavity length variation range of $\lambda_C//4$, and the curves are shown in figure 6:

$$R_a = V_2/V_1 = k_1 + k_2\cos\left(4\pi L/\lambda_C\right) \quad (3)$$

where, λ_C is the center wavelength of the light source.

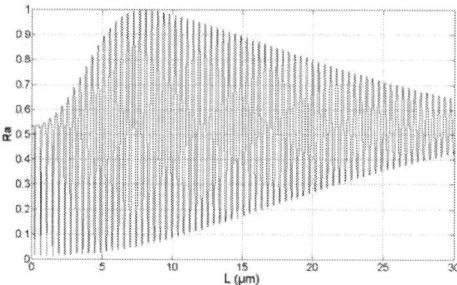

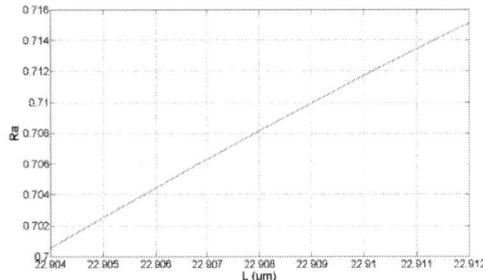

Figure 6. Theoretically calculated result of the ratio R_a.

IV. EXPERIMENT AND RESULTS

The variation of the light power of the narrowband was measured in experiments, while the sensor sensed the sound signals provided by a sound source with the frequency of 1000 Hz, with the sound intensity of 1Pa and 10Pa. Then the amplitudes of the AC output of the photo-electricity conversion circuit after the PD2 were 0.3 mV and 2.9 mV separately. When the sound intensity is 10Pa, there is a slot to leak sound, so the output declined slightly. And it could be considered that the sensitivity of the sensor is about 0.3 mV/Pa without amplifier.

When the cavity length is 22.908 μm, the light powers of the narrowband and broadband are measured separately, they are15.8 μW and 22.3 μW, then R_a=0.7085. The experimental result is almost equal to the theoretical result R_a=0.7081.

V. CONCLUSION

In this paper, we described a diaphragm-based F-P sensor for acoustic detection, and an outlook of the compensation technique for fluctuation of static pressure. The acoustic sensor was tested, and the sensitivity is about 0.3 mV/Pa without amplifier. The ratio R_a was measured, and highly consistent with the theoretical value, so the compensation technique for fluctuation of static pressure was accomplished, and then the sensor is applicable to underwater acoustic detection.

REFERENCES

[1] J. H. Cole, R. L. Johnson, and P. G. Bhuta. Fiber-optic detection of sound. J. Acoust. Soc. Am., vol. 62, pp. 1136–1138, 1977.

[2] J. A. Bucaro, H. D. Dardy, and E. F. Carome. Fiber-optic hydrophone. J. Acoust. Soc. Am., vol. 62,pp. 1302–1304, 1977.

[3] G. D. Peng and P. L. Chu, "Optical Fiber Hydrophone Systems," in Fiber Optic Sensors, edited by S. Yin, P. B. Ruffin, and F. T. S. Yu, 2nd ed. (CRC Press, Boca Raton, FL, 2008), Chp. 9, pp. 369–373.

[4] H. Xiao, P.G. Duncan, J. Deng, et al. Thin silica diaphragm-based SCIIB fiber optic acoustic sensors. Part of SPIE Conference on Harsh Environment Sensors, SPIE 1999, vol.3852, pp. 36–45.

[5] J. Wang, H. Xiao, J. Deng, R. May, A. Wang, Self-calibrated interferometric/intensity-based (SCIIB) optical fiber pressure sensor, SPIE 1998, vol.3538-03.

[6] Y. Ge, M. Wang, H. Yan. Optical MEMS pressure sensor based on a mesa-diaphragm structure. Optics Express, vol.16, pp.21746–21752, 2008.

[7] Y. Ge, M. Wang, X. Chert, H. Rong. An optical MEMS pressure sensor based on a phase demodulation, Sensors and Actuators A: Physical, vol. No. 143, pp. 224–229, 2008.

[8] V. Bhatia, K. A. Murphy, et al. Multiple strain state measurements using conventional and absolute optical fiber-based extrinsic Fabry–Perot interferometric strain sensors, Smart Master, Struct., vol. 4, pp. 240–245, 1995.

[9] V. Bhatia, K. A. Murphy, et al. Recent developments in optical-fiber-based extrinsic Fabry–Perot interferometric strain sensing technology, Smart Mater. Struct., vol. 24, no.4, pp. 246–251, 1995.

[10] T. Liu, G.F. Fernando, et al., Simultaneous strain and temperature measurements in composites using extrinsic Fabry–Perot interferometric and intrinsic rare-earth doped fiber sensors, Sensors and Actuators, vol. 80, pp. 208–215, 2000.

[11] C. I. Lin, F.G. Tseng. A micro Fabry–Perot sensor for nano-lateral displacement sensing with enhanced sensitivity and pressure resistance. Sensors and Actuators A: Physical, vol. 114, pp. 163–170, 2004.

[12] W. Wang, J. Luan, et al. The Optical Acoustics Sensor Based on Si-MEMS Sensitive Structure. Journal of CAEIT, vol.6, no.3, pp. 320–323, 2011.

[13] O. Kilic, M. Digonnet, G. Kino, and O. Solgaard. External fibre Fabry–Perot acoustic sensor based on a photonic-crystal mirror. Meas. Sci. Technol., vol. 18, pp. 3049–3054, 2008.

[14] O. Kilic, M. Digonnet, G. Kino, and O. Solgaard. Asymmetrical spectral response in fiber Fabry–Perot interferometers. J. Lightwave Technol., vol. 28, pp. 5648–5656, 2009.

[15] O. Kilic, M. Digonnet, G. Kino, and O. Solgaard. Miniature photonic-crystal hydrophone optimized for ocean acoustics. J. Acoust. Soc. Am., vol. 129, no. 4, pp. 1837–1850, 2011.

[16] G. Zhang, Q. Yu. Shide Song. An investigation of interference/intensity demodulated fiber-optic Fabry–Perot cavity sensor. Sensors and Actuators A, vol. 116, pp. 33–38, 2004.

A Vehicle Laser Doppler Velocimeter Configured With Three Transmitting Beams

Meng Shanshan

Institute of Electrical Engineering, Chinese Academy of Sciences
Beijing, China
mss@mail.iee.ac.cn

Abstract—**A laser Doppler velocimeter (LDV) with three transmitting beams is designed to simultaneously measure the three components of vehicle's velocity, which cannot be achieved in conventional LDV with single or double beam. The system configuration and design method of the LDV is proposed, It is based on all fiber-optics and the signal collecting and processing is accomplished by FPGA and DSP. Some key technologies, including antenna layout, signal process and software design, are introduced. The experimental results indicated that the LDV was able to provide the real-time three-dimensional velocity components for vehicle navigation systems.**

Keywords- laser Doppler velocimeter; vehicle navigation; three-dimensional velocity measurement

I. INTRODUCTION

The automobile navigation system cannot only realize the automobile orientation and direction, but also ensure and strengthen the rapid maneuverability of the automobile weapon system. The land combat vehicles are usually equipped with automobile navigation system. As the rapid expansion of the width and depth of war-related area of the modern war, the navigation system is playing a more and more important role in modern wars. The essential factor of automobile navigation is the measurement and collection of velocity information. Laser Doppler Velocimeter (LDV) has a good application prospect in terms of automobile navigation with its advantages such as non-contact measurement, fast dynamic response, high-precision measurement and high anti-jamming ability.

As the gradual maturity of the Laser Doppler speed measurement technology, many universities and scientific research institutions inland and abroad have a rather deep research with regard to the LDV based on the single or double beam configuration. In the year 2009, Chinese Engineering Physics Research Institute developed the LDV that could measure the velocimeter ranging from 1-39mm/s[1]. In the year 2010, the National University of Defense Technology developed the LDV based on Janus configuration that has a relative velocity measurement accuracy of more than 1%[2]. In the year 2011, the 27th Research Institute of China Electronics Technology Group Corporation developed the laser radar that was applied to the space environment firstly all over the world, which could acquire the real-time parameter of the relative position and speed of the flight. In the year 2011, NASA reported the multi-radar system that can be used to airborne accurate navigation and autonomous landing, and the multi-

radar system can realize the real-time velocity vector measurement of the aircraft as accurate as 0.01m/s[3]. Until now, there has not been a report with regard to the Laser Doppler Three-dimensional velocimeter inland.

II. VELOCITY MEASUREMENT PRINCIPLE

At present, the working principle of velocimeter is mostly based on Doppler Effect. For wave propagation in all forms, all the movement of wave source, receiver, propagation medium, middle reflector or scatterer can cause the change of wave frenquency.

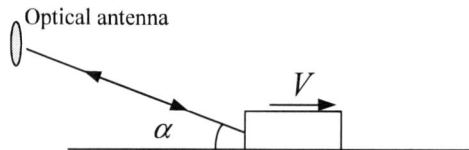

Figure 1. Principle of Velocity Measurement by Doppler Effect

As shown in Fig. 1, the target speed is represented as V, the angle between the target moving direction and the laser emitted by optical antenna is represented α, then the Doppler frenquency received by the optical antenna is:

$$f_d = 2\frac{v\cos\alpha}{\lambda} \tag{1}$$

In the equation, λ represents wavelength. We can learn from the Eq. (1) that when the Doppler frequency of echo signal is measured and the angle α between the beam and target moving speed is defined, the target moving speed can be retrieved.

$$v = \frac{\lambda f_d}{2\cos\alpha} \tag{2}$$

The Doppler frequency of three different beam directions can be achieved through three optical antenna, and then the moving speed of the target in three coordinate directions can be deduced based on Eq. (2). This is the velocity measurement principle of three-dimensional LDV.

III. VELOCIMETER SYSTEM DESIGN

The target of system design is to provide three-dimensional velocity parameter to automobile navigation system. The high

978-1-4799-1215-5/13 $31.00 © 2013 IEEE

mobility and complicated motion environment (as vibration and different ground reflection etc.) of the ground vehicle require that the optical system of LDV be stable and reliable. Therefore the all-fiber relative optical path has been chosen to realize light field wave front match easily and to eliminate the influence of wave front errors on detection performance. Furthermore, the all-fiber relative optical path has other advantages such as low price, small structure, light weight and convenient installation debugging [4].

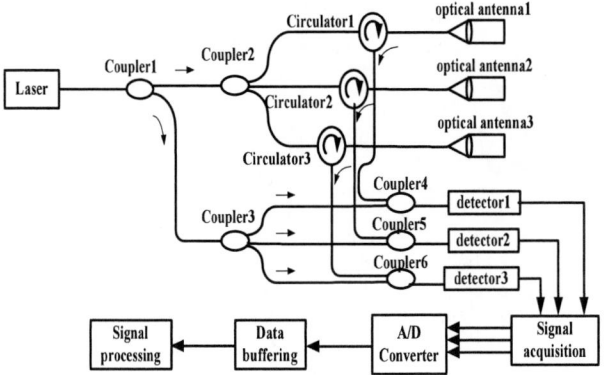

Figure 2.   LDV system principle Schematic diagram

The transmitter in the system adopts the transmitting way of laser continuous wave. It takes fiber collimator as system optical antenna which act as transceiver and receiver, finishes alignment of outgoing beam and couples the Doppler return light signal to the optical system. It is equipped with three optical antenna and presets the installing angle of the antenna. The circulator is set to isolate emission ray and echo signal ray. Based on the operation wavelength and dynamic range of velocity measurement, the band width of photo detector and the adoption of photodiode is confirmed. The signal processing adopts FPGA to process high speed data cache and provide clock signal to A/D converter, and A/D sampling rate is set based on actual velocity measurement range. After the algorithm processing by DSP, the calculating result is sent to host computer through serial port and displayed on LCD.

Three –dimensional LDV system principle block diagram is as shown in Fig. 2. In the figure, the ray emitted by laser device is separated into two beams by the coupler. One of them is turned into three ray after entering coupler 2, and radiates the detected target by three optical antenna after going through three circulator. The other is separated into three local oscillator ray after entering coupler3, and then combines with echo signal ray received by optical antenna inside coupler 4, 5 and 6. After that, it enters three photo detectors whose output signal enters signal acquisition and processing circuit, and then the relative velocity of detecting target in three directions is deduced.

IV.   ANTENNA LAYOUT

As to the three-dimensional laser velocimeter, the antenna layout is comparatively complicated to measure the relative velocity of the vehicle in three coordinate directions. It requires

individual setting of the angels concerning antenna layout. The details will be discussed below.

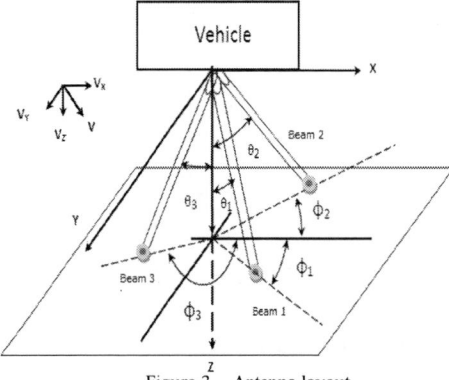

Figure 3.   Antenna layout

Assuming the Installation position of three antenna is as shown in Fig. 3. The headstock orientation is X-axis, and the vertical line is Z-axis, then rectangular coordinate system (x, y, z) and spherical coordinate system(r, θ, φ) is built. The angle between antenna I and Z-axis is θ_i. The horizontal rotary angle is φ_i. According to Doppler Effect, assuming that the vehicle has no pitching and axial rotation, then the Doppler frequency shift of beam i is:

$$f_i = \frac{2}{\lambda}(V_x \sin\theta_i\cos\varphi_i + V_y\sin\theta_i\sin\varphi_i + V_z\cos\theta_i) = \frac{2}{\lambda}(a_iV_x + b_iV_y + c_iV_z)$$

(3)

In Eq. (3), $i=1,2,3$, $a_i = \sin\theta_i\cos\varphi_i$, $b = \sin\theta\sin\varphi$, $c_i = \cos\theta_i$, based on the three Doppler frequency shift, the following equation system is built:

$$\begin{cases} f_1 = \frac{2}{\lambda}\vec{V}\cdot\vec{A}_1 = \frac{2}{\lambda}(a_1V_x + b_1V_y + c_1V_z) \\ f_2 = \frac{2}{\lambda}\vec{V}\cdot\vec{A}_2 = \frac{2}{\lambda}(a_2V_x + b_2V_y + c_2V_z) \\ f_3 = \frac{2}{\lambda}\vec{V}\cdot\vec{A}_3 = \frac{2}{\lambda}(a_3V_x + b_3V_y + c_3V_z) \end{cases}$$

(4)

In Eq. (4), i equals 1, 2, 3. The three equations of component value of velocity vector can be deduced by solving the above equation system:

$$\begin{cases} V_x = \frac{\lambda}{2}\dfrac{(c_3b_2 - c_2b_3)f_1 + (c_1b_3 - c_3b_1)f_2 + (c_2b_1 - c_1b_2)f_3}{D} \\ V_y = \frac{\lambda}{2}\dfrac{(a_3c_2 - a_2c_3)f_1 + (a_1c_3 - a_3c_1)f_2 + (a_2c_1 - a_1c_2)f_3}{D} \\ V_z = \frac{\lambda}{2}\dfrac{(b_3a_2 - b_2a_3)f_1 + (b_1a_3 - b_3a_1)f_2 + (b_2a_1 - b_1a_2)f_3}{D} \end{cases}$$

(5)

In Eq. (5), $D = a_3(-b_2c_1 + b_1c_2) + a_2(b_3c_1 - b_1c_3) + a_1(-b_3c_2 + b_2c_3)$.

V.   DESIGN OF SIGNAL PROCESSING CIRCUIT

As to the three-dimensional laser velocimeter, the antenna layout is comparatively complicated to measure the relative

978-1-4799-1215-5/13 $31.00 © 2013 IEEE   171

velocity of the vehicle in three coordinate directions. It requires individual setting of the angels concerning antenna layout. The details will be discussed below.

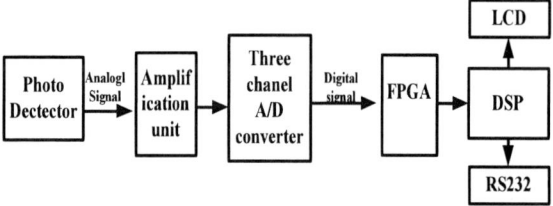

Figure 4. Signal processing circuit digram

As shown in Fig. 4, the preamplifier circuit adopts Integrated Operational Amplifier design, A/D conversion adopts three pieces of twelve units high-speed AD conversion chip AD9236 to conduct high speed sampling to three analog signal separately. The digital signal after conversion is exported into asynchronous FIFO (first in first out) memory of FPGA (here we adopts the Cyclone II series EP2C8Q208 chip of Altera Corporation) chip through parallel output. Besides, FPGA provides clock signal to AD9236 to control conversion rate of AD. The data bus between FPGA and DSP (TMS320C6747 of TI Corporation is selected) is linked to data port of three asynchronous FIFO inside the FPGA at one end, and the other end is linked to 16 units data port of external memory interface EMIFA.

The connection between FPGA and DSP is as shown in Fig. 5. When the memory of three FIFO inside the FPGA is full, FPGA would send interruption signal INT to DSP, which enters interrupt service subroutine and put RD_EN_x (x=1, 2, 3) as 1, and then the data inside three FIFO would be sent to data bus through parallel scheme. After DSP reads data according to its clock signal RD_CLK, the corresponding Doppler frequency would be concluded through spectrum extraction algorithm, and the velocity information would be demodulated at last.

At the moment, the signal processing of Radar Doppler velocity measurement system is mostly going on inside frequency domain, mainly taking FFT (Fast Fourier Transform, FFT) as natural frequency estimation algorithm. The frequency can be extracted from low signal-to-noise ratio signal efficiently by adopting FFT technology to process Doppler signal. This system adopts TMS320C6747 chip of TMS320C6747 of TI Corporation, whose dominant frequency is 300MHz. The Experiment Result Shows that this chip can finish 1024 FFT calculation and satisfy the system requirement of operating precision and speed.

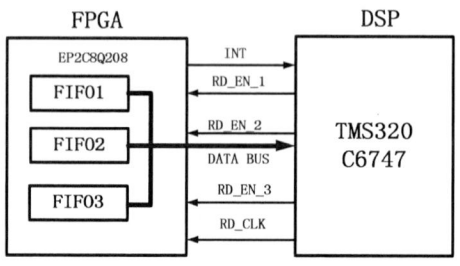

Figure 5. Interface between FPGA and DSP

VI. EXPERIMENT

A. Testify Frequency extraction algorithm using signal generator

The AFG3102 signal generator of Tektronix Corporation is selected as signal source in the experiment. Load the three series of signals produced by two signal generators to analog signal port of signal processing circuit to replace the output signal of photodetector to conduct the experiment. The expeiment photo is as shown in Fig. 6.

Figure 6. Experiment of Frequency extraction

In order to evaluate the accuracy of frequency extraction algorithm, we choose six series of signal whose separate frequency is: 0.5MHz, 1MHz, 5MHz, 10MHz, 20MHz and 30MHz. Take each frequency for 100 times. The detailed data are as shown in table 1.

TABLE I. FREQUECY THROUGH CCS

Frequency of input signal (MHz)	Mean of measurement (Hz)	root-mean-square error	Mean error(Hz)	Standard error (Hz)	Relative error (%)
0.5	500480.8	642.22	480.8	427.9	0.0962
1	1001057	1000	1057	877.3	0.1057
5	5004831	2236.07	4831	4607.5	0.0966
10	10010073	3162.28	10073	8434.9	0.1007
20	20020937	4472.14	20937	17209.4	0.1047
30	30033210	5477.23	33210	23506	0.1107

B. Field experimental test

In the actual application of three-dimensional LDV, the output frequency of the detector has a louder noise compared to signal generator. Therefore, external field experiment is needed to tesify the performance of prototyping system and velocity measurement algorithm under low-noise circumstance. In the external field experiment, microwave velocimeter and three-dimensional LDV are installed at the same time. With the velocity value measured by microwave velocimeter as reference, we take collimator as optical antenna, and install three antenna to the antenna platform, and then the mesured velocity vector is synthesized to work out the radial velocity of the vehicle. The experiment equipment is installed on the automobile running on expressway.

The data of a 60s experiment are shown as in Fig. 7, from which we can see that the the data between radar velocity measurement sampling prototype and microwave velocimeter are consistent with regard to velocity trend, with a great curve change of the radar velocity measurement sampling prototype.

The reason of this is that the beam of radar velocity measurement sampling prototype is narrow and a slight vibration of the vehicle can cause a major change of the beam angle. Meanwhile, the microwave velocimeter has a broad transmitted beam which is immune to this kind of slight vibration. Therefore, when the vibration detected by tri-beam laser velocimetry radar prototype in three directions is accumulated, the velocity wave has a major fluctuation, which states that there is a good relativity between the velocity measurement result of this laser volocimeter and target velocity. It also tesifies the comparatively high practicability and reliability of the prototype back-wave signals detection and processing circuit designed in this paper.

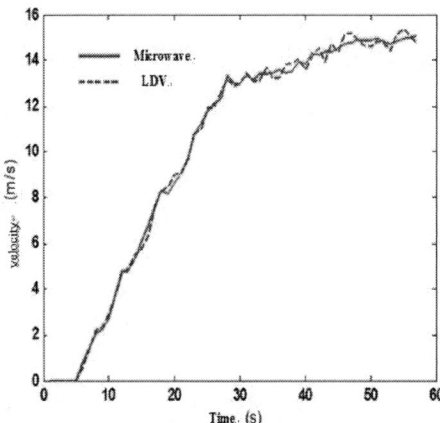

Figure 7.   Field experiment velocity curve

VII.   CONCLUSION

Aiming at the defect that the traditional single beam and double beam LDV cannot measure the vehicle velocity signal at various directions simultaneously, this paper has designed a three-dimensional LDV. The system design, antenna layout, signal processing circuit and software design have been introduced in detail, the frequency extracting algorithm has been testified by experiment. Through experimental verification, velocity measurement function of this prototype is qualified.

REFERENCES

[1]   ZHANG Zhao-yun, GAO Yang, ZHAO Xing-hai, ZHAO Xiang, Laser Doppler velocity measurement based on self-mixing effect [J]. *Infrared And Laser Engineering*, 2009. 38(Supplement): 343-346,2009.

[2]   ZHOU Jian,LONG Xing-wu, Laser Doppler velocimeter based on Janus configuration [J]. *Journal of Optoelectronics.Laser*, 2011. 22(2): 266-270,2011.

[3]   D. Pierrotteta, F. Amzajerdianb, L. Petwayb, etl. Navigation Doppler Lidar sensor for precision altitude and vector velocity measurements: flight test results. Proc. of SPIE Vol. 8044（2011）, doi: 10.1117/12.886826.

[4]   Ma Zongfeng, Zhang Chunxi ,Wang Xiaxiao, Wang Jiqiang, All-fiber laser radar at 1.55μm for speed measurement[J]. *Journal of Beijing University of Aeronautics and Astronautics*,34(05): 596-597,2008.

Investigation of focal ratio degradation caused by stress in Large-Core Astronomical Fibers

Yunxiang Yan, Ruichen WANG, Weimin Sun*, Yongjun Liu*

Key Laboratory of In-Fiber Integrated Optics, Ministry of Education, Harbin Engineering University
liuyj@hrbeu.edu.cn or sunweimin@hrbeu.edu.cn

Abstract—The focal ratio degradation (FRD) properties of optical fibers in the Large Sky Area Multi-Object Fiber Spectroscopy Telescope (LAMOST) caused by different stress were investigated by employing an Energy Distribution Method (EDM). We analyzed the FRD in two conditions with uniform stress and non-uniform stress. The qualitative and quantitative analyses on the output light spot from the optical fibers show that the output focal ratio is stable with uniformly compressed while FRD becomes more serious with the increasing non-uniform stress.

Keywords-optical fiber; focal ratio degradation; stress

I. INTRODUCTION

Optical fibers have been used in astronomical spectroscopy for years for its high transmission efficiency [1]. The LAMOST can observe 4000 celestial bodies at the same time by transmitting the light to the optical spectrum analyzer (OSA) via 4000 optical fibers. Transmission and focal ratio degradation have an important influence on the performance of the optical fiber telescope. The FRD is invariant in the ideal straight fiber without considering the absorption. While in the practical application, the bending [2, 3], the dispersion [4] and the stress [3] will cause the FRD. WANG GANG, et al. presented the qualitative analyses on the influence of the stress [5]. And in this paper, we tested the feasibility of Energy Distribution Method (EDM) and the reported the quantitative analyses in different conditions with different stresses.

II. EXPERIMENTS

In this paper, we choose the fiber used in LAMOST to investigate the FRD caused by stress. It is a large-core multimode fiber with a core of 320μm. The experiments include 2 parts. First, we test the feasibility of Energy Distribution Method. Then, we research the influence on FRD caused by uniform stress and non-uniform stress.

A. Testing Energy Distribution Method and Setting the input Light Source System

Fig. 1 is the schematic of the experiment system. We choose the LED as the light source to simulate the light of stars. According to the definition

$$F/\# = f/D, \qquad (1)$$

while the distance between the aperture (f) and the focal point of the lens (e) is $f=100.0\pm1.0$mm, and the diameter of

the aperture is $D=20.00\pm0.01$mm. The input focal ratio is $F/\#=5.00\pm0.05$.

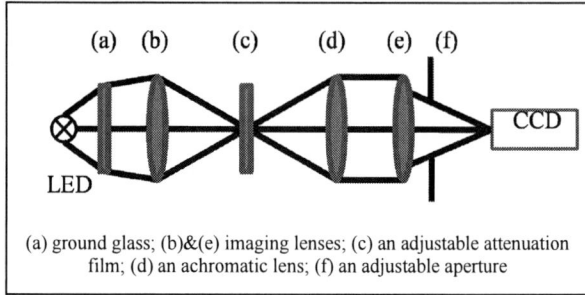

(a) ground glass; (b)&(e) imaging lenses; (c) an adjustable attenuation film; (d) an achromatic lens; (f) an adjustable aperture

Figure 1. The schematic of the experiment system

We use EDM to process the images of output spots taken by CCD to acquire the input focal ratio. EDM calculates the radium of the output spot by searching a circle inside which the encircled energy (EE) is a certain percent energy, for example 90%, of the total energy. We chose a series of output spots separately to give a series of diameters [6]. The results are as Fig. 2 shows

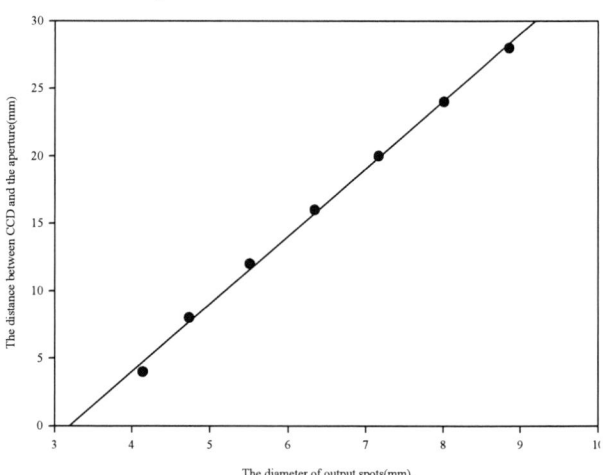

Figure 2. The fitting curve of the output spots' diameters with input focal ratio of $F/5.00$. According to the definition $F/\#=f/D=\Delta f/\Delta D$, the slop of the fitting curve is the focal ratio

The focal ratio of the light source is shown in Tab. I.

TABLE I.   THE FOCAL RATIO OF INPUT LIGHT

Theoretic Value	Focal Ratio of Input Light System		
	EE85	*EE90*	*EE95*
5.00±0.05	5.17	5.04	4.99

The results show that EDM with EE95 is feasible and better for the input light source system with the input focal ratio of *F*/5.00.

B. The Influence on FRD with Different Stresses

In the experiment, we qualitatively and quantitatively analyzed the influence on FRD with two styles of stresses, uniform stress and non-uniform stress. The experiment system was built as Fig. 3 shows. The input light system was the same as in Fig. 1. The large-core fiber (b) is fixed on an X-Y-Z axis platform (a), and then the fiber connects to the receptor, a CCD camera, on the other end. A computer is used to process the data using EDM.

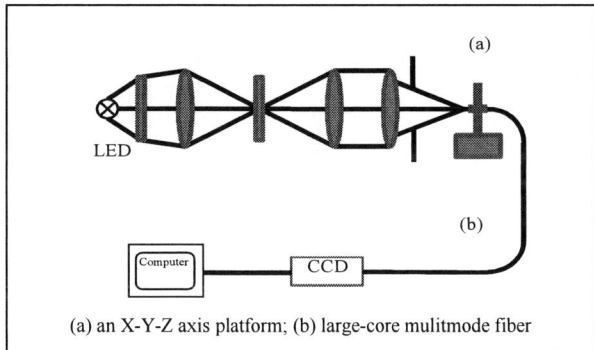

(a) an X-Y-Z axis platform; (b) large-core mulitmode fiber

Figure 3.   The schematic of the experiment system

1) Influence on FRD with uniform stress.

As is shown in Fig. 4, two fibers were placed in parallel on a substrate (1cm (L) ×1.0cm (W) ×1.0cm (H)) made of PTFE to set the uniform stress from 0N to 100N with 20N step on the fiber separately in 30cm and 50cm away from the input end of the fiber, so each of the fiber bears half the stress every 10N from 0N to 50N. In order to make sure the fiber is fixed, we regulate the platform scale up and down to set different stresses during the experiment. The resolution of the platform scale is 1g, and acceleration of gravity is chosen as $g=10m/s^2$.

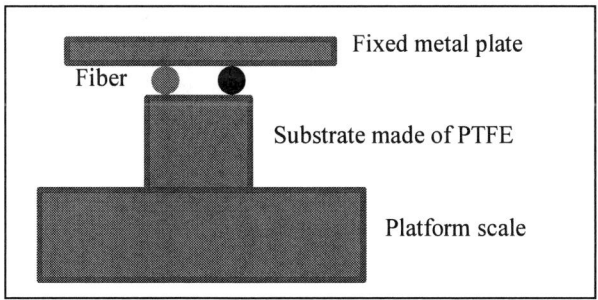

Figure 4.   Two fibers were placed in parallel on a substrate made of PTFE.

The output focal ratio of fibers is shown in Tab. II and Fig 5 shows the trend of the output focal ratio.

TABLE II.   THE FOCAL RATIO OF UNIFORM STRESS CONDITION

Distance	Uniform Stress					
	0N	*10N*	*20N*	*30N*	*40N*	*50N*
30cm	4.73	4.69	4.76	4.72	4.74	4.72
50cm	4.74	4.69	4.70	4.69	4.73	4.71

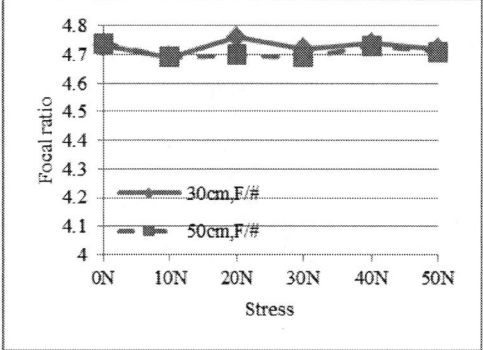

Figure 5.   The trend of output focal ratio with increasing uniform stress

Fig. 6 shows some of the output spots taken by CCD. The upper 3 images are the spots changing with increasing distance. The lower 3 images are the output spots taken in the same position with different stresses.

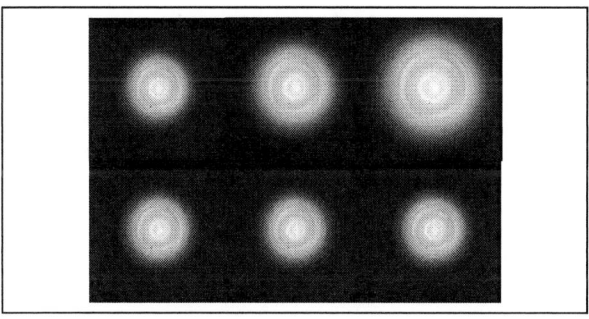

Figure 6.   The output spots taken by CCD in uniform stress condition. The upper 3 images show the diameters' change of the output spots with increasing distance between CCD and the aperture. The lower 3 images are taken in the same position, and the diameters of the output spots are relatively the same.

It's reasonable to notice that the focal ratio are 4.73 and 4.74 without stress, both less than the input focal ratio of 4.99 since the existence of FRD in fibers caused by the absorption, the dispersion and some other factors as in [2-4]. And the output focal ratio is relatively stable with the increasing uniform stress. The output spots are basically the same size though fibers bear on different stresses.

2) Influence on FRD with non-uniform stress.

In LAMOST, the location of fiber ends is controlled by a double-axis rotation device. The fibers would twist and touch the main axis when fiber rotates, so the fiber would be under non-uniform stress.

To investigate the influence, we tested the FRD caused by non-uniform stress bearing on the fiber in 30cm away from the input end. The experiment system is almost the same as in Fig. 4, and we use a rubber substrate instead as Fig. 7 shows.

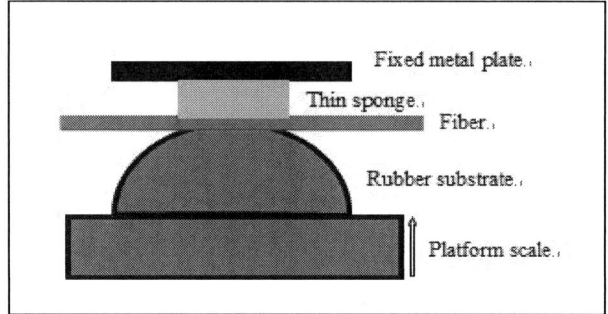

Figure 7.   Using a rubber substrate to set non-uniform stress on the fiber.

In this case, only one fiber was fixed on the substrate, so the stress was also set every 10N from 0N to 50N. The output focal ratio is shown in Tab. III.

TABLE III.   THE FOCAL RATIO OF NON-UNIFORM STRESS CONDITION

Distance	Non-uniform Stress					
	0N	10N	20N	30N	40N	50N
30cm	4.72	4.64	4.54	4.43	4.30	4.01

Fig. 8 shows the trend of the output focal ratio decreasing with the increasing non-uniform tress. So the FRD is more serious than that in uniform stress condition.

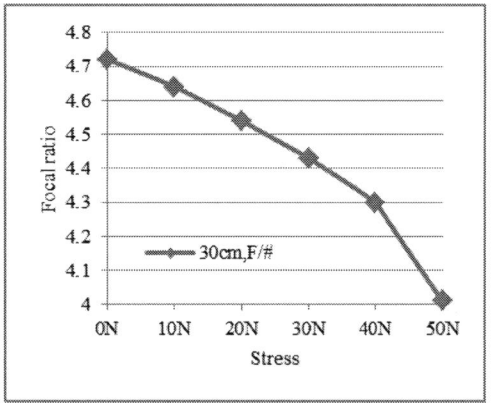

Figure 8.   The curve shows that the output focal ratio decrease with the increasing non-uniform stress.

And Fig. 9 shows some of the output spots taken by CCD in non-uniform stress condition.

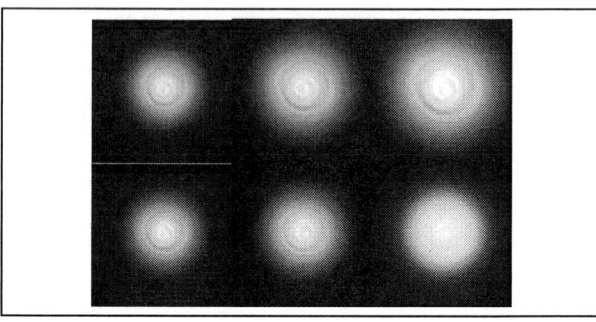

Figure 9.   The output spots taken by CCD in non-uniform stress condition. Compared with the output spots in Fig. 6 in uniform stress condition, the diameters of the upper 3 images change much faster, so the output focal ratio is smaller with the same input focal ratio of $F/5.00$. And in the lower 3 images, the spots are more dispersed with the increasing stress, though they are taken in the same position, showing that the FRD is more serious.

III.   CONCLUSION

In this paper, we confirmed the feasibility of the EDM in investigating the influence on FRD caused by stress. The output focal ratio decreases from 4.99 to ~4.73 for the existence of FRD even without any stress. And the output focal ratio is relatively stable in uniform stress condition while decreasing fast with the increasing non-uniform stress, and the output spots are more dispersed, decreasing the density of the output spot's energy and making FRD more serious, which is unfavorable for spectrum analysis.

ACKNOWLEDGMENT

This work is supported by the Natural Science Foundation of China (No. 11078009 & No. 61307005), 111 projects (B13015), to the Harbin Engineering University and the Opening Project of Key Laboratory of Astronomical Optics & Technology, Nanjing Institute of Astronomical Optics & Technology, Chinese Academy of Sciences.

REFERENCES

[1]   J.R.P Angel, M.T. Adams, T.A. Boroson, et al, "A very large optical telescope array linked with fused silica fibers," Astrophysical Journal (S0004-637X), 218(1), 1977, pp. 776-782.

[2]   D. Yan, Z. Bing, "Analysis of focal ratio degradation of light-transmitting optical fiber in LAMOST," Journal of University of Science and Technology of China, vol. 37(6), 2007, pp. 606-611.

[3]   W. Xiaoke, G. Jian, G. Pengcheng, et al, "A fiber feed system for a multiple object Doppler instrument at Sloan Telescope," Pro. Of SPIE (S0277-786X), 6269, 2006, pp.62692T-1-9.

[4]   C.L. Poppet, J.R. Allington-Smith, "Fibre systems for future astronomy: anomalous wavelength–temperature Effects," Monthly Notices of the Royal Astronomical Society (S0035-8711), 379(1), 2007, pp.143-150.

[5]   W. Gang, J. Xiaojun, L. Men, "The Performance of Fiber Optics for LAMOST," ACTA ASTROPHYSICA SINICA, vol. 20 Supp., 2000, pp.65-70.

[6]   J.L. Xue, "The performance measurement and analysisi of the fiber used in lLAMOST,"[M] Harbin Engineering University, China, 2013, pp. 42–45.

Broadband optical beam power splitter for wavelength dependent light circuits on silicon substrates

Zhiyong Li [1]*, Jiejiang Xing [1], Biao Yang [1] and Yude Yu [1]

1 State Key Laboratory on Integrated Optoelectronics, Institute of Semiconductors, Chinese Academy of Sciences, P.O. Box 912,
Beijing, 100083, China
* lizhy@semi.ac.cn

Abstract—**A broadband optical splitter based on silicon-on-insulator (SOI) is proposed for uniform beam power splitting in the wavelength from 1450 nm to 1650 nm, which has low wavelength dependence of less than 0.2dB in the arrange of about 100 nm. The novel types of splitters are predicted to have a great potential in the field of broadband light circuits for wavelength-divided multiplexing (WDM).**

Keywords-beam power splitter; broad-band; silicon photonics; light circuits

I. INTRODUCTION

Recently, silicon photonics has become an attractive technology for chip-scale optical integration due to its compatibility with standard CMOS technology at high reliability and low cost. In many multi-port optical devices such as optical switches and cross-bar meshes, optical power splitters are the most important building blocks for kinds of optical components. An optical beam power splitter can be developed by many waveguide configurations [1], including directional coupler (DC) [2] and multimode interference (MMI) [3], as shown in Fig. 1. However, most of these waveguides are wavelength dependent or have narrow operation bandwidth, which is a big stone on the way of wavelength division multiplexing (WDM) [4] light circuits on SOI substrate. So, a broadband 3dB splitter with high uniformity attracts the attention of many researchers [5]. In this paper, a silicon-based 2×2 broadband splitter is proposed, which has an adiabatic mode evolution region consisting of two tapered waveguides with the simultaneously varied distances between the two waveguides.

Among of uniform splitters reported by other research teams, broadband performance normally require two waveguide regions, the first one works for tapered coupling by varied distance with the constant waveguide widths along the light propagation region in the splitter [1]. And it is followed by the tapered waveguides with the constant distance for the enhanced mode transition. These designs lead to so much long splitter and large excess loss. Therefore, adiabatic couplers with simultaneously tapered both waveguides widths and distances should be focused for the tapered coupling and mode profile, as shown is Fig. 2 (a). As the coupled-mode theory [6] predicts, this adiabatic coupler has two system modes, both even and odd mode, in these strongly coupled optical

waveguides. According to simulation, even mode and odd mode can carry the same optical power, which is 3dB beam splitting. The novel splitter proposed here has high performance in splitting uniformity and broad bandwidth due to sufficiently adiabatic mode evolution, which also means the shorter device length and less excess loss compared to the previous works.

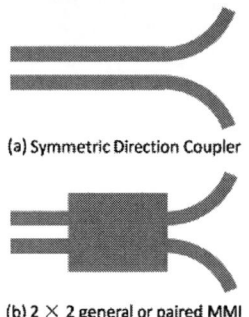

Figure 1. Comparison of beam splitters, including (a) a symmetric direction coupler (DC), (b) a 2x2 general interference (GI) or paired interference (PI) multimode interference (MMI)

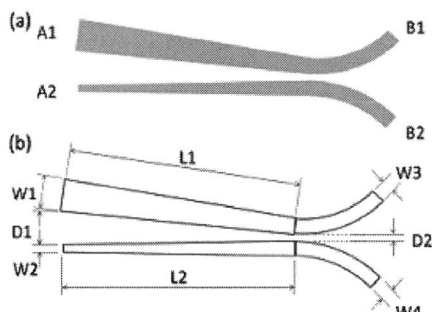

Figure 2. Proposed 2x2 adiabatic splitters, (a) the schematic diagram of the novel design, (b) geometric parameters of transition regions in the adiabatic coupling waveguides

II. DEVICE DESIGN

The schematic diagram of the novel design is shown in Fig. 2. The proposed 2x2 adiabatic splitter have two different-size optical waveguides, which are A1-B1 and A2-B2, placed

978-1-4799-1215-5/13 $31.00 © 2013 IEEE

closely to each other in the same plane. The width of waveguide A1-B1 varies from W1 at the starting end to W3 at the exit end, and the width of waveguide A2-B2 begins with W2 and ends at W4. As the baseline of simulation, W1 is 600 nm which is linearly tapered to 500 nm (W3=W4), with W2 of 400 nm. The distance between two waveguides is linearly tapered from D1 of 1100 nm to D2 of 200 nm. In the reference model, the adiabatic region length (L1=L2) is set as 300 μm. All waveguides are built on SOI substrates, which have the top silicon thickness of 340 nm and the buried silica thickness of about 2 μm. Assumed as etching depth of 260 nm, an optical rib waveguide is defined and simulated by 3 dimension beam propagation method (3D-BPM). The electric field distribution of single mode in the waveguide is shown in Fig. 3 (only TE polarization is considered in this work) and the effective index of it is estimated at the value of about 2.9.

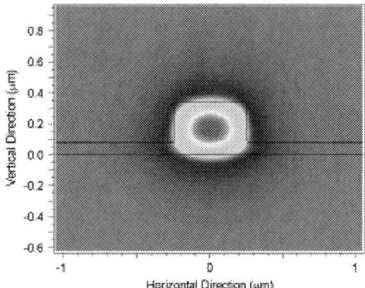

Figure 3.   An eigenmode (TE) profile of silicon optical rib waveguides, with the width of 500 nm and the etched depth of 260 nm

III.   SIMULATION RESULTS

It can be proven by simulation that, when gap width (D2) at the narrowest location varies from 100 to 300 nm, high uniformity of less than 1 dB can be obtained. Considering the fact that 180 nm CMOS fabrication can be accessed to, the gap width of about 200 nm is adopted in this design. This width value make the non-uniformity of beam splitting less than 1 dB, as shown in Fig. 4 (a). For the width W3 (or W4) at the exit end of the coupler waveguides, the calculated results show a lower non-uniformity (~ 0.2 dB) compared with that of gap width D2, as shown in Fig. 4 (b). For uniformity operation, the exit end widths of coupler waveguides are suggested as about 300 nm, which can be further optimized. Fig. 4 also shows that the mode evolution region length L1 (or L2) has the effect on non-uniformity. For more than 300 μm, splitting uniformity can catch up to 0.2 dB and the proposed splitter is fabricated with a large fabrication tolerance of more than 1 μm, which can be improved further in the future. Fig. 5 shows the simulated propagation mode distribution for the reference model of the calculated adiabatic couplers. The beam power ratio between the two waveguides of the splitter is much approximate to 3dB, with small non-uniformity of about 0.17 dB in the wavelength from 1450 nm to 1650 nm.

IV.   CONCLUSION

An adiabatic 2×2 3dB splitter is theoretically demonstrated. This splitter has its mode evolution and coupling tapered simultaneously. Simulation results show a high uniformity of less than 0.2 dB for a large operation wavelength range of more than 100 nm.

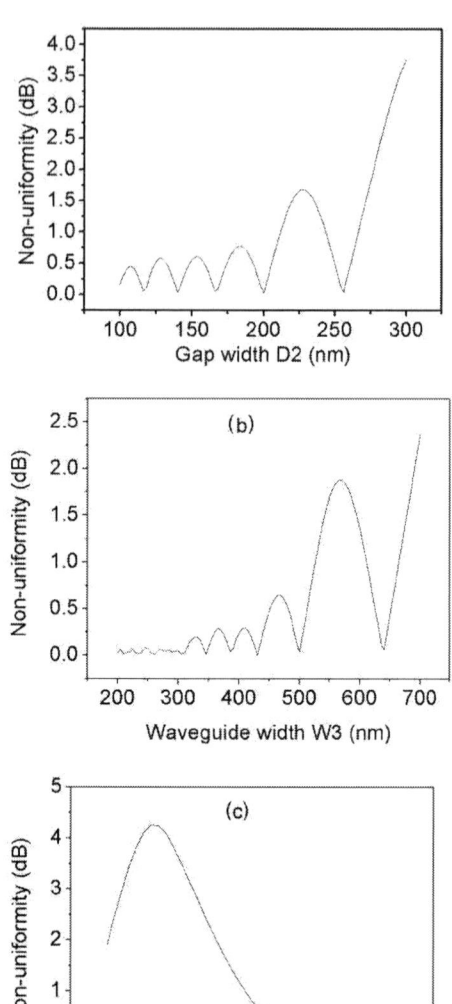

Figure 4.   Calculated non-uniformity of beam power splitting for (a) varied gap widths D2 at the narrowest location, (b) varied rib waveguide widths at the exit end, and (c) varied coupling region lengths L1 (or L2) at the starting end

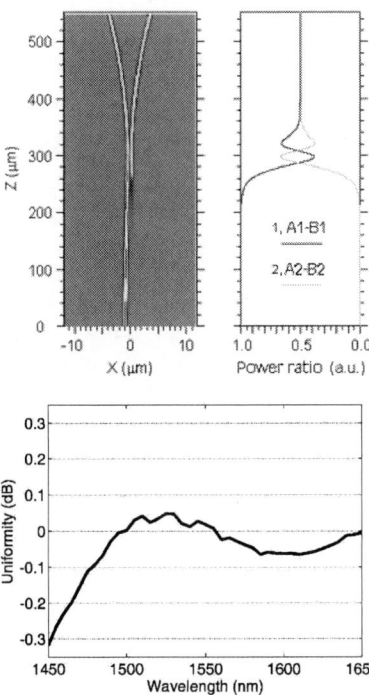

Figure 5. Propagating mode distribution and weak wavelength dependent splitting of the novel optical beam power splitter as the reference model, W1 = 600 nm, W2 = 400 nm, W3 = W4 = 500 nm, D1 = 1100 nm, D2 = 200 nm, L1 = L2 = 300 μm

ACKNOWLEDGMENT

This work is supported by the National Basic Research Program of China (Grant No. 2011CB301701), the Knowledge Innovation Program of the Chinese Academy of Sciences (Grant No. KGCX2-EW-102), and the National Natural Science Foundation of China (Grant No. 61275065).

REFERENCES

[1] J. J. Xing, K. Xiong, H. Xu, Z. Y. Li, X. Xiao, J. Z. Yu, and Y. D. Yu, "Silicon-on-insulator-based adiabatic splitter with simultaneous tapering of velocity and coupling," Optics Letters, Vol. 38, pp. 2221-2223. 2013.

[2] A. Prinzen, M. Waldow, and H. Kurz, "Fabrication tolerances of SOI based directional couplers and ring resonators," Optics Express, Vol. 21, pp. 17212-17220. 2013.

[3] H. Shahoei, D.-X. Xu, J. H. Schmid, J. P. Yao, "Photonic Fractional-Order Differentiator Using an SOI Microring Resonator With an MMI Coupler," IEEE Photonics Technology Letters, Vol. 25, pp.1408-1411. 2013.

[4] D. X. Xu, A. Densmore, P. Waldron, J. Lapointe, E. Post, A. Delâge, S. Janz, P. Cheben, J. Schmid, and B. Lamontagne, "High bandwidth SOI photonic wire ring resonators using MMI couplers," Optics Express, Vol. 15, pp. 3149-3155. 2007.

[5] M. R. Watts, W. A. Zortman, D. C. Trotter, R. W. Young, and A. L. Lentine, "Low-Voltage, Compact, Depletion-Mode, Silicon Mach–Zehnder Modulator," IEEE J. Sel. Top. Quantum Electron. Vol. 16, pp. 159-164. 2010.

[6] A. Yariv, "Coupled-mode theory for guided-wave optics," IEEE Journal of Quantum. Electronics, Vol. QE-9, pp. 919-933. 1973.

The influence on FRD with different encircled energy

Ruichen Wang[1], Yunxiang Yan[1], Yongjun Liu[1*], Hongquan Zhang[2], Weimin Sun[1*]

1 Key Lab of In-fiber Integrated Optics, Ministry Education of China, Harbin Engineering University, Harbin, China
2 The 49th Research Institute of China Electronics Technology Group Corporation, Harbin, China
liuyj@hrbeu.edu.cn or sunweimin@hrbeu.edu.cn

Abstract—**An energy distribution method was used to calculate the focal ratio, verifying the feasibility of this method by testing of optical fiber's input and output focal ratio to determine the appropriate encircled energy (EE) in practical application. This method had good repeatability, and it provided an effective way for quantitative research of the focal ratio degradation (FRD).**

Keywords-component: optical fiber; focal ratio degradation; method; quantitative research.

I. Introduction

Because of the high transmission efficiency, fiber is used universally in the field of astronomy [1], such as Integral field spectroscopy and Multi-object spectroscopy [2]. There are two main factors influencing the performance of Multi-object spectroscopy technology, namely optical fiber's characteristic of transmittance and focal ratio degradation (FRD) [3]. The transmittance of fiber can be optimized by adjusting fiber materials and some other factors. However, FRD is very complicated and accurate FRD measurements are proven to be challenging, and repeatability of the results from group to group has been elusive [4].

II. Energy Distribution Method

Due to focal ratio degradation not only makes the output spot of the fiber aggravating the degree of dispersion, it also seriously causes the loss of energy, which is very bad for information analysis in the spectrometer, also undesirable in the astronomical use. Therefore, we put forward the energy distribution method to measure the focal ratio.[5]

The basic principle we used is to measure the spot size in different position apart from the focal point, and acquire a series of image data, and then calculate the focal ratio by using an energy distribution method. The spot size is figure out by processing the CCD's spot image. Theoretically, the size of the light spot contains 100% of incident light energy within a circular, but in fact, mainly because of two reasons we can't get the corresponding circle containing 100% energy: 1. The experimental instrument cannot image ideally, so it is hard to measure the faint halos associated with optical phenomenon such as diffraction and aberration; 2. The environmental noise light and dark current of the measurement instruments cause the existence of background noise all the time.

First we take an image of a light spot by CCD, then use the mass center method to solve the gray value center of the whole image matrix, set this center as the center of a circle,

and choose variables *r*, then calculate the sum of the energy within the radius *r* of the circle. When *r* increases to *R*, the sum of the energy equals to a certain percentage of the total energy contained in whole image matrix, we identify the 2*R* with the spot diameter *D*. We named the energy within the radius *r* encircled energy (EE).

Our work is to find out the appropriate EE of the effective energy corresponding to the theoretical value.

III. Experimental Data

Experimental schematic is shown in Fig.1. We choose a white-light LED as the light source to simulate the light of stars. The light emitted from the LED diffusion through the frosted glass (a), then the beam of light focused by lens (b). We put an adjustable attenuation film (c) in focus point behind the lens (b) to adjust the light intensity. After being transmitted through a collimator (d), the light is converged again through a lens (e). An iris-diaphragm (f) is placed close to the lens (e) to change the aperture *D*. The CCD can be moved back and forth along the optical axis direction to capture the light spot image in different location away from focus point.

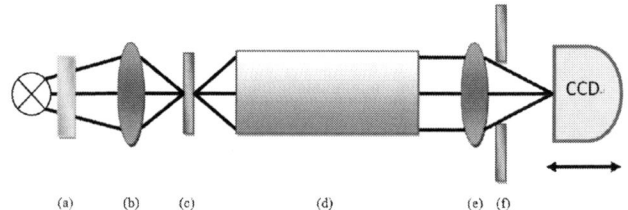

Figure 1. Experimental schematic

A. Theoretical Calculation

According to the definition, the distance between aperture and focal point is f=100.0$\pm$1.0mm, and we use iris-diaphragm to set the diameter of the aperture D=20.00$\pm$0.01mm, so the input focal ratio is $F\#$=5.00$\pm$0.05.

978-1-4799-1215-5/13 $31.00 © 2013 IEEE

B. measuring and processing

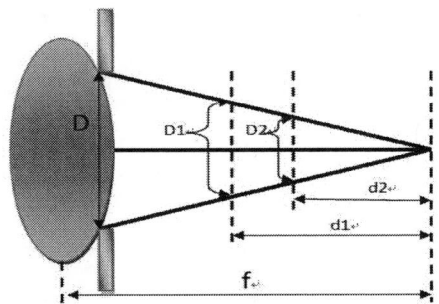

Figure 2. Measuring method

As is shown in Fig. 2, when the distance between CCD receiving screen and the focus point is d_1, the light spot diameter is D_1 (the diameter is calculated by energy distribution method in chapter II); When the distance is d_2, we gets D_2; ...... When distance is d_n, the diameter is D_n. Do $d \sim D$ function, according to the definition of focal ratio *F/#*:

$$F\# = f/D \qquad (1)$$

And the geometric relationship：

$$d_1/D_1 = d_2/D_2 = \ldots = d_n/D_n = f/D \qquad (2)$$

Namely the focal ratio *F#* equals to the slope of the function.

We use different EE percent value to process the light spot image collected by CCD separately, filter out the result with large deviations, and find that when the encircled energy divide by the total energy equal to 85% (EE85), 90% (EE90), 95% (EE95), these three sets of data are approximately consistent with the theoretical value, the graph of *d-D* function are shown in Fig. 3.

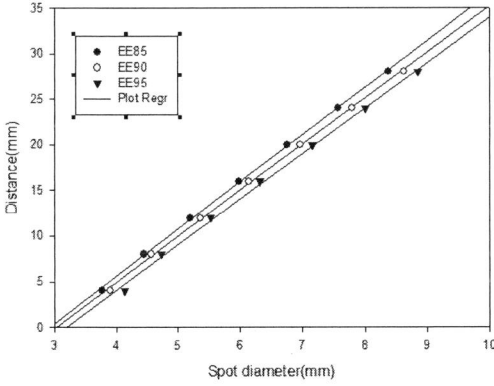

Figure 3. Measuring method

After many repeated experiments, we have tested analysis the three values of EE. As is shown in table 1, the data is stable, and mean variance is small, so we think the Energy distribution method is feasible. Among the three, the data of EE95 is closer to the theoretical value (5.00)

compared to EE85, EE90, so the value of EE95 is most appropriate in practical application.

TABLE I.   RESULT OF REPEATED EXPERIMENT

EE	$F\#_{1ST}$	$F\#_{2ND}$	$F\#_{3RD}$	$F\#_{4TH}$	$F\#_{5TH}$	$\overline{F\#}$	σ
85	5.16	5.15	5.17	5.18	5.17	5.17	0.010
90	5.02	5.03	5.04	5.04	5.04	5.03	0.008
95	4.97	4.97	5.00	5.00	5.00	4.99	0.014

IV.   OPTICAL FIBER TESTING

We have picked the suitable measurement method, and our purpose is mainly investigating the FRD properties of the optical fiber. So we had a long-length optical fiber for testing experiments by using the Large-Core Astronomical Fibers.

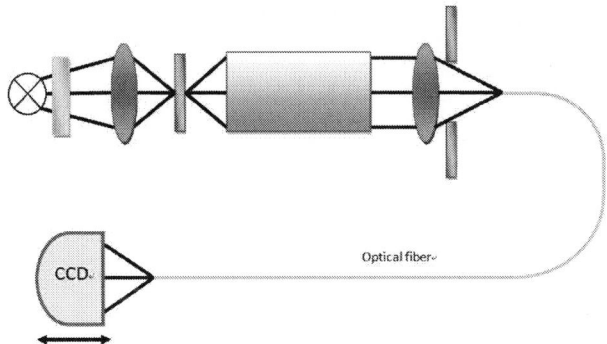

Figure 4.   Experimental schematic of optical fiber testing

The forepart of the experiment instruments are same as before, and the two-meter freely bending fiber is placed on the experimental platform. The incident end of the fiber is directed at the focus of the last lens, and CCD is placed behind the exit end of the optical fiber to capture the light spot image. We still use the energy distribution method to calculate light spot diameter, and one group of spot image are shown in figure 5.

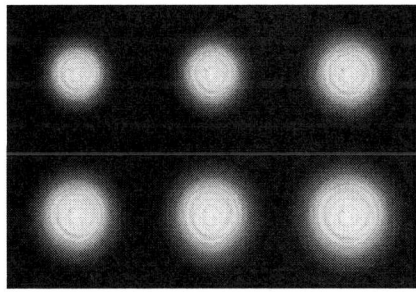

Figure 5.   One group of output spot image

We processed the image data with EE85, EE90, EE95 separately, and obtained the results are shown in table II. It agrees well with our prior testing results of optical fiber, which indicates we can use this method for quantitative research of the FRD in optical fiber.

TABLE II.   RESULT OF FIBER OPTIC TESTING

	EE85	EE90	EE95
Focal ratio	4.67	4.54	4.49

V.   CONCLUSION

In this paper, we test and verify the feasibility of the Energy distribution method in investigating the focal ratio, and draw the conclusion that EE95 is the most appropriate in practical application. This method has good repeatability. Although this method cannot eliminate the elusive results from group to group completely, it provides an effective way for quantitative analysis of the FRD in optical fiber from different sources.

ACKNOWLEDGMENT

This work is supported by the National Natural Science Foundation of China (No.61077047,No.61307005), 111 project (B13015), to the Harbin Engineering University and the Opening Project of Key Laboratory of Astronomical Optics & Technology, Nanjing Institute of Astronomical Optics & Technology, Chinese Academy of Sciences.

REFERENCES

[1] J.R.P Angel, M.T. Adams, T.A. Boroson, et al, "A very large optical telescope array linked with fused silica fibers," Astrophysical Journal (S0004-637X), 218(1), 1977, pp. 776-782.

[2] Sun Weimin, Xue Jinlai, Yu Haijiao, Yan Yunxiang, Liu Xiaoqi, Jiang Yu," The influence on output spot of large core fibers in noncentral incidence" Proceedings ofSPIE,Vol.1,2012,pp.261-264.

[3] Ramsey Lawrence W, "Focal ratio degradation in optical fibers of astronomical interest", Astronomical Society of the Pacific,(S1050-3390),1988,3:26-39.

[4] Jeremy D. Murphy, Phillip J. MacQueen, Gary J. Hill, et al, "Focal Ratio Degradation and Transmission in VIRUS-P Optical Fibers", Advanced Optical and Mechanical Technologies in Telescopes and Instrumentation, Proc. of SPIE Vol. 7018, 70182T, (2008),doi: 10.1117/12.788411.

[5] J.L. Xue, "The performance measurement and analysisi of the fiber used in lLAMOST,"[D] Harbin Engineering University, China, 2013, pp. 46-47.

Supermodes Coupling Characterizaion of Optical Fiber Brush

Xiaoliang Zhu[1,2], Qi Yan[1], Haijiao Yu[1], Weimin Sun[1]*

1 Harbin Engineering University, Harbin, China
2 Zhejiang Gongshang University, Hangzhou, China
zhuxiaoliang@mail.zjgsu.edu.cn

Abstract—A special kind of fiber device which we called it as optical fiber brush which would like to be used in astronomy is introduced in our paper. In this paper, we investigate its fabricating technique and the supermodes coupling characterization. In order to investigate and compare the divergence angle of different supermodes, near-field and far-field distributions of them are also calculated. Comparing with the far-field mode radius, we find the in-phase supermode has the least divergence angle and the best beam quality. The research could be helpful to improve the beam quality and to transmit stars' images in astronomy.

Keywords-supermode; optical fiber brush; microstructured fiber; astronomy

I. INTRODUCTION

Fiber is used in astronomy to capture the light of star and transfer it to the fiber spectrometer. But there is the problem to transfer the circle spot of the star image to a slit in the spectrometer. How to achieve the best beam quality and improve the optical information integration is the key.

Optical fibers which have multiple cores in one cladding would become very important for various applications in the fields of optical communication [1, 2], optical sensors [3-5], multiple optical couplers [6, 7]. Also, the multicore fiber (MCF) has attracted considerable attention in the astronomical field because of the opportunity of integration and achieving enhanced power [8, 9].

We have introduced a novel multicore fiber device and called it as the optical fiber brush (OFB) in our previous paper [10]. The light propagation characteristics was analyzed in our previous work. However, to our knowledge, the supermodes of the astronomical optical fiber brush has not been investigated. As the cores are close enough in OFB, the core array generates various supermodes depending on the relative phase difference between neighboring cores owing to evanescent coupling in adjacent cores. The in-phase supermode (all cores have the same phase) has the best beam quality and is the desired operation mode.

In this paper, the fabrication technology of the fiber brush is presented firstly. And then the super-mode analysis for describing the beam quality of the OFB is accomplished by neglecting coupling effects between nonadjacent cores.

II. FABRICATION OF ASTRONOMICAL OPTICAL FIBER BRUSH

The structure of an ideal astronomical optical fiber brush are illustrated in Fig. 1. Multiple doped cores arranged on a linear array in the common cladding have the same radius and doping concentration.

The astronomical optical fiber brush was manufactured by inserting fiber cores into microstructure fiber preform in a linear array as shown in Fig. 2(a). And we heat and stretch the complex fiber preform into a fiber taper as shown in Fig. 2(b). Then we cut the fiber taper and the micrograph of the fabricated fiber brush as shown in Fig. 2 (c). The shape of the cores is slightly close to rectangle because of the imperfect control technique of temperature and velocity in the practice fabrication process.

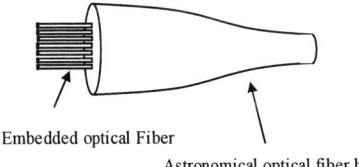

Embedded optical Fiber

Astronomical optical fiber brush

Figure 1. Structure of astronomical optical fiber brush

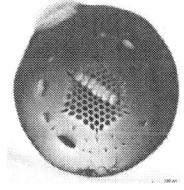

(a) Insert fiber cores into a microstructure fiber

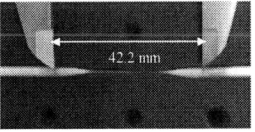

(b) The shape of the taper

National Natural Science Foundation of China (No. 61307005, No. 11078009)

(c) The cross section of the fiber brush

Figure 2. Fabrication of astronomical optical fiber brush

The modes exists in the fiber taper are mainly the guided core mode, only a small part of the optical power is lost into leakage mode and radiating out of the fiber. And the tapering shape was slow change of formation and of structure as it progress. Thus we can use guided mode theory to establish a simple theoretical model to describe these processes.

III. SUPERMODES ANALYSIS FOR OPTICAL FIBER BRUSH

The cross section of the OFB to be treated in this paper is shown in Fig. 2(c). Each core only supports a single mode for the wavelength λ=0.98 μm and has the same physical parameters approximately. The radius of each core is r=3.5 μm, and the distance between two adjacent cores is d=7.5 μm. The refractive index of cores and the cladding are n_1=1.4600 and n_2=1.4572, respectively. The propagation constant and coupling coefficient between two adjacent cores in all the cores are identical.

We use SMF which connected to 980 nm LD as an exciter to excite the middle core of OFB. The other cores in OFB are excited by evanescent wave owing to the adjacent cores are close enough. The supermode theory analysis is presented as follows.

Considering the coupling effects between nonadjacent cores are considered to be quite weak in the first-order approximation, we can neglect them. According to the coupled-mode theory and the mode field superposition theory [11], we can calculate the incident power of each core is 0, 0, 0, 1, 0, 0, 0 for the cores which are linearly distributed from left to right in OFB. Coupling mode equation can be solved as an eigenvalue and eigenvector problem with the initial condition. Based on the linear superposition of supermodes, the amplitude distribution of the supermode is written as [12]

$$E^v(x,y,z) = \left[\sum_m E_m^v E_m(x,y)\right]\exp(\lambda_v z) \qquad (1)$$

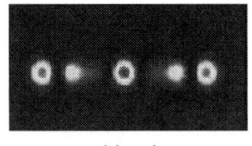

(a) v=1

(b) v=2

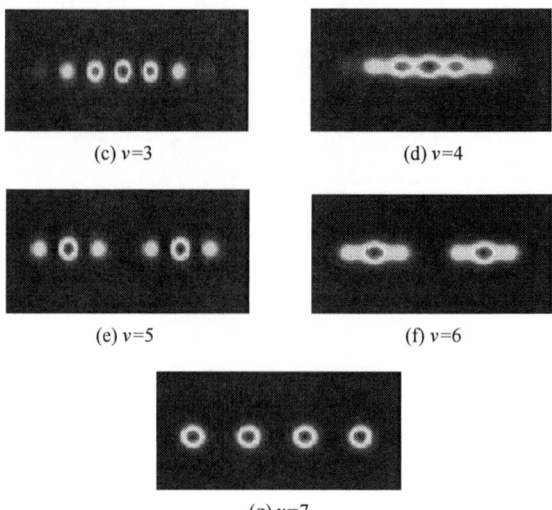

(c) v=3

(d) v=4

(e) v=5

(f) v=6

(g) v=7

Figure 3. Intensity distribution of 5 supermodes excited by fusion splice

where $E^v(x,y,z)$ is the amplitude distribution of the vth supermode. E_m^v denotes the mth element of the eigenvector E^v. λ_v is the vth eigenvalue of coupling mode equation. $E_m(x,y)$ is the transverse field in the mth core. And the transverse field in a single mode fiber can be expressed as

$$E(x,y) = \sqrt{\frac{2}{\pi}}\frac{1}{\omega_0}\exp(-\frac{x^2+y^2}{\omega_0^2}) \qquad (2)$$

where ω_0 is the mode field diameter and can be obtained by using the invariability of propagation constant.

As the output field distribution of OFB is the linear superposition of each supermode. The supermodes for OFB are calculated and illustrated in Fig.4. The modes are numbered as an integral v. When v=4, the supermode is in-phase supermode (all cores have the same phase).

IV. COMPARISON OF BEAM QUALITY

Once the near field is known, the far field distributions according to each supermodes can be calculated using diffraction equation as in (3).

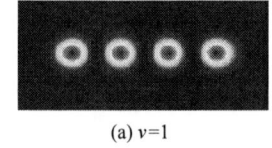

(a) v=1

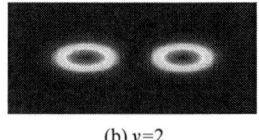

(b) v=2

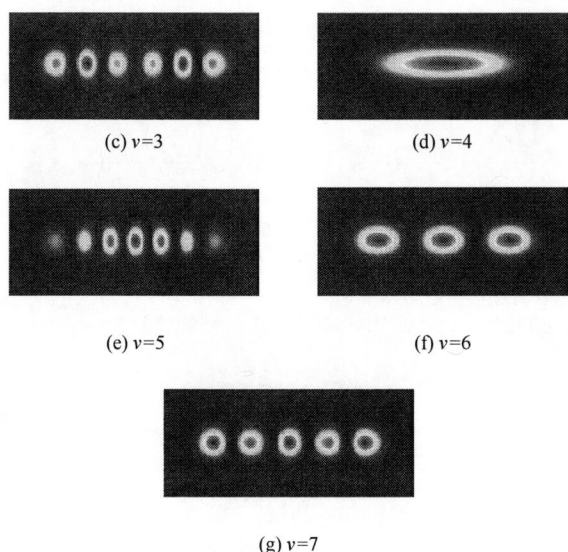

(c) $v=3$ (d) $v=4$

(e) $v=5$ (f) $v=6$

(g) $v=7$

Figure 4.   Far field intensity distribution of supermodes excited middle core

$$E_l^v(x_l,y_l,l)=\left(\frac{e^{-jkl}}{j\lambda l}\right)\times\int\limits_{-\infty}^{\infty}\int\limits_{-\infty}^{\infty}E^v(x_0,y_0,z_0)\cdot e^{\frac{-jk}{2l}\left[(x-x_l)^2+(y-y_l)^2\right]}dx_0dy_0$$

(3)

where x_0, y_0 and z_0 are the coordinate positions at the end of the fiber, and x_l and y_l are the coordinate positions in propagation distance l of light field. $E^v(x_0,y_0,z_0)$ was given by (1). Fig. 4 shows the far-field patterns of supermodes at l=75 μm.

We defines the width of maximum intensity/exponent as the field-radius of supermode. The far field mode radius of supermodes excited by middle core can be calculated when l=75 μm, and are 27.34 μm, 23.80 μm, 28.53 μm, 17.60 μm, 29.12 μm, 26.31 μm and 28.10 μm respectively. It is clearly that the in-phase supermode4 has the best beam quality and least divergence angle.

V. CONCLUSION

Considering the promising application in the field of astronomy, we have investigated the fabricating technique and fabricated a special kind of fiber which we called it as OFB successfully. It can link to single core single mode fiber directly and transfer Gauss mode to linear mode field. The supermodes coupling characterization of OFB was also investigated in this paper. We investigate the supermode characterizations including near-field and far-field. Comparing with the far-field mode radius, we find the in-phase supermode

has the least divergence angle. Thus, if we could adjust the OFB output in phase, the beam quality will be improved and can be used to transmit stars' images in astronomy. And we would like to do this work in the future.

ACKNOWLEDGMENT

This work is supported by the National Natural Science Foundation of China (No. 61307005, No. 61077047), 111 project (B13015) to the Harbin Engineering University and the Opening Project of Key Laboratory of Astronomical Optics & Technology, Nanjing Institute of Astronomical Optics & Technology, Chinese Academy of Sciences.

REFERENCES

[1] S. Inao, T. Sato, H. Hondo, M. Ogai, et. al., "High density multicore-fiber cable," in Proc. International Wire & Cable Symposium (IWCS), pp. 370-384, 1979.

[2] G. Le Noane, D. Boseher, P. Grosso, J. C. Biseul, C. Botton, "Ultra high density cables using a new concept of bunched multicore monomode fibers: a key for the future FTTH networks," in Proceedings of the International Wire & Cable Symposium (IWCS), pp. 203-209, 1994.

[3] J. R. Dunphy, G. Meltz, M. M. El Leil Abou, and E. Snitzer, "Twin-core fiber-optic sensor for simultaneous temperature and strain measurement," in Proc. 3rd Conf. on Opt. Fiber Commun., pp. 58–60, January, 1984.

[4] M. J. Gander, D. Macrae, E. A. C. Galliot, R. McBride, J. D. C. Jones, P. M. Blanchard, J. G. Burnett, A. H. Greenaway, and M. N. Inci, "Two-axis bend measurement using multicore optical fiber," Opt. Commun., vol. 182, pp. 115-121, August, 2000.

[5] W. N. MacPherson, M. Silva-Lopez, and J. S. Barton, et. al, "Tunnel monitoring using multicore fiber displacement sensor," in Proc. 17th Inter. Conf. on Opt. Fiber Sensors, pp. 274–277, May, 2005.

[6] J. P. Donnelly, N. L. DeMeo, Jr., and G. A. Ferrante, "Three-guide optical couplers in GaAs," J. Lightw. Technol., vol. 1, LT-1, pp. 417-424, June 1983.

[7] H. A. Haus and L. Molter-Orr, "Coupled multiple waveguide systems," IEEE J. Quantum Electron., vol. QE-19, pp. 840-844, May, 1983.

[8] Leon-Saval SG, Birks TA, Bland-Hawthorn J, et al., "Multimode fiber devices with single-mode performance," OPTICS LETTERS, vol. 30: 2545-2547, October, 2005.

[9] Joss Bland-Hawthorn, Pierre Kern, "Astrophophotonics: a new era for astronomical instruments," OPTICS EXPRESS, vol. 17, 1880-1884, March, 2009.

[10] Sun Weimin, Yu Haijiao, Yan Qi, Tian Fengjun, Liu Xiaoqi, Jiang Yu, Huang Zhongjun, Hu Zhongwen, "Light propogation in a fiber-brush-shape converter," Proc SPIE Int Soc Opt Eng, vol. 8450: 845014, September, 2012.

[11] Xiaoliang Zhu, Libo Yuan, Jun Yang, and Shouxiu Cao, "Coupling model of standard single-mode fiber and capillary fiber," Appl. Opt., vol. 48, pp. 5624-5628, October, 2009.

[12] J. A. Besley, J. D. Love, "Supermode analysis of fibre transmission," IEE Proc. –Optoelectron., vol. 144, pp. 411-419, December, 1997.

Tapering technique for an embedded microstructure fiber device

Weimin Sun[1]*, Qi Yan[1], Haijiao Yu[1], Xiaoliang Zhu[12]

1 Key Lab of In-fiber Integrated Optics, Ministry Education of China, Harbin Engineering University, Harbin, China
2 Zhejiang Gongshang University, Hangzhou, China
sunweimin@hrbeu.edu.cn

Abstract—**In order to further improve the optical efficiency of the optical fiber image slicer, we designed an embedded microstructure fiber slicer. To reduce the transmission loss, we designed and fabricated a large-heating-zone moving planner-flame taper system to manufacture the fiber slicer. We studied the influence of the heating-zone, reasonable heating-zone could improve the device quality. As the heater is mobile and there is air holes in the perform, the heating temperature must be in accurate range.**

Keywords-component; embedded microstructure fiber devices; photonic lantern; optical fiber brush; astronomy

I. INTRODUCTION

As the technology improves, looking for the earth-like exoplanets and the extraterrestrial life arouses more and more attention. One of the important methods for astronomical observations is spectrum analysis technique[1]. High resolution astronomical spectrograph is one of the most important instruments for spectrum analysis. A high resolution astronomical spectrograph has to sample the stellar image with a narrow slit, with the consequent rejection of most the light. The problem can be alleviated by image slicer[2]. Traditional image slicers composed of lenses and mirrors, it has the disadvantage of that, all the parts require high machining accuracy, complex structures, complex structures, and are hard to install[3-5]. An optical fiber image slicer not only effectively improves the light power captured by high-resolution spectrographs, but also has the advantages of simple structure, easy fabrication and easy installation[6]. In order to further improve the optical efficiency of the optical fiber image slicer, we designed an embedded microstructure fiber devices.

This embedded microstructure fiber device has two special structures: photonic lantern and optical fiber brush. Light receiving efficiency of the traditional optical fiber image slicer is limited by the fiber cladding and the gaps between the embedded fibers[6]. Fig. 1 shows the input end of traditional optical fiber image slicer. We made a large-core plastic-cladding fiber, because it can receive more light and the plastic cladding can be dissoluble in acetone. We inserted 7 none-cladding fibers into a pure silica capillary tube and then fused and tapered down into a solid glass element. Fig. 2 shows the sketch of the photonic lantern. The solid glass element has the same waveguide structure as the common optical fiber, and the core of it has the same refractive index, so that the photonic lantern can receive more light.

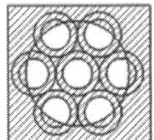

Figure 1. The input end of traditional optical fiber image slicer

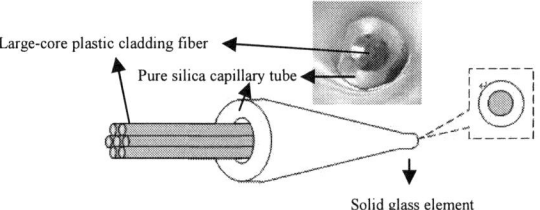

Figure 2. The sketch of the photonic lantern

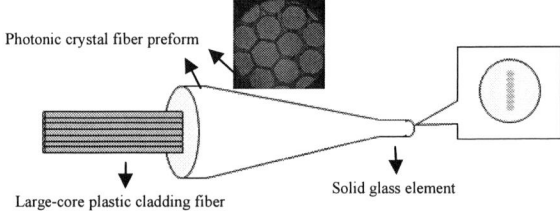

Figure 3. The structure of fiber brush

The aim of the optical fiber brush was fabricated to couple the light signal into the spectrometer. The fiber brush is made by tapering a photonic crystal fiber preform inserted with seven none-cladding fibers. The seven fibers are inserted into a straight row of air holes of a photonic crystal fiber preform. At the taper end, the fiber brush has linear multi-cores. Fig. 3 shows the structure of fiber brush. When light propagates along the taper, it will couple among the cores. Compared to the optical fiber bundle, the fiber brush has better linearity and flatness, and makes it easier to match the slit of spectrograph.

II. MANUFACTURING METHOD OF EMBEDDED MICROSTRUCTURE OPTICAL FIBER DEVICE

At beginning, we used some commercial equipments to try two different tapering methods. First, we used the KF-FBT optical fiber taper system. Fig. 4 shows the photonic lantern made by this system. The device has a good tapered section, but has the gap between the optical fibers. The burner of this taper system is fixed, so that the burner size limits the heating zone and the preform size. Due to the lack of oxygen mixing with hydrogen, flame temperature is limited, the fibers can't fuse together.

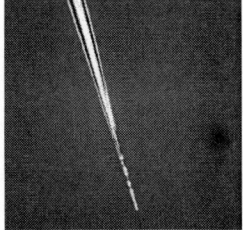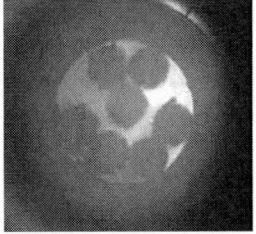

Figure 4. The photonic lantern made by KF-FBT optical fiber taper system

Then we used an optical fiber drawing tower. In order to have a good tapered section, the preform must have large enough diameter. We inserted 5 fibers into a photonic crystal fiber preform, and then inserted the preform into a pure silica tube with 7.5cm outer diameter. Fig. 5 shows the device made by the optical fiber tower. In the taper end, the 5 cores are arranged in a line, and the fibers and preform are fused together. The taper shape is limited by the heater size of the drawing tower. Through the taper section length of 3.5cm, the outer diameter of the preform reduces to 200μm. The taper shape changes dramatically, and the light leak out seriously, so this fiber brush is very lossy.

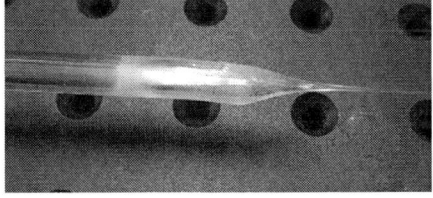

Figure 5. The device made by the optical fiber tower

According to the local-mode theory of slowly varying fibers[7], if the fiber is slowly varying, the modes in the taper are accurate approximations to local modes and the local modes propagated without losing significant power. Based on the conservation of material, the diameter of the preform and the taper end is a definite value, in the tapering process, the larger the heating-zone is, the more slowly the fiber vary, the smaller the loss of the device is.

The existing equipment can't meet this requirement, so we designed and fabricated a large-heating-zone moving planner-flame taper system. Fig. 6 shows the sketch and photo of the large-heating-zone moving planner-flame taper system. The most specialty of the system is the adjustable large heating zone formed by the moving heater whose speed and distance are adjustable. Blow holes of the burner are in a straight row to form a planner flame, and so the width of the flame is decreased to help control the shape of the taper. The heating temperature could be adjusted by controlling the gases. For the adjustable heating zone, the taper is lengthened, and as a result, the loss of the device is reduced.

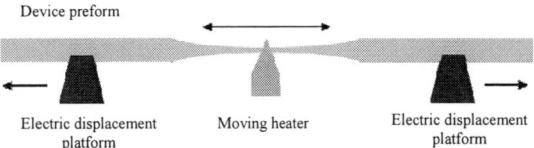

(a) sketch map

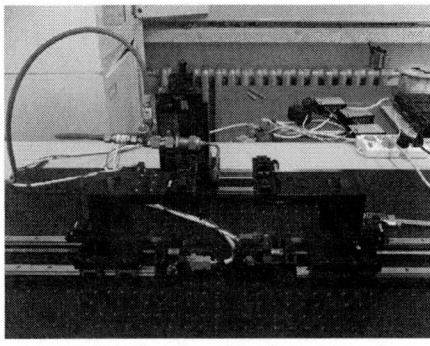

(b) photo

Figure 6. The large-heating-zone moving planner-flame taper system

III. THE EFFECT OF THE HEATING-ZONE

Using the large-heating-zone moving planner-flame taper system, we made two photonic lanterns with different heating-zone. The taper length is 34mm and 7mm, the loss test results are shown in Table 1 and Table 2. Comparing the test results of two devices, the heating-zone influences the loss of the device, the larger the heating-zone is, the smaller the loss is.

TABLE I.   THE LOSS TEST RESULTS OF THE PHOTONIC LANTERNS WITH 34MM TAPER LENGTH

Fiber port	P1 (μW)	P0 (μW)	Loss (dB)
1	0.12	/	/
2	0.09		
3	0.15		
4	0.15		
5	0.11		
6	0.14		
7	0.10		
Sum	0.86	1.08	0.99

TABLE II.   THE LOSS TEST RESULTS OF THE PHOTONIC LANTERNS WITH 7MM TAPER LENGTH

Fiber port	P1 (μW)	P0 (μW)	Loss (dB)
1	0.10	/	/
2	0.11		
3	0.09		
4	0.09		
5	0.12		
6	0.07		
7	0.13		
Sum	0.71	1.27	2.53

National Natural Science Foundation of China (NO.11078009 and 61307005)

Compared with the photonic lantern, the optical fiber brush has more stringent requirement to heating-zone. The preform of the brush has many air holes. Because of the mechanical motion, the heating-zone ends have more heating time than the middle part, so that air holes collapse faster at the ends. When the heating-zone is too large, the heating-zone will form a fusiform section. Fig. 7 shows the fusiform section of the device for too large heating-zone. We should control the heating-zone size in a certain range to ensure a good taper section. Fig.8 shows the optical fiber brush made with reasonable heating-zone.

Figure 7.   The fusiform section of the device for too large heating-zone

Figure 8.   The optical fiber brush made with reasonable heating-zone

IV.   THE EFFECT OF HEATING TEMPERATURE

When using the large-heating-zone moving planner-flame taper system, the heating temperature is a key factor. The heating temperature could be adjusted by controlling the proportion of gases. If the heating temperature is too low, the fibers and perform can't fuse together. If the heating temperature is too high, the moving of the heater will cause the air holes uneven subsidence. It is easy to form bubbles in the taper. Bubble led to uneven refractive index distribution and shape irregular, the light in the core is scattered, and then increase the device loss. When a bubble is in the taper section of the photonic lantern, the loss is 5.28dB. Fig. 9 shows the scattered light caused by the bubble in the taper.

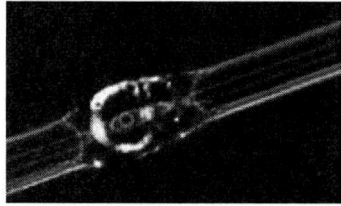

Figure 9.   The scattered light caused by the bubble in the taper

Because the preform of brush is larger than the preform of lantern, so it needs larger flame and higher temperature. More air hole in the brush preform more easily lead to uneven subsidence. The heating temperature range is very small. Fig.

10 shows the uneven subsidence of the optical fiber brush for a high temperature.

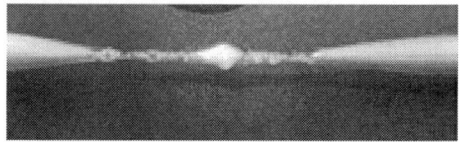

Figure 10. the uneven subsidence of the optical fiber brush for a high temperature

V.   CONCLUSION

According to the local-mode theory of slowly varying fibers, we designed a large-heating-zone moving planner-flame taper system for the production of embedded microstructure fiber devices. Heating-zone influences the loss of the device, the larger the heating-zone is, the smaller the device loss is. The heating temperature is a key factor for the device quality. Compare with the photonic lantern, the optical fiber brush has higher requirements to the taper parameters.

ACKNOWLEDGMENT

This work is supported by the National Natural Science Foundation of China (NO.11078009 and 61307005); 111 project (B13015), to the Harbin Engineering University and the Opening Project of Key Laboratory of Astronomical Optics & Technology, Nanjing Institute of Astronomical Optics & Technology, Chinese Academy of Sciences.

REFERENCES

[1]   Joss Bland-Hawthorn, Pierre Kern, "Astrophotonics: a new era for astronomical instruments," OPTICS EXPRESS, Vol.17, No3, pp.1180-1184. 2009.

[2]   Fancisco Diego , "Confocal image slicer," Applied Optics, Vol.32, NO 31, pp. 6284-6287, NOVEMBER, 1993.

[3]   Bowen I S, "The Image-Slicer a Device for Reducing Loss of Light at Slit of Stellar Spectrograph," The Astrophysical Journal, Vol.88, NO 2, pp.113-124,1938,

[4]   T. Walraven, J. H. Walraven, "Some features of the Leiden radial velocity instrument," Conference on Auxiliary Instrumentation for Large Telescopes, 2 nd, Geneva, Switzerland, pp 175, 1972.

[5]   E. H. Richardson, "An Image Slicer for Spectrographs," Publications of the Astronomical Society of the Pacific, pp.436-437, 1966.

[6]   L. B. Li, Y. T. Zhu, and Z. W. Hu, "Analysis of the Characteristics of Optical Fibre Image Slicers," Astronomical research and technology, 6, pp.220-227, 2009.

[7]   A. W. Snyder, J. D. Love, " Optical waveguide theory," Chapman and Hall, London and New York, pp. 407-412, 1983.

Effect of oil molecular contamination on the space infrared optical system

Chunlian Lu*, He Lv, Shijie Wang, Cuiling Wang, Weimin Sun*

Key Lab of In-fiber Integrated Optics, Ministry Education of China
Harbin Engineering University
Harbin, China
luchunlian@hrbeu.edu.cn or sunweimin@hrbeu.edu.cn

Abstract—Space optical systems are usually exposed to the external space environment. Therefore, space contamination has a bad influence on the space optical systems. Infrared optical system is the most part of space optical systems. In this paper, the contamination source and principle of space contamination were analyzed. The experiment was designed to observe effects of oil molecular contamination. Then, the results of thin film interference were analyzed. We could use the results of research to choose proper detection wavelength on the space infrared optical system or get the mass of contamination. At last, we could draw the conclusions about contamination effects of oil molecular contamination on the space infrared optical system.

Keywords-space contamination; oil molecular; infrared optical system; thin film interference

I. INTRODUCTION

With the development of science and technology, many countries have made progress in aerospace and other military affairs. Including the development of space technology such as satellites for countries to increase military strength provides an important guarantee. Space optical systems are the important parts of space technology. Therefore, optical systems directly affect the performance of the whole systems. In recent years, the space optical contamination has been researched at home and abroad [1-5]. Including Midcourse Space Experiment MSX, it has installed pollution control system. Our country also has begun to focus on space optical system of spacecraft contamination effects [6-7].

Space optical systems are the important parts of spectral imaging in space communications and detections. The degree of space contamination has an influence on spectral imaging quality of space optical systems [8]. Currently, a large number of space passive-contamination will cause serious pollution for optical system, increases stray light, reduces optical transmittance and has an effect on detection sensitivity of the entire system. Among the space contamination, oil, atomic oxygen erosion products such as gas molecules in this class of molecules spacecraft contaminant is the main source of common contamination.

II. THE RESEARCH OF SPACE CONTAMINATION

A. Classification of Space Contamination

In general, the space contamination is divided into two main parts [9-11]:

By the propulsion system, attitude control system ejected propellant impurities, incomplete combustion of the propellant, oxidants such as particulate matter deposited on the surface of spacecraft cause pollution. It is known as particle contamination.

The high vacuum outgassing products of spacecraft materials, such as atomic oxygen erosion product gas molecules deposited on the surface in the form of spacecraft contamination, called molecular contamination.

B. Principle of Space Contamination

When a beam of light is incident into the surface of oil film molecular contamination, some of the light is reflected, part of the light is absorbed and the other light is transmitted. According to the law of conservation of energy,

$$\rho + \alpha + \tau = 1 \qquad (1)$$

Where ρ is reflectivity, α is the rate of absorption, τ is the transmittance.

When the contaminants deposited on the surface of infrared optical system, the transmission of optical surfaces will change. Because reflection and absorption change the results, the particulate contamination of optical systems will degrade the detection ability. We call it the optical damage. Currently, a large number of space passive-contamination will cause serious pollution for optical systems, increase stray light, reduce optical transmittance.

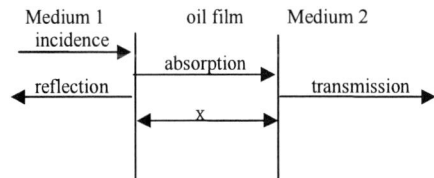

Figure1. The relationship of different light

III. SPACE CONTAMINATION EXPERIMENT

A. Method

In order to analyze oil paint contaminants damage degree of the entire optical system, the experiments were tested. In the experiment, we used transmission decay factor to stand for the optical damage factor. Changes in spectral transmittance could be measured with a spectrophotometer. In order to simulate the optical lens, polyester film was chosen to be the substitute of optical lens.

Space optical contamination of the surface mass was m. S was contamination area. V was contamination volume. l was thickness of contamination material. d was density. D was defined as mass thickness, which was the mass of contamination in unit area. So we had

$$D=m/S=dV/S=dlS/S=dl$$

(2)

$$D= \qquad\qquad dl$$

(3)

In the experiment, silicone oil was used as a contamination source and polyester film was chosen to be the substitute of optical lens. The contamination degree was expressed by mass thickness with the unit $\mu g/mm^2$. Electronic precise balance was used to measure the mass thickness of contaminant and spectrophotometer was used to measure the transmittance of samples at each specific wavelength.

B. Results

In the near-infrared, λ was from $1.4\,\mu m$ to $3.0\,\mu m$. The function of between the optical damage factor and sample plate without contamination was given as follow figure 2. The function of between the optical damage factor and oil molecular contamination was given as follow figure 3.

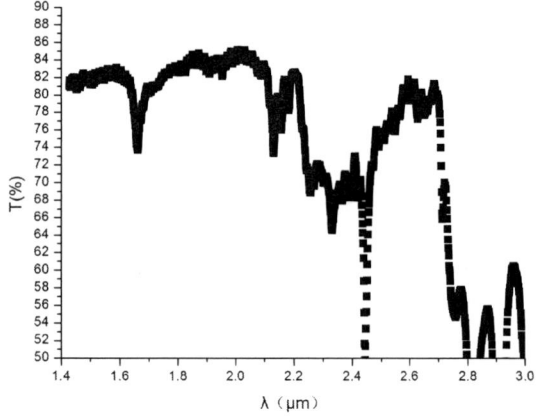

Figure2.The relationship of sample between transmittance and wavelength without contamination

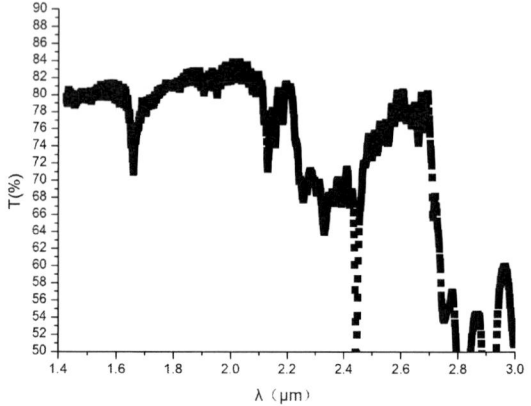

Figure3. The relationship of sample between transmittance and wavelength with contamination

C. Discussion

As we can see from the figure 2 and figure 3, when the contaminants deposited on the surface of infrared optical system, the transmission of optical surfaces will change. We could see transmittance which was with contamination decreased about 2% from figure 2 to figure 3.

The film obtained by the experimental data can be seen contaminated. Although generally contaminated with the mass thickness increased, the transmittance decreased, can be found through the graph, the relationship between the two figures were not strictly met the law of diminishing. There was little fluctuation curve. Through the analysis, it should be made by thin contaminant film surface of the interference result. Then, the analysis of thin film interference would be considered.

IV. ANALYSIS OF THIN FILM INTERFERENCE

A. Principle

When the contaminant is oil molecular, thin film interference will occur in the oil film layer. The schematic is given as figure 4.

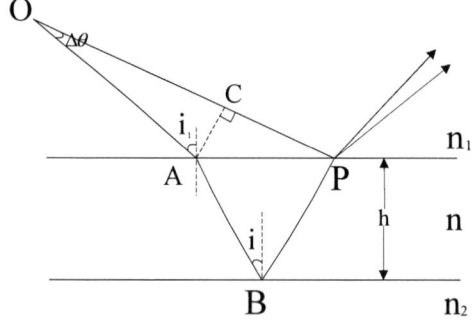

Figure4. The schematic of thin film interference

$$\Delta L(P)=(OABP)-(OP)=(OA)-(OP)+(ABP) \qquad (4)$$

Due to the film is thin, A is close to P, the angle $\Delta\theta$ is very small. So, as an approximation, AC is perpendicular to OP.

$$(OA)-(OP)\approx-(CP)=-n_1APsini_1 \qquad (5)$$

$$=-nAPsini=-n(2htani)sini \qquad (6)$$

$$=-2nhsin2i/cosi \qquad (7)$$

What is more, $\qquad (ABP)=2(AB)\approx2nh/cosi \qquad (8)$

So, $\Delta L(P)\approx2nhcosi \qquad (9)$

Where, i is inclination angle of light in the thin film.

Actually, the most used way is normal incidence. The incident light and the reflected light are everywhere perpendicular to the film surface. At that time, i=0, $\Delta L=2nh$

When $\Delta L=(2k+1)\lambda/2$, the thin film interference will be minimum.

$$h=(2k+1)\lambda/4n, \qquad (10)$$

B. Analysis of results

In the experiment, dimethyl silicone oil was used as oil molecular. The density of it was $0.97g/cm^3$ and its refractive index was 1.56. According to the formula $D = dh$, the density of the molecular film d, the thickness of contamination h, the above formula (10) was substituted into the minimum interference occurred could be obtained when the film mass thickness met proper wavelength. In terms of the mass thickness of two samples, we chose two wavelengths to research thin film interference as follow Table 1. The mass thickness that was $0.328\mu g/mm^2$ was corresponding to the wavelength of 2108nm. The mass thickness that was $0.410\mu g/mm^2$ was corresponding to the wavelength of 2632nm.

TABLE I.   THE TRANSMITTANCE OF DIFFERENT SAMPLES FOR SPECIFIC WAVELENGTH (%)

Mass thickness($\mu g/mm^2$) \ Wavelength(nm)	2108	2632
0.082	83.57	79.29
0.287	82.56	78.31
0.328	82.72	78.10
0.410	81.09	78.84
0.492	81.06	77.11
0.574	80.52	76.88

From the results of Table 1, we could see different results for specific wavelength. In general, the transmittance decreased with the increase of contamination for the same wavelength. When the surface of oil film occurred thin film interference, the effect of contamination would be changed. If the mass thickness was corresponding to proper wavelength for thin film interference, the result would bring about the peak of transmittance. According to the research of experiments, we could get the proper wavelength for infared detections when we knew the mass thickness of oil molecular contamination. On the contrary, if the mass thickness of oil molecular contamination was given, we could select the proper wavelength.

V. CONCLUSION

In this paper, the classification of space contamination and principle of space contamination were given at first. Then, we gave the research of space contamination experiment. Consider that when transmittance decay to a certain value, the image quality of optical system became bad. So we could use the relationship as the references of space system design. The results of near-infrared were shown. We could find the characteristics of sample plates in near-infrared light. Finally, the analysis of thin film interference was presented. Two samples were selected to verify the principle of thin film interference. From the research of thin film interference, the proper detection wavelength could be chosen for contamination. We could also get the mass thickness of contamination from detection wavelength of minimum interference.

ACKNOWLEDGMENT

This work is supported by the National Natural Science Foundation of China (No. 61307005) .This work was supported by the 111 project (B13015), to the Harbin Engineering University. This paper is funded by the International Exchange Program of Harbin Engineering for Innovation-oriented Talents Cultivation.

REFERENCES

[1] B. A. Banks, K. K. De Groh, S.K. Miller "Low earth orbital atomic oxygen interactions with spacecraft materials". NASA TM-220042213233, 2004, pp. 7-11

[2] Kim K. G., Bruce A. B., Dever J.A., et al. "NASA Glenn Research Center's Materials International Space Station Experiments (MISSE 1–7)".NASA / TM－2008－215482, 2008, pp. 1-39

[3] M. Krasoweki, L. Greer, J. Flatico, et al. "A Hardware and Software Perspective of the Fifth Materials on the International Space Station Experiment (MISSE－5)." NASA TM－2005－213840, 2005, pp.1-20

[4] E. M. Silverman"Space environmental effects on spacecraft: LEO materials selection guide", NASA CR-4661, 1995, pp. 88-92

[5] J. G. Funk, J. W. Strickland, J. M. Davis "Materials and Processes Technical Information System (MAPTIS) LDEF materials database." NASA CP－3194, 1993, pp. 1201-1222

[6] X. X. Yuan, C. L. Zhou, D. S. Yang, "The contamination control of aerospace optical system", Spacecraft Environment Engineering vol. 29(2), 2012, pp. 168-172

[7] C. L. Zhou ,"Contamination control through entire process of spacecraft development." Spacecraft Environment Engineering vol. 22(6), 2005, pp. 335-341

[8] C. L. Lu, X. Zhao, Y. P. Zhou, "Analysis of the contamination affection on the space optical system," Journal of Harbin Institute of Technology, vol. 37(2), 2005, pp. 223-226

[9] X. H. Zhao, Z.G. Shen, Y. S. Xing, "Effect of Molecular Contamination on Spacecraft" Acta Aeronautica et Astronautica Sinica vol. 30(1), 2009, pp. 159-164

[10] Y. J. Wang, Y. M. Wang "Effects of Solar UV Radiation on Space Optical Films" Chin. J. Space Sci. vol. 29(2), 2009, pp. 222-228

[11] Z. L. Jiao, D. S. Yang, H.W. Pang "Experimental study on optical effects of molecular contamination". Spacecraft Environment Engineering vol. 26(1), 2009, pp. 17-20

Effect of selective saturated absorption on Atomic Line Filter

Shuangqiang Liu*, Weimin Sun

Harbin Engineering University, Harbin, China
liukunwu@hrbeu.edu.cn

Abstract—We report an atomic filter based on optical pumping induced anisotropy in Doppler-broadened two-level atoms. Atomic line filters (ALF) in previous research used a selective pump to populate atoms on special Zeeman sub-levels to obtain optical anisotropy because of dipole selection rules. Unlike them, we proposed a novel method to get circular anisotropy realized by selectively saturated absorption in rubidium. .

Keywords-atomic filter; anisotropy; rubidium;

I. Introduction

It is well known that ALFs with their merits of high transmission, narrow bandwidth, excellent out-of-band rejection, and wide-field of view have played a key role in the free-space laser communication [1] and lidar systems [2-4]. Since the early 1990s, several types of ALF have been studied and applied to many practical systems, such as Faraday anomalous dispersion optical filter (FADOF) [5-10], induced-dichroism-excited atomic line filter (IDEALF) [13]. We have investigated IDEALF both on theory and experiment [14-16].

FADOF and IDEALF are both consisted by an atomic vapor between two crossed polarizers, and their transmission are both resulted from the polarization plane rotation of the linearly polarized probe. In the case of FADOF, the rotation is caused by the resonance enhancement and high dispersion of the Faraday Effect near the narrow absorption line, while circular dichroism and birefringence induced by the selective optical pumping cause the rotation of polarization in the case of IDEALF.

The IDEALFs which have been reported [17-20] use a laser pump to populate atoms on special energy levels, so that only one sigma component of the probe undergoes absorption while the other one can pass though the medium freely because of dipole forbidden. In this paper, we report an atomic filter based on laser induced circular dichroism realized by "selectively" saturated absorption in rubidium. We believe this work will be useful for polarized spectroscopy and atom-photon interaction experiments, and expand our knowledge of ALF.

II. Theory Model

We show the diagram of our considered two-level system relevant to ^{87}Rb D_2 line in Fig. 1(a). A σ^+ polarized strong coherent pump filed with the Rabi frequency Ω_c is detuned from the $5S_{1/2}(F=2) - 5P_{3/2}(F=3)$ transition by Δ_c in the pumping process, while a linearly polarized field with the Rabi frequency Ω_p also probes $5S_{1/2}(F=2) - 5P_{3/2}(F=3)$

transition by Δ_p. As is shown in our previous work [21], the σ^+ polarized pump will create a net population movement from low magnetic sublevel to high magnetic sublevel, resulting in an accumulation of population on $5S_{1/2}(F=2, m_F=+2)$ and $5P_{3/2}(F'=3, m_{F'}=+3)$. When the pump get saturated, only these two sublevels have atoms populated on in the steady state, while other sublevels of $5S_{1/2}(F=2)$ and $5P_{3/2}(F'=3)$ are all empty (shown in Fig. 1(b)). According to the selection rule for electric-dipole transition, when a linearly polarized probe which can be decomposed into two beams of equal amplitude and opposite circular polarization passing though the atoms, only the σ^- probe component undergoes absorption while the σ^+ component passes through the atoms freely because of the saturated absorption. On the other hand, the saturated absorption also changes the refractive index of the σ^+ probe component. So the medium exhibits an optical anisotropy consists of circular dichroism and birefringence which displays the differences in the absorption and refractive index between the two components of the probe field, respectively. An atomic line filter consists of an atomic vapor between two crossed polarizers, therefore the atoms interact with neither component of the probe when the light is far from resonant with the $5S_{1/2} - 5P_{3/2}$ transition, and hence the probe is blocked by the crossed analyzer for the unchanged polarization. For the light near resonant, the probe's polarization was altered for optical anisotropy, leading it can pass through the filter, and that is the physical mechanism of the atomic filter.

We take the probe's two components separately in order to analyze the pump-probe process. For the σ^- component, there are four sublevels ($m_F=0, +2$ of the ground state and $m_{F'}=+1, +3$ of the excited state) related to the pump and probe, therefore it can be treated as a four-level system shown in Fig. 1(c). $m_F=0 \rightarrow m_{F'}=+1$ and $m_F=2 \rightarrow m_{F'}=+3$ are coupled by circular polarized pump with detuning Δ_c, while $m_F=2 \rightarrow m_{F'}=+1$ is coupled by the σ^- component probe with detuning Δ_p. In the interaction picture, the Hamiltonian with the rotating-wave and electric-dipole approximations can be expressed as

---

This work was supported by Doctoral Fund of Ministry of Education of China, China Postdoctoral Science Foundation and Heilongjiang Postdoctoral Science Foundation.

$$H = -\frac{\hbar}{2}\begin{bmatrix} 0 & \Omega_c & 0 & 0 \\ \Omega_c^* & -2\Delta_c & \Omega_p & 0 \\ 0 & \Omega_p^* & -2(\Delta_c - \Delta_p) & \Omega_c \\ 0 & 0 & \Omega_c^* & -2(2\Delta_c - \Delta_p) \end{bmatrix} \quad (1)$$

We take these Rabi frequencies as real for simplicity, and then describe the quantum dynamics of the four-level atomic system by the Liouville equation,

$$\frac{\partial \rho}{\partial t} = \frac{1}{i\hbar}[H_I, \rho] - \frac{1}{2}(\Gamma \rho + \rho \Gamma) \quad (2)$$

The full statements of (2) are expressed as follows:

$$\dot{\rho}_{12} = \frac{i}{2}\Omega_c(\rho_{22} - \rho_{11}) - \frac{i}{2}\Omega_p \rho_{13} + [i\Delta_c - \frac{1}{2}(\Gamma_{21} + \Gamma_{23})]\rho_{12}$$

$$\dot{\rho}_{13} = \frac{i}{2}\Omega_c \rho_{23} - \frac{i}{2}\Omega_c \rho_{14} - \frac{i}{2}\Omega_p \rho_{12} + i(\Delta_c - \Delta_p)\rho_{13}$$

$$\dot{\rho}_{14} = \frac{i}{2}\Omega_c \rho_{24} - \frac{i}{2}\Omega_c \rho_{13} + [i(2\Delta_c - \Delta_p) - \frac{1}{2}\Gamma_{43}]\rho_{14}$$

$$\dot{\rho}_{22} = \frac{i}{2}\Omega_c(\rho_{12} - \rho_{21}) + \frac{i}{2}\Omega_p(\rho_{32} - \rho_{23}) - (\Gamma_{21} + \Gamma_{23})\rho_{22}$$

$$\dot{\rho}_{23} = \frac{i}{2}\Omega_c \rho_{13} - \frac{i}{2}\Omega_c \rho_{24} + \frac{i}{2}\Omega_p(\rho_{33} - \rho_{22})$$
$$- [i\Delta_p + \frac{1}{2}(\Gamma_{21} + \Gamma_{23})]\rho_{23}$$

$$\dot{\rho}_{24} = \frac{i}{2}\Omega_c \rho_{14} - \frac{i}{2}\Omega_c \rho_{23} + \frac{i}{2}\Omega_p \rho_{34}$$
$$- [i(-\Delta_c + \Delta_p) + \frac{1}{2}(\Gamma_{21} + \Gamma_{23} + \Gamma_{43})\rho_{24}$$

$$\dot{\rho}_{33} = \frac{i}{2}\Omega_c(\rho_{43} - \rho_{34}) + \frac{i}{2}\Omega_p(\rho_{23} - \rho_{32}) + \Gamma_{23}\rho_{22} + \Gamma_{43}\rho_{44}$$

$$\dot{\rho}_{34} = \frac{i}{2}\Omega_c(\rho_{44} - \rho_{33}) + \frac{i}{2}\Omega_p \rho_{24} + (i\Delta_c - \frac{1}{2}\Gamma_{43})\rho_{34}$$

$$\rho_{44} = \frac{i}{2}\Omega_c(\rho_{34} - \rho_{43}) - \Gamma_{43}\rho_{44}$$

$$\rho_{11} + \rho_{22} + \rho_{33} + \rho_{44} = 1$$

(3)

Here $\Gamma_{ij}(i = 2, 4, j = 1, 3)$ is the spontaneous decay rates of the excited state $|i\rangle$ to the ground state $|j\rangle$. The most important ingredient of (3) is the presence of the matrix element ρ_{23}, which indicates the coherence between the levels $|3\rangle$ and $|2\rangle$, and it develops as a result of the coherent driving of the two allowed transitions. By solving (3) in the steady state, we can obtain the numeric result of ρ_{23} and then the linear susceptibility of the atomic system for the σ^- probe can be written as

$$\chi_- = \frac{2\mu_{23}^2 N}{\hbar \Omega_p \varepsilon_0} \rho_{23} \quad (4)$$

Where N is the density of the atoms, μ_{23} denotes the matrix element that couples the two correlative hyperfine sublevels, and those numerical values can be obtained by solving the relevant equations in [22] and [23].

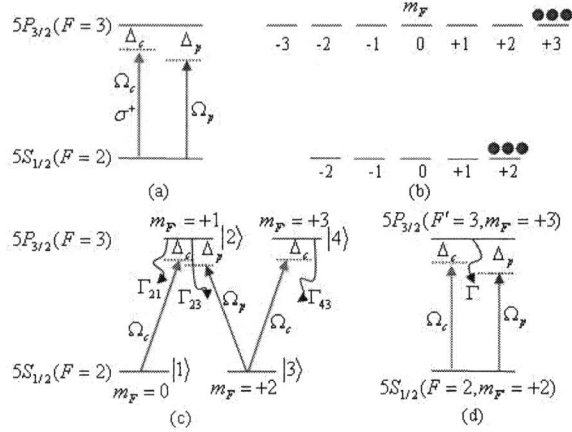

Figure 1. (a) Schematic diagram of a two-level ^{87}Rb atom system with a ciuclarly polarized pump and a linearly polarized probe. (b) Schematic diagram of atomic population on Zeeman sub-levels. (c) Schematic representation of the σ^- probe component passing though the system, there are four sub-levels correlated to the pump-probe process, so it can be trated as a four-level system. (d) Schematic representation of the σ^+ probe component passing though the system, which can be trated as a two-level system.

For the σ^+ component probe, there are only two Zeeman sub-levels which attach themselves to the pump-probe process, therefore the system can be equivalent to a two-level system with a strong pump and a weak probe (shown in Fig. 1(d)), and the effective susceptibility is well known from the work of Mollow [24] and Agarwal [25],

$$\chi_+ = \frac{N\mu_{34}^2}{\hbar \varepsilon_0} \cdot \frac{\Gamma_{43}^2 + \Delta_c^2}{(2\Omega_c^2 + \Gamma_{43}^2 + \Delta_c^2)(\Delta_c + \Delta_p + i\Gamma)} \times [1 -$$
$$\frac{2\Omega_c^2(\Delta_p + 2i\Gamma_{43})(\Delta_p - \Delta_c + i\Gamma_{43})/(\Delta_c - i\Gamma_{43})}{(\Delta_p + 2i\Gamma_{43})(\Delta_p + \Delta_c + i\Gamma_{43})(\Delta_p - \Delta_c - i\Gamma_{43}) - 4\Omega_c^2(\Delta_p + i\Gamma_{43})}]$$
(5)

The main limitation for laser spectroscopy of gas samples is Doppler broadening. It is well known that a pump beam and a probe beam that counter-propagate along the same path in an absorption cell can effectively decrease the impact of Doppler broadening [18, 19]. In order to reduce the influence of the Doppler frequency shift to the lowest, we assume that the pump passing on the opposite direction and then Δ_c and Δ_p in (3) and (5) can be separately substituted by $\Delta_c - k_c v$ and $\Delta_p + k_p v$, where v is the atomic velocity, k_c and k_p are the wave number of the pump and probe, respectively. The linear susceptibility $\chi(v)$ which is averaged over the Doppler distribution of velocities as [14]

$$\chi_\pm(v) = \frac{1}{\sqrt{\pi}u}\int_{-\infty}^{+\infty}\chi_\pm e^{-v^2/u^2}dv \quad (6)$$

978-1-4799-1215-5/13 $31.00 © 2013 IEEE

Here u is the most probable velocity which is defined as $u = \sqrt{2k_B T / M}$.

The transmission of the probe beam passing through the crossed polarizer beside the cell is given by [26]

$$T = \frac{1}{2}\exp(-\alpha L)[\cosh(\Delta\alpha L) - \cos(2\rho L)] \quad (7)$$

Here L is the length of the atomic vapor. $\alpha = \omega_p \operatorname{Im}(\chi_+ + \chi_-)/(2c)$, $\Delta\alpha = \omega_p \operatorname{Im}(\chi_+ - \chi_-)/(2c)$ and $\rho = \omega_p \operatorname{Re}(\chi_+ - \chi_-)/(4c)$ are the mean absorption coefficient, circular dichroism, and rotary power, respectively.

III. NUMERICAL RESULTS AND DISCUSSION

On the beginning, in order to show anisotropy in the atomic system, we present numerical results for the susceptibility of each component of the probe by evaluating (6) for different Rabi frequency of the pump. We use typical parameter for ^{87}Rb D_2 transition: $\Gamma_{43} = 6.066 \times 2\pi \times 10^6$ Hz , $\Gamma_{21} = \Gamma_{43} \times 2/5$, $\Gamma_{23} = \Gamma_{43} \times 1/15$ [27], $N = 1.57 \times 10^{10} cm^{-3}$, u=239.4m/s (at room temperature 300K). We show respectively in Fig. 2 and Fig.3 the behavior of each probe's component assuming that the pump is in resonance. In each figure, (a) is the imaginary part of susceptibility which refers to absorption, while (b) is the real part that refers to dispersion. In figure 2, the imaginary part of χ^+ shows the absorption of the probe decrease with the increase in the Rabi frequency of the pump beam, and even a typical Lamb dip appears because of saturating effect. The real part of χ^+ exhibits abnormal dispersion when the pump isn't big enough, and the slope at the center decrease with the pump intensity increase, and it change to normal dispersion when a Lamb dip comes forth.

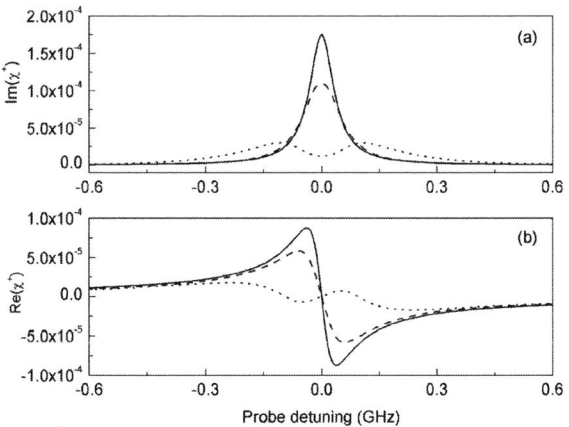

Figure 2. The imaginary (a) and real (b) parts of susceptibility χ of the σ^+ component as a function of the probe detuning in different pump Rabi frequencies. $\Omega_c = 10MHz$ (solid), $\Omega_c = 100MHz$ (dash), $\Omega_c = 300MHz$ (dot).

It is shown Figure 3 that there are three maximum point of imaginary part of χ^- at a certain pump, and this is because of Aulter-Townes Splitting under strong pump condition.

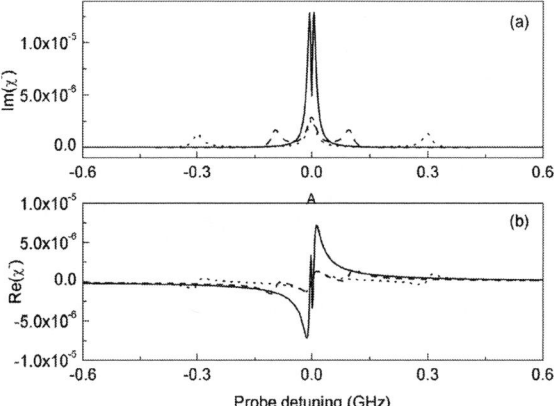

Figure 3. The imaginary (a) and real (b) parts of susceptibility χ^- of the σ^- component as a function of the probe detuning in different pump Rabi frequencies. $\Omega_c = 10MHz$ (solid), $\Omega_c = 100MHz$ (dash), $\Omega_c = 300MHz$ (dot).

By comparing Figure 2 and Figure 3, it is palpable that the susceptibility of σ^+ is an order of magnitude that that of σ^- around the frequency center, which is mainly in respect that dipole matrix elements for σ^+ transition $m_F = +2 \rightarrow m_{F'} = +3$ is much bigger than that of σ^- transition $m_F = +2 \rightarrow m_{F'} = +1$ [27]. Therefore the differences in the absorption and dispersion will respectively lead to circular dichroism and birefringence. The optical anisotropy consisted of them will result in the diversification of the probe field's polarization, and then its transmission. By solving (7), the calculated transmission as a function of the detuning of the probe from the atomic transition is shown in Figure 4.

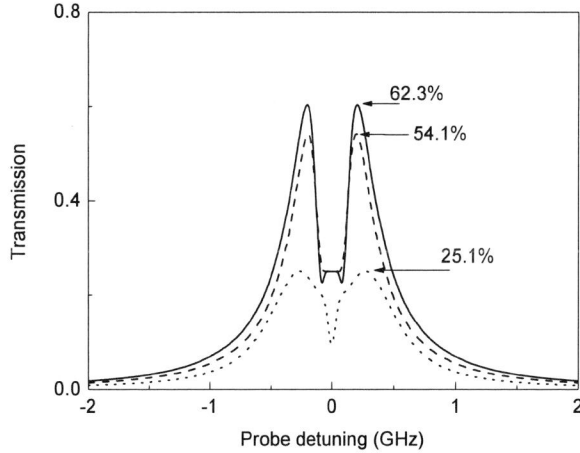

Figure 4. Transmission as a function of the probe detuning in different pump Rabi frequencies. $\Omega_c = 10MHz$ (solid), $\Omega_c = 100MHz$ (dash), $\Omega_c = 300MHz$ (dot), cell length L=2cm, other paramenters are the same as Figure 2 and 3.

Figure 4 intuitively lays out the filtering ability of the two-level system. Even at room temperature, there are two transmission peaks of 62.3%. When the pump intensity increase, the susceptibility of the σ^+ decrease because of selective saturation absorption, and the disparity of $\chi_\pm$ reduces, and then the optical anisotropy becomes inconspicuous, hence the transmission

IV. CONCLUSION

In conclusion, we present an atomic line filter in Doppler broadened a two-level system, and further a novel method to get optical anisotropy realized by selectively saturated absorption in rubidium. Numeric results show selective optical pumping does have influence on the filter characteristic. Moreover, the pumping scheme used can be extended to other energy levels or other alkali atomic vapors. We believe this work will be useful for polarized spectroscopy and atom-photon interaction experiments, and expand our knowledge of ALFs.

REFERENCES

[1] J. Tang, Q. Wang, Y. Li, L. Zhang, J. Gan, M. Duan, J. Kong and L. Zheng, "Experimental study of a model digital space optical communication system with new quantum devices," Appl. Opt. **34**, 2619-2622 (1995)

[2] H. Chen, M. A. White, David. A. Krugger, and C. Y. She, "Daytime mesopause temperature measurements with a sodium-vapor dispersive Faraday filter in a lidar receiver," Opt. Lett. **21**, 1093-1095 (1996)

[3] C. Fricke-Begemann, M. Alpers, and J. Höffner, "Daylight rejection with a new receiver for potassium resonance temperature lidars," Opt. Lett. **27**, 1932-1934 (2002).

[4] J. Höffner and C. Fricke-Begemann, "Accurate lidar temperature with narrowband filters," Opt. Lett. **30**, 890-892 (2005).

[5] J. Menders, K. Benson, S. H. Bloom, C. S. Liu, and E. Korevaar, "Ultranarrow line filtering using a Cs Faraday filter at 852 nm," Opt. Lett. **16**, 846-848 (1991)

[6] D. J. Dick and T. M. Shay, "Ultrahigh-noise rejection optical fitler," Opt. Lett. **16**, 867-869 (1991)

[7] R. I. Billmers, S. K. Gayen, M. F. Squicciarini, V. M. Contarino, W. J. Scharpf, and D. M. Allocca, "Experimental demonstration of an excited-state Faraday filter operating at 532 nm," Opt. Lett. **20**, 106-108 (1995).

[8] Y. Peng, "Transmission characteristics of an excited-state Faraday optical filter at 532 nm," J. Phys. B **30**, 5123 (1997).

[9] Y. Zhang, X. Jia, Z. Ma, and Q. Wang, "Potassium Faraday optical filter in line-center operation," Opt. Commun. **194**, 147 (2001).

[10] Y. Zhang, X. Jia, Z. Ma, and Q. Wang, "Optical filtering characteristic of potassium Faraday optical filter," IEEE. J. Quantum Electron. **37**, 372 (2001).

[11] L. Zhang and J. Tang, "Experimental study on optimization of the working conditions of excited state Faraday filter," Opt. Commun. **152**, 275 (1998).

[12] S. K. Gayen, R. I. Billmers, V. M. Contarino, M. F. Squicciarini, W. J. Scharpf, G. Yang, P. R. Herczfeld and D. M. Allocca, "Induced-dichroism-excited atomic line filter at 532 nm," Opt. Lett. **20**, 1427-1429 (1995).

[13] Yufeng Peng, Wenjin Zhang, Liang Zhang, Junxiong Tang, Opt. Commun. 282 (2009) 236.

[14] Shuangqiang Liu, Yundong Zhang, Hao Wu, and Ping Yuan, "A large scale tunable atomic filter and its performance on the co-propagating and cunter-propagating pump fields," Journal of the optical society of the America B-optical physics. Vol. 28, Iss. 5, pp. 1104~1110 (2011).

[15] Shuangqiang Liu, Yundong Zhang, Hao Wu, and Ping Yuan, "Gain assited large-scale tunable atomic filter based on birefringence by double selective optical pumping", Optics Communications, Vol. 284, Iss. 18, pp. 4180~4184 (2011).

[16] Shuangqiang Liu, Yundong Zhang, Hao Wu, and Ping Yuan, "Ultra-narrow band atomic filter based on optical pumping induced dichroism realized by selectively saturated absorption", Optics Communications, Vol. 285, Iss. 6, pp. 1181-1184 (2012).

[17] S. K. Gayen, R. I. Billmers, V. M. Contarino, M. F. Squicciarini, W. J. Scharpf, G. Yang, P. R. Herczfeld and D. M. Allocca, Opt. Lett. **20** (1995) 1427.

[18] L. D. Turner, V. Karagnanov, and P. J. O. Teubner, "Sub- Doppler bandwidth atomic atomic optical filter," Opt. Lett. 27, 500-502 (2002)

[19] Alessandro Cerè, Valentina Parigi, Marta Abad, Florian Wolfgramm, Ana Predojević, and Morgan W. Mitchell, Opt. Lett. 34 (2009) 1012.

[20] Z. S. He, Y. D. Zhang, H. Wu, P. Yuan, S. Q. Liu, "Theoretical model for an atomic optical filter based on optical anisotropy," J. Opt. Soc. Am. B. **26**, 1755-1759 (2009).

[21] Shuangqiang Liu, Yundong Zhang, Daikun Fan, Hao Wu, and Ping Yuan, "The selective optical pumping process in Dopper-broadened atoms," Applied Optics. Vol. 50, Iss. 11, pp. 1620~1624 (2011).

[22] D. M. Brink and G. R. Satchler, Angular Momentum (Oxford, 1962).

[23] R. Loudon, The Quantum Theory of Light, 2nd ed. (Oxford University Press, 1983).

[24] B.R. Mollow, Phys. Rev. A **5**, 2217 (1972).

[25] G. S. Agarwal and Tarak Nath Dey, "Slow light in Doppler-broadened two-level systems," Phy. Rev. A, **68**, 063816 (2003)

[26] P. Yeh, "Dispersive magnetooptic filters," Appl. Opt. **21**, 2069-2075 (1982).

[27] D. A. Steck, "Rubidium 87 D Line Data," http://steck.us/alkalidata/rubidium87numbers.1.6.pdf

Photorefractive effect in relaxor ferroelectric $0.88Pb(Zn_{1/3}Nb_{2/3})O_3$-$0.12PbTiO_3$ single crystal

Hong Jia[1]*, Yang Li[2], Jun Li[2], Hao Tian[2], Liang Sun[3]

1College of Electrical Engineering, Suihua University, Suihua City, Heilongjiang Province, China
2 Department of Physics, Harbin Institute of Technology, Harbin, Heilongjiang Province, China
3 Department of Physics and Electronic Engineering, Yibin University, Yibin, Sichuan Province, China
jiahong21@163.com

Abstract—**Photorefractive properties of $0.88Pb(Zn_{1/3}Nb_{2/3})O_3$-$0.12PbTiO_3$ single crystal were investigated by the two-wave coupling experiment. The maximal gain coefficient is 5.5cm⁻¹, the effective trap density is $1.21 \times 10^{16}cm^{-3}$, and the normalized photorefractive response time under 1W/cm² illumination is 1.48s at the wavelength of He-Ne laser. The dominant charge carrier was identified as electron from the direction of two-wave coupling energy transfer.**

Keywords-PZN-PT; photorefractive effect; two-wave coupling

I. INTRODUCTION

Ferroelectric single crystals with high electro-optic response have been widely used in advanced photonic and microelectronic devices such as high speed light modulators, parametric oscillators, and nonlinear frequency converters [1-3]. High electro-optic coefficients will allow of smaller size devices and lower operating voltages. Most oxygen-octahedral ferroelectrics, which exhibit excellent electromechanical properties, also have outstanding optical properties [4-6]. $(1-x)Pb(Zn_{1/3}Nb_{2/3})O_3$-$xPbTiO_3$ (PZN-xPT) single crystals have a morphotropic phase boundary (MPB) between the rhombohedral and tetragonal phases (0.08 < x < 0.10) [7]. Because of extremely high dielectric and piezoelectric constants near the MPB compared to lead zirconium titanate (PZT) ceramics [7,8], PZN-xPT single crystals are being considered as one of the most promising materials for next generation of electromechanical transducers in a broad range of advanced applications, such as medical ultrasound imaging and underwater communication. Recently, there is an increasing interest in their optical properties. PZN-xPT single crystals have been found to exhibit outstanding optical properties [9-11]. PZN-xPT single crystals are expected to be promising materials for optical devices.

As compared with rhombohedral and morphotropic phase boundary crystals, tetragonal PZN-xPT single crystals possess the optimal transmission properties [12]. For tetragonal $0.88Pb(Zn_{1/3}Nb_{2/3})O_3$-$0.12PbTiO_3$ (PZN-0.12PT) single crystal, high optical transmittance was found in the wavelength range from 0.45 to 5.5μm, large EO coefficients were observed, low-frequency value of the linear electro-optic coefficient $\gamma_c=\gamma_{33}-(n_o/n_e)^3\gamma_{13}$ of poled PZN-0.12PT single crystal was found to be equal to 165 pm/V at the wavelength of 633nm, which is a rather high value [13]. Thus far, the photorefractive effects were seldom studied for PZN-xPT single crystals. Here we report the photorefractive effect in PZN-0.12PT single crystal.

II. EXPERIMENT

A. Sample preparation

The PZN–xPT single crystals used in this study were grown using a high temperature flux method [10]. For tetragonal PZN-0.12PT single crystals at room temperature, the spontaneous polarization is along <001> direction. The (001) surfaces were determined using an orientation device equipped with x-ray diffractometer. The samples were poled along the <001> direction under an electric field of 1kV/mm for 10 minutes at about 120°C in silicone oil, then slowly cooled to room temperature while maintaining half of the applied electric field. By this procedure, the samples were poled into single domain crystals. The surfaces for light transmission were polished to optical quality.

B. Two-wave coupling

The photorefractive properties were investigated at 632.8nm wavelength by two-wave coupling measurement with an extraordinary polarization [14]. The experimental setup is shown in Fig. 1. The bisected line of signal and reference lights was in the [100] direction, the grating wave vector K of interference fringes were oriented along the [001] axis (c-axis) of the crystal. In this geometry, $\gamma_{eff} = \gamma_{33}$. When measuring the gain coefficient Γ, the intensity ratio of the reference light and signal light was 500:1. The grating formation rate is defined to be the inverse of the time for the amplified light in two-wave coupling to reach (1-1/e) of the saturated value. And it was determined as a function of the total light intensity on the crystal.

This work is supported by the National Natural Science Foundation of China, grant #11074059.

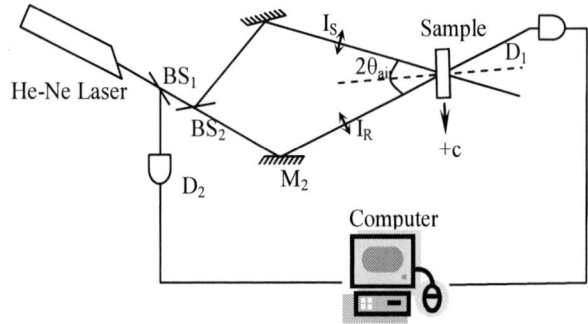

Figure 1. Experimental configuration of the two-wave coupling measurement. M_1, M_2: mirrors; BS_1, BS_2: beam splitters; D_1, D_2: detectors; I_R: reference light; I_S: signal light

III. RESULT AND DISCUSSION

Gain coefficients of two-wave coupling Γ were calculated from the ratio of the output signals with and without amplification. In Fig. 2, it is shown as a function of external crossing angles 2θ. The maximum value of gain coefficient Γ is 5.5cm^{-1} at the crossing angle of $20°$. The gain coefficient Γ can be written in the form

$$\Gamma=[A\sin\theta/(1+B^{-2}\sin^2\theta)](\cos2\theta_i/\cos\theta_i) \quad (1)$$

where Γ is the external half angle, θ_i is the internal half angle between the two incident laser beams [15]. Over the range of external crossing angles 2θ used here ($0<2\theta<60°$), the internal crossing angle $2\theta_i$ in Eq. (1) was less than $22°$, and the factor $(\cos2\theta_i/\cos\theta_i)$ varied by less than 6% from unity. In Eq. (1) we have assumed that the electron-hole competition factor $\xi(k)$ is constant with k in order to simplify the data analysis. The parameter A is proportional to the effective electro-optic coefficient γ_{eff},

$$A =\gamma_{eff}\xi(K)\frac{8\pi^2 n^3 k_B T}{q\lambda^2}=\frac{\delta\Gamma}{\delta\theta}\bigg|_{\theta=0} \quad (2)$$

A is determined by the slope of the plot of Γ versus 2θ near $\theta=0$. The parameter B is related to the effective photorefractive charge density N_{eff},

$$B = \frac{\lambda K_0}{4\pi}=\frac{q\lambda}{4\pi}\sqrt{\frac{N_{eff}}{\varepsilon\varepsilon_0 k_B T}}\approx\sin\theta_{peak} \quad (3)$$

B is determined by $2\theta_{peak}$, the crossing angle at which the gain reaches its maximum value.

The dominant photorefractive charge carrier in PZN-0.12PT crystals was determined by comparing the direction of two-beam coupling to the direction of the positive c-axis of the crystals. The direction of two-wave coupling gain is toward the positive poling electrode c face, thus the dominant charge carrier is identified as electron [16].

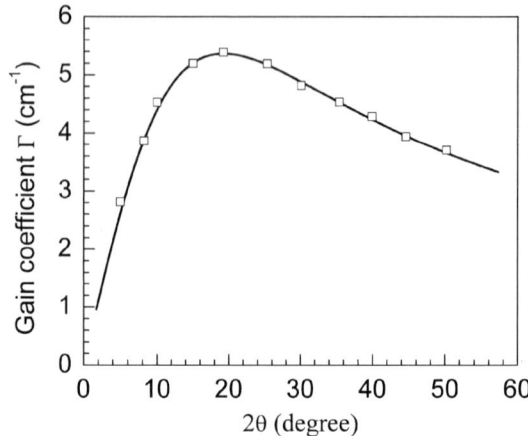

Figure 2. Gain coefficient of two-wave coupling at the wavelength of 632.8nm as a function of external crossing angle 2θ. The solid curve is the optimal fitting result of Eq. (1)

Fig. 3 shows the grating formation rate $1/\tau$ as a function of the light intensity at wavelength of 632.8nm and the grating period Λ of 1.9µm. The photorefractive response time τ at the intensity of 1W/cm^2 is 1.48s. The grating formation rate can be written as

$$\frac{1}{\tau}=\frac{\sigma}{\varepsilon\varepsilon_0}=\frac{\sigma_d}{\varepsilon\varepsilon_0}+\frac{\sigma_{ph}}{\varepsilon\varepsilon_0}=\frac{\sigma_d}{\varepsilon\varepsilon_0}+\frac{e\mu\tau_R\alpha\lambda}{\varepsilon\varepsilon_0 hc}I_0 \quad (4)$$

where σ_d is the dark conductivity, σ_{ph} is the photoconductivity, α is the absorption coefficient at wavelength λ, e is the elementary charge, h is the Planck's constant, c is the light velocity in vacuum, μ is the carrier mobility, τ_R the carrier lifetime, and I_0 is the light intensity [15,16]. The absorption coefficient α is 7.6cm^{-1} at 623.8nm wavelength. From Fig. 3, the characteristic parameters were obtained such that $\sigma_d=4.1\times10^{-13}(\Omega\text{cm})^{-1}$, $\sigma_{ph}=4.3\times10^{-11}(\Omega\text{cm})^{-1}$ at the light intensity of 1W/cm^2, and the product of the carrier mobility and carrier lifetime $\mu\tau_R=1.1\times10^{-11}\text{cm}^2/\text{V}$.

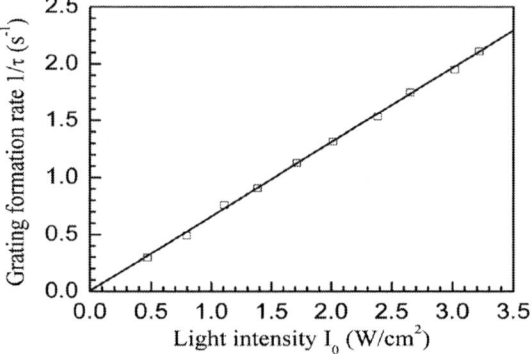

Figure 3. Grating formation rate $1/\tau$ as a function of the light intensity I_0. Solid curve is the linear fitting result of Eq. (4)

IV. CONCLUSION

We have studied the photorefractive effects in PZN-0.12PT single crystal. By the two-wave coupling measurement, we observed a large effect comparable with that of SBN single crystals. The maximal gain coefficient is 5.5cm^{-1}, the effective trap density is 1.21×10^{16}cm^{-3}, and the normalized photorefractive response time under 1W/cm^2 illumination is 1.48s at the 632.8nm laser wavelength. The dominant charge carrier was identified as electron. By adding appropriate transition metal ions as the photorefractive centers, the photorefractive proprieties are likely to be enhanced.

REFERENCES

[1] Y. Li, J. Li, Z. X. Zhou, R. Y. Guo, and A. S. Bhalla, "Low-frequency-dependent electro-optic properties of potassium lithium tantalate niobate single crystals," Europhys. Lett., vol. 102, p. 37004, May 2013.

[2] C. J. He, Z. X. Zhou, and Y. W. Liu, "Dendrite crystal morphology evolution mechanism of β-BaB2O4 crystal," Sci. China Ser. E, vol. 52, pp. 1703-1706, June 2009.

[3] C. P. Hu, H. Tian, B. Yao, Z. X. Zhou, and D. Y. Chen, "Large quadratic electro-optic effect in K0.99Li0.01Ta0.60Nb0.40O3 single crystal," Curr. Appl. Phys., vol. 13, pp. 785-788, June 2013.

[4] C. J. He, Y. G. Zhang, L. Sun, J. M. Wang, T. Wu, F. Xu, C. L. Du, K. J. Zhu, and Y. W. Liu, "Electrical and optical properties of Nd3+-doped Na0.5Bi0.5TiO3 ferroelectric single crystal," J. Phys. D: Appl. Phys., vol. 46, p. 245104, June 2013.

[5] C. J. He, F. Xu, J. M. Wang, and Y. W. Liu, "Refractive index dispersion of relaxor ferroelectric 0.9Pb(Zn1/3Nb2/3)O3-0.1PbTiO3 single crystal," Cryst. Res. Technol., vol. 44, pp. 211-214, February 2009.

[6] C. J. He, H. B. Chen, F. Bai, Z. B. Fan, L. Sun, F. Xu, J. M. Wang, Y. W. Liu, and K. J. Zhu, "Composition and orientation dependence of high electric-field-induced strain in Pb(In1/2Nb1/2)O3–Pb(Mg1/3Nb2/3)O3–PbTiO3 single crystals," J. Appl. Phys., vol. 112, p. 126102, December 2012.

[7] M. Jin, J. Y. Xu, X. H. Li, H. Shen, and Q. B. He, "Growth defects in a PZNT93/7 crystal prepared by directional solidification," Cryst. Res. Technol., vol 43, pp. 1074-1077, October 2008.

[8] C. J. He, W. P. Jing, F. F. Wang, K. J. Zhu, and J. H. Qiu, "Full tensorial elastic, piezoelectric and dielectric properties characterization of [011]-poled PZN-9%PT single crystal," IEEE Trans. Ultrason. Ferroelectr. Freq. Control, vol. 58, pp. 1127–1130, June 2011.

[9] E. W. Sun, Z. Wang, R. Zhang, and W. W. Cao, "Reduction of electro-optic half-wave voltage of 0.93Pb(Zn1/3Nb2/3)O3-0.07PbTiO3 single crystal through large piezoelectric strain," Opt. Mater., vol. 33, pp. 549-552, January 2011.

[10] C. J. He, W. P. Jing, K. J. Zhu, and J. H. Qiu, "Linear electro-optic properties of orthorhombic PZN-8%PT single crystal," IEEE Trans. Ultrason. Ferroelectr. Freq. Control, vol. 58, pp. 1118–1121, June 2011.

[11] C. J. He, X. D. Fu, F. Xu, J. M. Wang, K. J. Zhu, C. L. Du, and Y. W. Liu, "Orientation effect on bandgap and dispersion behavior of 0.91Pb(Zn1/3Nb2/3)O3-0.09PbTiO3 single crystals," Chin. Phys. B, vol. 21, p. 054207, May 2012.

[12] C. J. He, F. Xu, J. M. Wang, C. L. Du, K. J. Zhu, and Y. W. Liu, "Composition dependence of dispersion and bandgap properties in PZN-xPT single crystals," J. Appl. Phys., vol. 110, p. 083513, May 2011.

[13] C. J. He, H. B. Chen, L. Sun, J. M. Wang, F. Xu, C. L. Du, K. J. Zhu, and Y. W. Liu, "Effective electro-optic coefficient of (1–x)Pb(Zn1/3Nb2/3)O3–xPbTiO3 single crystals," Cryst. Res. Technol., vol 47, pp. 610-614, June 2012.

[14] H. Jia and L. Sun, "Photorefractive effect of ion-doped Pb(Mg1/3Nb2/3)O3-PbTiO3 single crystals (in Chinese)," Sci. Sin.-Phys. Mech. Astron., vol. 42, pp. 333–338, March 2012.

[15] C. J. He, Z. X. Zhou, D. J. Liu, X. Y. Zhao, and H. S. Luo, "Photorefractive effect in relaxor ferroelectric 0.62Pb(Mg1/3Nb2/3)O3-0.38PbTiO3 single crystal," Appl. Phys. Lett., vol. 89, p. 261111, December 2006.

[16] Y. W. Liu, Y. Y. Zhou, M. J. Zhu, and C. J. He, "Experimental investigation of the photochromic effect and two-color holographic recording in near-stoichiometric LiNbO3:Fe:Mn crystals," J. Nonlinear Optic. Phys. Mat., vol. 21, p. 1250052, December 2012.

Dependence of Cs Atomic Clock CPT Signal on Magnetic Intensity

Hongsong Mei, Qiang Huang, Junhai Zhang, Weimin Sun and Zongjun Huang

College of Science
Harbin Engineering University
Harbin, China, 150001
Email: jhzhang@hrbeu.edu.cn

Abstract—This work demonstrates the CPT (Coherent Population Trapping) signal of Cs atomic clock can be affected by magnetic field theoretically. The amplitude and the width of the resonance line of the CPT signal are the functions of the magnetic intensity. When both the coherent lasers have a specific rate of intensity the amplitude of the CPT signal keeps the most stable. Compared with four lambda models in theory, research shows that the scheme of the translation from ground states $|F_g=4$, $m_F=0\rangle$ and $|F_g=3$, $m_F=0\rangle$ to excited state $|F_e=4$, $m_F=-1\rangle$ is the most suitable for Cs CPT atomic clock.

Keywords-CPT; Magnetic field; Wave function

I. INTRODUCTION

Atomic clock based on CPT has been researched in both theory and experiment for many years because it has smaller dimension and consumption. Frequency stability is the most important for atomic clock, temperature and light intensity affect the signal of CPT of atomic clock have be reported[1, 2]. Early work reported in based of ^{87}Rb and ^{133}Cs [3, 4], later work has also been done using ^{85}Rb [5]. The CPT signal is a function of the buffer-gas pressure that has been discussed [6]. The scientific works showed that laser light beam transverse variation could affect the shape of the CPT resonance line [7, 8]. The advantage of the weaker spin-exchange relaxation increasing signal amplitude has been also researched [9, 10].

In this paper we analyzed the CPT spectrum of Cs atomic D_1 line relative to clock transition in the case of magnetic field, and discussed the effect of the field on the signal amplitude at the four lambda models. The results show that for the lambda scheme involving the ground states $|F_g=4$, $m_F=0\rangle$, $|F_g=3$, $m_F=0\rangle$ and the excited state $|F_e=4$, $m_F=-1\rangle$ the amplitude can not depend on field when both the coherent lasers have a specific rate of intensity. Based on the most stable CPT, the width of the resonance can keep the most stable.

II. ANALYSIS IN THEORY

Considering Zeeman Effect, the total Hamiltonian of Cs atomic in magnetic field B is the sum of the non-perturbed atomic Hamiltonian H_0 and the Zeeman Hamiltonian, and is described by:

This work was supported by the Aviation Science Foundation of China (201207P6002) and Fundamental Research Founds of Central University (HEUCF20111111).

$$H_z = \left(-\frac{\mu_B}{\hbar} \right) B \cdot (L + g_s S + g_I I) \quad (1)$$

Where B is the static magnetic field, L, S and I denote the orbital, electron spin and nuclear spin angular moments respectively, g_S and g_I are the electron and nuclear spin Landé factors, μ_B and $\hbar$ are the Pohl magneton and the Planck constant. In the case of the unperturbed representation, the diagonal matrix element of the Hamiltonian become:

$$\langle F, m_F | H | F, m_F \rangle = E_0(F) - \mu_B g_F m_F B_z \quad (2)$$

Where $E_0(F)$ is the energy of the sublevel $|F, m_F\rangle$ and g_F is the associated Landé factor, and we neglect completely the influence of nonlinear Zeeman Effect on the shift of the hyperfine structure. The off-diagonal matrix element is the following formulation [11]:

$$\langle F-1, m_F | H | F, m_F \rangle = -\frac{\mu_B}{2} (g_J - g_I) B_z \cdot \left(\frac{[(J+I+1)^2 - F^2][F^2 - (J-I)^2]}{F} \right)^{1/2}$$
$$\left(\frac{F^2 - m_F^2}{F(2F+1)(2F-1)} \right)^{1/2} \quad (3)$$

The wave function of the quantum state in the field can be expressed in terms of the unperturbed atomic state vectors [12]:

$$|\psi(F_g, m_g)\rangle = \sum_g C_{gg'}(B, m_g) |F_g', m_g\rangle \quad \text{Ground states} \quad (4)$$

$$|\psi(F_e, m_e)\rangle = \sum_g C_{ee'}(B, m_e) |F_e', m_e\rangle \quad \text{Excited states} \quad (5)$$

The modified transfer coefficients are expressed as:

$$a[\psi(F_e, m_F); \psi(F_g, m_g)] = \sum_{F_e', F_g'} C_{F_e F_e'} a(F_e', m_e; F_g', m_g; q) C_{F_g F_g'} \quad (6)$$

Where $a(F_e', m_e; F_g', m_g; q)$ is the unperturbed transfer coefficients, q=-1, 0, 1.

We can calculate the square of the modified transfer coefficients in field for eight transitions of Cs atomic D_1 line relative to clock transition, the results are shown in Fig.1.

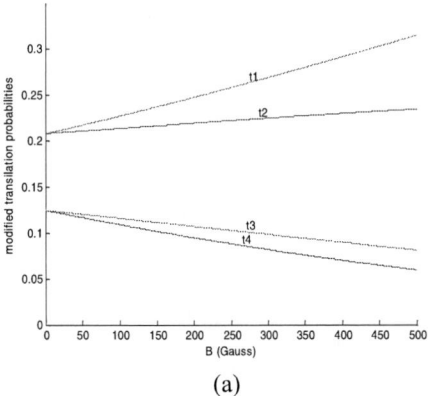

(a)

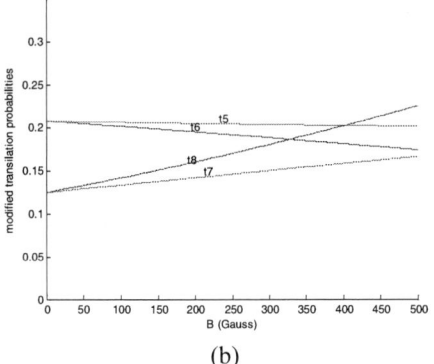

(b)

Figure 1. Fig.1. (a) modified translation probabilities versus the magnetic flux intensity for the left-hand circularly polarized laser

t1: $|F_g=3,m_F=0>\rightarrow|F_e=4,m_F=1>$,
t2: $|F_g=4,m_F=0>\rightarrow|F_e=4,m_F=1>$,
t3: $|F_g=3,m_F=0>\rightarrow|F_e=3,m_F=1>$;
t4: $|F_g=4,m_F=0>\rightarrow|F_e=3,m_F=1>$.

Figure 2. (b) modified translation probabilities versus the magnetic flux intensity for right-hand circularly polarized laser

t5: $|F_g=3,m_F=0>\rightarrow|F_e=4,m_F=-1>$,
t6: $|F_g=4,m_F=0>\rightarrow|F_e=4,m_F=-1>$,
t7: $|F_g=3,m_F=0>\rightarrow|F_e=3,m_F=-1>$,
t8: $|F_g=4,m_F=0>\rightarrow|F_e=3,m_F=-1>$.

Based on the above analysis, we find that the modified translation probabilities are the functions of magnetic intensity; taking it into account, we should redefine the transitional Rabi frequencies as:

$$\omega_{R1} = \frac{d_{\mu'm}E_1}{\hbar}a_{\mu'm}(B) \qquad (7)$$

$$\omega_{R2} = \frac{d_{\mu m}E_2}{\hbar}a_{\mu m}(B) \qquad (8)$$

Where d_{im} is the electric dipole moment for the unperturbed translation from level i to m (i=μ or μ'), $a_{im}(B)$ is the modified transfer coefficient from level i to m (i=μ or μ'). E_1 and E_2 denote the amplitudes of the laser radiation field.

Atomic clock based on the CPT is usually composed of three levels lambda models, as shown in Fig. 2. Here $|1>$, $|2>$ and $|3>$ represent the ground states $|F_g=3, m_F=0>$ and $|F_g=4,$

$m_F=0>$ and excited state respectively, Δ_0 is the frequency detuning of single photon transition, ω_1 and ω_2 are the angle frequency of both coherent lasers.

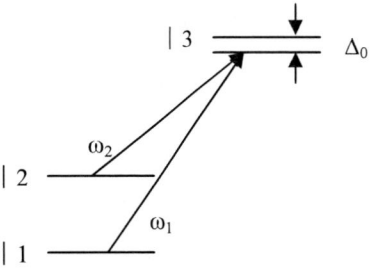

Figure 3. Fig.2: schematic diagram of the relative energy levels of cesium clock based on the CPT

The rate equation for the population of the levels for the coherence in the ground state is obtained from density-matrix:

$$\frac{\partial \rho}{\partial t} = \frac{i}{\hbar}[\rho, H] \qquad (9)$$

Where ρ is the density-matrix element and H is the total Hamiltonian which involves Zeeman Effect. Taking RWA into account, we take the solution of the off-diagonal matrix element as the form:

$$\rho_{12} = \delta_{12}e^{i(\omega_1-\omega_2)t} \qquad (10)$$

$$\rho_{13} = \delta_{13}e^{i\omega_1 t} \qquad (11)$$

$$\rho_{23} = \delta_{23}e^{i\omega_2 t} \qquad (12)$$

It is the reason that the two ground states are in a close proximity compared with the excited states, we take the spontaneous emission of the excited states to the two ground states as the same, which means both the two ground states obtain half of the spontaneous emission of the excited state, and then it will be a closed translation. And for the three levels system the Bloch equations are [1]:

$$\frac{d}{dt}\rho_{33} = -\omega_{R1}\operatorname{Im}\delta_{13} - \omega_{R2}\operatorname{Im}\delta_{23} - \Gamma\rho_{33} \qquad (13)$$

$$\frac{d}{dt}\rho_{22} = \omega_{R2}\operatorname{Im}\delta_{23} + \frac{\Gamma}{2}\rho_{33} - \frac{\gamma_1}{2}(\rho_{22}-\rho_{11}) \qquad (14)$$

$$\frac{d}{dt}\rho_{11} = \omega_{R1}\operatorname{Im}\delta_{13} + \frac{\Gamma}{2}\rho_{33} - \frac{\gamma_1}{2}(\rho_{11}-\rho_{22}) \qquad (15)$$

$$\frac{d}{dt}\delta_{13} = -[\frac{\Gamma}{2} + i(\omega_1-\omega_{31})]\delta_{13} + \frac{i\omega_{R1}}{2}(\rho_{33}-\rho_{11}) + i\omega_{R2}\delta_{12}$$
$$(16)$$

$$\frac{d}{dt}\delta_{23} = -[\frac{\Gamma}{2} + i(\omega_2-\omega_{32})]\delta_{23} + \frac{i\omega_{R2}}{2}(\rho_{33}-\rho_{22}) + i\omega_{R1}\delta_{21}$$
$$(17)$$

$$\frac{d}{dt}\delta_{12} = -[\gamma_2 + i(\omega_{12}-\omega_1+\omega_2)]\delta_{12} + \frac{i\omega_{R1}\delta_{32}}{2} - \frac{i\omega_{R2}\delta_{13}}{2}$$
$$(18)$$

$$\rho_{11} + \rho_{22} + \rho_{33} = 1 \qquad (19)$$

978-1-4799-1215-5/13 $31.00 © 2013 IEEE

Where Γ is the decay rate of $|3>$, γ_1 is the decay rate of the ground state, γ_2 is the decay rate of the coherence. We define $f = \dfrac{\omega_{R1}}{\omega_{R2}}$, $\Gamma_P = \dfrac{\omega_{R2}^2}{2\Gamma}$. Under the conditions that

$$\Delta_0 = \omega_1 - \omega_{31} = \omega_2 - \omega_{32} = 0$$ and steady solutions straightforward algebra gives the excited state population as:

$$\rho_{33} = \frac{\omega_{R2}^2}{\Gamma^2} \left\{ \frac{1+f^2}{2} + \frac{(f^2-1)^2 \Gamma_P}{2[\gamma_1 + (1+f^2)\Gamma_P]} - \frac{2f^4 \Gamma_P [\gamma_2 + (1+f^2)\Gamma_P]}{[\gamma_2 + (1+f^2)\Gamma_P]^2} \right\}$$

$$(20)$$

The amplitude of the CPT signal as a function of the magnetic intensity can be expressed by:

$$A_{33} \propto a_{23}^2(B) - \left(\frac{d_{13}E_1}{d_{23}E_2} \right)^2 a_{13}^2(B)$$

$$(21)$$

We should take the stability of the amplitude of CPT into account, so we make a variation for the formulation (21), we get:

$$\delta A_{33} \propto \frac{\partial}{\partial B} \left[a_{23}^2(B) - \left(\frac{d_{13}E_1}{d_{23}E_2} \right)^2 a_{13}^2(B) \right] \delta B \qquad (22)$$

To get the most stable clock transition signal, we can have:

$$\frac{\partial}{\partial B} a_{23}^2(B) - \left(\frac{d_{13}E_1}{d_{23}E_2} \right)^2 \frac{\partial}{\partial B} a_{13}^2(B) = 0$$

$$(23)$$

It is found that the four lambda models of the transition relative to the CPT clock are adapted to the formulation (23) according to the figure 1. That means the intensity of the two coherent lasers must be in specific ratio, the magnetic field will not have affect on amplitude of the CPT signal. The width of the resonance line is given by:

$$\nu_{1/2} = \frac{\gamma_2 + \left[\left(\frac{d_{13}E_1}{d_{23}E_2} \right)^2 a_{13}^2(B) + a_{23}^2(B) \right] \frac{(d_{23}E_2)^2}{2\Gamma}}{\pi} \qquad (24)$$

We can see that:

$$\frac{\partial}{\partial B} [\beta a_{13}^2(B) + a_{23}^2(B)] \neq 0 \qquad (25)$$

Where $\beta = \left(\dfrac{d_{13}E_1}{d_{23}E_2} \right)^2$.

So if we want to get the most stable atomic clock, we should prefer:

$$\min \left\{ \left| \frac{\partial}{\partial B} [\beta a_{13}^2(B) + a_{23}^2(B)] \right|_{B_0} \right\} \qquad (26)$$

Under the condition of the formulation (23) we can have:

$$\min \left\{ \left| \frac{\partial}{\partial B} [\beta a_{13}^2(B) + a_{23}^2(B)] \right|_{B_0} \right\} = \min \left\{ 2 \left| \frac{\partial}{\partial B} [a_{23}^2(B)] \right|_{B_0} \right\}$$

$$(27)$$

Comparing the modified translation probabilities in figure 1, we find the best scheme is t5 → translation from

$|F_g=3, m_F=0>$ and $|F_g=4, m_F=0>$ to $|F_e=4, m_F=-1>$ based on the formulation (27).

III. CONCLUSIONS

We have analyzed how the magnetic field affects the CPT signal of the Cs atomic and worked out the formulation of the CPT signal in magnetic field. At the same time we discussed the stability of the amplitude and the width of the resonance line of the CPT signal of Cs atomic in magnetic field. This work implied that we could select the specific ratio of the intensity of the two coherent lasers to get the most stable amplitude of the CPT signal of Cs atomic and the most stable width of the resonance line is the scheme of the translation from $|F_g=3, m_F=0>$ and $|F_g=4, m_F=0>$ to $|F_e=4, m_F=-1>$.

References

[1] [1] J. Vanier "Atomic clock based on coherent population trapping: a review," Appl. Phys. B, vol.81, pp.421-442, 2005

[2] [2] S. Knappe, J. Kitching, L. Hollberg et al "Temperature dependence of coherent population trapping resonances," Appl. Phys. B, vol.74, pp.217–222, 2002

[3] [3] N. Cyr, M. Tetu and M. Breton "All-optical microwave frequency standard: a proposal" IEEE Trans. Instrum. Meas., vol42, pp.640-649, 1993

[4] [4] F. Levi, A. Godone and J. Vanier "Cesium microwave emission without population inversion" Ultrasonics, Ferroelectrics and Frequency Control, IEEE Transactions vol.46, pp.609-615, 1999

[5] [5] M. Merimaa, T. Lindvall, I. Tittonen et al "All-optical atomic clock based on coherent population trapping in 85Rb" JOSA B, vol.20, pp.273-279, 2003

[6] [6] A. B. Post, Y. Y.Jau, N. N. Kuzma et al "End resonances for atomic clocks" 2004, in Program of the IEEE International Ultrasonics, Ferroelectrics, Abstracts booklet, pp.341, 2004

[7] [7] F. Levi, S. Micalizio, A. Godone et al "Realization of a CPT Rb maser prototype for Galileo" Frequency Control Symposium and PDA Exhibition Jointly with the 17th European Frequency and Time Forum, 2003 Proceedings of the IEEE International, pp.22-26, 2003

[8] [8] Taichenachev.A.V, Tumaikin A M, Yudin V I et al "Nonlinear resonance line shapes: Dependence on the transverse intensity distribution of a light beam" Phys. Rew. A, vol.69 024501, 2004

[9] [9] Y. Y. Jau, A. B. Post, N. N. Kuzma et al "The physics of miniature atomic clocks: 0-0 versus "end" resonances" Frequency Control Symposium and PDA Exhibition Jointly with the 17th European Frequency and Time Forum, 2003, Proceedings of the IEEE International, pp.33-36, 2003

[10] [10] Y. Y. Jau, A. B. Post, N. N. Kuzma et al "Intense, narrow atomic-clock resonances" Phys Rev. Lett. vol.92, 110801, 2004

[11] [11] P. Tremblay, M. Michaud, M. Levesque et al "Absorption profiles of alkali-metal D lines in the presence of a static magnetic field" Phys. Rev. A, vol.42, pp.2766-2773, 1990

[12] [12] R. Nibedita, M. Pattabiraman and C. Vijayan "Low filed Zeeman magnetometry using Rubidium absorption spectroscopy" Journal of Physics: Conference Series, vol.80, 012035, 2007

978-1-4799-1215-5/13 $31.00 © 2013 IEEE

Bistability of laser-induced thermal radiation in rare earth doped solids

Jing Dai, Hong Li, Zhenguo Zhang, Xinlu Zhang*, Li Li

Key Lab of In-Fiber Integrated Optics of Ministry of Education, and College of Science,
Harbin Engineering University, Harbin 150001, China
* zhangxinlu1@yahoo.com.cn

Abstract—This paper presents the unusual bistable thermal radiation in rare earth doped solids under intense laser excitation. A theoretical model involving the thermal radiation balance is established to discuss the intrinsic bistability of laser-induced thermal radiation. The origin of thermal radiation bistability is attributed to the nonlinear response of radiation quantum efficiency to local temperature.

Keywords-bistability; thermal radiation; rare earth material

I. INTRODUCTION

Since the middle of last century, optical bistability has been keeping much attractive to researchers due to the potential applications in all-optical switch and optical logical system. In recent years, the bistability phenomena on the luminescence and thermal radiation from rare earth materials were increasingly concerned [1-4]. Taking into account the special electronic structures of rare earth ions, the thermal radiation in rare earth materials exhibits special properties which are different from the ordinary materials [5-11]. Redmond et al. investigated the fluorescence quenching and the bistable blackbody radiation emerged in the rare earth nanomaterials doped Er/Yb ions [8,9]. They believed that the fluorescence reabsorption by Er impurities is mainly responsible for the bistable radiation. In this paper, we theoretically study the laser-induced thermal radiation characteristics in Tm,Yb doped material. The thermal radiation-balanced model is built to discuss the characteristics of the radiation bistability induced by intense laser excitaion.

II. THEORETICAL MODEL

Here in this work, we concern the Tm,Yb doped system under 980 nm laser excitaion. The energy level structures and energy transfer processes in Tm,Yb system can be found in literature [3-5]. Tm ions are introduced to emit upconversion luminescence (UL). The sensitizer Yb ions are to resonantly absorb 980 nm pump photons, then to efficiently transfer energy to Tm ions. Laser photons at 980 nm are resonantly absorbed by the Yb ions ($^2F_{7/2} \rightarrow ^2F_{5/2}$). The energy transfer of Yb-to-Tm ions has been confirmed by highly efficient upconversion blue (480 nm, $^1G_4 \rightarrow ^3H_6$ transition) spectra measurements [3-5]. The entire quenching of upconversion spectra have been proved at adequate laser excitation, which indicates efficient thermal phonon emission [10-11].

Let us start with a thermal radiation-balanced equation needed firstly. It can be written as the following form [9],

$$Q_{in} = Q_{out} \qquad (1)$$

where Q_{in} is the total absorbed energy, Q_{out} is the total out-exchange energy. Q_{in} can be taken as the sum of non-resonantly absorbed intensity by the background scattering and resonantly absorbed intensity by rare-earth ions. So Q_{in} can be expressed as

$$Q_{in} = (\alpha_S + \alpha_R)I_p \qquad (2)$$

where a_S is the background absorption coefficient, a_R is the resonance absorption coefficient, I_p is the applied pump intensity. Considering the luminescence of rare earth materials, Q_{out} is described as the sum of three parts, including the luminescence radiation I_l, blackbody radiation I_B, and the thermal conduction I_c.

$$Q_{out} = I_l + I_B + I_c \qquad (3)$$

The luminescence intensity I_l from excited Tm ions can express in equation (3) as

$$I_l = \alpha_R I_P \frac{v_f}{v_p n} q_t q_r(T) q_e \qquad (4)$$

where v_f is the average fluorescence frequency radiated from Tm ions, v_p is the frequency of pump. n denotes the number of Yb ions required for upconversion transfer to one Tm ion. q_t is the quantum efficiency of Yb-Tm energy transfer, $q_r(T)$ is the radiation quantum efficiency of Tm ions, q_e is the external quantum efficiency of fluorescence radiation. Note that $q_r(T)$ is a temperature dependence term, it can be written as

$$q_r(T) = r_{21}/(r_{21} + r_{nr}(T)) \qquad (5)$$

where r_{21} is the radiation rate determined by the intrinsic construction of rare-earth solids. $r_{nr}(T)$ is the non-radiation transition rate dependent strongly on local temperature, and can be written as [11]

$$r_{nr}(T) = r_{nr}(0)[1 - \exp(\frac{-hc}{\lambda kT})]^{-N} \qquad (6)$$

where $r_{nr}(0)$ is the non-radiation transition rate at the temperature of absolute zero, h is the Planck constant, c is the light velocity in vacuum, λ is the radiation wavelength of Tm ions, k is the Boltzmann constant, T is the local temperature, N is the multiphonon number emitted by non-radiation decay.

Based on the Stefan-Boltzmann's law, the blackbody radiation intensity I_B can be written as

$$I_B = \sigma(T^4 - T_0^4) \qquad (7)$$

where σ is the Stefan-Boltzmann constant. T_0 is the ambient temperature. The thermal conduction intensity I_c is written as

$$I_c = K(T)\chi(T - T_0) \qquad (8)$$

where $k(T)$ is the thermal conduction rate generally related with temperature. But in the range of high temperature, $k(T)$ can be approximately regard as a constant [9]. χ with the unit of m^{-1} is the geometry factor. Incorporating the above equations (4)-(8), we develop the model of laser-induce blackbody radiation in Tm,Yb solid system with the thermal balance condition $Q_{in}=Q_{out}$ [9],

$$(\alpha_s + \alpha_R)I_p$$
$$= \alpha_R I_P \frac{v_f}{v_p n} q_t q_e \frac{r_{21}}{r_{21} + r_{nr}(0)[1 - \exp(\frac{-hc}{\lambda kT})]^{-N}} \qquad (9)$$
$$+ \sigma(T^4 - T_0^4) + K(T)\chi(T - T_0)$$

Based on the theoretical modeling, we can discuss the bistable blackbody radiation through the numerical simulations based on the sufficient spectra analyses.

III. RESULTS AND DISCUSSION

Simulation of bistability mode of blackbody radiation are shown in Fig. 1. The intensity of blackbody radiation has a strong threshold effect with the growth of laser intensity, following an up-jump at a critical high threshold I_{p0}; then decreasing the laser intensity even below the up-jump threshold, the intensive blackbody radiation keeps emerging. It does not suddenly vanish from sight until a critical lower threshold I_{p1} reaches. Thereby the bistability of the laser-induced thermal radiation is performed in modeling.

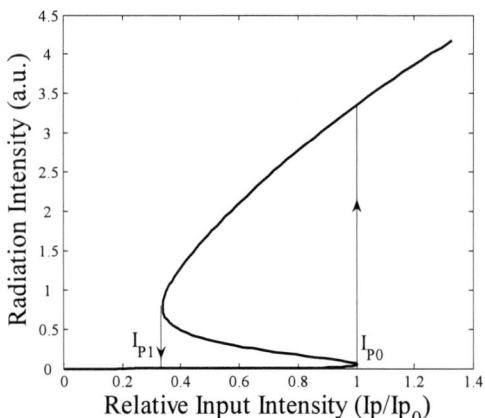

Figure 1. Simulation of bistability mode of blackbody radiation

During the operation processes of thermal radiation bistability, the hysteresis of the local thermal accumulation (or the local temperature) takes place inevitably. This can be revealed by the bistable hysteresis of radiation quantum efficiency shown in Fig. 2. The relationship between the radiation quantum efficiency of Tm ions and the temperature is given from Eqs. (5)-(6). Generally, the laser-induce local temperature changes proportionately with the absorbed pump power by changing the pump excitation intensity. When the local temperature is low, the radiation quantum efficiency approaches to unity. With the further increasing of the temperature, it drops abruptly to be close to zero. This means that the changes in the local temperature can result in the strong transition of radiation quantum efficiency.

Following the radiation quantum efficiency decreases suddenly, non-radiation quantum efficiency increases rapidly, as show in Fig. 3. The pump photon energy is absorbed by the electrons of Tm and emits thermal phonons through non-radiation decay. The emitted phonons are the energy particles to produce thermal radiation. Non-radiation transition becomes more efficiency with the enhancement of input intensity, which produces a lot amount of thermal phonons.

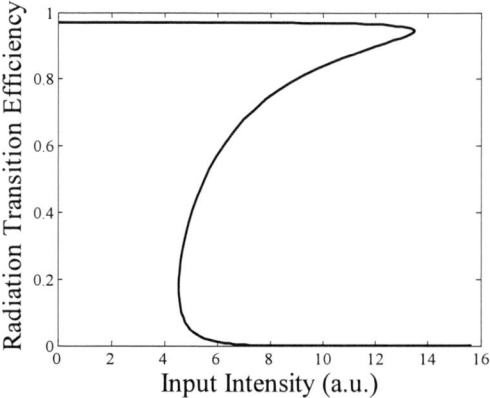

Figure 2. The bistable hysteresis of radiation quantum efficiency

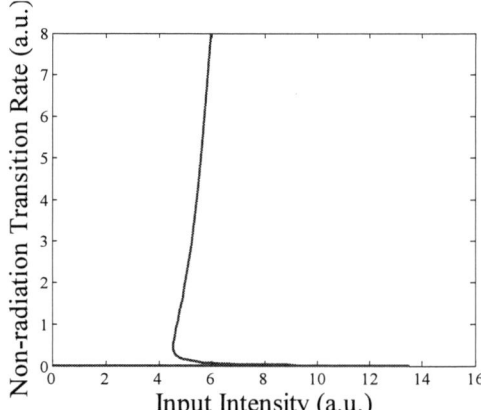

Figure 3. The bistable hysteresis of non-radiation quantum efficiency

The microscopic interpretations above indicate that the local temperature becomes increasingly high with the increasing of pump intensity. When the input intensity approaches I_{p0}, the local temperature leads to a nonlinear abrupt transition of radiation quantum efficiency. It has an intense influence on blackbody thermal radiation in Tm,Yb system, which results in the remarkable rise of the thermal radiation. It can be expected that the incandescent white light appeared suddenly in experiments. When the laser excitation is reduced from a high pump intensity, the local temperature of Tm,Yb solid decreases consequently. However, in the falling process, at the up-jump critical value of I_{p0}, the local temperature is still higher than that in the pump rising process. It is believed that the previous local thermal accumulation can not release synchronously to the surroundings and will impose an effect on the thermal radiation later. Thus the thermal radiation intensity is delayed to decrease abruptly. As a consequence, the unusual bistability characteristics of the thermal radiation are observed.

In order to further justify the vital role of the dependence of radiation quantum efficiency on the local temperature in the bistability mode, we can assume it constant, that is, $q_r(T)=q_0<1$. As show in Fig. 4, no bistability phenomenon in the thermal radiation is observed. This indicates that the temperature dependence of radiation quantum efficiency is responsible for the radiation bistability in Tm,Yb doped system.

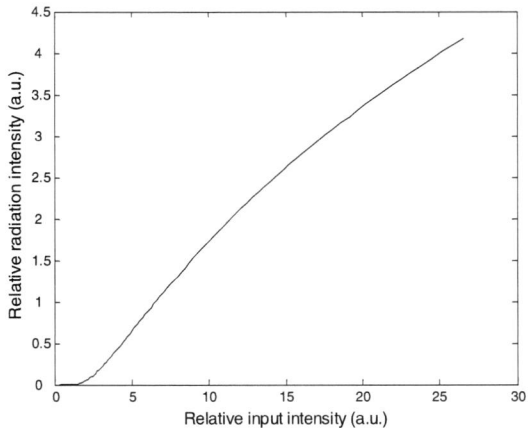

Figure 4. Blackbody radiation curve with a constant $q_r(T)=q_0$

We also carried out some simulations to measure the role of rare earth dopants on the occurrence of thermal radiation bistability. If the solid host material is not doped with any rare earth ions, the luminescence radiation from active ions can not happen. Thus the blackbody thermal radiation model can be rewritten as

$$\alpha_s I_p = \sigma(T^4 - T_0^4) + K(T)\chi(T - T_0) \qquad (10)$$

By incorporating with the Planck blackbody radiation law, the numerical result is obtained in Fig. 5. There is no bistable hysteresis phenomenon in the undoped host materials. The thermal radiation intensity is just monotonously increasing with the laser excitation intensity. In the undoped host material, the laser-induced role is as same as the thermal excitation due to background absorption. When the wavelength of the

material is ascertained, the thermal radiation intensity is only determined by the sample temperature. The simulation in Fig. 5 indicates the blackbody radiation itself does not exhibit the bistability feature. Therefore, it can be concluded that the introduction of rare earth ions into the host material is essential for the bistability mode of the thermal radiation.

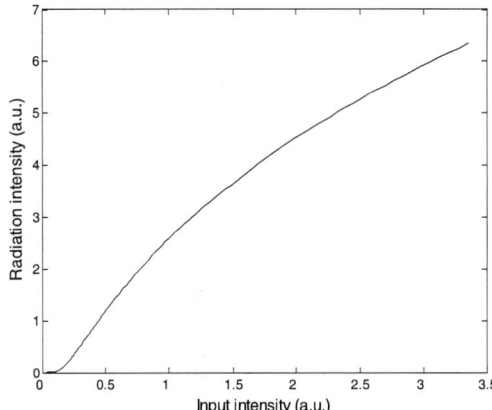

Figure 5. Blackbody radiation curve in an undoped host material

The bistable blackbody radiation due to the strongly temperature dependence of radiative quantum efficiency is explained as above. The rare earth doped materials must be chosen to achieve the bistable blackbody radiation. In a codoped system, one type of Tm ions is taken as activator; the other type of Yb ions is taken as sensitizer to transfer energy to activator efficiently. In addition, the multi-phonon number N implies the avalanche heating capacity in the bistability mode. A large amount of thermal phonon emission can accelerate the rapid growth of thermal accumulation or temperature. Our many simulations also show that the larger the multi-phonon number N, the better the bistability mode with both lower thresholds and wider bistable region. This suggests that it is favorable, for improving bistability mode, to choose the activator ion with wide energy gap to yield a large phonon number. If one wants to observe bistable thermal radiation in rare earth material, the laser excitation intensity must reach the critical threshold.

IV. CONCLUSION

In conclusion, the bistability phenomena of the laser-induced thermal radiation in Tm,Yb doped materials has been presented theoretically. Based on the thermal radiation-balanced model, the unusual bistability of the laser-induced thermal radiation is performed in modeling. The simulation shows that the radiation quantum efficiency exhibits the remarkable bistable hysteresis, accompanying with the operation of the thermal radiation bistability mode. The necessary condition for the thermal radiation bistability to occur has been discussed, and its origin is attributed to the intrinsically nonlinear response of radiation quantum efficiency to the laser-induced local temperature.

ACKNOWLEDGMENT

This work is supported by the National Natural Science Foundation of China (grants 11204048 and 61275138), and Heilongjiang Postdoctoral Science Foundation (LRB11190), the Fundamental Research Funds for the Central Universities.

REFERENCES

[1] M. P. Hehlen, A. Kuditcher, S. C. Rand and S. R. Lüthi, Site-selective, intrinsically bistable luminescence of Yb^{3+} ion pairs in CsCdBr3, Phys. Rev. Lett. 82(15): 3050-3053 (1999).

[2] A. Kuditcher, M. P. Hehlen, C. M. Florea, K. W. Winick and S. C. Rand, Intrinsic bistability of luminescence and stimulated emission in Yb- and Tm-doped glass, Phys. Rev. Lett. 84(9): 1898-1901 (2000).

[3] L. Li, H. Li, X. L, Zhang, Y. F. Peng, M. Nie, B. Jiang, X. W. Zhang and R. M. Li, Observation of bistable upconversion emission in Tm,Yb codoped yttria nanocrystal, Proc. of SPIE, 7846: 78460Y (2010).

[4] D. X. Wang, M. Nie, Z. G. Zhang, H. Li, L. Li, X. L. Zhang, X. W. Zhang and R. M. Li, Intrinsic bistability effect of Yb-sensitized Tm upconversion emission in zirconium dioxide nanocrystal, Proc. of SPIE, 8335: 833505 (2012).

[5] C. Y. Cao, X. M. Zhang, M. L. Chen, W. P. Qin and J. S. Zhang, Up-conversion fluorescence and thermal optical bistability in Gd_2O_3: $10\%Yb^{3+}$, $0.5\%Tm^{3+}$, Phys. B 405: 3685-3689 (2010).

[6] D. R. Gamelin, S. R. Lüthi and H. U. Güdel, The role of laser heating in the intrinsic optical bistability of Yb^{3+}-doped bromide lattices, J. Phys. Chem. B 104: 11045-11057 (2000).

[7] M. A. Noginov, M. Vondrova and B. D. Lucas, Thermally induced optical bistability in Cr-doped colquiriiteb crystals, Phys. Rev. B. 65(3): 035112 (2001).

[8] S. M. Redmond, S. C. Rand and S. L. Oliveira, Bistable emission of a black-body radiator, Appl. Phys. Lett. 85(23): 5517-5519 (2004).

[9] S M. Redmond, Luminescent instabilities and non-radiative processes in rare earth systems, Ph.D dissertation (University of Michigan, Ann Arbor, Michigan, 2003).

[10] S. M. Redmond, S. C. Rand, X. L. Ruan and M. Kaviany, Multiple scattering and nonlinear thermal emission of Yb^{3+}, Er^{3+}:Y_2O_3 nanopowders, J. Appl. Phys. 95(8): 4069-4077 (2004).

[11] X. L. Ruan and M. Kaviany, Enhanced nonradiative relaxation and photoluminescence quenching in random, doped nanocrystalline powders, J. Appl. Phys. 97(10): 104331 (2005).

Enhanced laser cooling of Tm-doped solids by upconversion pumping

Jing Dai, Li Li, Xinlu Zhang*

Key Lab of In-Fiber Integrated Optics of Ministry of Education, and College of Science,
Harbin Engineering University, Harbin 150001, China
* zhangxinlu1@yahoo.com.cn

Abstract—**We present a physical scheme of upconversion-assisted anti-Stokes fluorescence cooling in Tm doped low phonon materials. The upconversion pumping from the metastable level to the upconversion excited level is introduced to produce extra anti-Stokes cooling cycles which can improve the laser cooling performance. Both the limitation of the cooling cryogenic temperature and the range of the cooling wavelength can be extended notably in contrast to the traditional anti-Stokes cooling.**

Keywords-laser cooling; rare earth solid; anti-Stokes fluorescence

I. INTRODUCTION

Laser cooling of solids is an interesting, vagarious phenomenon of nonlinear photo-thermal transition dynamics in the field of photon-matter interaction. It is generally based on the principle of anti-Stokes fluorescence [1], where the emitted photons have a mean frequency higher than that of the absorbed photons. Laser cooling of solids offers a feasible approach to all-solid-state optical refrigeration technique. It has potential applications in the fields of integrated optoelectronic circuit and aerospace detecting devices due to the pollution-free, vibrationless, no electro-magnetic radiation. Since the first experiment of laser cooling in Yb doped fluorozirconate glass was demonstrated by Epstein et al. in 1995 [2], laser cooling has experienced rapid progress in various rare-earth (Yb, Tm and Er) doped solids [3-9]. The cooling cryogenic temperature by individual anti-Stokes fluorescence has culminated to an absolute temperature 155 K in 2010 [3] and 119 K in 2013 [4] in Yb doped LYF crystal by Seletskiy et al.. Also, the net laser cooling by about 40 K in a semiconductor using group-II–VI cadmium sulphide nanoribbons has been first demonstrated by Zhang et al. in 2013 [10].

Following the successful progress in experiment, the fundamental theories with new cooling schemes or physical mechanisms need to be developed to provide guidance for enhancing laser cooling performance [11, 12]. In this paper, we present a physical scheme of upconversion-assisted anti-Stokes fluorescence cooling in Tm doped low phonon materials. The idea of upconversion cooling comes from the proposal related to Er doped system by Garcia-Adeva et al. [13], but the physical construction of upconversion cooling in Tm doped system is distinct with that because of the unique electronic structure of Tm ion. The upconversion pumping from the metastable level to the upconversion level in Tm doped system is introduced to produce extra anti-Stokes cooling cycles to contribute to the cooling capacity. A six-level model for optical

refrigeration is built to discuss the improvement of the laser cooling performance by upconversion pumping. The simulations show that both the limitation of the cooling temperature and the range of the cooling wavelength can be extended notably in contrast to the traditional anti-Stokes cooling.

II. THEORETICAL MODEL

Here we consider a Tm doped lithium yttrium fluoride low phonon crystal (Tm:YLF) under both the ground-state pumping (λ_{p1} pumping) the excited-state pumping (λ_{p2} pumping). The energy level diagram and energy transition processes involved in laser cooling are shown in Fig. 1. The ground-state manifold consists of two closely spaced levels (n_0 and n_1) with an energy separation δ_1. The metastable manifold consists of two levels (n_2 and n_3) with an energy separation δ_2. The upconversion excited manifold consists of two levels (n_4 and n_5) with an energy separation δ_3 as well. The six-level energy model for optical refrigeration is adopted to develop the theoretical model of the upconversion-assisted anti-Stokes fluorescence cooling, which mainly consists of the traditional anti-Stokes cooling cycle ($n_1 \rightarrow n_2$ pumping, $n_3 \rightarrow n_0$ radiation) and the upconversion cooling cycles ($n_3 \rightarrow n_4$ pumping, $n_5 \rightarrow n_2$ and $n_5 \rightarrow n_0$ radiations).

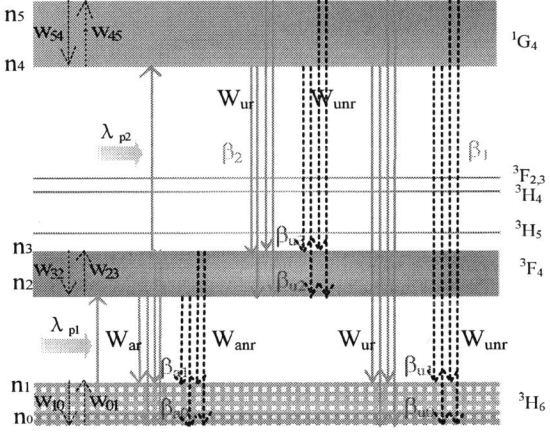

Figure 1. The six-level energy diagram for optical refrigeration consisting of the traditional anti-Stokes cooling cycle (λ_{p1} pumping) and the upconversion cooling cycles (λ_{p2} pumping) in low dopant Tm;YLF crystals.

The rate equations governing the density populations (n_i, i=0, 1, 2, 3, 4, 5) on the six levels can be written as

978-1-4799-1215-5/13 $31.00 © 2013 IEEE

$$\frac{dN_1}{dt} = -\frac{I_1\alpha_{r1}}{h\gamma_{p1}} + \beta_{a1}(N_2 + N_3)R_{af} + \beta_{u1}(N_4 + N_5)R_{uc} \tag{1}$$
$$- w_{10}(N_1 - N_0 e^{-\delta_1/K_bT})$$

$$\frac{dN_2}{dt} = \frac{I_1\alpha_{r1}}{h\gamma_{p1}} + \beta_{u2}(N_4 + N_5)R_{uc} - N_2 R_{af} \tag{2}$$
$$+ w_{32}(N_3 - N_2 e^{-\delta_2/K_bT})$$

$$\frac{dN_3}{dt} = -\frac{I_2\alpha_{r2}}{h\gamma_{p2}} - N_3 R_{af} + \beta_{u3}(N_4 + N_5)R_{uc} \tag{3}$$
$$- w_{32}(N_3 - N_2 e^{-\delta_2/K_bT})$$

$$\frac{dN_4}{dt} = \frac{I_2\alpha_{r2}}{h\gamma_{p2}} - N_4 R_{uc} + w_{54}(N_5 - N_4 e^{-\delta_3/K_bT}) \tag{4}$$

$$\frac{dN_5}{dt} = -N_5 R_{uc} - w_{54}(N_5 - N_4 e^{-\delta_3/K_bT}) \tag{5}$$

$$N_t = N_0 + N_1 + N_2 + N_3 + N_4 + N_5 \tag{6}$$

where γ_{p1} is the ground-state pumping frequency with the intensity I_1 and γ_{p2} the upconversion pumping frequency with the intensity I_2. α_{r1} and α_{r2} are the near-resonant absorption coefficients to the λ_{p1} and λ_{p2} pump lasers, respectively. R_{af} is the decay rate of the metastable $^3F_4$ level, and R_{uc} the decay rate of the upconversion $^1G_4$ level. β_{a0} and β_{a1} are the branch ratios from the metastable level $^3F_4$ to the Stark sublevels n_0 and n_1, respectively. We also denote the branch ratios β_{u0}, β_{u1}, β_{u2} and β_{u3}, respectively, corresponding to the transitions from the upconversion $^1G_4$ level to the Stark sublevels n_0, n_1, n_2 and n_3. The weighting factors (w_{10}, w_{32}, w_{54}) in the electron-phonon interaction terms maintain the Boltzmann distribution among each manifold at quasi equilibrium. N_t is the total concentration of Tm ions.

The net cooling power density removed from the sample can be deduced by using the thermal-radiation balance equation,

$$P_{cool} = P_{rad} - P_{abs}$$

In the system associated with upconversion pumping, the absorbed pump power per volume by the sample is

$$P_{abs} = (\alpha_{r1} + \alpha_b)I_1 + (\alpha_{r2} + \alpha_b)I_2$$

where α_b is the background absorption coefficient. The out-radiated fluorescence power per volume is deduced to be

$$P_{rad} = \eta_e(N_2 + N_3)W_{ar}h\gamma_{af} + \eta_e(N_4 + N_5)W_{ur}(\beta_1 h\gamma_{uc1} + \beta_2 h\gamma_{uc2})$$
$$= \eta_{ext1}(\frac{I_1\alpha_{r1}}{\gamma_{p1}} - \beta_2\frac{I_2\alpha_{r2}}{\gamma_{p2}})\gamma_{af} + \eta_{ext2}\frac{I_2\alpha_{r2}}{\gamma_{p2}}(\beta_1\gamma_{uc1} + \beta_2\gamma_{uc2})$$

where $\eta_{ext1} = \eta_e W_{ar}/(\eta_e W_{ar} + W_{anr})$ and $\eta_{ext2} = \eta_e W_{ur}/(\eta_e W_{ur} + W_{unr})$. And the mean fluorescence frequencies for the six-level upconversion cooling model can be given by

$$h\gamma_{af} = E_{12} + \delta_1\beta_{a0} + \delta_2 N_3/(N_2 + N_3) \tag{7}$$
$$= E_{12} + \delta_1\beta_{a0} + \delta_2/[1 + (1 + R_{af}/w_{32})e^{\delta_2/k_BT}]$$

$$h\gamma_{uc1} = E_{34} + \delta_2\beta_{u2} + \delta_3 N_5/(N_4 + N_5) \tag{8}$$
$$= E_{34} + \delta_2\beta_{u2} + \delta_3/[1 + (1 + R_{uc}/w_{54})e^{\delta_3/k_BT}]$$

$$h\gamma_{uc2} = E_{12} + E_{34} + \delta_1\beta_{u0} + \delta_2 + \delta_3 N_5/(N_4 + N_5) \tag{9}$$
$$= E_{12} + E_{34} + \delta_1\beta_{u0} + \delta_2 + \delta_3/[1 + (1 + R_{uc}/w_{54})e^{\delta_3/k_BT}]$$

According to the Eqs. (7)-(9), one can numerically estimate the temperature-dependent shift of the mean fluorescence wavelength during the laser cooling process.

Once the net cooling power and the absorbed power are known, one can calculate the cooling efficiency of the upconversion-assisted laser cooling system by the relationship

$$\eta_{cool} = P_{cool}/P_{abs}$$
$$= \frac{\eta_{ext1}(\frac{\alpha_{r1}}{\gamma_{p1}} - \beta_2\frac{\mu\alpha_{r2}}{\gamma_{p2}})\gamma_{af} + \eta_{ext2}\frac{\mu\alpha_{r2}}{\gamma_{p2}}(\beta_1\gamma_{uc1} + \beta_2\gamma_{uc2})}{(\alpha_b + \alpha_{r1}) + (\alpha_b + \alpha_{r2})\mu} - 1 \tag{10}$$

where $\beta_1 = \beta_{u2} + \beta_{u3}$ and $\beta_2 = \beta_{u0} + \beta_{u1}$, and the ratio $\mu = I_2/I_1$. Note that the absorption coefficients $\alpha_{r1}(\lambda, T)$ and $\alpha_{r2}(\lambda, T)$ are dependent on the pumping wavelength and the temperature of the sample. They can be expressed approximately under low pump excitation to be

$$\alpha_{r1} = \frac{\sigma_{12}(\lambda_{p1}, T)}{1 + e^{\delta_1/k_BT}}(N_0 + N_1) \simeq \frac{\sigma_{12}(\lambda_{p1}, T)N_t}{1 + e^{\delta_1/k_BT}} \tag{11}$$

$$\alpha_{r2} = \frac{\sigma_{34}(\lambda_{p2}, T)(N_2 + N_3)}{1 + e^{\delta_2/k_BT}} \simeq \frac{\sigma_{34}(\lambda_{p2}, T)}{1 + e^{\delta_2/k_BT}}\frac{I_1\alpha_{r1}}{h\gamma_{p1}\beta_{a1}R_{af}} \tag{12}$$

In the condition of $\mu = 0$ without the upconversion pumping, the cooling efficiency of Eq. (10) is reduced to be the standard formulation of the traditional anti-Stokes cooling [3],

$$\eta_{cool} = \eta_{ext1}\frac{\alpha_{r1}}{\alpha_b + \alpha_{r1}}\frac{\gamma_{af}}{\gamma_{p1}} - 1 \tag{13}$$

Comparing the upconversion cooling efficiency of Eq. (10) with the traditional cooling efficiency of Eq. (13), one can find that the upconversion pumping makes important effects on the laser cooling performance. This is mainly due to the notable contributions of the upconversion cooling cycles to the net cooling power. On the basis of the theoretical model developed above, one can discuss numerically the upconversion-assisted cooling performance by using the experimental spectra data [14-15]. Here we put emphasis on the improvement of the minimum achievable cryogenic temperature (MAT).

III. RESULTS AND DISCUSSION

We first calculate the temperature-dependent shift of the mean fluorescence wavelengths of λ_{af}, λ_{uc1} and λ_{uc2}. This is useful to determine the effective pumping wavelengths λ_{p1} and λ_{p2} to perform the net cooling. In a wide temperature range of 300 K to 80 K, the maximum of the mean fluorescence wavelength λ_{af} is close to about 1800 nm as shown in Fig. 2. Similarly, the simulation of Fig. 3 shows that the mean wavelength of λ_{uc1} shifts to a maximum wavelength of about

651 nm along with the reduction of temperature to 80 K. In addition, the other upconverison fluorescence produced by $^1G_4 \rightarrow {}^3H_6$ transition approaches to a longer wavelength of about 477.5 nm during the cooling process, as shown in Fig. 4. These calculations indicate the requirement for the pumping wavelengths, that is, $\lambda_{p1} > 1800$ nm and $\lambda_{p2} > 651$ nm for their individual anti-Stokes radiation.

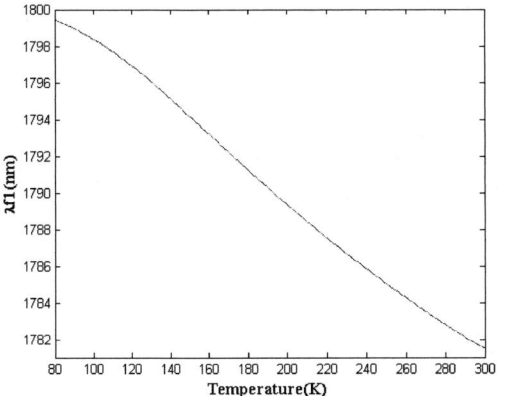

Figure 2.   The temperature-dependent shift of the mean wavelength of anti-Stokes fluorescence λ_{af}.

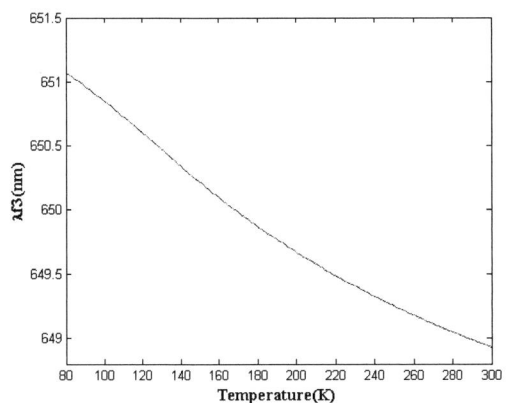

Figure 3.   The temperature-dependent shift of the mean wavelength of one upconversion fluorescence λ_{uc1}.

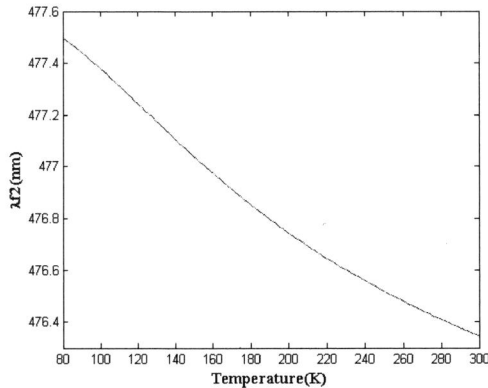

Figure 4.   The temperature-dependent shift of the mean wavelength of the other upconversion fluorescence λ_{uc2}.

In order to evaluate the minimum achievable temperature (MAT), the cooling efficiencies versus the pump wavelength and the sample temperature are simulated by using the experimental spectra data [14-15]. Fig. 5 shows the traditional anti-Stokes cooling efficiency in a high purity Tm:YLF crystal with $\alpha_b = 0.04$ m^{-1} under the condition of $\mu = 0$. The minimum cooling temperature of 132 K can be achieved theoretically at the pumping wavelength of $\lambda_{p1} = 1845$ nm. In contrast, Fig. 6 shows the simulation of upconversin-assisted anti-Stokes cooling efficiency with the same absorption of $\alpha_b = 0.04$ m^{-1} under the condition of $\mu = 0.5$. The minimum achievable temperature of 114 K are theoretically predicted in modeling at the pumping wavelength of $\lambda_{p1} = 1795$ nm. A notable reduction by about 18 K in the minimum achievable temperature is demonstrated through the enhancement of upconversion cooling cycles to laser cooling capacity. In terms of the temperature-dependent shift of the mean fluorescence wavelength λ_{af} shown in Fig. 2, the net cooling of the traditional individual anti-Stokes scheme is not able to implement under such a pump wavelength of $\lambda_{p1} = 1795$ nm. The heating effects induced at $\lambda_{p1} = 1795$ nm can be found in Fig. 5. The comparative results imply that the upconversion cooling cycles exhibit the powerful capacity of removing the thermal power from the sample. And besides, the range of pump wavelength λ_{p1} covering the net cooling region is notably extended by the cooperating upconversion pumping. The net laser cooling of Tm doped solids is also possibly achieved even at the pump wavelength of λ_{p1} where the traditional anti-Stokes fluorescence cooling is unable. The outspread of the cooling wavelength is another interesting feature for the upconversion-assisted cooling scheme, which is mentioned analytically in the proposal related to Er system by Garcia-Adeva et al [13]. Here the important feature is demonstrated numerically in a different system of upconversion-assisted anti-Stokes fluorescence cooling of Tm doped solid.

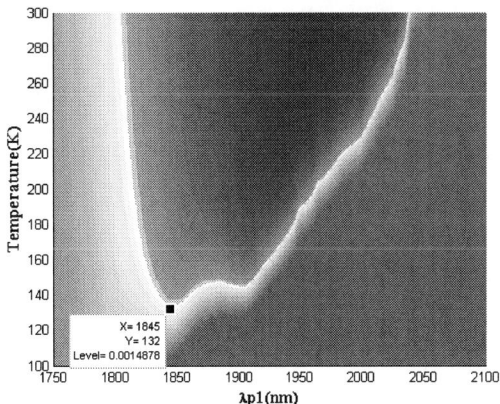

Figure 5.   Surface map of individual anti-Stokes cooling efficiency versus pump wavelength λ_{p1} and sample temeperature T with $\alpha_b = 0.04$ m^{-1}. Blue and red regions correspond to the cooling ($\eta_{cool} > 0$) and heating ($\eta_{cool} > 0$) respectively, separated by a white borderline ($\eta_{cool} = 0$).

978-1-4799-1215-5/13 $31.00 © 2013 IEEE   209

Fig. 7 shows the simulation of the upconversion-assisted anti-Stokes cooling efficiency versus upconversion pumping λ_{p2} and sample temperature with the same absorption of α_b=0.04 m^{-1} under the condition of μ=0.5. The net cooling for the fixed ground-state pumping wavelength of λ_{p1}=1845 nm is enabled when the upconversion pumping wavelength of λ_{p2} is larger than the critical wavelength of 657 nm. Moreover, the change in the minimum achievable temperature varies slightly with the adjustment of the upconversion pumping wavelength.

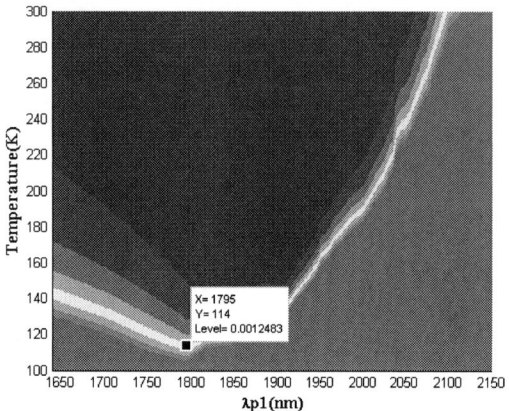

Figure 6. Surface map of upvonersion-assisted anti-Stokes cooling efficiency versus pump wavelength λ_{p1} and sample temeperature T with α_b=0.04 m^{-1}. Blue and red regions correspond to the cooling (η_{cool}>0) and heating (η_{cool}>0) respectively, separated by a white borderline (η_{cool}=0).

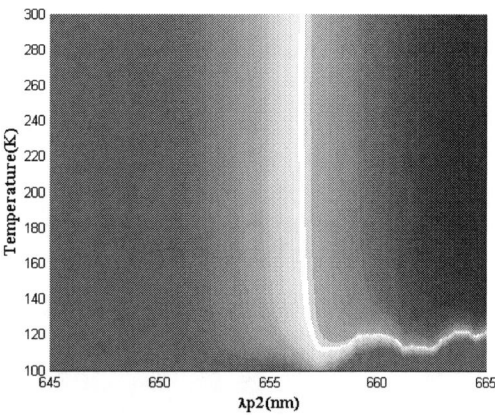

Figure 7. Surface map of upvonersion-assisted anti-Stokes cooling efficiency versus upconversion pump λ_{p2} and sample temeperature T with α_b=0.04 m^{-1}. Blue and red regions correspond to the cooling (η_{cool}>0) and heating (η_{cool}>0) respectively, separated by a white borderline (η_{cool}=0).

IV. CONCLUSION

In conclusion, the physical scheme of upconversion-assisted anti-Stokes fluorescence cooling in Tm doped low phonon materials has been presented. The six-level model for upconversion-assisted optical refrigeration is developed based on the experimental spectra data, to discuss the improvement of the laser cooling performance by upconversion pumping. The upconversion pumping from the metastable $^3F_4$ level to the upconversion excited $^1G_4$ level can produce extra anti-Stokes cooling cycles which can powerfully enhance the laser cooling capacity. The simulations demonstrated that in contrast to the traditional anti-Stokes cooling, the minimum achievable cryogenic temperature can be further pulled down and the range of traditional ground-state pump wavelength covering the net cooling region can be notably extended by the cooperating upconversion pumping.

ACKNOWLEDGMENT

This work is supported by the National Natural Science Foundation of China (grants 11204048 and 61275138), and Heilongjiang Postdoctoral Science Foundation (LRB11190), the Fundamental Research Funds for the Central Universities.

REFERENCES

[1] P. Pringsheim, Zwei bemerkungen uber den unterschied von lumineszenz und temperaturstrahlung, Z. Phys. 57: 739-746 (1929).

[2] R. I. Epstein, M. I. Buchwald, B. C. Edwards, T. R. Gosnell and C. E. Mungan, Observation of laser-induced fluorescent cooling of solid, Nature 377: 500-503 (1995).

[3] D. V. Seletskiy, S. D. Melgaard, S. Bigotta, A. D. Lieto, M. Tonelli and M. Sheik-Bahae, Laser cooling of solids to cryogenic temperatures, Nature Photonics 4: 161-164 (2010).

[4] S. D. Melgaard, D. V. Seletskiy, A. D. Lieto, M. Tonelli and M. Sheik-Bahae, Optical refrigeration to 119 K, below National Institute of Standards and Technology cryogenic temperature, Opt. Lett. 38(9): 1588-1590 (2013).

[5] D. V. Seletskiy, S. D. Melgaard, R. I. Epstein, A. D. Lieto, M. Tonelli and M. Sheik-Bahae, Local laser cooling of Yb:YLF to 110 K, Opt. Express 19(19): 18229-18236 (2011).

[6] C. W. Hoyt, M. Sheik-Bahae, R. I. Epstein, B. C. Edwards and J. E. Anderson, Observation of anti-Stokes fluorescence cooling in thulium-doped glass, Phys. Rev. Lett. 85(17): 3600-3603 (2000).

[7] J. Fernandez, A. J. Garcia-Adeva, R. Balda. Anti-Stokes laser cooling in bulk erbium-doped materials. Phys. Rev. Lett. 97(3): 033001 (2006).

[8] A. R. Albrecht, M. Ghasemkhani, J. G. Cederberg, D. V. Seletskiy, S. D. Melgaard and M. Sheik-Bahae, Progress towards cryogenic temperatures in intra-cavity optical refrigeration using a VECSEL, Proc. of SPIE 8638: 863805 (2013).

[9] W. Patterson, S. Bigotta, M. Sheik-Bahae, D. Parisi, M. Tonelli and R. I. Epstein, Anti-Stokes luminescence cooling of Tm3+ doped BaY2F8, Opt. Express 16(3): 1704-1710 (2008).

[10] J. Zhang, D. H. Li, R. J. Chen and Q. H. Xiong, Laser cooling of a semiconductor by 40 kelvin, Nature 493: 504-508 (2013).

[11] X. L. Ruan and M. Kaviany, Advances in laser cooling of solids, J. Heat Transfer 129: 3-10 (2007).

[12] S. V. Petrushkin, V. V. Samartsev, Laser cooling of solids (Cambridge International Science Pub. Ltd, 2009).

[13] A. J. Garcia-Adeva, R. Balda, J. Fernandez, Upconversion cooling of Er-doped low phonon fluorescent solids, Phys. Rev. B 79: 033110 (2009).

[14] B. M. Walsh, N. P. Barnes and B. D. Bartolo, Branching ratios, cross sections, and radiative lifetimes of rare earth ions in solids: Application to Tm3+ and Ho3+ ions in LiYF4, J. Appl. Phys. 83(5): 2772-2787 (1998).

[15] J. W. Szela and J. I. Mackenzie, Excited-state absorption measurements of Tm3+-doped crystals, Proc. of SPIE, 8433: 84331O (2012).

The techniques of fixed pattern noise reduction for high speed digital CMOS image sensor

Na Zhang *, Haiyong Zheng

School of Information Science and Engineering, Ocean University of China, Qingdao, China
baiquanbaiquan@126.com

Abstract—**Techniques of fixed pattern noise reduction for CMOS image sensor are presented. Double sampling reusing the existing switch-capacitor amplifier of column ADC, double resets for 5-T active pixel, and negative offset storage of the cyclic ADC, are proposed to reduce the fixed pattern noise caused by the threshold voltage's mismatch of the amplifying and transmitting transistor of the pixel, and the offset of the amplifier in ADC, respectively. Experimental results show that the typical offset voltages observed are 0.97mV, which is about 1LSB, that indicate that these techniques are efficient without additional processes and devices.**

Keywords-CMOS image sensor; fixed pattern noise; double sampling; column parallel

I. INTRODUCTION

High frame rate is one of the three trends in development of CMOS image sensor[1]. In present, three structures are mainly used by high frame-rate CMOS image sensor: analog CMOS image sensor, digital pixel CMOS image sensor, column parallel CMOS image sensor. Because of integrating analog-to-digital converter, clock generator, LVDS, and some other modules, column parallel CMOS image sensor has the advantages of ease use and system integration, which also has high fill factor, high image quality and can be easily expanded [2].

There are some noises in CMOS image sensor, which reduce the image quality, for example, photon shot noise, dark current shot noise, 1/f noise, fixed pattern noise, and so on[3]. The fixed pattern noise(FPN) which is fixed in a spatial caused by the mismatch of the transistors' threshold voltage due to semiconductor technology, influences the image quality greatly[4]. There are two fixed pattern noises in column parallel CMOS image sensor: pixel FPN because of the mismatch of the transistors' threshold voltage in pixel, column FPN caused by the offset of the column amplifiers.

Double sampling [5] can remove the FPN caused by the amplifying transistor of pixel, but can't reduce the ones introduced by the pixel's transmitting transistor and the amplifier's offset. In this paper, some techniques have been presented, which can cancel the fixed pattern noise caused by all these reasons, without additional devices.

II. THE STRUCTURE OF CMOS IMAGE SENSOR

To meet high frame requirement, the designed column parallel CMOS image sensor is comprised of pixel array, cyclic ADC, bias circuit, PLL, readout circuit, LVDS and logic control circuit, shown in Fig.1. In each column, there is one ADC. The ADCs in the odd and even columns lay on the upside and downside of the pixel array respectively.

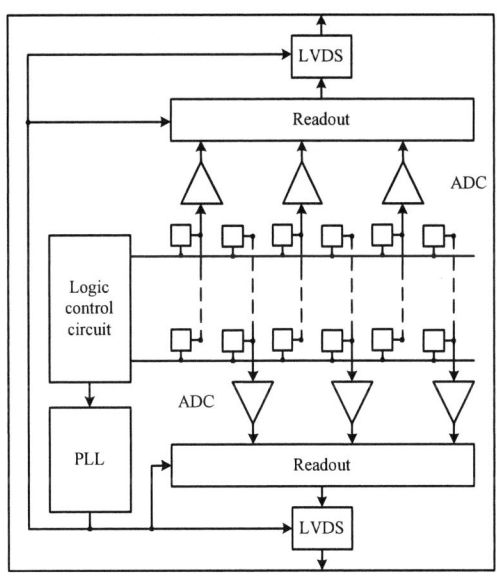

Figure 1. Structure of high frame rate column parallel CMOS image sensor

III. DESIGN OF FIXED PATTERN NOISE REDUCTION

A. Double sampling technique

Double sampling technique needs a sample-hold circuit. In this paper, the existing switch-capacitor amplifier of column ADC is reused to realize the double sampling, without additional circuits.

The cyclic ADC, shown in Fig.2, is comprised of sub-ADC, two stage register, sub-DAC, logic correction and switch-capacitor amplifier module. Fig. 3 shows the switch-capacitor amplifier module schematic, which samples and holds the residue voltage, multiplies it by 2 and adds reference voltage.

According to the timing diagram of switch-capacitor amplifier module shown in Fig. 4, in clock phase B, the residue voltage of the amplifier's output is

$$V_{out}(1) = V_{reset} - V_{signal} - V_R \qquad (1)$$

978-1-4799-1215-5/13 $31.00 © 2013 IEEE

Where $V_{out}(1) = V_{OM}(1) - V_{ON}(1), V_R = V_{RM} - V_{RN}$.

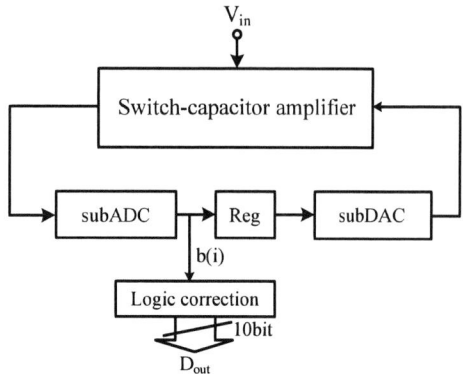

Figure 2.   Structure of RSD cyclic ADC

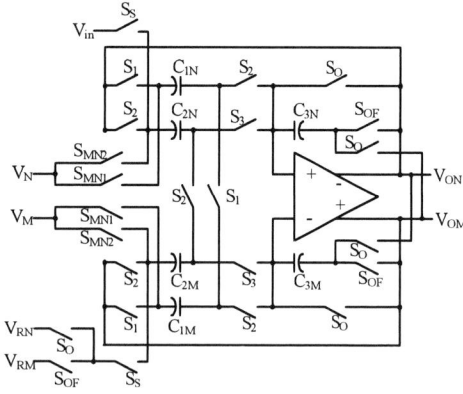

Figure 3.   Switch-capacitor amplifier module schematic diagram

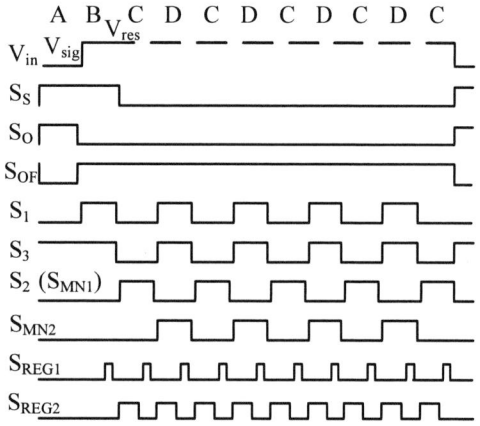

Figure 4.   Timing diagram of switch-capacitor amplifier

Considering the FPN V_{F1} caused by the amplifying transistor M3, the pixel's readout signal offsets by V_{F1} when transmitting through transistor M3. Thus formula (1) can be written as

$$V_{out}(1) = (V_{reset} - V_{F1}) - (V_{signal} - V_{F1}) - V_R$$
$$= V_{reset} - V_{signal} - V_R \qquad (2)$$

Where the FPN caused by the threshold voltage's mismatch of the amplifying transistor of the pixel, has been cancelled by double sampling of the cyclic ADC.

B. Negative offset storage technique

The mismatch of the devices because of the uncertainty of the technology can bring dc offset to the column processing circuit and cause the column FPN which introduces stripes onto the captured image. Unfortunately, compared to the pixel FPN, the column FPN is often more noticeable to the human eye and it is more difficult to be removed through circuitry solutions. The main source of the column FPN in column parallel CMOS image sensor is the offset voltage of the amplifiers in ADC.

If only the column amplifier's offset voltage is considered, the input referred offset voltage is

$$V_{OS}(1 + 2 + 2^2 + \cdots + 2^{n-1})/2^{n-1} \approx 2V_{OS} \qquad (3)$$

In this paper, a negative offset storage technique has been presented to reduce this offset. In clock phase A, capacitor C_{3M} and C_{3N} are crossed over the amplifier's positive and negative terminals. In clock phase B, the residue voltage of the amplifier is

$$V_{out}(1) = V_{reset} - V_{signal} - V_R - V_{OS} \qquad (4)$$

In the following clock phase C and D, the residue voltage is

$$V_{out}(i+1) = 2V_{out}(i) - b(i) \times V_R + V_{OS}$$
$$i = 1,2,\cdots,9 \qquad (5)$$

According to (4), the offset of $V_{out}(1)$ is $-V_{OS}$. In (5), it is multiplied by 2 and added by V_{OS}, thus there is also $-V_{OS}$ offset in $V_{out}(2) \sim V_{out}(10)$. The offset voltage hasn't been amplified and accumulated, so the input referred offset voltage is

$$-V_{OS}/2^{n-1} \approx 0 \qquad (6)$$

C. Technique of double resets for 5-T pixel active pixel

For high speed application, global shutter that all the pixels integrate simultaneously and 5-T active pixel with memory node in it (fig. 5), are adopted.

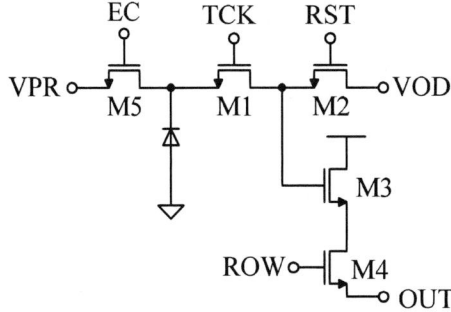

Figure 5.   Structure of 5-T active pixel

At the beginning of the integration, the photodiode of the pixel is reset by high voltage of VPR. However, to improve the dynamic range of the pixel, the high voltage of TCK is generally increased, higher than the voltage of EC. So the reset of the pixel is incomplete and fixed pattern noise caused by the threshold voltage (V_{T1}) deviation of the transmitting transistor M1 is introduced. This FPN can't be eliminated by double sampling, but only by pinned diode where additional technology processes and more cost are needed.

In this paper, a double resets technique is proposed to reduce this FPN, shown in. Fig. 6. At first, VPR is low and EC becomes high to turn on the exposure-controlling transistor M5, which makes the photodiode reset to VPR voltage. Then, VPR becomes high voltage, extra electrons flow into VPR port, and photodiode is reset to the difference of the high voltage of EC and the threshold voltage of transistor M5. After integration, TCK turns high, the signal is read out through M1, the reverse node of photodiode rise to the difference of high voltage of TCK and threshold voltage of M1. Hence, the transmitting signal of the pixel is

$$V_{out} = (V_{H,TCK} - V_{TH1}) - (V_{H,EC} - V_{TH5}) + V_{sig}$$
$$= V_{sig} + (V_{H,TCK} - V_{H,EC}) + (V_{TH5} - V_{TH1}) \qquad . \qquad (7)$$

Where $V_{H,TCK}$, $V_{H,EC}$ are respectively the high voltage of TCK and EC, V_{sig} is the integration signal.

According to (7), because of the mismatch of the transistors, the value of $V_{TH5} - V_{TH1}$ varies between different pixels. However, in single pixel, the uniformity of threshold voltage deviation due to the neighbor of transistor M1 and M5 in the layout, greatly decreases the deviation of the $V_{TH5} - V_{TH1}$ relatively to V_{TH1}. Hence to the whole pixel array, the deviation of the pixel readout voltage is reduced heavily, and the FPN is almost eliminated by adding one reset step, without additional technology processes which cuts down the chip's cost.

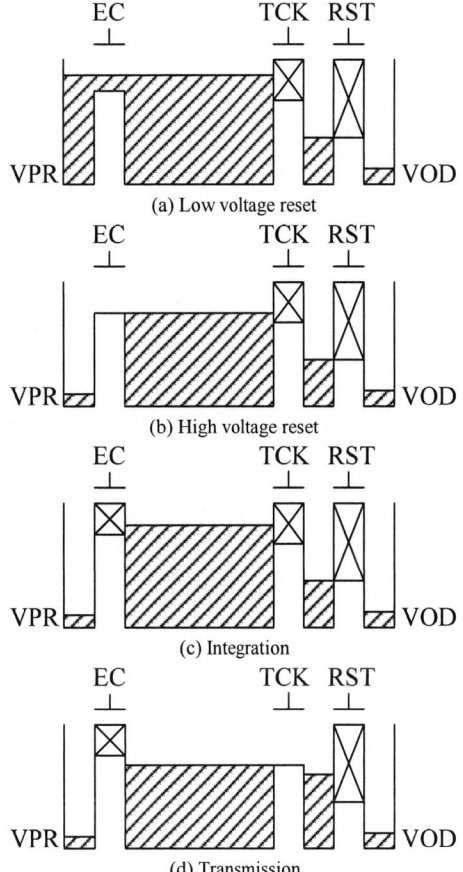

Figure 6.   Barrier diagram of double resets for 5-T pixel active pixel

IV.   SIMULATION RESULTS

An ADC based on the proposed techniques was implemented using the SMIC 0.18μm 1P3M salicide process. The converter achieves 10-b linearity at a sampling rate of 454KS/s. Fig.7 shows the residue voltage of the amplifier under different conditions.

According to Fig.7, without offset cancellation, the amplifier's offset has been brought in every cycle, and been amplified in the following cycles. But with offset cancellation, the offset hasn't been amplified cycle by cycle.

Fig. 8 shows the plots of linearity. The measured DNL and INL are 0.74LSB/-0.25LSB, and 1.1LSB/-1.2LSB respectively. Typical offset voltages observed are ± 0.97mV, which is about 1LSB. The results indicate that the techniques proposed in this paper can eliminate the FPN effectively.

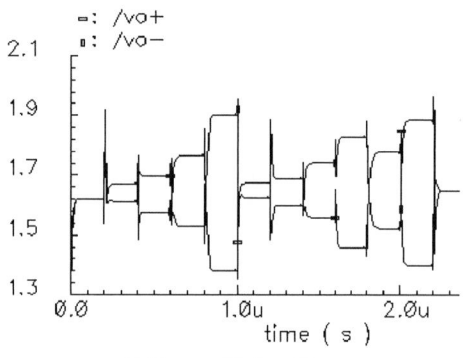

(a) Non-considering offset

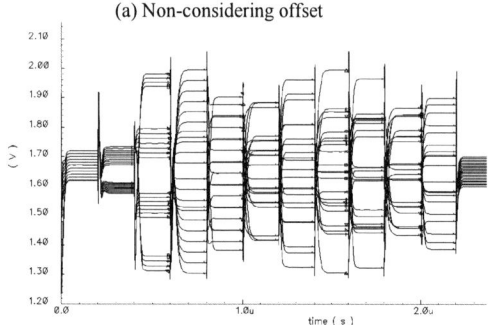

(b) Considering offset but without offset cancellation

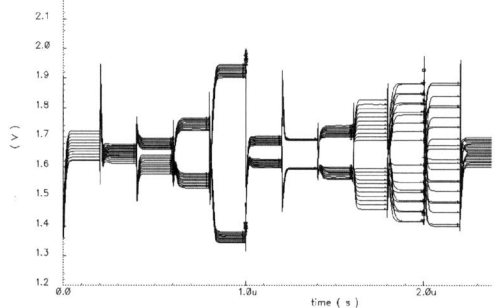

(c) Considering offset and with offset cancellation

Figure 7.   Diagram of residue voltage of the amplifier

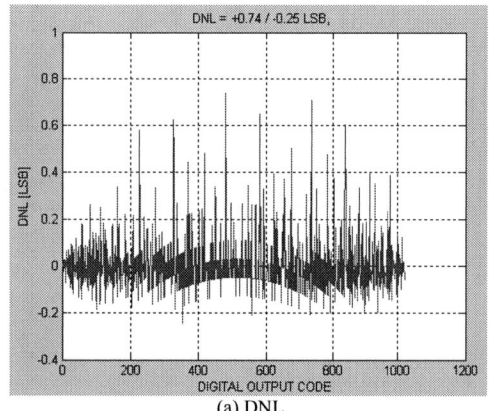

(a) DNL

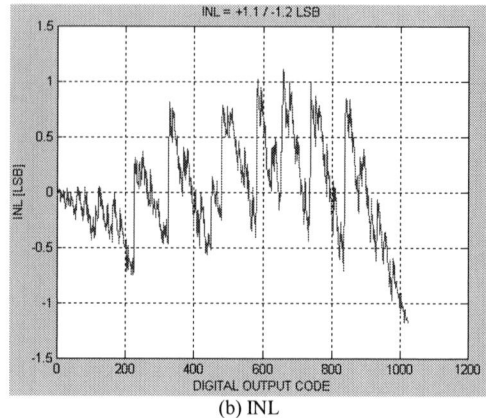

(b) INL

Figure 8.   DNL and INL of the proposed ADC

V.   CONCLUSION

For high speed column parallel CMOS image sensor, three techniques are proposed in this paper to reduce the fixed pattern noise. Experimental results justify that the FPN in the CMOS image sensor has been almost eliminated without additional processes and devices.

ACKNOWLEDGMENT

This work is supported by the Project of Science and Technology Program for Basic Research of Qingdao (No. 13-1-4-249-jch).

REFERENCES

[1]   A. El Gamal and H. Eltoukhy, "CMOS image sensors," IEEE J. Circuits and Devices, Vol. 21, No 3, pp. 6–20, JUNE, 2005.

[2]   Min-Seok Shin, Jong-Boo Kim, Min-Kyu Kim, Yun-Rae Jo and Oh-Kyong Kwon, "A 1.92-Megapixel CMOS image sensor with column-parallel low-power and area-efficient SA-ADCs," IEEE Transactions on Electron Devices, Vol. 59, No 6, pp. 1693–1700, APRIL, 2012.

[3]   Takeyuki Fujii, Shoichi Suzuki and Shinichiro Saito, "Noise evaluation standard of image sensor using visual spatio-temporal frequency characteristics," SPIE Proc. Multimedia Content and Mobile Devices, pp. 86671I-1–86671I-15, MARCH, 2013.

[4]   Xiao-fen JIA and Bai-ting Zhao, "Noise analysis and column FPN suppression technology," International Journal of Computer Science, Vol. 10, No 1, pp. 208–211, JANUARY, 2013.

[5]   Y. Nitta, et al. "High-speed digital double sampling with analog CDS on column parallel ADC architecture for low-noise active pixel sensor," ISSCC Dig. Tech. Papers, pp. 500–501, FEBRUARY , 2006.

Polarization Characteristics and High Birefringence for Chiral Photonic Crystal Fiber with Squeezed Triangular Lattice

She Li[*12], JunQing Li[2], DunLiang Ren[1], Lin Zhang[1], Hui Liu[1], HongXin Shi[1]

College of Science, Heilongjiang University of Science and Technology, Harbin 150022, China

[2]Department of Physics, Harbin Institute of Technology, Harbin 150001, China

Lishe1979@163.com

Abstract-This paper propose a highly birefringent and circular polarization-maintaining chiral photonic crystal fiber with squeezed air holes and triangular-lattice. By a chiral plane-wave expansion method, the modal birefringence and polarization of the fundamental modes are studied. The calculated results show that the feeble squeezing together with trivial chirality can obtain a highly modal birefringence around 1550 nm wavelength. Furthermore, the corresponding polarizations are computed and discussed in detail.

Key word: chiral medium; photonic crystal fiber; polarization

I. INTRODUCTION

In recent decades, a novel kind of fibers, named photonic crystal fiber (PCF) that can guide light in a defect surrounded by a regular lattice of holes or rods, even air hole, attracts a great deal of attention of researchers duo to some novel characteristics, such as endless single-mode guidance[1-2], tailorable dispersion and nonlinearity[3-6], and polarization characteristics[7-8]. Generally, PCFs can be divided into two types from the light-guiding mechanism. The first is named index-guiding PCFs[9] which are based on modified total internal reflection (TIR) and first realized in experiment. The second kind of PCFs are known as photonic bandgap (PBG) effect[10] and allow the light propagation in a low-index core, even in hollow core.

Commonly, for a perfectly traditional PCF, the linearly polarized (LP) modes are supported. However, because of the imperfection[7] in manufacture process, the holes in PCFs cannot be kept perfectly circular and lead to complicated characteristics of polarization. To eliminate the imperfection, it is necessary to need to designing and producing maintain-polarization (highly birefringent) fiber. At present, there are two conventional methods[11-13] to achieve large birefringence in index-guiding PCFs studied in theory and experiment. The first method is stress induction as

conventional fibers; the second is to design asymmetry microstructure near the fiber core. The birefringence in the second method is higher than in that of first method. In addition, there are some reports that can achieve much higher birefringence in PBG fiber with noncircular air core.

The both kinds of PCFs mentioned lead to linear birefringence (LB)[11-13]. There is also another method that dielectrically chiral medium is introduced into PCF, named chiral PCF[14] which gives rise to circular birefringence (CB). Chirality is a symmetry in which an object cannot be mapped into its mirror image with the use of only rotation or translation and can be found in nature in structures that span many length-scales, named asymmetry of material. When both asymmetries exist, a kind of novel phenomenon maybe emerges.

Generally, there are two kinds of chiral fiber (including chiral PCFs): structurally chiral fiber[15-19] and dielectrically chiral fiber[14,20-22]. In the structurally chiral fiber a helical structure or a twisted structure, whose size is comparable with interested wavelength, is introduced into core or cladding; such a kind of chiral PCF can be used to design circular-polarization components[17-18], such as polarization-selective filter, resonator. The dielectrically chiral PCF is based on the waveguide structures of ordinary PCF, in which the composing medium is partly or entirely replaced by chiral medium; for instance, the chiral molecules is absorbed into the background, such PCF is converted into chiral one which can be used to design circular-polarization-maintain PCF.

In this paper, we propose a novel polarization-maintain PCF that is composed of dielectrically chiral background, a cladding with squeezed-triangular-lattice and elliptical air holes. The effect of dielectric chirality on polarization states and mode indices of guided modes is studied by a chiral plane-wave expansion (PWE) method[14]. Mode index divergence between right-handed circular polarized (RCP)

978-1-4799-1215-5/13 $31.00 © 2013 IEEE   215

and left-handed circular polarized (LCP) states is still demonstrated in structurally asymmetry PCF. Chirality's effect on mode index and CB in such a PCF is found to be similar to that in bulk chiral media.

II. THEORETICAL MODEL

The diagram of originally chiral PCF that possesses is shown in Fig.1, in which the dark background indicates chiral medium, the regular array of holes are air ones, r indicates the original radius of holes, X_0 and Y_0 respectively denote the horizontally and vertically original distance for the two nearest holes , ε_1, ξ_1 and ε_2, ξ_2 respectively represent the relative permittivity and chirality parameter of the background and air holes.

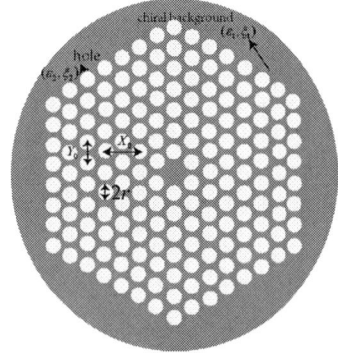

Figure 1. Diagram of chiral PCF without squeezed holes and lattice.

To describe the dielectrically chiral medium that fills the background of chiral photonic structure, we adopt the Drude-Born-Fedorov's material equations as construction relations, which can be expressed as

$$\boldsymbol{D} = \varepsilon_0 \varepsilon_r (\boldsymbol{E} + \xi \nabla \times \boldsymbol{E}) \ \boldsymbol{B} = \mu_0 \mu_r (\boldsymbol{H} + \xi \nabla \times \boldsymbol{H}),$$

where ξ denotes the chirality parameter, which can determine the specific rotary power δ through $\delta = -\xi k_0^2 n_0^2$, in which k_0 and n_0 represent the wavenumber in vacuum and the refractive index for chiral medium, respectively. In simulation, the relative permittivity of background ε_1, original distance X_0, Y_0 and radius of the holes r are respectively chosen to be 2.25, 3.81μm, 2.2μm and 0.9167μm. In this letter, when the cross-section of PCF is extruded horizontally, the corresponding X_0, Y_0 and the major semi-axis A and minor semi-axis B can be set to be eX_0, Y_0 and er, r, respectively, where e indicates the squeezing degree.

To describe the polarization distribution, the third localized Stocks parameter s_3 is adopted, which can be used determine whether a polarization is purely circular ($s_3 = \pm 1$) or purely linear ($s_3 = 0$). s_3 ranges from -1 to +1, of which the negative (positive) sign indicates the left-handed (right-handed) polarized states. In order to more accurately describe the polarization in the core, the third Stocks parameter for the core is redefined as

$$S_3 = \iint_{S_{core}} s_3 \left| \mathrm{E}^2 \right| dS \Big/ \iint_{S_{core}} \left| \mathrm{E}^2 \right| dS ,$$

where dS denotes an area element in the core, s_3 and $\left| \mathrm{E}^2 \right|$ indicate the corresponding localized third Stokes parameter and intensity of light for dS.

III. RESULTS AND DISCUSSION

The circular polarized states are supported by the chiral PCF with a circular solid core and equilateral triangular lattice, and the corresponding field distributions are similar to that of achiral PCF. However, the corresponding polarization distributions are completely different. Here the modified PWE method[14] is also adopted. Fig.2 exhibits the field intensity distributions and polarization distributions of fundamental modes for the achiral and chiral PCF, where the squeezing degree is chosen to be $e=0.9$, the corresponding wavelength in vacuum is 1.5μm and the chirality are selected as $\xi=0$ and $\xi=-0.0003$, respectively.

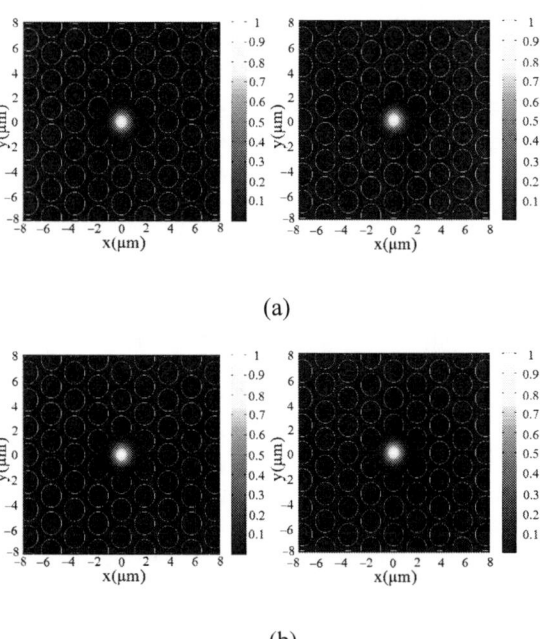

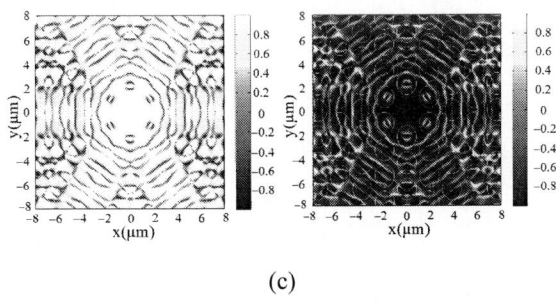

(c)

Figure 2. The field intensity distributions and the corresponding polarization distributions for fundamental modes in chiral PCF. (a) and (b) are field intensity distributions of PCFs with the same strength of squeezing degree (*e*=0.9), where the chirality is chosen to be ξ=0 (achiral case) and ξ=-0.0003 (chiral case). (c) is the polarization distributions corresponding to (b).

From Fig.2 (a) and (b), one knows the field intensity distributions of fundamental modes for squeezed chiral and achiral case are also alike; however, the corresponding polarization distributions are completely different. Fig.2(c) shows the complicated polarization distributions corresponding to Fig.2(b), nevertheless the modal polarizations in the core are almost identical for each mode, from which one can find that in the area of the core both of $|S_3|$ approach to 0.98 which corresponds to nearly circular polarization (elliptical polarization, EP). And it is well known that the squeezed achiral PCF supports the linearly polarized states. For the perfectly circular holes and equilateral triangular lattice, the corresponding polarized states are almost perfectly circular polarized (CP). It is clear that the distribution of mode field is mainly dependent on the structure of PCF, while the polarization distribution is closely related to the strength of chirality and of the structural asymmetry.

Fig.3 compares the effect of chirality and the strength of squeezing on S_3 with varied wavelength, in which the five types of line correspond to ξ=-0.0001, ξ=-0.0003, ξ=-0.0005, ξ=-0.001, ξ=0, respectively. Here we define the cases $S_3 \geq 0.99$ and $S_3 \leq -0.99$ as purely RCP and LCP states and $|S_3| \leq 0.01$ as linearly polarized states. From Fig.3 one can find that for all chiral cases the $|S_3|$ decrease with increase of wavelength. This is because in the shorter wavelength region the more intensity of light is concentrated in the chiral core and the effect of chirality is stronger. Meanwhile, with increase of wavelength, the intensity of light reduces in the chiral core, which indicates that the effect of structural asymmetry changes stronger. So the decrease of the $|S_3|$ gradually indicates the polarized

states of guided modes convert from CP to EP. For example, for ξ=-0.001, the two $|S_3|$ almost equal to unity from 0.9μm to 2μm, which can be viewed as perfectly CP. While for ξ=-0.0005, ξ=-0.0003 and ξ=-0.0001 correspond to CP from 0.9μm to 1.7μm, 1.5 μm and 1.1μm. Furthermore, with increase of wavelength or weakening of chirality, the polarized states may eventually convert from EP into LP. For achiral case, as we all know the both $|S_3|$ are almost equal to zero.

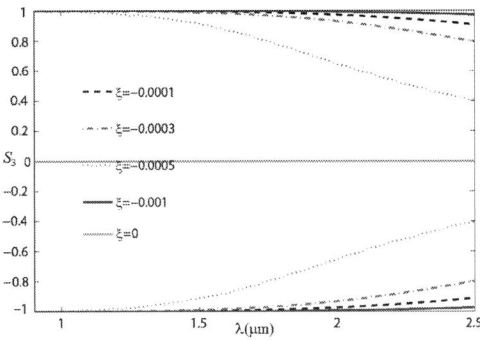

Figure 3. The dependence of S_3 on wavelength with different chirality.

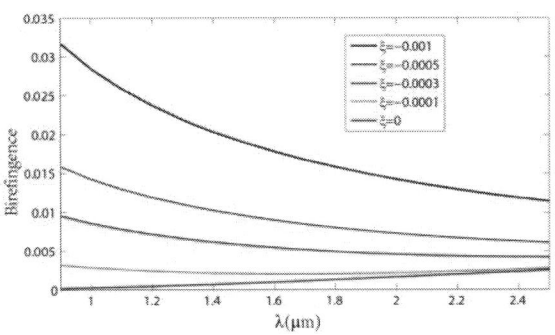

Figure 4. EB of fundamental modes in PCFs with different chirality.

As is well known that the PCFs squeezed for holes and lattices result in LB; while the chirality of medium give rise to CB. When the both exist, a larger birefringence with EP emerges, shown in Fig.4. Fig.4 demonstrates the variation of birefringence with wavelength for different chirality, from which one can see the EB is larger with increase of the chirality in the background. When the chirality gets to a modest value, the EB can almost be converted into CB entirely. Comparing to Fig.3 one can find the purely CB corresponding to wavelength regions and chirality are from 0.9μm to 2μm, 1.7μm, 1.5μm, 1.1μm and the and chirality ξ=-0.001, ξ=-0.0005, ξ=-0.0003, ξ=-0.0001, respectively. In addition, the birefringence is larger in the short wavelength region thank to the more intensity of light concentrated in the chiral core and stronger chiral effect of medium. For achiral case, the birefringence rises with increase of wavelength duo to the light of intensity of light decreasing in the chiral core and the larger structural asymmetry. This is the reason that the birefringence for $\xi = -0.0001$ decreases from $0.9\mu m$ to $1.7\mu m$ and increases from $1.7\mu m$ to $2.5\mu m$. For $1.55\mu m$, the birefringence can achieve 10^{-2} for $\xi = -0.001$, which is much larger than that for achiral case, from which we can know the larger chirality the PCFs, the larger the CB or EB.

IV. CONCLUSION

In summary, we investigated a kind of dielectrically chiral and squeezed chiral PCFs. Owing to the chirality of the background and the squeezed holes and lattice, the PCFs can support complicated polarized modes, from CP ones to LP ones operating in optical fiber communication band. When the chirality is relatively weak, the polarized states are almost EP and with increase of chirality, those become CP, for which the birefringence correspond to EB and CB. The stronger the chirality, the larger the EB or CB will be. The moderate chirality is necessary to obtain the more purely CP and larger birefringence for squeezed PCF. These unique properties promise some applications in optical fiber communication, such as EP or CP maintaining PCF, lasers and sensors.

ACKNOWLEDGMENT

This work is supported by the National Natural Science Foundation of China (grant No. 60977032, 11104049), the Science and Technology Programs of Hei-longjiang Province Educational Committee of China (Grant No.12531573) and Institute of higher education the Twelfth Five Plan Project of Heilongjiang (Grant No. HGJXHB2110884).

REFERENCES

[1] J. C. Knight, T. A. Birks, P. S. Russell, and D. M. Atkin, All-silica single-mode optical fiber with photonic crystal cladding, Opt. Lett. 21(19):1547- 1549, 1996.

[2] T. A. Birks, J. C. Knight, and P. S. Russell, Endlessly single-mode photonic crystal fiber, Opt. Lett. 22(13): 961-963, 1997.

[3] W. H. Reeves, D. V. Skryabin, F. Biancalana, J. C. Knight, P. St. J. Russell, F. G. Omenetto, A. Efimov and A. J. Taylor, Transformation and control of ultrashort pulses in dispersion-engineered photonic crystal fibres, Nature. 424:511-515, 2003.

[4] G. S. Wiederhecker, C. M. B. Cordeiro, F. Couny, F. Benabid, S. A. Maier, J. C. Knight, C. H. B. Cruz and H. L. Fragnito. Field enhancement within an optical fibre with a subwavelength air core, Nature Photonic. 1:115-118, 2007.

[5] S. G. Yang, Y. J Zhang, L. N. He and S. Z. Xie, Broadband dispersion-compensating photonic crystal fiber, Opt. Lett. 31(19):2830-2832, 2006.

[6] D.V. Skryabin, F. Biancalana, D.M. Bird, and F. Benabid, Effective Kerr Nonlinearity and Two-Color Solitons in Photonic Band-Gap Fibers Filled with a Raman Active Gas, Phy. Rev. Lett. 93(14): 143907, 2004.

[7] R. J. Kruhlak, G. K. L. Wong, J. S. Y. Chen, S. G. Murdoch, R. Leonhardt, and J. D. Harvey, Polarization modulation instability in photonic crystal fibers, Opt. Lett. 31(10): 1379-1381, 2006.

[8] M. Y. Chen and Y. K. Zhang,Improved design of polarization-maintaining photonic crystal fibers, Opt. Lett. 33(21): 2542-2544, 2008.

[9] S. E. Barkou, J. Broeng, and A. Bjarklev, Silica–air photonic crystal fiber design that permits waveguiding by a true photonic bandgap effect, Opt. Lett. 24(1): 46-48, 1999.

[10] R. F. Cregan, B. J. Mangan, J. C. Knight, T. A. Birks, P. St. J. Russell, P. J. Roberts and D. C. Allan, Single-mode photonic band gap guidance of light in air, Science. 285(3):1537-1539, 1999.

[11] Ortigosa-Blanch, J. C. Knight, W. J. Wadsworth, J. Arriaga, B. J. Mangan, T. A. Birks, and P. St. J. Russell, Highly birefringent photonic crystal fibers. Opt. Lett. 25(18):1325-1327, 2000.

[12] M. J. Steel and R. M. Osgood, Jr. Elliptical-hole photonic crystal fibers. Opt. Lett. 26(4): 229-231, 2001.

[13] Z. J. He, Highly birefringent extruded elliptical-hole photonic crystal fibers with single defect and double defects, Chin. Opt. Lett. 7(5):387-389, 2009.

[14] J. Q. Li, Q. Y. Su, and Y. S. Cao, Circularly polarized guided modes in dielectrically chiral photonic crystal fiber, Opt. Lett. 35(16):2270-2272, 2010.

[15] A. Argyros, M. Straton, A. Docherty, E. H. Min, Z. Y. Ge, K. H. Wong, F. Ladouceur, and L. Poladian, Consideration of chiral optical fibres, Front. Optoelectron. China. 3(1): 67–70, 2010.

[16] V. I. Kopp and A.Z. Genack, Chiral fibres : adding twist, nature photonics. 5:470-472, 2011.

[17] V. M. Churikov, V.I. Kopp, and A. Z. Genack, Chiral diffraction gratings in twisted microstructured fibers, Opt. Lett. 35(3):342-344, 2010.

[18] G. K. L. Wong, M. S. Kang, H. W. Lee, S. Burger, L. Zschiedrich, F. Biancalana, and P. St.J. Russell, Strongly twisted solid-core PCF: a one-dimensional chiral metamaterial, Frontiers in Optics. 2011,9,30.

[19] D. Shemuly, Z. M. Ruff, A. M. Stolyarov, G. Spektor, S. Johnson, Y. Fink and O. Shapira, Asymmetric wave propagation in planar chiral fibers. Opt. Express. 21(2):1465-1472, 2013.

[20] R. C. Qiu and I. T. Lu, Guided waves in chiral optical fibers, J. Opt. Soc. Am. A. 11(12):3212-3219, 1994.

[21] Y.S. Cao, J. Q. Li, and Q. Y. Su, Guided modes in chiral fibers, J. Opt. Soc. Am. B. 28(2):319-324, 2011.

[22] J. Q. Li, S. Li, X. O. Wang, Y. D. Zheng, and C. F. Li, Self-induced optical rotation of solitons in a chiral fibre, Chin. Phys. Lett. 21(4):675-678, 2004.

A real non-contact remote trapping single fiber tweezers

Yu Zhang*, Zhihai Liu, Jun Yang, Libo Yuan

Key Lab of In-fiber Integrated Optics, Ministry Education of China, Harbin Engineering University, Harbin, China
zhangy0673@163.com

Abstract—A real non-contact two-core single fiber optical tweezers is proposed. It is fabricated by the fiber grinding and polishing technology. The yeast cells trapping performance realized by this single fiber optical tweezers with a special designed truncated cone tip fiber probe is demonstrated and investigated. The distribution of the output optical field emerging from the truncated cone fiber tip is simulated by using the FDTD (finite domain time difference) method. The axial and transverse optical trapping forces are also simulated with the FDTD method. This new single fiber optical tweezers can realize truly non-contact remote optical trapping of micro particles and provides a new application for the biological and medical researching fields.

Keywords—single fiber optical trapping; non-contact remote optical trapping; two-core fiber; truncated cone fiber tip

1. INTRODUCTION

The optical trapping method proposed by Ashkin [1] is considered as a useful technique to manipulate or probe most biological specimens or living objects because of the non-contact character of light [2-3]. The traditional lens optical tweezers utilize high numerical aperture objective focusing light beams to trap the micro particle, despite the relevant results achieved in many fields, the bulky structure and expensive cost still limits their applications in several environments. The realization of the optical tweezers based on optical fibers turns this device into a miniaturized tool, suitable for many relevant applications, like in vivo biological operations [4-5]. The early fiber optical tweezers include multi optical fibers, using two or more optical fiber guiding light beams crossing prorogating getting high optical intensity gradient distribution to manipulate the micro-particle. Then as the need for further, the single fiber optical tweezers turns out [6-8], which is more simple and practical to be a tool to research the bio-cells, but the disadvantage of the single fiber optical tweezers is the 3-D trapping has been achieved through highly tapered fibers: in this case, the trapping point gets very close to the fiber tip, which means it will not be really non-contact manipulation, making the tweezers quite unpractical for real applications. Therefore Minzioni [9] presents an improved single-like fiber optical tweezers, which is composed of multi optical fibers bundled into a capillary to form an optical tweezers probe realizing micro-particle trapping. It is both a "single" optical tweezers and can realize trapping micro-particle non-contact. But the manufacture of this optical tweezers probe needing high accuracy manipulation device to arrange and assemble multi fibers into a capillary, the processing for this optical tweezers is much complicated and hard to fabricate. However we propose and experimentally demonstrate a new simple two-core single fiber optical tweezers, whose size is as a normal single mode fiber and this tweezers also can realize micro-particle really remote non-contact trapping.

II. THE TWO-CORE FIBER PROBE

The new two-core single optical fiber tweezers are fabricated by grinding and polishing technology. Two laser beams propagate through two cores in the two-core fiber respectively and focus with high NA in front of the fiber tip, Fig.1 shows the two-core fiber profile, the two cores distribute in a line passing through the center of the fiber profile, and the distance between two cores is 62μm. The core diameter is 3.7μm, and the surrounding cladding diameter is 125μm just like the normal single mode single core fiber. The core index is 1.4681, and the cladding index is 1.4632.

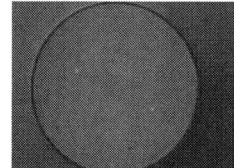

Figure 1.   The profile of two-core fiber

Unlike the heating and drawing manufacture techniques, the twin-core fiber optical tweezers was manufactured by grinding and polishing the two-core fiber to form a truncated cone tip. Fig.2 shows the cross-section schematic diagram of the two-core fiber truncated cone fiber probe used in our experiment. In order to generate a large enough optical trapping force, the grinding angle α of the tapered tip is selected as 22° [10], and h is the distance between the particle and the fiber. In the truncated cone zone, the two beams coming from two cores were guided in the tapered angle, total refracted on the interface AB and CD between the fiber and the medium respectively, then refracting out of the fiber from BC and focusing on one point E to trapping the micro-particle. By using this two beams combination technique, a strong enough gradient force potential well is obtained for micro-particles trapping in three dimensions in front of the two-core fiber tip.

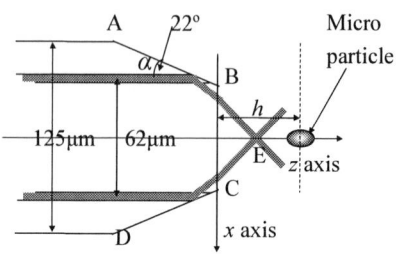

Figure 2.   The schematic diagram of grinded tip of four-core fiber

It can be explained that two laser beams with a large crossing angle (which can be concluded as high numerical aperture) can form a single gradient force trap, which is exactly the situation in our case. The distribution of the optical field emerging from the fiber probe used in our experiment was numerically calculated based on the FDTD method. In the FDTD calculation, a continual Gaussian beam come from the laser diode light source with wavelength 980nm propagating down to up along the truncated cone fiber tip. The refractive index of the surrounding medium (water) is 1.33. The grid step is chosen as 0.05μm in the FDTD calculation. The simulation result, optical field distribution of E_x, E_z and H_y are shown in Fig. 3.

As shown in Fig. 3, the laser beam emerging from the fiber tip is focused at a position where the intensity of the laser beam is the highest, and the focal zone is about 18~22μm away from the fiber tip plane.

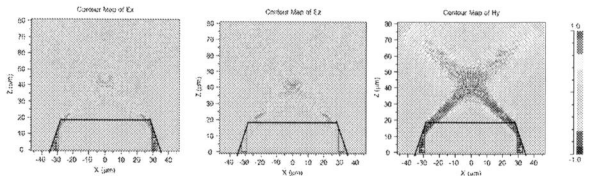

Figure 3.   The outputting optical filed of grinded fiber tip

III. EXPERIMENT SETUP AND RESULTS

The two core fiber optical tweezers experiment setup and its working principle are described as Fig.4. A laser diode (LD) with wavelength at 980nm is used as the light source with its power to be tuned from 0 to 120 mW by adjusting the driving current of the LD.

The pigtail fiber of the laser diode is single mode fiber. In order to launch the optical power into the two cores of two-core fiber, we are heating and drawing the fiber at the splicing point after spliced the single mode single core fiber and the two-core fiber, thus a couple zone between the single mode single core and two-core single mode fiber is formed [11]. By this way, the input power is coupled and shared in the two cores and guided the two beams to the two-core fiber tapered end.

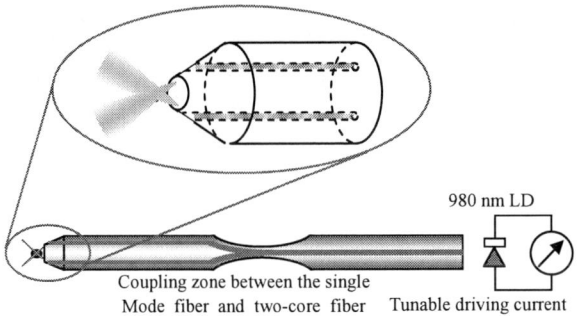

Figure 4. Sketch of two-core fiber optical tweezers experiment setup

The trapping of yeast cells in water was realized based on our new truncated cone tip two-core fiber optical tweezers system mentioned above. Under the trapped status, we could freely move the trapped yeast cell to the forward and backward or right and left directions. The trapping force is also can be controlled by adjusting the optical power of the laser source. The photo of experiment video is shown in Fig.5, and according to the ratio relation of the photo, we can calculate the distance between the trapped particle and the tip of the fiber is about 20μm.

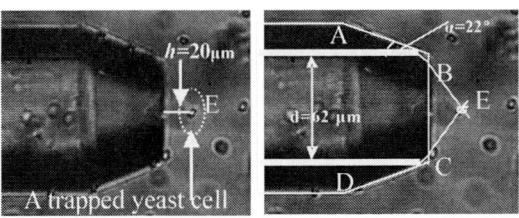

Figure 5. Images of the grinded tip of the two-core fiber.

IV. OPTICAL TRAPPING FORCE ANALYSIS

Light force will be produced when light refracted and reflected at the interface between two different media, which acted on the particle within a laser trap results from the change in momentum of light due to refraction or absorption.

The optical momentum density is normal used to analysis the optical trapping force, and it is given by

$$g = \frac{nS}{c^2} \qquad (1)$$

Where n is the medium index, S is the Poynting Vector, and c is the speed of light.

The yeast cells are normally not perfectly spherical, therefore, according to reference [12] and Fig.6, A is a point on the border of the ellipsoidal yeast cell, and A_1 is a point nearby A outside of the cell, A_2 is a point nearby A inside of the cell. g_{A1} and g_{A2} are the momentum density of A_1 point and A_2 point, respectively. Then the time average change of momentum density is

$$< \Delta g > = < g_{A1} > - < g_{A2} > \qquad (2)$$

Based on momentum conservation theory, the force exerted on per unit area at point A can be given by:

$$F_A = < \Delta g > \cdot ds = \{< g_{A1} > - < g_{A2} >\} \cdot ds \qquad (3)$$

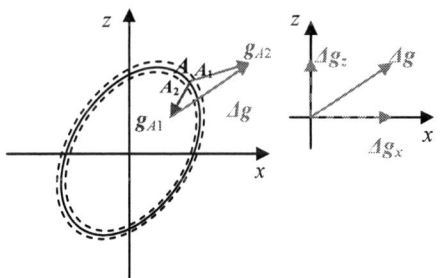

Figure 6. The optical momentum density on the surface of sphere

We projected F_A along x direction (F_{Ax}) and z direction (F_{Az}), respectively, and then the axial and transversal trapping force exerted on the microsphere can be given by:

978-1-4799-1215-5/13 $31.00 © 2013 IEEE

$$\vec{F}_x = \int_s F_{Ax} \cdot ds \qquad \vec{F}_z = \int_s F_{Az} \cdot ds \qquad (4)$$

We build the truncated cone modal in *Rsoft®* software, using *FullWAVE* method getting the E_x, E_z and H_y components of output optical field (TM mode) to compute the optical momentum density g, (supposing the trapped yeast cell is ellipsoidal, with the long axis 4μm, short axis 2.5μm), and finally getting the yeast cell axial and transversal trapping force of this grinded truncated cone tip optical fiber tweezers, shown as Fig.7 and Fig.8.

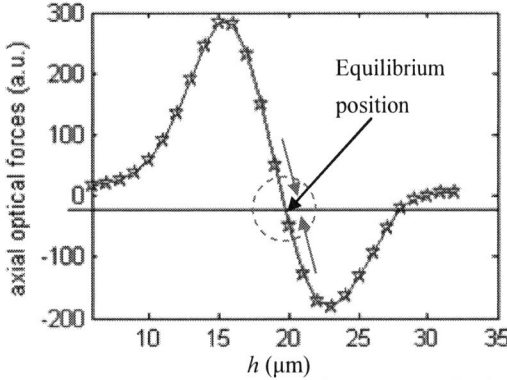

Figure 7.   The simulation results of the axial trapping force

From Fig. 7, we can see that near the tip (say, in the range of 0~20μm), the axial trapping force is larger than zero, which means the direction of the trapping force is the same with the beams propagating direction, and this force will pushing the particle away from the fiber tip; in the range of 20~28μm, the axial trapping force is less than zero, which means the direction of the trapping forces is opposite to the beam propagating direction, and the forces will pull the particle near to the fiber tip; therefore on the position about 20μm, the forces working on the particle is zero, the particle will be equilibrium, which means the particle will be trapped stably on the position of 20μm. Compare with Fig.5, in the real photo of experiment result, the distance of the particle trapped stably is also about 20μm, which is consistent with the simulation result in Fig.7; in the rang of 28μm to more farther, the scattering

force plays a major role, which will pushing the particle moving away from the fiber tip.

When the distance between the particle and the fiber tip is in the range of 0~20μm, the particle is pushed towards the beam propagation direction, and will be propelled into the trapping zone; when the distance is in the range of 20~28μm, the particle is pulled reverse to the beam propagation direction, which will also propel the particle into the trapping zone.

The transverse trapping forces are also calculated as shown in Fig.8 (keeping the axial distance between the particle and the fiber tip is 20μm). According to the Fig.8, when the distance between the particle and the fiber axis is 0, the transverse trapping force is 0, which means only the particle on the fiber axis, the particle will be in equilibrium, once the particle moves away from the equilibrium position, a negative transverse trapping force will work, which will pull the particle back to the equilibrium position, and the effective range is about 15μm. When the distance is larger than 15μm, the scattering force will play a major role, which will push the particle away from the fiber.

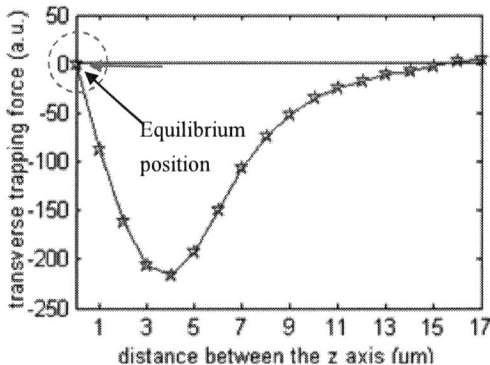

Figure 8.   The simulation results of the transverse trapping force (h=20μm)

V. CONCLUSION

In conclusion a real non-contact trapping micro particles two-core single fiber optical tweezers is proposed. Both theoretical and experiential results show that this new two-core single fiber optical tweezers with a truncated cone tip can realize the real non contact optical trapping and manipulating of the micro particles immersed in fluid medium.

Compared with traditional lens optical tweezers and early multi fibers optical tweezers, this new single fiber optical tweezers is more simple and suitable to manipulate practical micro particles, such as bio cells.

Compared with other single optical fiber tweezers, which normally have a tapered or hemispherical lens fiber tip, this new single two-core fiber optical tweezers with a truncated cone tip can realize micro particle remote non-contact trapped.

The theoretical analysis and controlling method of using this new two-core fiber optical tweezers trapping and rotating micro particles will be carried out after the further experiment.

ACKNOWLEDGMENT

This work is supported by the National Natural Science Foundation of China (Grants No. 61077062, 61177081, 61107069, 41174161, 11204047, 61227013, 61205027 and 61275087), and partially supported by the following grants: Ministry of Science and Technology of China (Grants No. 2010DFA22770), Research Fund for the Doctoral Program of Higher Education of China (Grants No. 20112304110017), Post - Doctor Research Fund of Heilongjiang Province of China (Grants No. LBH Q10147), and Fundamental Research Funds for Harbin Engineering University of China.

REFERENCES

[1] A. Ashkin, J. M. Dziedzic, J. E. Bjorkholm, and Steven Chu , "Observation of a single-beam gradient force optical trap for dielectric particles", Opt. Lett., vol. 11, pp. 288-290, May, 1986.

[2] A. Ashkin, "Optical trapping and manipulation of neutral particles using□lasers", Proc. Natl. Acad. Sci. USA, vol. 94, pp. 4853-4860, May, 1997.

[3] A. Ashkin, "History of optical trapping and manipulation of small-neutral particle, atoms, and molecules", IEEE J Sel. Top. Quant., vol. 6, pp. 841-856, Nov-Dec, 2000.

[4] M. Ikeda, K. Tanaka, M. Kittaka, M. Tanaka, and T. Shohata, "Rotational manipulation of a symmetrical plastic micro-object using fiber optic trapping", Opt. Commun., vol. 239, pp. 103-108, Sep, 2004.

[5] C. J McMullin, H. P. Lee, and E. R. Lyons, "Demonstration of trapping, motion control, sensing and fluorescence detection of polystyrene beads in a multi-fiber optical trap", Opt. Express, vol. 13, pp. 2634-2642, Apr, 2005.

[6] A. Constable, J. Kim, J. Mervis, F. Zarinetchi, and M. Prentiss, "Demonstration of a fiber-optical light-force trap", Opt. Lett., vol. 18, pp. 1867-1869, Nov, 1993.

[7] L. B. Yuan, Z. H. Liu, J. Yang, C. Y. Guan, "Twin–core fiber optical tweezers", Opt. Express, vol. 16, pp. 4551, Mar, 2008.

[8] Z.H. Liu, C.K. Guo, J. Yang, and L.B. Yuan, "Tapered fiber optical tweezers for microscopic particle trapping: fabrication and application" Opt. Express, vol. 14, pp. 12510–12516, Nov, 2006.

[9] P. Minzioni, F. Bragheri, C. Liberale, E. D. Fabrizio, and I. Cristiani, "A Novel Approach to Fiber-Optic Tweezers: Numerical Analysis of the Trapping Efficiency", IEEE J Sel. Top. Quant., vol. 14, pp. 151-157, Jan-Feb, 2008.

[10] Y. Zhang, Z.H. Liu, J. Yang, L.B. Yuan, "A non-contact single optical fiber multi-optical tweezers probe: Design and fabrication", Opt. Commun., vol. 285, pp. 4068-4071, Sep, 2012.

[11] L.B. Yuan, Z.H. Liu, J. Yang, C.Y. Guan, "Bitapered fiber coupling characteristics between single-mode single-core fiber and single-mode multicore fiber", Appl. Opt., vol. 47, pp. 3307-3312, Jun, 2008.

[12] H. Yang, G.Y. Feng, S.H. Zhou, H.Y. Ding, "A new method based on conservation momentum to calculate the trapping force using FDTD algorithm", Proc. SOPO, 19-21 June, 2010, pp.1-4.

978-1-4799-1215-5/13 $31.00 © 2013 IEEE

Pencil-lead-break transient detecting by phase- optical time domain reflectometer based on coherent detecting

Yuelan Lu

Key Lab of In-fiber Integrated Optics, Ministry Education of China, Harbin Engineering University, Harbin, China
Luyuelan1968@163.com

Abstract—**Pencil-lead-break transient detecting is demonstrated by phase- optical time domain reflectometer based on coherent detecting and data processing. By using coherent detecting technology, combined with Erbium-doped Optical Fiber Amplifier and high power load Acousto-optical Modulators (AOM), we improves signal detecting sensitive and Signal-noise-ratio by 2~3dB coherent detection advantage over direct detection of phase-OTDR. Location of pencil-lead-break transient signal was detected with higher spatial resolution in time domain by averaged traces transiently. Using varying power with time of the location, frequency information can be obtained by fast Fourier transforms (FFT) analyzer. With loop geometry fiber, as short as 0.25cm fiber can give effective signal over 12-km long single-mode fiber and as high as 50 kHz frequency component can be detected.**

Keywords- vibration; coherent detection; pencil-lead-break

I. INTRODUCTION

In recent years, structural health monitoring has become a new field in civil, mechanical and aerospace engineering. Many public civil structures are facing a problem of aging. It has been reported that 40% of the bridges in Canada are 50 years old , while the number in United States is over 50% and 42% of all bridges are structurally deficient [1]. However, the structural response under impact wave has been recognized as a global damage detection method [2]. As early as 1980, 1-D optical Fourier spectrum analyzer system is used to detect fatigue cracks, as small as 0.5mm fatigue cracks can be tested[3]. Since 1989, the Bragg grating has become the dominant sensor to monitor strains and deformation [4-5]. One remarkable advantage of distributed optical fiber sensors is that the desired measurement can be obtained in a distributed way and spatial profiling can easily be accomplished Distributed sensor corresponds to optical time domain reflectometry (Rayleigh scattering and Fresnel scattering) [6-7]. Other detecting method such as Fabry–Perot interferometric sensor, a phase compensating technique is developed in applying the sensor to detecting an acoustic emission signal [8].Nelson H discovered that breaking of a pencil-lead on the surface of an aircraft panel would act as a simulated source of an acoustic emission signals. Since that time, pencil-break has been used to adequately replicate the growth and propagation of a crack. Pencil-lead-break monitoring has been proposed and demonstrated in this paper based on a coherent detection phase-sensitive optical time domain reflectometer. By using the data processing of moving averaging method and fast Fourier transforms (FFT), the information of location and frequency of signal can be obtained transiently.

II. SIGNAL PROCESSING METHOD

A CW light with an optical frequency of f is divided by a 3dB coupler into two parts, a signal and a local oscillator (LO). The basic principle of coherent detection consists of combining the optical signal coherently with a local oscillator (LO) before it falls on the photo detector. The signal is modulated to a pulse with a Δf frequency shift by an acousto-optic modulator (AOM). The Rayleigh backscattered signal $E_b(t)\exp j(2\pi(f+\Delta f)t+\phi(t))$ will be mixed with the LO light $E_{LO}(t)\exp j(2\pi ft)$ by a 3dB coupler (shown in Fig.1). The mixed signal is then launched to a balanced detector. The detected current is proportional to the optical power:

$$I(t) \propto E_{Lo}(t)^2 + E_b(t)^2 + 2E_{Lo}(t)E_b(t)\cos\theta(t)\cos(2\pi\Delta ft+\varphi(t)) \quad (1)$$

Where $\theta(t)$ and $\varphi(t)$ are the relative polarization angle and phase between the backscattered signal and the LO light, respectively. The AC component of (1) can be obtained by using heterodyne detection, where the frequency difference of two inputting signals to mixer is Δf.

In order to reduce the amplitude fluctuation of traces due to phase noise [9], we adopt moving averaging and moving differential method as well as to increase frequency response in crack detection. Supposed there are N raw traces set $r=\{r_1,r_2,r_3,...r_i...,r_N\}$, where r_i means the ith raw trace. If moving averaging number is M, the averaged traces set $R=\{R_1, R_2,R_3, ...R_i...R_K\}$, where $K=N-M+1$; and:

$$R_i = \frac{1}{M}\sum_{l=i}^{l=i+M-1} r_l \quad , l \in [1,N], i \in [1, N-M+1] \quad (2)$$

Because the adjacent traces in R is very close we could not use the adjacent differential method to figure out the crack signal. Considering the values of pulse width and the lasting time of pencil-lead-break vibration, we choose $R_r = R_{\text{int}(i/2M)\times M+1}$ as the moving references, and then the differential traces can be gotten by using:

$$\Delta R = \{\Delta R_1, \Delta R_2,...,\Delta R_i,...,\Delta R_J\} \quad (3)$$

978-1-4799-1215-5/13 $31.00 © 2013 IEEE

Where $\Delta R_i = R_i - R_r$, $J=K-1=N-M$. For this method, even a great average number the traces number decrease from N to $N-M$, which means frequency response does not decrease too much with the average number. So we can increase the SNR by using this method without loss of too much higher frequency components.

III. A COHERENT PHASE-OPTICAL TIME DOMAIN REFLECTOMETER

The setup of phase-OTDR with coherent detection is shown by Fig.1. Both the probe light and the local oscillation are generated from the same laser source to ensure the coherent interaction. Output CW laser from ECL is split through a coupler: 50% of the light is launched into the AOM through a polarization controller while 50% is launched into the other path. AOM modulates CW to pulses and meanwhile shifts the probe frequency by 200MHz with respect to the local oscillator, so that heterodyne coherent detection can be achieved. (If the frequency of the laser source is f, then the frequency of probe pulse becomes $f + \Delta f = f + 200MHz$. The AOM is pulsed continuously with a repeat rate of 10 kHz;

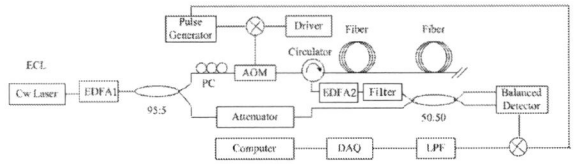

Figure 1.   Experimental setup of coherent Phase-OTDR

IV. RESULTS AND DISCUSSIONS

Loops of fiber are glued at an Al plate of thickness 1.58mm. They are located around 880m along the testing fiber with the total fiber length of 1.2Km. HB 2 pencils were used to perform pencil lead break experiment adjacent to a point of the circumference of fiber loop. Because the vibration signal of pencil-lead-break is an impact signal, so its frequency components are broad even as high as 5MHz. Optical pulses have been put into single mode fiber at a rate of 10 kHz (0.1ms). We took waveforms for averaging by moving averaging and moving difference method to give enough SNR and high frequency response. 50000 raw waveforms were recorded and saved on the memory board to test pencil-lead-break signal. By subtracting electrical signal within two channels of the balanced detector, DC component will be gotten rid off. And then through a mixer, the interference signals are mixed with 200MHz reference sine electrical signal generated by the other channel of the function generator. The profiled heterodyne coherent signal is extracted after a low pass filter. The final signal is recorded by a DAQ card, and saved in computer for trace analysis and FFT processing by labview program. The overlapped consecutive waveforms can give the location of crack signal as Fig.2. Power spectrum of Fig.2 shows that there is an obvious peak at 880m along the fiber, which means that there should be a vibration occurs. So we are confident that the location of pencil-lead-break is 880m along the fiber. For three pencil-lead-break experiments, we have accomplished pencil-lead-break with once, twice and three times. Using the power of the location, we have gotten power of pencil-lead break varying with time as Fig.3（a-c）. Fig.3 a~c show once, twice and three responses impact which means different pencil-lead-break events have occurred in time domain, each event lasts about 0.01s. For some cases, not only the location information but also the frequency information of the crack is required. In order to analyze the frequency of the event, a power spectrum analysis was performed by FFT. By using FFT, we have obtained the power spectrum in frequency domain as Fig.4 shown. Because there are many peaks in the graph, which means there are frequency components in the pencil-lead-break impact signal.

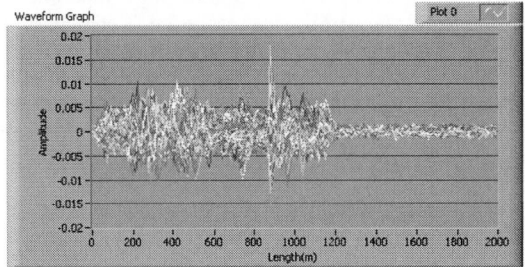

Figure 2.   Power spectrum distribution along fiber

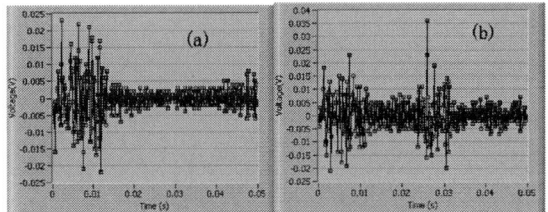

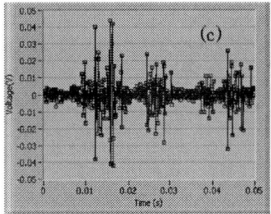

Figure 3.   Response impact showing mutiple pencil-lead-break events in the time domain

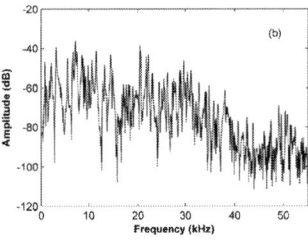

Figure 4.   Power spectrum distribution in the frequency domain

The use of coherent detecting combined with new data processing of moving averaging and moving differential method is very effective in enhancing the performance of phase-OTDR. The sensitivity of phase-OTDR can be enhanced and higher SNR will be obtained. Frequency information of the crack can be obtained by power spectrum of same position transiently. With loop geometry fiber, as short as 0.25cm fiber can get effective signal, and as high as 50 kHz frequency component can be detected

V. CONCLUSION

We demonstrated pencil-lead-break detecting by phase-OTDR using coherent detecting technology experimentally for the first time. For this, we proposed a practical moving averaging and moving differential method to fast realize the averaged waveforms of phase-OTDR to improve SNR while not excessively degrading the spatial resolution. Also we experimentally demonstrated the SNR 2~3dB improvement of coherent phase-OTDR compared with direct phase-OTDR.

ACKNOWLEDGMENT

The author thanks Xiaoyi Bao, Yongkang Dong and ziyi zhang for their contributions to this paper.

REFERENCES

[1] J. M. Stallings, J. W. Tedesco, M. El-Mihilmy, and M. McCauley, "Field performance of FRP bridge repairs," Journal of Bridge Engineering, Vol. 5,pp. 107-113, 2000.

[2] S. W. Doebling, C. R. Farrar, M. B. Prime, and D. W. Shevitz, "Damage identification and health monitoring of structural and mechanical systems from changes in their vibration characteristics: A literature review," Los Alamos National Laboratory, Tech. Rep. LA pp.130-137 1996.

[3] B. J. Pernick and J. Kennedy, "Optical method for fatigue crack detection,"APPL .OPT, Vol. 19, pp. 3224-3229,September 1980.

[4] S M Melle; K Liu; R M Measures, "Practical Fiber-Optic Bragg Grating Strain Gauge," APPL .OPT, Vol. 32, pp.3601-3609, 1993.

[5] C. S. Baldwin, A. J. Vizzini, "Acoustic emission crack detection with FBG " Proceedings of SPIE Vol. 5050, pp.133-143,2003.

[6] Duckey Lee, Hosung Yoon, Na Young Kim, Hansuek Lee, and Namkyoo Park, "Analysis and Experimental Demonstration of Simplex Coding Technique for SNR Enhancement of OTDR,"IEEE LTIMC 2004 pp. 118-122, [Digest Lightwave Technologies in Instrumentation & Measurement Conference October 2004].

[7] I. Alasaarela, P. Karioja, and H. Kopola, "Comparison of distributed fiber optic sensing methods for location and quantity information measurements," Optical Engineering, Vol. 41, pp.181-189 ,2002.

[8] Dae-Hyun Kim, Bon-Yong Koo, Chun-Gon Kim,and Chang-Sun Hong, "Damage detection of composite structures using a stabilized extrinsic Fabry–Perot interferometric sensor system," Smart Mater. Struct. Vol.13, pp. 593–598, 2004.

[9] Hisashi Izumita, Shin-ichi Furukawa, Yahei Koyamada, and Izumi Sankawa, "Fading Noise Reduction in Coherent OTDR," IEEE Photonics. Technol . LETT, Vol. 4,pp.201-203,February 1992.

Anti-Relaxation Coating of the Vapor Cell

Xie Xin ; Liu Yunhui ; Bai Yuanyuan ; Xu Yunfei

Physics Department of Zhejiang University

xuyf@zju.edu.cn

Abstract—**In order to achieve the maximum possible polarization lifetime of the alkali atoms in the vapor cell, we discussed several spin-relaxation mechanisms and anti-relaxation methods. We found that octadecyltrichlorosilane (OTS) acts as an anti-relaxation coating for alkali atoms, with good effect in high-temperature conditions.**

Keywords- spin-relaxation; anti-relaxation; OTS

I. INTRODUCTION

Alkali metal atoms are useful for the application of atomic magnetometers because they have a single unpaired electron in the outer shell that can be easily manipulated. Atomic magnetometers work by exploiting the energy structure of the ground and excited states to polarize the atoms and measure the magnetic field. What can influence the sensitivity of an atomic magnetometer includes the following three aspects: the lifetime of the excited state, pressure broadening due to collisions with other gas species, and Doppler broadening due to thermal motion of the alkali atoms [1]. Among the three, perhaps the most important method for optimizing the magnetometer sensitivity is to maximize the spin polarization lifetime.

II. SPIN RELAXATION

A. Wall Collisions

When an alkali atom encounters the bare glass surface of the cell wall, it becomes adsorbed into the surface for a finite period before being ejected back into the cell volume. During this time, the atom experiences the electromagnetic interaction. Although the adsorption period is short, it can be measured in the range of 0.1-10 us [3]. By the time the atom leaves the surface, its spin direction becomes completely randomized. Wall collisions therefore completely depolarize the atoms and can dominate all the other spin-relaxation mechanisms unless they are suppressed.

B. Spin-Exchange Collisions

With high alkali densities, spin-exchange collisions between alkali atoms are the dominant cause of spin relaxation. In such a collision, the direction of the electron spins of the two atoms can be reversed while the total spin is conserved. The process may be represented symbolically by the following:

$$A(\uparrow) + B(\downarrow) \Rightarrow A(\downarrow) + B(\uparrow)$$

The arrows refer to spin-up and spin-down. Spin-exchange collisions are sudden with respect to the hyperfine interaction and do not affect the nuclear spin of the colliding atoms. The overall effect of spin-exchange collisions is to redistribute the atoms among the atomic Zeeman sublevels and preserve the total atomic spin. As the magnetic field approaches zero, spin-exchange no longer affects the polarization lifetime. [6]

C. Spin-Destruction Collisions

Spin-destruction collisions may occur between alkali atoms or with the buffer and quenching gases. Such collisions between alkali atoms transfer spin angular momentum to the rotational angular momentum of the colliding pair of atoms. The process may be represented symbolically by the following:

$$A(\uparrow) + B(\downarrow) \Rightarrow A(\downarrow) + B(\downarrow)$$

Spin-destruction collisions do not preserve the total spin of the alkali ensemble. Alkali-alkali spin-destruction collisions do not play an important role in determining the polarization lifetime unless spin-exchange relaxation is suppressed.

Collisions with the buffer or the quenching gases can also destroy the alkali polarization. In the cell, with a noble gas species for the purpose of slowing alkali diffusion to walls, the buffer gas density is typically chosen so that the relaxation due to spin-destruction collisions with noble gas atoms is comparable to the other dominant spin-relaxation mechanisms.

D. Magnetic Field Gradients

With the presence of a magnetic field gradient, atoms in different parts of the cell experience different local fields and therefore precess at different frequencies. In a coated cell without buffer gas, the alkali atoms freely diffuse in the entire volume, average the whole magnetic field and suppress the effect of the field gradients in an example of motional narrowing. However, in the cell where the atoms diffuse very slowly, they only remain in a small region of the cell during a single coherence lifetime and only experience a single value for the magnetic field strength. In the system, the gradient broadening is given by the spread of precession frequencies throughout the cell. In general, the broadening due to a magnetic gradient depends on the extent to which atoms diffuse with the gradient. Operation in the system of slow diffusion can allow direct measurement of the magnetic field gradient across the cell.

E. Reservoir Effect

In the presence of condensed alkali metal atoms in the glass surface, atoms in the cell transfer between condensed state and gaseous state. In the process, the whole distribution of polarization state in the cell is destroyed. A method for suppressing reservoir effect is by using buffer gas.

III. ANTI-RELAXATION METHODS

Alkali spins depolarize immediately after colliding with the glass walls of the vapor cell. Wall collisions dominate all the other spin-relaxation mechanisms unless they are suppressed. The two common methods for suppressing wall depolarization are the use of anti-relaxation surface and buffer gas.

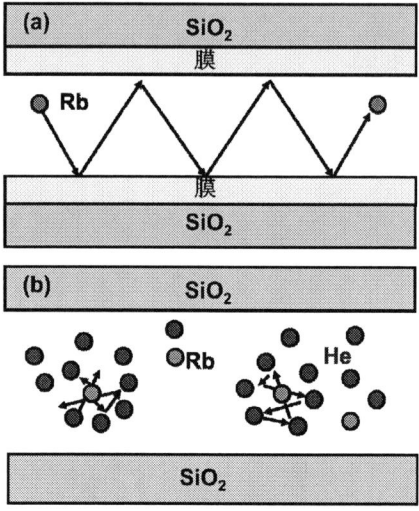

Figure 1. (a) In the cells, surface coatings prevent the alkali atoms from approaching the glass wall. (b) In the cells with buffer gas, the atoms require a long period of time to diffuse from the center of the cell to the wall.

Surface coatings are molecules that cover the surface and prevent the alkali atoms from approaching the glass wall. These coatings must be chemically inert and ideally allow the atoms to collide with the surface many times without depolarizing. The most well-known effective coating is paraffin, which melts at temperatures of about 60-80°C [2]. Above these temperatures, paraffin cannot work. Therefore, at a high temperature, effective coatings were previously unavailable before the discovery of the effectiveness of OTS up to 170ºC [4].

With high buffer gas pressure, the atoms experience diffusive motion, increasing the time to reach the cell wall. In the aspect of spin-destruction collisions, the buffer gas density should be typically chosen so that the relaxation due to spin-destruction collisions with noble gas atoms is comparable to the other dominant spin-relaxation mechanisms.

Anti-relaxation coatings have numerous advantages over buffer gas, including larger optical rotation signals, lower laser power requirements, and lack of spin-destruction effects [1]. The demonstrations of high-density magnetometers, including the RF magnetometers and the SERF magnetometer, used cells filled with buffer gas because there was no available coating for operation at a high temperature [8] [9]. However, now Seltzer demonstrated that OTS acted as an effective anti-relaxation coating at temperatures up to about 170º C.

IV. OTS COATING

OTS is the first known anti-relaxation coating for high-temperature application of polarized alkali vapor. It is an organic molecule that resembles paraffin, with eighteen carbon atoms each attached to two or three hydrogen atoms, and a silane end group that reacts with the glass surface and binds to it. The form of an OTS coating molecule after being attached to the surface is illustrated in Figure 2 and figure 3.

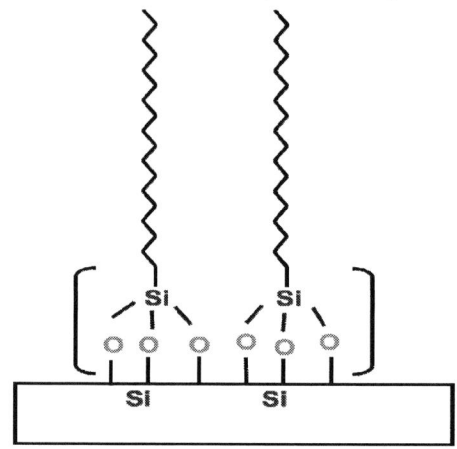

Figure 2. Attachment of an OTS molecule to the glass surface in anhydrous conditions.

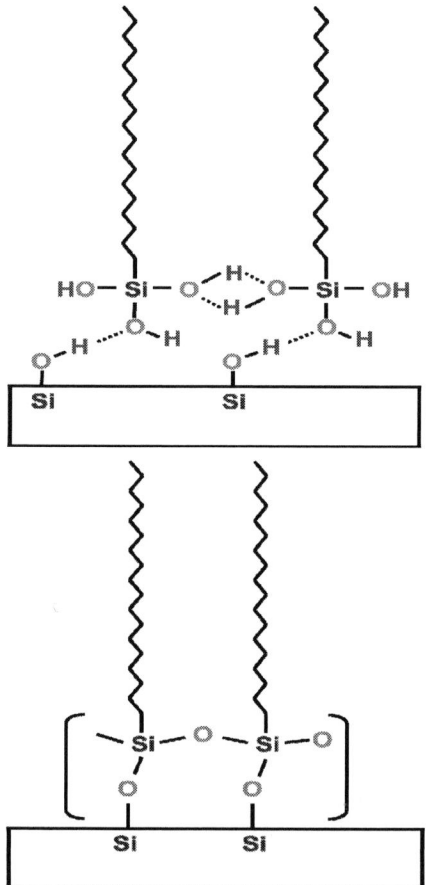

Figure 3. Attachment of an OTS molecule to the glass surface in the presence of water conditions

Seltzer found that OTS acted as an anti-relaxation coating for alkali atoms at temperatures below 170° C, which allowed them to collide with a glass surface up to 2 000 times before depolarizing [1]. We did an experiment to evaluate the anti-relaxation effect of OTS.

V. COATING PROCEDURE

We operated in ambient atmosphere that contained moisture, which led to cross-linking between OTS molecules (Figure 3), as opposed to the dry environment necessary for the formation of monolayer films (Figure 2).

First, we cleaned our cells with piranha solution, which removed all organic material from the surface of the glass so that the reaction with OTS could occur. Next, we rinsed the cell by filling them first with deionized water and then with methanol. We then dried out the cells by baking in vacuum at a temperature of at least 100 °C for one hour. The cells were then returned to the fume hood for OTS coating. The OTS solution was formed. The cells were exposed to the solution for five minutes, during which time OTS molecules were attached to the glass. The presence of water led to cross-linking between OTS molecules. Finally, we filled the OTS-coated cell with rubidium which acted as the source of alkali vapor during the experiment.

A perfect coating can suppress spin-relaxation so that the atoms in the cell can achieve long polarization lifetime. The coating quantity was determined by polarizing the atoms with a strong, circularly polarized pump beam and then quickly turning off it, and detecting changes in the polarization state with a weak, polarized probe beam whose optical pumping effect could be ignored [5]. When we turn off the optical pumping, we subsequently measured T, the time constant with which the longitudinal spin polarization decays. By reducing the intensity of the probe beam to the extent that attenuation no longer increased the observed polarization lifetime, we ensured that absorption of probe photons was slow enough not to affect T1. We compared the vapor cell that was coated with the one that contained the buffer gas to prove the efficiency of OTS.

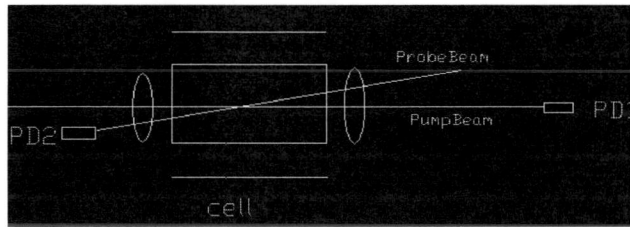

Figure 4. Experimental schematic of anti-relaxation coating characterization.

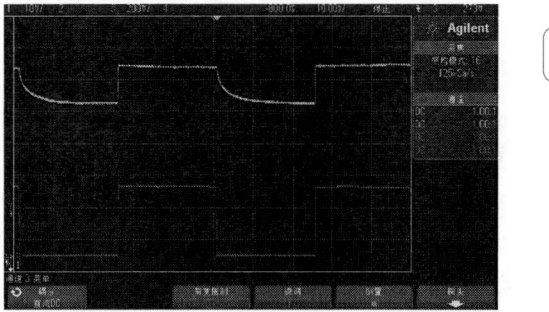

Figure 5. the signal that Photodiode 1 and 2 obtained in the experment.

VI. THE ANTI-RELAXATION EFFECT OF OTS

The figures are obtained from our experimental signals. We dealt with the data obtained. Most of the observations described below were made.

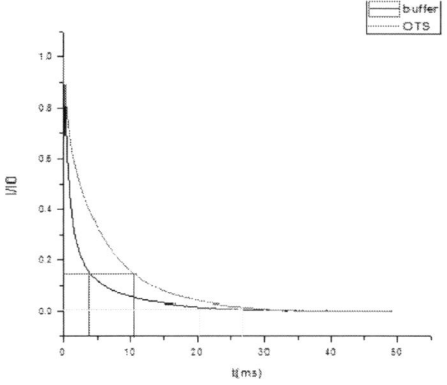

Figure 6. comparison between two curves that were obtained from the signals of probe beam at 60° C.

TABLE I.   TIME OF SPIN-RELAXATION AT SOME TEMERATURES

temperatures (°C)	40	50	55	60	80
$T_{\#1}(ms)$	5.90	4.50	8.84	5.86	2.03
$T_{\#2}(ms)$	2.99	2.72	2.36	1.11	0.40

$T_{\#1}(ms)$ the cell which was coated;

$T_{\#2}(ms)$ the cell which contained buffer gas

Figure 6 displays the signal curve of the probe beam. It is demonstrated that the OTS coating is more efficient than the buffer gas as shown in Figure 6. Our results were summarized in Table I. Besides the comparison between OTS coating and buffer gas, the anti-relaxation effect of OTS and buffer are also observed at different temperatures.

VII. CONCLUSIONS

According to the experiment, an interesting dependence of OTS coating efficiency on temperature has observed. To be more specific, OTS coating is more efficient than the buffer gas at high temperatures.

REFERENCES

[1] Seltzer S J Developments in Alkali-Metal Atomic Magnetometry 2008

[2] M.V.Balabas,T.Karaulanov,M.P.Ledbetter and D.Budker,Polarized Alkali-Metal Vapor with Minute-Long Transverse Spin-Relaxation Time,Phys.rev.lett.105

[3] **H. N. de Freitas, M. Oria, and M. Chevrollier (2002)**. Spectroscopy of cesium atoms adsorbing and desorbing at a dielectric surface. Applied Physics B **75** (6), 703–709 http://dx.doi.org/10.1007/s00340-002-1029-y. 64

[4] S. J. Seltzer, D. M. Rampulla, S. Rivillon-Amy, Y. J. Chabal, S. L. Bernasek, and M. V. Romalis, J. Appl. Phys. 104, 103116 (2008)

[5] W.Frazen, Phys. Rev. 115, 850 (1959)

[6] **W. Happer and A. C. Tam (1977)**. Effect of rapid spin exchange on the magnetic resonance spectrum of alkali vapors. Physical Review A **16** (5), 1877–1891. http://link.aps.org/abstract/PRA/v16/p1877. 61, 161, 163, 166

[7] H. G. Dehmelt, Phys. Rev. 105, 1487 (1957); J. Opt. Soc. Am.SS, 335 (1965)

[8] **I. M. Savukov, S. J. Seltzer, M. V. Romalis, and K. L. Sauer (2005)**. Tunable Atomic Magnetometer for Detection of Radio-Frequency Magnetic Fields. Physical Review Letters. **95** (6), 063004. http://link.aps.org/abstract/PRL/v95/e063004. 5, 69, 125, 131, 132, 170,264

[9] **S. J. Seltzer and M. V. Romalis (2004)**. Unshielded three-axis vector operation of a spin-exchange relaxation-free atomic magnetometer. Applied Physics Letters **85**(20), 4804–4806 http://link.aip.org/link/?APL/85/4804/1.166

Research on the measurement range of particle size with laser backscattering based on PT algorithm

Jian Xing[1]*, Yuandong Sui[1], Weimin Sun[1]

1 Key Lab of In-fiber Integrated Optics, Ministry Education of China, Harbin Engineering University, Harbin, China
xingniat@sina.com

Abstract—During the test of particle size with backscattering, the measurement range of particle size based on PT independent model algorithm. R-R particle size distribution function was simulated and analyzed as target for determining the particle size range. Simulation experiments illustrate that the particle size distribution can be retrieved very well in the range from 0.5μm to 15μm at relative refractive index *m*=1.57 in the visible-infrared spectral region, and the measurement range of particle size will vary with the varied wavelength range and relative. It provided a theory basis for practical application.

Keywords-PT algorithm; Mie scattering; backscattering; particle size distribution

I. INTRODUCTION

Light backscattering method is one technology of light scattering particle size measurement, and it has more development space and application potential because of the simple and co-rotating measuring device (Fig 1). It is also not influenced by the extreme particles concentration compared with the transmission method. The forward scattering method has been widely used at present, but the method of backscattering is seldom reported. Specially, the range of particle size is different according to laser source and refraction, etc. The theory of laser backscattering particle size measuring has been derived based on Mie scattering theory and simulation calculated through independent model algorithm in this paper. It has also laid a solid foundation for subsequent development and application of experimental device.

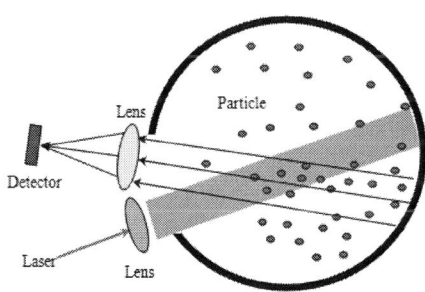

Figure 1. Sketch diagram of light backscattering theory

II. INDEPENDENT MODEL ALGORITHM

According to the Mie scattering theory, when a one-way single color light beam with a wavelength for λ and intensity for I_0 irradiates to the suspended particles distance for L, the incident light will be scattered in all directions because of the scattering effect. Some of the scattered light which come a way back nearly 180 degrees can be conducted to the photodetector by the reflection type fiber-optic probe. Every particle system has it's own range of size distribution, so it should keeps to the following relation on the premise that the particles meet the condition of irrelated single scattering[1]:

$$I_{sca} = I_{inc}\frac{3\lambda^2}{4\pi^3 L^2}N_D\int_{D_{min}}^{D_{max}}(i_1+i_2)\frac{f(D)}{D^3}dD \quad (1)$$

Where, I_{sca} stands for scattering intensity; I_{inc} stands for incident intensity; N_D stands for the total number of particles in the particle system; i_1 and i_2 stand for scattering intensity function, it relates to the particle complex refractive index and scattering angle and scale parameter and can be worked out through the Mie scattering theory; $f(D)d(D)$ stands for volume frequency distribution, that is the volume frequency of particles which diameter in the range from D to $D+d(D)$ in unit volume. It is an unknown quantity waiting for solving in the particle size measurement.

Formula (1) is the Fredholm Integral Equation of the First Kind, it is on the left of the measured values and the right of the particle size distribution waiting for solving that represented by $f(D)$ in the integral sign. There isn't a theoretical solution for this kind of equation at present, so direct solution for formula (1) could be very difficult and then discretization is a must, that is it should be divided into a series of size fraction:

$$I(\lambda)_{sca}=I(\lambda)_{inc}\frac{3\lambda^2}{4\pi^3 L^2}N_D\sum_{j=1}^{S}C_j\left[i_1\left(\lambda,m,\theta,\overline{D}\right)+i_2\left(\lambda,m,\theta,\overline{D}\right)\right]\frac{f(\overline{D}_j)}{\overline{D}_j^{\,3}} \quad (2)$$

Where, S stands for the numbers of classification of particle size range from D_{min} to D_{max}; C_j stands for numerical integration coefficient; $f_j(\overline{D}_j)=\int_{D_j}^{D_{j+1}}f(D)dD$, $\overline{D}$ stands for the halfway point of subinterval $[D_j,D_{j+1}]$. It should be measured with multiple wavelengths simultaneously because the particle size distribution in formula (2) has S unknown variables, so a system of linear equations can be acquired from it[2]:

$$\mathbf{E = Af} \quad (3)$$

Where, $\mathbf{E}=[(I_{1sca}/I_{1inc});\cdots;(I_{Usca}/I_{Uinc})]$; U stands for wavelength number; $\mathbf{A}=[A_{MN}]$ is a $U{\times}S$ step weighting matrix, one of the elements of A_{MN} is:

$$A_{MN}=\frac{3\lambda_M^2}{4\pi^3 L^2\overline{D}_N^{\,3}}N_DC_J\left[i_1\left(\lambda_M,m,\theta,\overline{D}_N\right)+i_2\left(\lambda_M,m,\theta,\overline{D}_N\right)\right]$$

$$(M = 1,\dots,U; N = 1,\dots,S) \qquad (4)$$

$\mathbf{f} = [f_1(\overline{D}_1); \cdots; f_S(\overline{D}_S)]$, where, M stands for wavelength number; N stands for the number of subinterval of division particle size.

In actual measurement, if the law of particle size distribution of the particle system is unknown, or the particle system size distribution can not be described simply by using any distribution, then the result received through independent model is not reliable. Among the dependent model inversion algorithm which has been put forward, limitation least square method raised by Phillips and Twomey has been most widely applied and the inversion result is also the most ideal. The least-squares solution of formula (3) is[3-4]:

$$\mathbf{f} = (\mathbf{A^T A})^{-1} \mathbf{A^T E} \qquad (5)$$

From a mathematical point of view, if weight matrix $\mathbf{A}$ is full-rank, the particle size distribution can be worked out through formula (5). But in fact, matrix $\mathbf{A}$ is highly ill-conditioned[5] and has a very large condition number, so it can't be solved by using formula (5). Phillips and Twomey imported a fairing matrix $\mathbf{H}$ and a fairing factor γ, then formula (5) is rewritten into[6]:

$$\mathbf{f} = (\mathbf{A^T A} + \gamma \mathbf{H})^{-1} \mathbf{A^T E} \qquad (6)$$

Where, γ stands for fairing factor, $\mathbf{H}$ stands for $N \times N$ step fairing matrix and it is defined as:

$$\mathbf{H} = \begin{bmatrix} 1 & -2 & 1 & 0 & 0 & & \\ -2 & 5 & -4 & 1 & 0 & 0 & \\ 1 & -4 & 6 & -4 & 1 & 0 & \\ 0 & 1 & -4 & 6 & -4 & 1 & 0 \\ & & & & \ddots & & \\ & & 0 & 1 & -4 & 5 & -2 \\ & & 0 & 1 & -2 & 1 \end{bmatrix} \qquad (7)$$

In this algorithm the choice of γ value is critical[7]: if γ=0, then formula (6) degenerate into the general algorithm of inverse matrix shape as formula (5), the resulting solution $f(D)$ will be in the form of considerable oscillation. The volatility can be reduced by increasing γ value gradually. If the γ value is very large, then the solution will be too fairing to avoid error. Therefore, how to select γ value has always been a popular concern. As one of the most commonly used methods is GCV (Generalized Cross Validation), γ value can be determined by solving the minimum through formula (8):

$$V(\gamma) = \frac{\frac{1}{S} \left\| [\mathbf{I} - \mathbf{K}(\gamma)] \mathbf{E} \right\|_2^2}{\left\{ \frac{1}{S} \mathrm{Trace}[\mathbf{I} - \mathbf{K}(\gamma)] \right\}^2} \qquad (8)$$

Where, $\mathbf{K}(\gamma) = \mathbf{A}(\mathbf{A^T A} + \gamma \mathbf{H})^{-1} \mathbf{A^T}$, $\mathbf{I}$ is a $M \times N$ step unit matrix.

III. NUMERICAL SIMULATION

In order to determine the range of particle size measurement of typical spherical particle system in visible light wave band, a large number of simulation experiments have been done in this section. At the beginning of numerical simulation, assuming a particle size distribution function $f(D)$ and selecting a set of measuring wavelength and a relative complex refractive index, put them into formula (1) to calculate a set of extinction values $(I/I_0)_{\lambda_1}, \cdots, (I/I_0)_{\lambda_U}$, then put this set of values into formula (2) as measuring values, the particle size distribution of the particle system was calculated through the appropriate inversion algorithm finally.

Assuming the particle size of polystyrene particle system obeys single peak R-R distribution which can be expressed as:

$$f(D) = 1 - \exp(-(\frac{D}{\overline{D}})^k) \qquad (9)$$

Let N_D=1, L=10mm, the wavelength of visible light wave band ranges from 0.4μm to 0.8μm, the particle size distribution inversion algorithm using PT algorithm in dependent model. The simulation software for this paper is Matlab 7.12. This paper made a study about particle size measurement range of polystyrene particle system dispersed in the water, the relative refractive index of this particle system in visible light wave band is 1.57. Table 1 lists the setting of parameters of inversing polystyrene particle system. The unit of characteristic parameters of R-R distribution is micrometer and the scattering angle is 175 degree.

TABLE I. PARAMETER SETTING FOR INVERSION OF POLYSTYRENE PARTICLE SYSTEM IN SINGLE PEAK

No.	Given $(\overline{D}, k)$	Given size range (D_{min}, D_{max})	Wavelength number U	Discretization S
(a)	(0.5,8)	(0.05,1)	40	40
(b)	(0.8,8)	(0.5,1)	40	40
(c)	(4.1,8)	(0.5,7)	40	80
(d)	(2,8)	(0.5,15)	40	80
(e)	(1,8)	(0.5,17)	40	80

The inversion results of different particle size measurement range are shown in Fig 2:

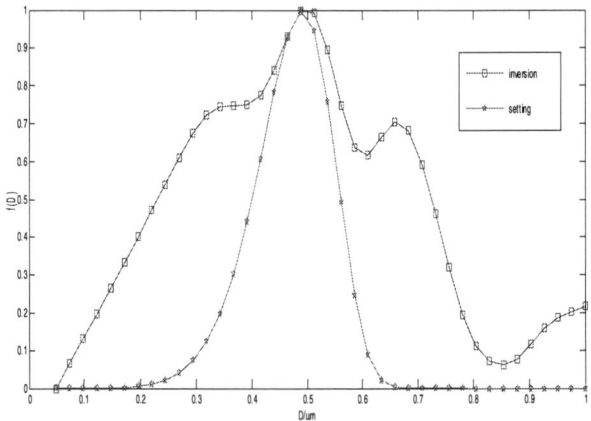

(a) Particle size distribution range 0.05μm -1μm

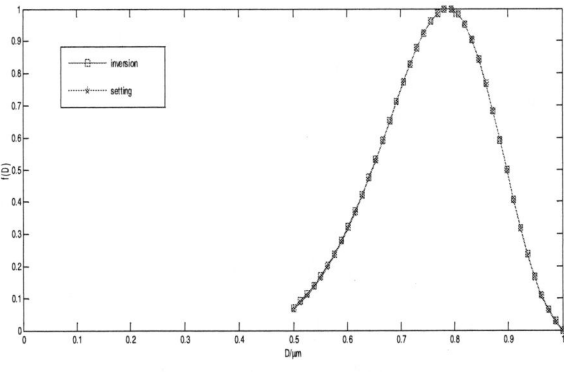

(b) Particle size distribution range 0.5μm -1μm

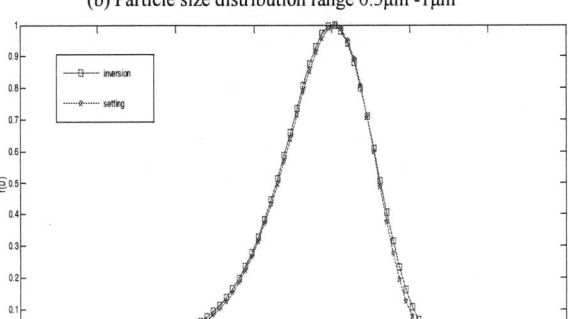

(c) Particle size distribution range 0.5μm -7μm

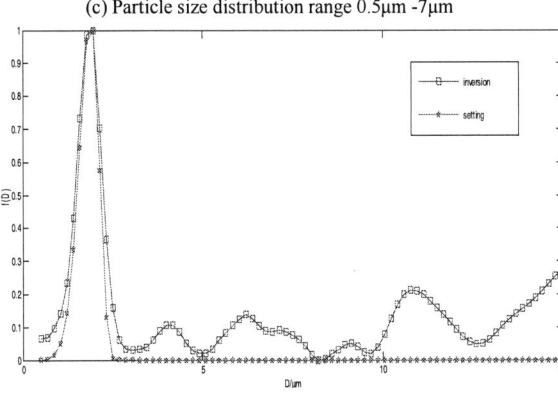

(d) Particle size distribution range 0.5 μm-15μm

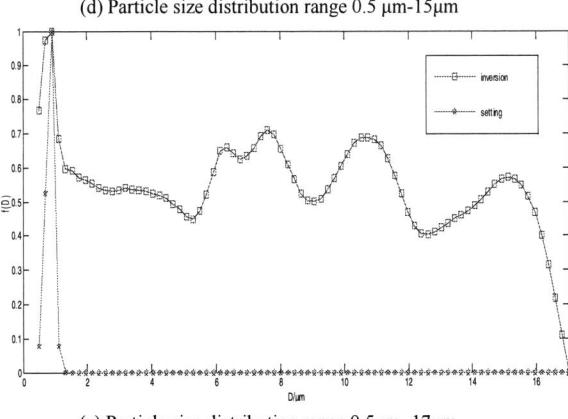

(e) Particle size distribution range 0.5μm -17μm

Figure 2.   Inversion result of R-R(m=1.57, ìλ=0.4μm -0.8μm)

It can be seen clearly from Fig 2 that the inversion curve shell basically tallies with the setting curve when the particle size range within 0.5μm - 15μm. While the lower limit is lower than 0.5μm, the inversion curve and the setting curve are put in a big difference as shown in Fig 2(a). However, inversion curve broadening was obtained when the ceiling is higher than 15μm as shown in Fig 2(e). Therefore, in order to obtain a better particle size distribution inversion result through independent model algorithm for polystyrene particle system dispersed in the water, the particle size measurement range in visible band must be limited within the scope of 0.5μm - 15μm.

IV.   CONCLUSION

In this paper, the particle size measurement range basic principle of visible laser backscattering method was introduced and simulated by PT algorithm. The solving method of particle size measuring through visible laser backscattering method in theory can boil down to the solving problem of Fredholm integral equation of the first kind, so it is necessary to determine upper and lower limit, namely the particle size measurement range, of the integral equation. PT algorithm in dependent model has been made by analyzing and comparing the inversion result in different wavebands of the particle system which obeys single peak R-R distribution in this article. Finally, the particle size measurement range is 0.5μm -15μm for polystyrene particle system dispersed in the water. It can be used to conduct the practical application.

REFERENCES

[1]   M. I. Mishchenko, Larry D. Travis, Andrew A. Lacis. Scattering, Absorption, and Emission of Light by Small Particles. Cambridge University Press. 2002:139.

[2]   Naining Wang. Optical Measurement Technique and Application of Particle Size. Atomic Energy Press. 2000:105-134.

[3]   M.Pahlow, D.Muller, M.Tesche. Retrieval of Aerosol Properties from Combined Multiwavelength Lidar and Sunphotometer Measurements. Applied Optics. 2006, 45(28):7429-7442.

[4]   M.L.Arias, G.L.Frontini. Particle Size Distribution Retrieved from Elastic Light Scattering Measurements by a Modified Regularization Method. Particle and Particle Systems Characterization. 2007, 23(5):374-380.

[5]   D.Muller, U.Wandingern, A.Ansmann. Microphysical Particle Parameters from Extinction and Backscatter Lidar Data by Inversion with Regularization:Theory. Applied Optics. 1999, 38(12):2346-2357.

[6]   Feng Xu, Xiaoshu Cai, Mingxu Su, Zhijun Zhao, Junfeng Li. The Study of Solving Particle Size Distribution through Dependent Model Algorithm. Chinese Journal of Lasers. 2004, 31(2):223-228.

[7]   Xiaoshu Cai. The Study of Total Light Scattering Particle Detection Technique and Application in Wet Steam Measurement. doctoral dissertation, Shang Hai: Shang Hai Machinery Institute, 1991.

Birefringence properties analysis of a novel three quasi-rectangular cores fiber

Fengjun Tian, Libo Yuan*

Key Lab of In-Fiber Integrated Optics of Ministry of Education, and College of Science, Harbin Engineering University, Harbin 150001, China
* tianfengjun0424@yahoo.com.cn

Abstract—**We demonstrate a three-core optical fiber with qusi-rectangular core. Due to fiber-core's geometry asymmetry the high birefringence properties are predicted and estimated. Based on the finite element method, the birefringence characteristics are analyzed numerically at 1550nm wavelength. The mode birefringence is calculated to be $B_m > 10^{-5}$. The origin of birefringence is attributed to the core's shape and the inner retained stress. By simulating inner self-stress distribution we discuss the contribution to mode birefringence.**

Keywords-rectangular cores fiber; three-core fiber; mode birefringence; numberical analyses

I. INTRODUCTION

Currently fiber optic interferometric structures have been widely applied in sensing field due to the high sensitivities that they exhibit on the measurement of a broad range of parameters. Especially, various in-fiber integrated interferometrics interest scientists due to its compactness and reliability [1-6]. Multi-core fibers (MCF) play an important role in the integrated interferometers, which have been widely researched in environmental sensing, curvature sensing and others [7-9]. These interferometers based on MCF are attractive for several reasons, including small size and application flexibility, as well as the presence of a reduced thermal sensitivity in view of the usually small difference of the thermo-optic coefficients of the fiber core/cladding due to integration in a fiber. However, interferometric fiber-optic sensors commonly consist of conventional low-birefringence single mode optical fiber (SMF) and fiber components. Consequently they typically exhibit signal fading owing to random fluctuations in the state of polarization (SOP) of the interfering beams [10]. In order to reduce influence of polarization-induced fading to interferometers [11-12], developing a kind of high birefringence MCF are challenges that we are confronted with now.

In this paper, we present a three-core fiber with quasi-rectangular cores (MRCF). Due to the rectangular core and geometric asymmetry of core/cladding high birefringence is predicted and characterized. This scheme is simple, and has some potential applications in in-fiber interferometers for reducing influence of polarization-induced fading.

II. DESIGN AND FABRICATION OF THE MRCF

The configuration of the MRCF is illustrated in Fig.1 (a), along with major waveguide parameters. The three rectangular cores are distributed circularly in the same cladding respectively. The length for rectangular cores is $l=10\mu m$, the width is $w=5\mu m$, and the distance from the center of fiber is $l=40\mu m$. The circular silica cladding diameter of MRCF is $R=125\mu m$. The mole fraction of GeO_2 in the core region is about 4%/mol. According to the Sellmeier dispersion formula, the refractive index of the cladding and the core is respectively designed as $n_{clad}=1.444$ and $n_{core}=1.450$ at wave length $\lambda=1.55\mu m$ respectively. The core-cladding index differences Δn is 0.006. Due to the rectangular core and geometric asymmetry of core/cladding, high birefringence has been theoretically predicted.

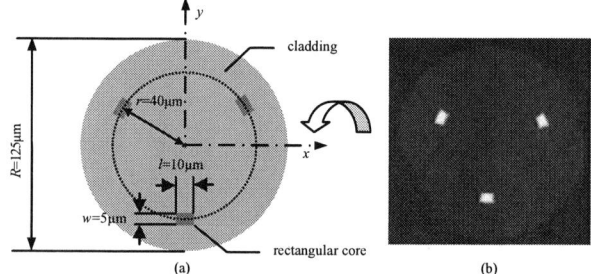

Figure 1. The three-core fiber with rectangular cores: (a) Schematic diagram, (b) Cross section

The fabrication technology of the MRCF includes preform preparation and fiber drawing. The preform was comprised of core rods and silica tube. The core rods were manufactured by modified-chemical-vapor deposition (MCVD ） and etching method. Then we make use of stack-method to organize the perform.

The fiber drawing tower with a vacuum function was used to draw the prepared preform. The drawing temperature was set at 2000℃ and feed speed was set at 500μm/min. The drawing speed was from 5 to 6 m/min. With the help of the vacuum the MRCFs have been successfully drawn. Fig.1 (b) shows the MRCF's cross section. The experimental results are well accordant with the configuration of the MRCF in Fig.1 (a).

III. THEORETICAL MODEL

Generally the mode birefringence B_m can be divided into three parts and expressed as:

978-1-4799-1215-5/13 $31.00 © 2013 IEEE

$$B_m = B_G + B_S + B_{S0} \qquad (1)$$

where B_G is the geometrical birefringence induced by the shape of core and cladding, B_S is the stress birefringence caused by the stress-applying parts in the fiber and B_{S0} is the self-stress birefringence originating from the thermal expansion difference of the rectangular cores and the cladding. For the proposed MRCF as shown Fig.1, introduction of the rectangular cores could cause special birefringence characteristics. There are mainly two factors to cause generation of mode birefringence: (1) the rectangular cores and the geometric asymmetry between the cores and the cladding (see Fig.1), (2) asymmetrical distribution of inner thermal residual stress around cores. This stress is named by self-stress, and results from mismatch in the thermal expansivity between the core and the cladding during the fabrication process of the MRCF. While cooling down from an approximately $1020\,°C$ high temperature to $20\,°C$ room temperature, it results in generation of thermally induced stresses in the MRCF. Moreover, because additional stress zone doesn't exist, the stress-applying birefringence B_S is equal to zero. Therefore the total mode birefringence in MRCF can be written as,

$$B_m = B_G + B_{S0} \qquad (2)$$

Note that B_G can't be simply plus together by B_{S0} in order to calculate correctly mode birefringence B_m in various wavelengths. Because both B_m and B_G are wavelength-dependent, and the self-stress directly changes material refractive index according to photo-elastic effect, B_{S0} isn't dependent on wavelength and doesn't directly contribute to B_m.

A lot of methods have been used to calculate the mode birefringence in PMF, such as complex variable method, finite element method [13-14] and infinitesimal element method etc. Finite element method (FEM) not only can respectively calculate the geometric birefringence and the complex stress distribution [11], but also can be use in combining operation of multiphysics. In this paper, we used FEM based on the multiphysics to study mode birefringence consisted of the geometric birefringence and self-stress birefringence in the MRCF.

In order to calculate mode birefringence of MRCF, we need to solve vector wave equation to get the difference of effective refractive index between two fundamental polarization modes. The third-order tensor permittivity of the fiber can be written as,

$$\varepsilon = \varepsilon_0 \begin{bmatrix} n_x^2 & 0 & 0 \\ 0 & n_y^2 & 0 \\ 0 & 0 & n_z^2 \end{bmatrix} \qquad (3)$$

where ε_0 is permittivity of vacuum, n_x, n_y and n_z are the refractive index respectively along the x, y and z axis. The wave equation with H_x and H_y is the following,

$$\begin{cases} \partial_y[\dfrac{1}{n_z^2}(\partial_x H_y - \partial_y H_x)] - \dfrac{1}{n_y^2}\partial_x(\partial_x H_x + \partial_y H_y) + k_0^2 n_{eff}^2 \dfrac{H_x}{n_y^2} = k_0^2 H_x \\ -\partial_x[\dfrac{1}{n_z^2}(\partial_x H_y - \partial_y H_x)] - \dfrac{1}{n_x^2}\partial_x(\partial_x H_x + \partial_y H_y) + k_0^2 n_{eff}^2 \dfrac{H_x}{n_x^2} = k_0^2 H_y \end{cases} \qquad (4)$$

where $k_0 = 2\pi/\lambda$, λ is free space wavelength. The n_{eff} is effective refractive index of polarization mode for solving. According to Galerkin method, we can get functional from Eq. (4). We combined FEM to get effective refractive index of two fundamental polarization modes. The mode birefringence is expressed as,

$$B_m = n_{eff}^y - n_{eff}^x \qquad (5)$$

In no stress conditions, $n_x = n_y = n_z = n_0$. The n_0 is the refractive index for a stress-free material. Combining Eq. (7) with FEM, we can get $B_m = B_G$.

When we consider exist of self-stress, the general linear stress-optical relation can be written, using tensor notation, as

$$\Delta n_{ij} = -C_{ijkl}\sigma_{kl} \qquad (6)$$

where $\Delta n_{ij} = n_{ij} - n_0 I_{ij}$, n_{ij} is the refractive index tensor, I_{ij} is the identity tensor, C_{ijkl} is the stress-optical tensor, and σ_{kl} is the stress tensor. Due to symmetry the number of independent parameters in the stress-optical tensor that characterizes this constitutive relation can be reduced. Because n_{ij} and σ_{kl} are both symmetric, $B_{ijkl} = B_{jikl}$ and $B_{ijkl} = B_{ijlk}$. In many cases it is possible to further reduce the number of independent parameters, and this model includes only two independent parameters, C_1 and C_2 that are first and second stress optical coefficient respectively. The stress-optical relation simplifies to

$$\begin{bmatrix} n_x \\ n_y \\ n_z \end{bmatrix} = n_0 - \begin{bmatrix} C_1 & C_2 & C_2 \\ C_2 & C_1 & C_2 \\ C_2 & C_2 & C_1 \end{bmatrix}\begin{bmatrix} \sigma_x \\ \sigma_y \\ \sigma_z \end{bmatrix} \qquad (7)$$

where $n_x = n_{11}$, $n_y = n_{22}$, $n_z = n_{33}$, $\sigma_x = \sigma_{11}$, $\sigma_y = \sigma_{22}$, and $\sigma_z = \sigma_{33}$. According to photo-elastic effect, the relation between the induced refractive index and the principle stress in x, y, and z axis direction can be written as,

$$\begin{cases} n_x = n_0 - C_1\sigma_x - C_2(\sigma_y + \sigma_z) \\ n_y = n_0 - C_1\sigma_y - C_2(\sigma_x + \sigma_z) \\ n_z = n_0 - C_1\sigma_z - C_2(\sigma_x + \sigma_y) \end{cases} \qquad (8)$$

where n_x, n_y and n_z are the transformational refractive index respectively along the x, y and z axis due to the inner self-stress. C_1 and C_2 are positive photo-elastic coefficients, σ_x, σ_y and σ_y are the principle stresses along the x, y and z axis, respectively. Therefore, the self-stress birefringence B_{S0} can be expressed as:

$$B_{S0} = n_x - n_y = -(C_1 - C_2)(\sigma_x - \sigma_y) \qquad (9)$$

Using the two parameters C_1 and C_2, we assumes that the nondiagonal parts of n_{ij} and σ_{kl} are negligible. This means that the shear stress corresponding to $\sigma_{12} = \tau_{xy}$ is neglected. In addition, the shear stresses corresponding to $\sigma_{13} = \tau_{xz}$ and $\sigma_{23} = \tau_{yz}$ are neglected by using the plane strain approximation. The plane strain approximation holds in a situation where the structure is free in the x and y directions but where the z strain is assumed to be zero. Moreover, according to above mentioned, B_{S0} doesn't directly contribute to B_m. We should substitute Eq. (8) into Eq. (3) and Eq. (4) to calculate effective refractive index of two fundamental polarization modes. From Eq. (5) we can get B_m with B_G and B_{S0} component.

IV. RESULTS AND DISCUSSION

According to Fig.1, we build a model of the MRCF to analyze its birefringence properties. Table 1 shows material parameters of the model, where n_{core} and n_{clad} is refractive index of Ge-doped core and silica cladding at λ=1550nm respectively, α_{Core} and α_{Clad} is thermal expansivity of core and cladding ($\alpha_{Core} = \alpha_{Clad} + 7.6 \times 10^{-5} \times m$, m=4% is Ge-doped mole fraction in core) respectively, E_{Core} and E_{Clad} is young's modulus of core and clad respectively, γ_{Core} and γ_{Clad} is Poisson's ratio of core and cladding respectively, ΔT is difference between operating temperature (20℃ room temperature) and reference temperature (1020 ℃ high temperature) during the drawing process of the MRCF. In addition, the length for rectangular cores is l=10μm, and the width is w=5μm. The circular silica cladding diameter of MRCF is R=125μm.

TABLE I.   MATERIAL PARAMETERS OF MRCF

n_{Core}	1.450029	ΔT	-1000℃
n_{Clad}	1.444024	E_{Core}, E_{Clad}	7830kg/mm^2
λ	1550nm	γ_{Core}, γ_{Clad}	0.186
α_{Core}	0.84×10^{-6}K^{-1}	C_1	0.7572×10^{-12}m^2/N
α_{Clad}	0.54×10^{-6}K^{-1}	C_2	4.1878×10^{-12}m^2/N

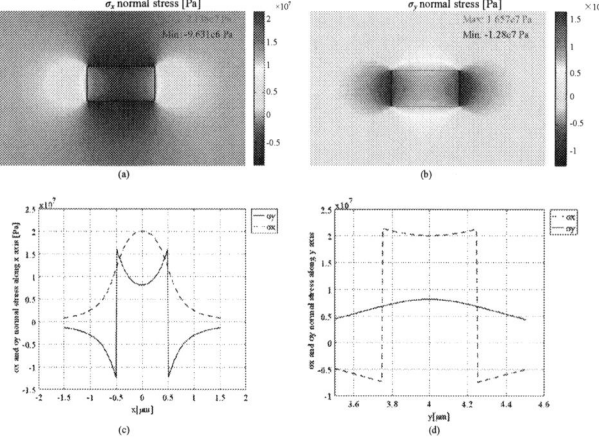

Figure 2.   Distribution of normal stress around core (a) σ_x (b) σ_y (c) cross-section plot along x axis (d) cross-section plot along y axis

To begin with, we respectively analysis the geometric birefringence and self-stress birefringence contribution to the mode birefringence at 1550nm wavelength. For geometry birefringence we got n^x_{eff}=1.446517, n^y_{eff}=1.446527 and B_G=-0.8×10^{-5}. Under considering the self-stress contribution to mode birefringence, a FEM method is initially use to calculate and plot distribution of inner thermal residual stress in MRCF. Fig.2 shows the distribution of normal stress σ_x and σ_y around rectangular core individually. The normal stresses σ_x and σ_y in core are about $(1.162 \sim 2.138) \times 10^7$ and $(0.683 \sim 1.657) \times 10^7$ Pa positive pulling stress respectively. There is about 4.8×10^6 Pa stress difference ($\sigma_x - \sigma_y$, $\sigma_x > \sigma_y$) in core. According to Eqs. (9), we obtain the self-stress birefringence of fiber core: $B_{S0} \approx 1.647 \times 10^{-5}$. In fact, B_{S0} distribution in core is no uniform, as shown Fig.3. Maximum is 4.981×10^{-5}, and minimum is -1.499×10^{-5}.

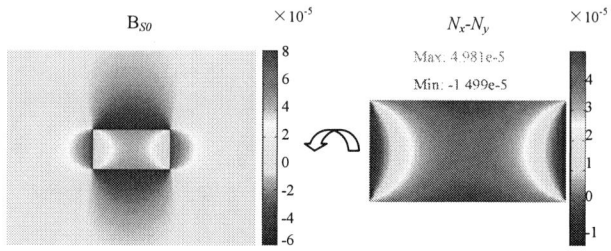

Figure 3.   Distribution of self-stress birefringence around core

Based on the result of self-stress distribution, the mode birefringence is estimated. The material index change induced by self-stress is got, and as an initial conditions we substitute it into Eq. (3) and Eq. (4) to calculate effective refractive index of two fundamental polarization modes. From Eqs. (5) we got B_m including B_G and B_{S0} component. Under considering B_{S0} the contribution to B_m, we got n^x_{eff}=1.446222, n^y_{eff}=1.446196 and B_m=-2.6×10^{-5}. Fig.4 shows the electric field distribution for two fundamental polarization modes HE_{11} parallel to x and y axes. By comparing them, we found that B_{S0} is a indispensable part of B_m, and this contribution is not negligible.

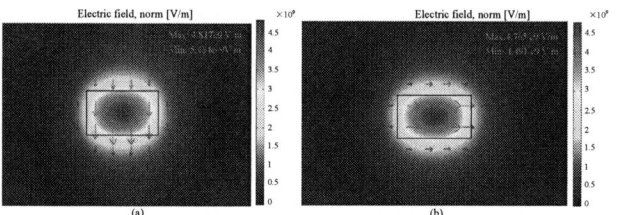

Figure 4.   The electrical field profiles of fundamental polarization modes along x and y axis respectively (a) x-polarization mode at 1550nm wavelength (b) y-polarization mode at 1550nm wavelengthθσμ

V. CONCLUSION

In conclusion, a kind of optical fiber with three quasi-rectangular cores is proposed and fabricated. Base on the finite element method, the theoretical model for the birefringence of MRCF is established. The birefringence characteristics are analyzed numerically at 1550nm wavelength. The mode birefringence is calculated to be $B_m > 10^{-5}$. Therefore, this rectangular cores fiber can be considered as a kind of high birefringence fiber. We expect that the MRCF has some potential applications in in-fiber interferometers with polarization maintaining.

ACKNOWLEDGMENT

This work is supported by the National Natural Science Foundation of China (grants 61205027). Supported by the 111 project (B13015), to the Harbin Engineering University. Partially supported by the Fundamental Research Funds for the Central Universities.

REFERENCES

[1] L. Yuan, J. Yang, Z. Liu, and J. Sun, "In-fiber integrated Michelson interferometer," Opt. Lett. **31**, 2692-2694 (2006).

[2] X. Daxhelet, J. Bures, R. Maciejko, "Temperature-independent all-fiber modal interferometer," Optical Fiber Technologgy, 1, 373-376 (1995).

[3] O. Duhem, J. F. Henninot, M. Douay, "Study of in fiber Mach-Zehnder interferometer based on two spaced 3-dB long period gratings surrounded by a refractive index higher than that of silica," Optics ommunications, 180, 255-262 (2000)

[4] L. B. Yuan, X. Wang, "Four-beam single fiber optic interferometer and its sensing characteristics," Sensors and Actuators 138, 9-15 (2007).

[5] L. B. Yuan, "Recent progress of multi-core fiber based integrated interferometers," in 2009 International Conference on Optical Instruments and Technology, Vol. 7508, 750802. (SPIE, 2009), pp. 1-8.

[6] Zhihai Liu, Fusen Bo, Lei Wang, Fengjun Tian, and Libo Yuan, "Integrated fiber Michelson interferometer based on poled hollow twin-core fiber," Opt. Lett. 36, 2435-2437 (2011).

[7] J.W. Arkwright, S. J. Hewlett, G. R. Atkins, and B. Wu, "High-isolation demultiplexing in bend-tuned twin-core fiber," J. Lightwave Technol. 14, 1740-1745 (1996).

[8] L. Yuan, J. Yang, and Z. Liu, "A compact fiber-optic flow velocity sensor based on a twin-core fiber michelson interferometer," IEEE Sens. J. 8, 1114-1117 (2008).

[9] O. Frazao, S. F. O. Silva, J. Viegas, J. M. Baptista, J. L. Santos, J. Kobelke, and K. Schuster, "All fiber Mach–Zehnder interferometer based on suspended twin-core fiber," IEEE Photon. Technol. Lett. 22, 1300-1302 (2010).

[10] A. D. Kersey, M. J. Marrone, and A. Dandridge, "Observation of input-polarization-induced phase noise in interferometric fiber-optic sensors," Opt. Lett. 13, 847-849 (1988).

[11] S. K. Sheem and T. G. Giallorenzi, "Polarization effects on single-mode optical fiber sensors," Appl. Phys. Lett. 35,914-917 (1979).

[12] D. W. Stowe, D. R. Moore, and R. G. Priest, "Polarizationfading in fiber interferometric sensors," IEEE J. Quantum Electron. 18, 1644-1647 (1982).

[13] Osakat, O., et al, "Stress analysis of optical fibers by a finite element method," IEEE J Quantum Elec, 17(10), 2123-2129, (1981).

[14] Stolenrh, "Calculation of stress birefringence in fibers by a finite element method," Journal of Lightwave Technology, 1(2), 297-301 (1983).

Coaxial Step Index Large Mode Area Fiber with low propagation loss

Souaci Farida, Li-Bo Yuan, Ya-Xian Fan[*]

Key Laboratory of Fiber Integrated Optics (MOE), Photonics Research Center, and College of Science,
Harbin Engineering University, Harbin, 150001, China
yxfan@hrbeu.edu.cn

Abstract—**A very simple coaxial step index optical fiber structure is introduced for large mode area (LMA) robustly-single-mode operation. Fundamental mode operation in the fiber is realized by the introduction of a narrow ring close to the core. Numerical results show that all the high order modes are coupled into the ring, while the fundamental mode dominates in the low numerical aperture core.**

Keywords- Coaxial step index structure, LMA fiber, low numerical aperture, narrow ring

I. INTRODUCTION

Over the last decade, high power fiber lasers have seen impressive progress in power-scaling. Multi-kilowatts of continuous-wave (CW) output power [1,2] and mega-watt levels of peak power in the ns-pulsed regime at 1.06 μm [3], with a high beam quality, have already been achieved in Yb-doped fiber lasers (YDFLs). However, major limitations in power scaling of fiber lasers are nonlinear effects in the forms of Raman, Brillouin and self-phase modulations. One efficient way to overcome these problems is using large-mode area (LMA) optical fiber. To ensure high beam quality, the LMA optical fiber should be single-mode guided. This is generally achieved by employing a multimode large-mode-area core design with a low numerical aperture (NA) in combination with bend-loss suppression of higher-order modes to promote single-mode operation [4, 5]. An alternative approach to achieve single-mode operation in a large core fiber is employing chirally-coupled-core (CCC) fiber structure. This structure provides efficient and highly selective coupling between higher order modes in the central straight core and the side-helix modes; and provide high loss for modes propagating in the helix cores, thus leading high loss on all coupled higher-order modes of the central core. A (CCC) fiber structure of 35 μm and 0.07 NA core has been designed and fabricated [6].

In this paper, we introduce a very simple design of LMA fiber with coaxial step index structure, which consists of a low numerical aperture large diameter core and a raised doped narrow ring close to the core. The single mode propagating in the large-core fiber is obtained by modulating the propagation loss difference between the fundamental mode (FM) and higher order modes (HOMs). It shows that the fiber can effectively operate in single mode at some or other desired wavelength (e.g. 1550 nm or 2020 nm) with a large effective mode area. As for other properties, such as polarization mode dispersion and chromatic dispersion, this coaxial fiber is alike to the conventional step-index fiber. It can also be flexibly

manufactured by conventional fiber manufacturing approach.

II. DESIGN OF FIBER

The fiber cross section and its refractive index profile are illustrated in Fig. 1. For example, the structure consists of a large diameter (42 μm, n_{core}=1.4771) core, a ring (width= 5 μm, distance to the core=3 μm, n_{ring} = 1.481) and cladding of index $n_{cladding}$ = 1.461. The value of numerical aperture of the fundamental mode at 1550 nm is 0.07.

We have simulated the propagation in the fiber and computed the modes using Beam Propagation Method (BPM).

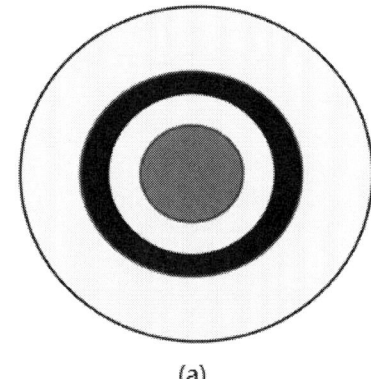

(a)

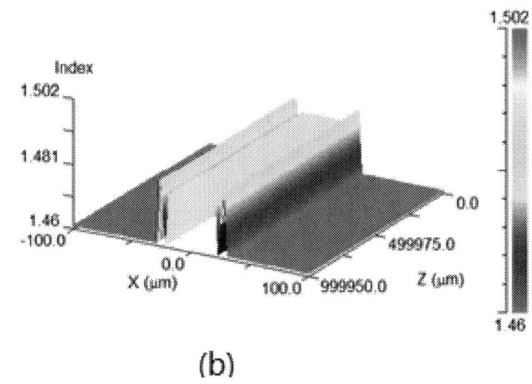

(b)

Figure 1. (a) Fiber cross-section of coaxial design, (b) Refractive index profile

We have simulated the propagation in the fiber and computed the modes using Beam Propagation Method (BPM). By proper choice of the ring width and its distance to the core, the higher order modes will undergo high rejection and propagate in the ring, allowing the fundamental mode to dominate in the core; which guarantee the single mode operation. Moreover, the tolerances for the absolute value of the ring refractive index are very tight. If the index of the ring is too high all the power is guided into this region. If the refractive index of the ring is too low, the ring has no impact on the core and the fiber is multimode.

III. . RESULTS AND DISCUSSIONS

The Beam Propagation Method (BPM) can predict from an incident field distribution within a structure. The main idea of this method is to divide a structure into "slices" elementary, spacing with ΔZ and then determine the scope of a given slice from the one before. However, the equations to solve are complex, which leads us to adopt certain approximations [7].

Fig. 2 and Fig. 3 show the computed fundamental and higher order modes respectively. The fundamental mode has a Gaussian-like mode field dominating in the fiber core, while the high order modes distribute in the narrow ring around the core.

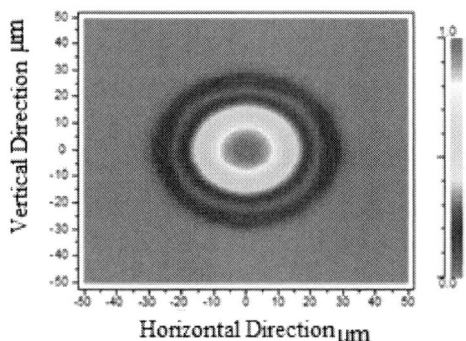

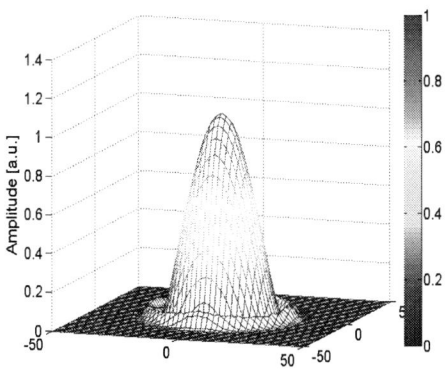

Figure 2.   Fundamental mode (FM)

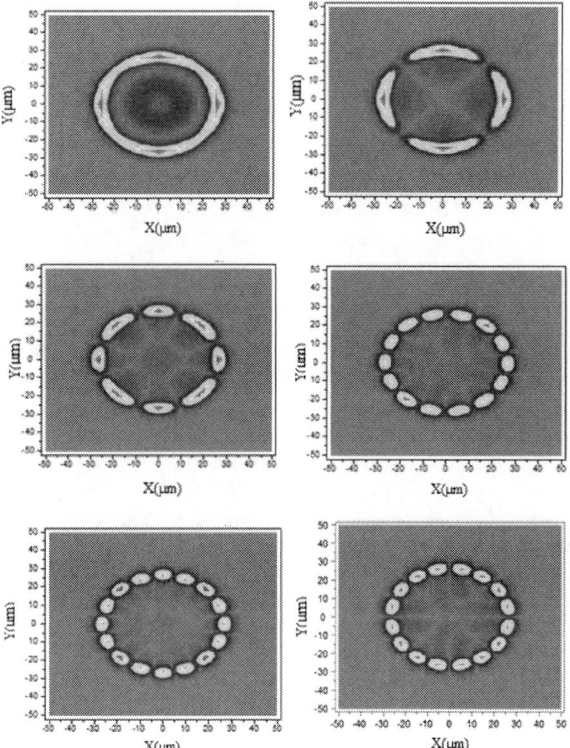

Figure 3.   Higher order modes (HOMs)

We used the imaginary distance BPM to compute the fiber modes. It should be noted that the imaginary distance BPM technique is formally equivalent to many other iterative mode solving techniques. The principle of this method, for simplicity, we consider 2D propagation of a scalar field, the incident field $\phi_{in}(x)$, can be expanded in the modes of the structure as

$$\phi_{in}(x) = \sum_m c_m \phi_m(x) \qquad (1)$$

The summation should consist of a true summation over guided modes and integration over radiation modes, but for brevity the latter is not explicitly shown. Propagation through the structure can then be expressed as

$$\phi(x,z) = \sum_m c_m \phi_m(x) e^{i\beta_m z} \qquad (2)$$

In each BPM-based mode-solving technique, the propagating field obtained via BPM is conceptually equated with the above expression to determine how to extract mode information from the BPM results. As the name implies, in the imaginary distance BPM the longitudinal coordinate z is replaced by z'=iz, so that propagation along

TABLE I.   PROPAGATION LOSS FOR FUNDAMENTAL MODE AND HIGHER ORDER MODES (DB/M)

Modes	FM	HOM1	HOM2	HOM3	HOM4	HOM5	HOM6
Propagation loss at 1550 nm (dB/m)	0.17	12.2	33.4	28.5	80	90	120
Propagation loss at 2020 nm (dB/m)	0.18	57.5	57.6	78.2	178.5	246.3	318.2

this imaginary axis should follow

$$\phi(x,z') = \sum_m c_m \phi_m(x) e^{\beta_m z'} \qquad (3)$$

The propagation implied by the exponential term in Eq.2 has become exponential growth in Eq.3, with the growth rate of each mode being equal to its real propagation constant. The essential idea of the method is to launch an arbitrary field, say a Gaussian, and propagate the field through the structure along the imaginary axis. Since the fundamental mode (m=0) has by definition the highest propagation constant, its contribution to the field will have the highest growth rate and will dominate all other modes after a certain distance, leaving only the field pattern . Consequently, the propagation constant can be obtained by the following expression

$$\beta^2 = \frac{\int \phi^* \left(\dfrac{\partial^2 \phi}{\partial x^2} + k^2 \phi \right) dx}{\int \phi^* \phi \, dx} \qquad (4)$$

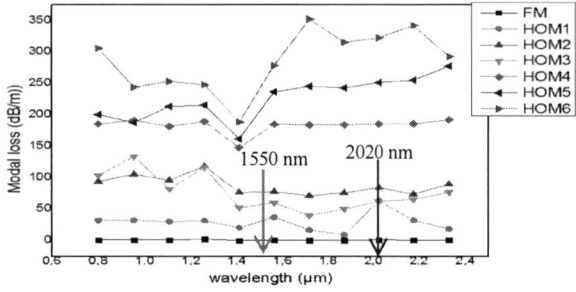

Figure 4.   Simulated modal loss using Beam Propagation Method

Higher order modes can be obtained by using an orthogonalization procedure to subtract contributions from lower order modes while performing the propagation [8]. An additional correction is added which removes the error due to the fact that we have solved for the eigenvalues of the paraxial equation. It is important to note that the imaginary distance BPM is not the same as the common technique of performing a standard propagation and waiting for the solution to reach steady state. The latter will only obtain the fundamental mode if the structure is single mode, and generally takes longer to converge. The imaginary distance BPM is closely related to the shifted inverse power method for finding eigenvalues and eigenvectors of a matrix.

In order to explore the single-mode operation, we need to compute the fiber propagation loss for FM and HOMs over a wide range of wavelengths; the results are presented in Fig. 4.

According to table 1, at 1550 nm all higher order modes have high loss (from >10 dB/m to >100 dB/m), while the predicted LP01mode loss is ~0.1 dB/m, i.e. expected fiber performance is effectively single-mode. Furthermore we have calculated the effective mode area and find that it is around 1018 μm^2, which represents an upper limit for coaxial step index fibers. And at 2020 nm, the higher loss (>57 dB/m) of high order modes have been obtained.

IV.   CONCLUSION

In conclusion, we have confirmed by numerical simulation a simple design of coaxial step index LMA fiber that effectively supports only a fundamental mode with core diameter of 42 µm (corresponding MFD at 1550-nm is 36 µm). The passive ring in cladding permits high suppression of HOMs. Such should be easy to fabricate and is expected to find applications in high power fiber lasers and amplifiers application [9-10], due to its effectively single-mode nature and large mode area of 1018μm^2.

ACKNOWLEDGMENT

This work was supported by the National Natural Science Foundations of China, under grant numbers 11074121 and 61290314, and partially supported by the 111 project (B13015), to Harbin Engineering University.

REFERENCES

[1] Jeong, Y., Sahu, J. K., Payne, D. N., and Nilsson, J., "Ytterbium-doped large-core fiber laser with 1.36 kW continuous-wave output power," Optics Express, 12: 6088 (2004).

[2] Jayanta K. Sahu ; Seongwoo Yoo ; Alexander J. Boyland ; Andrew S. Webb ; Mridu Kalita ; Jean-Noel Maran ; Yoonchan Jeong ; Johan Nilsson ; W. Andrew Clarkson ; David N. Payne "Fiber Design for high-power fiber lasers", Proc. SPIE 7195, Fiber Lasers VI: Technology, Systems, and Applications, 71950I (2009).

[3] Limpert, J., Schreiber, T., Liem, A., Nolte, S., Zellmer, H., and Tünnermann, A., "Megawatt peak power level fiber laser system based on compression in air-guiding photonic bandgap fiber," in Advanced Solid-State Photonics, OSA Technical Digest (Optical Society of America, 2004), paper MD1.

[4] N. G. R. Broderick, H. L. Offerhaus, D. J. Richardson, R. A. Sammut, J. Caplen, and L.Dong "Large Mode Area Fibers for High Power Applications", Optical Fiber Technology, 5:185-196 (1999).

[5] M. Y.Chen and Y.K.Zhang, "Bend insensitive design of large mode area microstructured optical fibers,"J. Lightw. Technol., 29: 2216 (2011).

[6] Chi-Hung Liu, Guoqing Chang, Natasha Litchinitser, and Almantas Galvanauskas, Doug Guertin, Nick Jabobson, Kanishka Tankala, "Effectively Single-Mode Chirally-Coupled Core Fiber", Conference Paper, Advanced Solid-State Photonics Vancouver, Canada January 28, (2007).

[7] M. D. Feit and J. A. Fleck, Jr, "Light propagation in gradedindex optical fibers", Applied Optics, Vol. 1 7, Issue 24, pp. 3990-3998 (1978)

[8] J.C. Chen and S. Jungling, "Computation of higher-order waveguide modes by the imaginary-distance beam propagation method", Optical and Quantum Electron. 26, S199 (1994).

[9] P. Kadwani, C. Jollivet, R. A. Sims, A. Schülzgen, L. Shah, and M. Richardson, "Comparison of higher-order mode suppression and Q-switched laser performance in thuliumdoped large mode area and photonic crystal fibers", OPTICS EXPRESS, 20(22): 24295 (2012).

[10] F. Jansen, F. Stutzki, C. Jauregui, J. Limpert, and A. Tünnermann, "High-power very large mode-area thulium-doped fiber laser", Optics Letters, 37(21): 4546, (2012).

The calculation of doped vanadium dioxide thin films

Xue-song Tian[13*], Qi Wang[2], Jian-feng Sun[2], Zhi-gang Fan[3]

1 College of Sciences, Heilongjiang University of Science and Technology, Harbin, China
2 National Key Laboratory of Science and Technology on Tunable laser, Harbin Institute of Technology, Harbin, China
3 Postdoctoral Research Station of Optical Engineering, Harbin Institute of Technology, Harbin, China
hit218@yeah.net

Abstract—**The vanadium oxide thin films are prepared on zinc selenide by DC magnet sputtering method. The components are gotten using The X-ray photoelectron spectroscopy (XPS). It's phase transition temperature can be changed by doping. Utilizing the Castep program package of the Material Studio simulation tool, based on local density function approximation and pseudo-potential method, optimization for the geometric structure of vanadium dioxides is accomplished with the BFGS calculate way. The phase transition temperature of the vanadium oxide thin films doped with Cr and Al is higher; and doped with W and F is lower.**

Keywords-component; vanadium dioxide; phase transition; doping

I. INTRODUCTION

The material vanadium dioxide(VO_2) undergoes a first order phase transition, from a high temperature metallic phase to a low temperature semiconducting phase. The underlying atomic and electronic transformations remain a topic of debate. The critical phase transition temperature, T_c, was first reported in 1959 [1] about 341 K. It's optical and electrical character changed abruptly after phase transition. The critical temperature, T_c, is also tuneable over a wide range by doping. Doped with different element changed the T_c higher or lower. It has broad applied foreground at laser protection, temperature sensor and optics storage[2-7].

The optical properties of vanadium dioxide thin films was calculated by utilizing the Castep program package of the Material Studio simulation tool. Based on the dynamic simulation process in Castep module, the electronic structure (energy band and state density), optical characteristics of crystal and the influence of components on phase-change temperature are calculated. doped with Cr Al W and F's affect on the phase transition temperature of the vanadium oxide thin films is also calculated.

II. EXPERIMETAL SYSTEM

The thin films were prepared on zinc selenide(ZnSe) by DC magnet sputtering. The V target's diameter is 60 mm and its thickness is 3 mm, the sputtering temperature was 450□, the distance between target and ZnSe is 79 mm. The other preparation parameters are: pressure before preparation 3.8×10^{-4} Pa, ratio of $O_2/Ar=1.8/16.3=0.11$(SCCM), pressure when sputtering 2.2 Pa, sputter current 0.5 A, sputter voltage 330 V, sputter power 165 W, sputter time 180 s. The thickness of the VO_2 is about 125 nm, measured by PGI1240 profile meter[8].

The spectral transmittances of the films were measured in the range 2.5 to 25 μm using a Nicolet 8700 IR spectrometer. The Nicolet 8700 IR spectrometer has heating accessories, the sample's temperature can be controlled; it also can decrease temperature using liquid nitrogen. In the optical character test before and after phase transition, using liquid nitrogen first, set the film's temperature at 5℃, begin measure. Then heating, waiting several minutes after the temperature raised 1℃ to make the film's temperature stably, rise the temperature and test the transmittance until phase transition. The transmittance changing of vanadium oxide make a whole Hysteresis cycle. Sample 1's transmittance from 5□ to 70□ at wavelength 2.5 μm~25 μm is shown in Fig.1. ZnSe base affect aside, at 10.6 μm (wavenumber 943), the transmittance is 73% at 5□, 12% at 68□. We can see from the data that the phase transition occur, the transmittance changed much, and the phase transition temperature is lower than 68□.

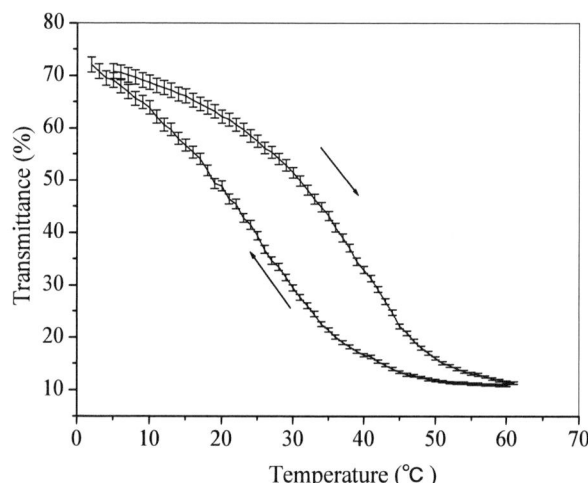

Figure 1. Transmittance of vanadium oxide at 10.6 μm before and after phase transition

III. CALCULATION

The material's forbidden band width is decided by its highest occupied molecular orbital(HOMO) and lowest unoccupied molecular orbital (LUMO). We can analysis the vanadium oxide's HOMO and LUMO to find the affect of doping. The calculation of HOMO and LUMO using module DMo 13. Figure 2 is the energy curve of vanadium oxide crystal model optimization. In the process of structure optimization, the energy of the system is significantly lower, and eventually reach a steady state.

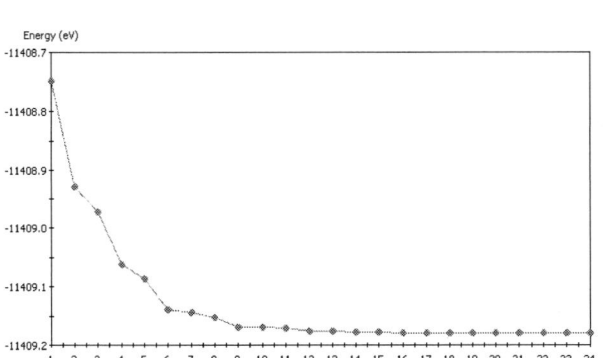

Figure 2. Vanadium oxide crystal model optimization energy curve

It is known by the valence band and the conduction band computed result, at different temperature, when the temperature is higher, the vanadium oxide's energy difference between the valence band and the conduction band is reduced. The higher temperature made the electron energy of the O's 2p orbital offset to conduction band. The V's orbital electron energy changed little.

The vanadium oxide's phase transition temperature can be changed by doping different element. Doping high valent state element can reduce the phase transition temperature; and doping low valent state element can increase the phase transition temperature. But study also show that doping can reduce the amount of change before and after phase transition.

To test and verify the stability of different doping structure, the structure's binding energy of gap doping and replace doping are all calculated. Table 1 is W's binding energy of gap doping and replace doping, it can be seen from the calculation, the binding energy of replace doping is higher than gap doping. The stability of replace doping is better than gap doping.

TABLE I.   TABLE I TUNGSTEN BINDING ENERGY OF GAP DOPING AND REPLACE DOPING

Doping methods	Binding energy （kJ/mol）
W2-V replace dope	-8821.879
W2-V gap dope	-9711.487
W1-V replace dope	-8953.678
W1-V gap dope	-9410.5

It can be seen from Figure 3. The VO_2 valence band and conduction band electronic density of states change curve before and after W doping. After W doping, the impurity energy level is introduced to the VO_2 valence band and conduction band, that make the energy difference between VO_2 valence band and conduction band decreased, reduces the activation energy in the band gap.

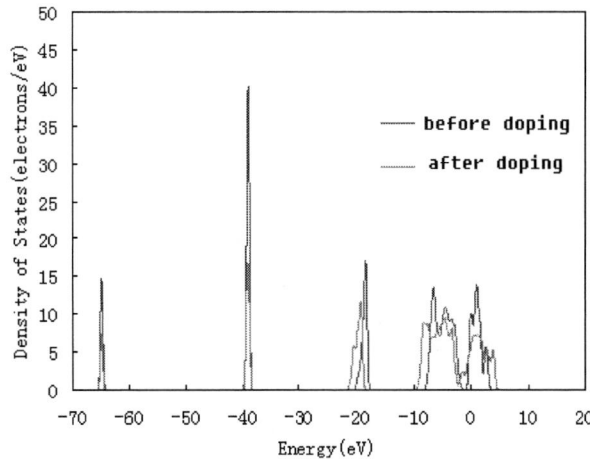

Figure 3.   The VO_2 valence band and conduction band electronic density of states change curve before and after W doping

It can be seen from the band gap changing before and after doping, the VO_2 doped element with low valence, as Cr and Al, band gap becomes bigger, can increase the vanadium oxide's phase transition temperature. the VO_2 doped element with high valence, as W and F, band gap becomes smaller, can decrease the vanadium oxide's phase transition temperature.

TABLE II. VO_2 ENERGY GAP CHANGING FEFORE AND AFTER DOPING

Dope	Valence band top （eV）	Conduction band bottom （eV）	Energy gap (eV)
VO_2	-1.774	-0.671	1.103
Al1	-1.473	-0.361	1.112
Cr1	-1.451	-0.325	0.826
F-1	-2.014	-1.028	0.986
W1	-1.761	-1.118	0.643
W2	-0.614	-0.492	0.122

IV. CONCLUSION

Vanadium oxide thin films is prepared on ZnSe by magnetron sputtering method. A spectral transmission study has been made from 2.5 μm to 25 μm. Then the transmission is calculated by utilizing the Castep program package of the Material Studio simulation tool. Doping with different element were calculated. The phase transition temperature of the vanadium oxide thin films doped with Cr and Al is higher; and doped with W and F is lower.

REFERENCES

[1] F.J.Morin, "Oxides Which Show a Metal-to-Insulator Transition at the Neel Temperature," Phys.Rev.Lett, vol. 3, No1, pp. 34-36, 1959

[2] Marvel, R. E., Appavoo, K, Choi, B. K et al, "Electron-beam deposition of vanadium dioxide thin films," APPLIED PHYSICS A-MATERIALS SCIENCE & PROCESSING, vol. 111, No3, pp. 975-981 2013

[3] Bonora, S, Beydaghyan, G, Hache, A et al, "Mid-IR laser beam quality measurement through vanadium dioxide optical switching," OPTICS LETTERS, vol. 38, No9, pp. 1554-1556, 2013

[4] Pergament, A, Stefanovich, G, Berezina, O et al, "Electrical conductivity of tungsten doped vanadium dioxide obtained by the sol-gel technique," THIN SOLID FILMS , vol. 531, pp. 572−576, 2013

[5] Allogho, Guy-Germain, Hamam, Habib, Beydaghyan, Gisia et al, "Continuously variable, electrically addressed beam splitter based on vanadium dioxide," APPLIED OPTICS, vol. 52, No2, pp. 241-247, 2013

[6] Bo Chen, DongfangYang, PaulA.Charpentier et al, "Al3+-doped vanadium dioxide thin films deposited by PLD," Solar Energy Materials & Solar Cells, vol. 93, pp. 1550–1554, 2009

[7] Agafonova, D. S, Grunin, V. K, Sidorov, A. I, "Attenuation modulation of guided modes in optical fibers with a coating based on vanadium dioxide," JOURNAL OF OPTICAL TECHNOLOGY, vol. 80, No1, pp. 1-6, 2013

[8] X. S. Tian, J. C. Liu, Q. Wang, "Components effect on vanadium oxide thin films phase transition character phenomenon observe," Laser Physics, vol. 18, No10, pp. 1207~1211, 2008

Linear polarization conversion in planar chiral metamaterial

Yiqun Xu, Xingchen Liu, Zheng Zhu, Zhengping Wang, Jinhui Shi

Key Laboratory of In-Fiber Integrated Optics of Ministry of Education, College of Science, Harbin Engineering University,
Harbin, China
shijinhui@hrbeu.edu.cn

Abstract: –We theoretically and numerically propose a kind of planar chiral metamaterial which consists of two layers of connected I-shape resonators arranged by a twist angle of 90°. Numerical simulation results show that our scheme can realize a polarization conversion for linearly polarized waves.

Keywords-component; Chiral metamaterials; resonators; polarization conversion

I. INTRODUCTION

Chiral metamaterials has attracted lots of attention since they can easily manipulate wave polarization [1-2]. Manipulating wave polarization is desired to achieve many devices like circular and linear polarizers [3-5], polarization rotators [6-7] and polarization spectrum filters [8-9]. Optical activity is one of the most distinguished properties in artificial chiral metamaterials that was widely studied [10-14]. In addition, asymmetric transmission of polarized light has been reported in chiral metamaterials. Asymmetric transmission in the chiral metamaterial is usually caused by the partial polarization conversion of the incident EM waves into one of the opposite polarization, which is asymmetric for the opposite directions of propagation [15]. Recently, several planar chiral metamaterial have been proposed in order to rotate the polarization plane of linearly polarized waves [16-19]. The chiral metamaterial was reported to reveal a dual-band AT effect for two orthogonal linearly polarized waves [19], distinct from the metamaterials previously reported that only show a single-band AT effect for one polarization. This phenomenon can also be explained by the de Hoop reciprocity as revealed by Jones matrix formulation [18, 20]. In the transfer-matrix representation of an optical system, the matrix for reverse transmission is the transpose of that of forward transmission. The system will give rise to asymmetric transmission of polarized light on condition that the system is chiral and anisotropic.

In this work, we propose a kind of chiral metamaterials which is composed of two layers of connected I-shape resonators arranged by a twist angle of 90°.The proposed scheme can perfectly achieve a conversion for the linearly polarized wave. Furthermore, the results show that the azimuth angle of the electromagnetic wave strongly depends on geometric parameters. We believe that our approach can efficiently modulate the polarization states.

II. POLARIZATION TRANSFORMER

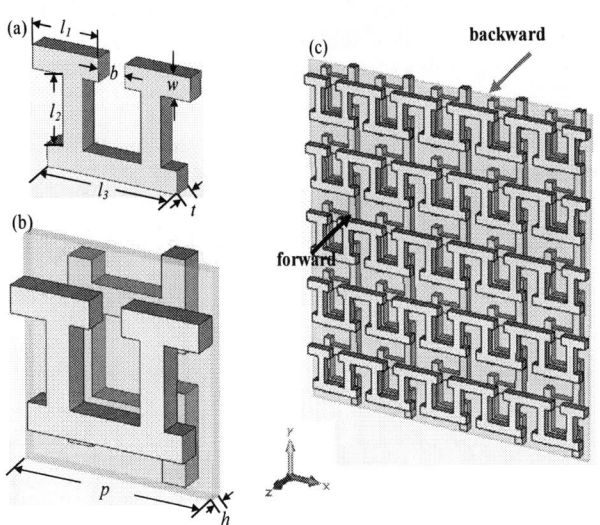

Figure 1. (a) Schematic of the single connected I-shape layer (b) Schematic of the unit cell in chiral metamaterial. (c) Metamaterial orientation with respect to the incident wave.

Figure 1(a) illustrates our proposed chiral metamaterials. The unit cell is composed of connected I-shape structure and a sandwiched dielectric spacer layer. The metallic layer on either side has the same pattern, but they are rotated with respect to each other by 90° around the z axis. In the theoretical model, the sandwiched dielectric layer is chosen as air, with a thicknesses of h=30nm. The connected I-shape structure is gold with the thickness of t=50nm. Other parameters are l_1=165nm, l_2=150nm, l_3=330nm and b=70nm. Figure 1(b) depicts one unit cell of our proposed structure that its dimension is 400×400 nm^2. Figure 1(c) show the metamaterial orientation with respect to the incident wave. We applied **T** matrice to analytically solve the polarization rotation problem of the present chiral metamaterials [17]. **T** matrice is the frequency-dependent Jones matrix, which depicts the complex amplitudes of the incident to the transmitted field:

$$\begin{pmatrix} t_x \\ t_y \end{pmatrix} = \begin{pmatrix} T_{xx} & T_{xy} \\ T_{yx} & T_{yy} \end{pmatrix}\begin{pmatrix} i_x \\ i_y \end{pmatrix} = \begin{pmatrix} A & B \\ C & D \end{pmatrix}\begin{pmatrix} i_x \\ i_y \end{pmatrix} = \mathbf{T}_{\text{lin}}^{f}\begin{pmatrix} i_x \\ i_y \end{pmatrix} \quad (1)$$

For convenience, we replaced T_{ij} by A, B, C, D. The superscript f indicates the forward propagation (along -z direction) and the subscript lin indicates the linear base (the base vectors are parallel to the coordinate axes). Then the transmission matrix for the propagation in the backward direction (+z direction) can be expressed as

$$\mathbf{T}_{\text{lin}}^{b} = \begin{pmatrix} T_{xx} & -T_{yx} \\ -T_{xy} & T_{yy} \end{pmatrix} \qquad (2)$$

The **T** matrix for a circular polarization base can be described as

$$
\begin{aligned}
\mathbf{T}_{\text{cir}}^{f} &= \begin{pmatrix} T_{++} & T_{+-} \\ T_{-+} & T_{--} \end{pmatrix} = \begin{pmatrix} a & b \\ c & d \end{pmatrix} \\
&= \frac{1}{2}\begin{pmatrix} A+D+i(B-C) & A-D-i(B+C) \\ A-D+i(B+C) & A+D-i(B-C) \end{pmatrix}
\end{aligned} \qquad (3)
$$

where we also replace the entries $T_{+/-}$ by a, b, c, d. The subscripts + and − represent the left-handed and right-handed polarized waves.

The azimuth rotation parameter ψ for the linear wave is defined as

$$\psi_x^b = -\frac{1}{2}\left[\arg a - \arg d\right] \qquad (4)$$

When a≠d, the optical activity will occur in the proposed double-layer structure.

This chiral structure has a C_2 symmetry with respect to the x or y axis. The T matrix obeys the form

$$T^{f} = \begin{pmatrix} A & B \\ -B & D \end{pmatrix} \qquad (5)$$

$$\mathbf{T}_{\text{cir}}^{f} = \frac{1}{2}\begin{pmatrix} A+D+2iB & A-D \\ A-D & A+D-2iB \end{pmatrix} \qquad (6)$$

In contrast to Eqs.(4), if T++•T--, optical dichroism appears in this structure.

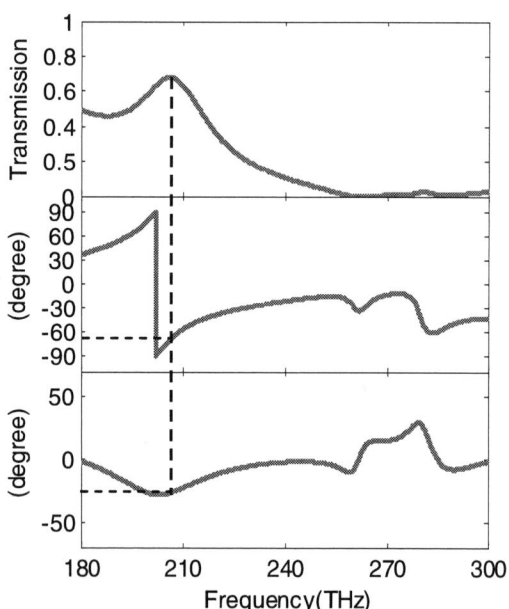

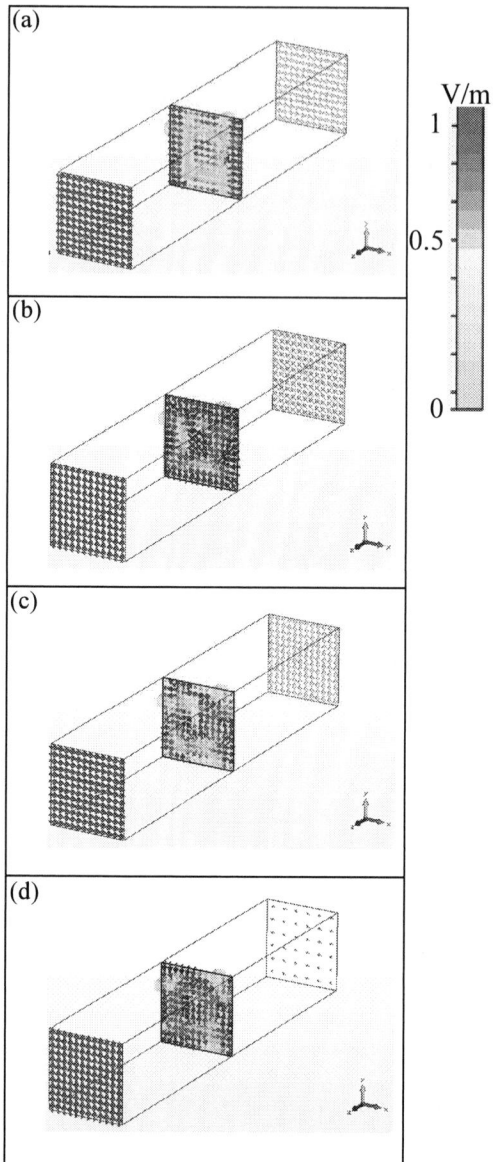

Figure 2. Calculated transmission coefficients (a), polarization rotation azimuth angle ψ (b) and ellipticity χ (c) of the transmitted wave when the incident wave is x-polarized.

Figure 3. The resonant modes for the linearly polarized wave at the resonant frequency.

III. SIMULATED RESULTS

We demonstrate the polarization rotation of the bilayered metamterial by using CST commercial simulation software. The periodic boundary condition is applied in the simulation. Usually the incident waves are *x*-polarized, propagating along the -z direction. The linear transmission coefficient is defined as $T = \left(\left|E_x^{\text{out}}\right|^2 + \left|E_y^{\text{out}}\right|^2\right)\Big/\left|E_x^{\text{in}}\right|^2$, shown in Figure 2(a).

978-1-4799-1215-5/13 $31.00 © 2013 IEEE

Figure 2(b) and 2(c) show the azimuth angle ψ and the ellipticity χ, which are defined as

$$\psi = \frac{1}{2}\tan^{-1}\left[\frac{2|t_{yx}|\cos\varphi}{|t_{xx}|\cdot(1-|t_{yx}|^2/|t_{xx}|^2)}\right] \quad (7)$$

$$\chi = \frac{1}{2}\sin^{-1}\left[\frac{2|t_{yx}|\sin\varphi}{|t_{xx}|\cdot(1-|t_{yx}|^2/|t_{xx}|^2)}\right] \quad (8)$$

Figure 2(a) show that the maximum transmission coefficient T is 0.68 at around 206.3THz, while the azimuth angle ψ and the ellipticity χ are around -60.99° and -26.65°, shown in Figure 2(b) and Figure 2(c). In order to observe the formation of polarization rotation, we have analyzed the electromagnetic wave evolution when propagating through the chiral metamaterial at about 204.3 and 260.7 THz. Figure 3 shows an electric field distribution for an *x*-polarized wave and *y*-polarized wave passing through the slab along the -z direction (The transmission of y-polarized incident wave is not shown in Fig.2). When the *x*/y-polarized wave incidents along the -z direction at about 204.3 THz [Fig. 3(a) and 3(b)], the fundamental electromagnetic resonance can be excited with a strong electromagnetic wave being coupled into the slab, and then the near-field coupling between the first and second layers results in an obvious transmitted wave, but the polarization direction has been rotated. In contrast, with a *y*-polarized wave along the -z direction at 260.7 THz [Fig. 3(d)], the metamaterial cannot be well excited due to its orientation. Meanwhile, *x*-polarized wave along the -z direction at 260.7THz can be converted to its cross-polarization. Therefore, the optical chiral metamaterial can achieve a dual-band polarization conversion for two orthogonal linearly polarized waves.

IV. CONCLUSIONS

To summarize, we proposed a chiral metamaterial constructed by two layers of connected I-shape resonators. The unit cells of two layers are arranged by a twist angle of 90°. The metamaterial can realize a dual-band polarization conversion for linearly polarized waves. Our findings are beneficial in designing polarization control devices.

ACKNOWLEDGMENT

The authors acknowledge financial support by the National Science Foundation of China under Grant Nos. 61201083, 61275094, U1231201, in part by the Natural Science Foundation of Heilongjiang Province in China under Grant No. LC201006, the China Postdoctoral Science Foundation under Grant Nos. 2012M511171 and 2013T60487, the Special Foundation for Harbin Young Scientists under Grant No. 2012RFLXG030, the Fundamental Research Funds for the Central Universities, and the 111 Project under Grant No. B13015.

REFERENCES

[1] M. Schäferling, D. Dregely, M. Hentschel, and H. Giessen, "Tailoring enhanced optical chirality: design principles for chiral plasmonic nanostructures," Phys. Rev. X, 2012, vol 2, pp.031010.

[2] E. Plum, X.-X. Liu, V. Fedotov, Y. Chen, D. Tsai, and N. Zheludev, "Metamaterials: optical activity without chirality," Phys. Rev. Lett., 2009, vol 102, pp. 113902.

[3] J. K. Gansel, M. Thiel, M. S. Rill, M. Decker, K. Bade, V. Saile, G. von Freymann, S. Linden, and M. Wegener, "Gold helix photonic metamaterial as broadband circular polarizer." Science, 2009, vol 325, pp. 1513–1515.

[4] Y. Zhao, M. A. Belkin, and A. Alù, "Twisted optical metamaterials for planarized ultrathin broadband circular polarizers." Nat. Commun., 2012, vol 3, pp. 870.

[5] J. Y. Chin, M. Lu and T. J. Cui, "Metamaterial polarizers by electric-field-coupled resonators," Appl. Phys. Lett., 2008, vol 93, pp. 251903.

[6] Y. Q. Ye and S. L. He, "90 degree polarization rotator using a bilayered chiral metamaterial with giant optical activity," Appl. Phys. Lett., 2010, vol 96, pp. 203501.

[7] M. Mutlu and E. Ozbay, "A transparent 90° polarization rotator by combining chirality and electromagnetic wave tunneling," Appl. Phys. Lett., 2012, vol 100, pp 051909.

[8] N. I. Zheludev, E. Plum, and V. A. Fedotov, "Metamaterial polarization spectral filter: isolated transmission line at any prescribed wavelength," Appl. Phys. Lett., 2011, vol 99, pp. 171915.

[9] J. H. Shi, H. F. Ma, W. X. Jiang, and T. J. Cui, "Multiband stereometamaterial-based polarization spectral filter," Phys. Rev. B, 2012, vol 86, pp. 035103.

[10] T. Q. Li, H. Liu, T.Li, S. M. Wang, F. M. Wang, R. X. Wu, P. Chen, S.N. Zhu and X. Zhang, "Magnetic resonance hybridization and optical activity of microwaves in a chiral metamaterial," Appl. Phys. Lett., 2008, vol 92, pp. 131111.

[11] M. Decker, R. Zhao, C. M.Soukoulis, S. Linden and M. Wegener, " Twisted split-ring-resonator photonic metamaterial with huge optical activity," Opt. Lett., 2010, vol 35, pp. 1593.

[12] M. Decker, M. W. Klein, M. Wegener and S. Linden, "Circular dichroism of planar chiral magnetic metamaterials," Opt. Lett., 2007, vol 32, pp. 856.

[13] S. V. Zhukovsky, A. V. Novitsky and V. M. Galynsky, "Elliptical dichroism: operating principle of planar chiral metamaterials," Opt. Lett., 2009, vol 34, pp. 1988.

[14] S. Engelbrecht, M. Wunderlich, AM. Shuvaev and A. Pimenov, " Colossal optical activity of split-ring resonator arrays for millimeter waves," Appl. Phys. Lett., 2010, vol 97, pp. 081116.

[15] V. A. Fedotov, P. L.Mladyonov, S. L. Prosvirnin, A. V. Rogacheva,Y. Chen and N. I. Zheludev, "Asymmetric propagation of electromagnetic waves through a planar chiral structure," Phys.Rev.Lett., 2006, vol 97, pp. 167401.

[16] C. Menzel, C. Helgert, C. Rockstuhl, E. B. Kley, A. Tunnermann, T. Pertsch and F. Lederer, "Asymmetric transmission of linearly polarized light at optical metamaterials," Phys. Rev. Lett., 2010, vol 104, pp. 253902.

[17] C. Huang, Y. J. Feng, J. M. Zhao, Z. B Wang and T. Jiang, "Asymmetric electromagnetic wave transmission of linear polarization via polarization conversion through chiral metamaterial structures," Phys Rev B., 2012, vol 85, pp. 195131.

[18] J. Han, H. Q. Li, Y. C. Fan, Z. Y. Wei, C. Wu, Y. Cao, X. Yu, F. Li and Z. S. Wang, "An ultrathin twist-structure polarization transformer based on fish-scale metallic wires," Appl. Phys. Lett., 2011, vol. 98, pp. 151908.

[19] J.H. Shi, X.C. Liu, S.W. Yu, T.T. Lv, Z. Zhu, H.F. Ma and T.J. Cui. "Dual-band asymmetric transmission of linear polarization in bilayered chiral metamaterial," Appl. Phys. Lett.,2013, vol 102, pp. 191905.

[20] R. J. Potton, "Reciprocity in optics," Rep. Prog. Phys., 2004, vol 67, pp. 717.

Beam cleanup of 20kW peak power laser pulses by SBS in 105μm large core diameter fiber with high beam quality ($M^2 \approx 1.5$)

C.Y. Zhu*, J.H. Zhang, Y.B. Yuan, D.X. Ba, J. Yan, Q.L. Gao and Z.W. Lu
Harbin Institute of Technology,
Harbin, China
zhuchy@hit.edu.cn

Abstract—**Stimulated Brillouin Scattering in multimode fiber leads to the event of beam cleanup, which has attractive potential applications in improving the beam quality of multimode fiber laser with large core diameter and high peak power. We report the experimental setup and result that using Stimulated Brillouin Scattering in 105μm core diameter GI fiber to achieve beam cleanup when the 20kW peak power pulsed pump is provided, and getting a high-quality output with $M^2 \approx 1.5$.**

Keywords- beam cleanup; SBS; 105μm GI multimode fiber

I. INTRODUCTION

Nowadays, along with increasing requirement of energy and peak power in fiber laser and amplifier field, the core diameter of transmission fiber becomes larger and larger. However, the beam quality in large core diameter fiber cannot be well guaranteed by existing technology. As the core diameter increases, the higher-order modes will increase rapidly in graded-index (GI) fiber. Through traditional way (e.g. winding method, etc.), the best available M^2 factor of output can reach less than 1.1 when the core diameter is 65μm, but if it grows to more than 100μm, the M^2 factor is difficult to drop to 5 or less, which has been become the important factors of limiting practical development in high power fiber field [1].

Beam cleanup might be a very promising new way to solve this problem by using nonlinear effects to realize in fiber. In 1993, H. Bruesselbach firstly reported that Stimulated Brillouin Scattering (SBS) in multimode fiber leads to the event of beam cleanup [2], which can purify multimode beam to fundamental beam, and several follow-up studies was evoked [3-8]. Lombard and Brignon achieved beam cleanup in GI multimode fiber by using a self-aligned Brillouin cavity, under the conditions of 65μm core diameter and 150W peak power, in 2006. Their work provides an important reference for deepening research of this technology [5]. This paper will show our study in larger core diameter and higher peak power conditions, which leads to the result that a beam composed by higher-order modes is purified to low-order mode output with $M^2 \approx 1.5$.

II. EXPERIMENTAL SETUP

Experiment is carried out by using a pulsed laser. Along with introducing a seed, Stimulated Brillouin amplification happens in GI fiber and reflects a Stokes pulse, which has the characteristic of beam cleanup. Our experimental setup is shown in Figure 1.

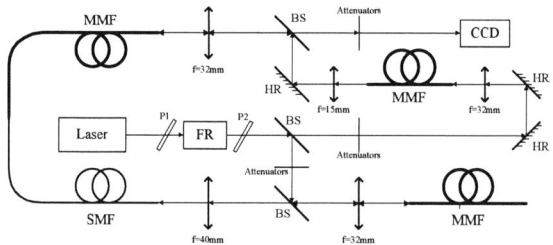

Figure 1. Experiment setup. MMF is multimode fiber; SMF is single-mode fiber; FR is Faraday isolators; BS is beamsplitter; HR is high reflection mirror.

A Q-switched pulsed Nd:YAG laser is provided with 1064μm wavelength, about 8ns pulse width, 10mJ single pulse energy, 1Hz repetition frequency and Gaussian beam output. After the light source, there is a Faraday isolators using for prevention of danger caused by reflected beam. The laser output is divided into two parts. One part constitutes the pump and another generates the seed. Through their interaction by Stimulated Brillouin amplification, the Stokes beam is brought out then be measured in CCD, obtaining its near-field distribution.

Stimulated Brillouin amplification requires the pump and Stokes to have specific frequency shift, which means an independent seed-produced component is necessary as the laser cannot be both pump and seed. So, one part of original laser is introduced into MMF1 previously, and a Stokes beam, which has a frequency shift, is reflected through self SBS. It is appropriate to be the seed with the specific frequency shift and then will take part in amplification process happened in MMF2, only if the MMF1 and MMF2 are identical in type.

978-1-4799-1215-5/13 $31.00 © 2013 IEEE

If the Stokes beam reflected from MMF1 is coupled into MMF2 through lens, using for pump, the experiment will not successfully develop since the mode excitation effect exists in multimode fiber. The mode excitation effect means that in multimode fiber as long as the position where a light enter the fiber in surface changes, the excited model groups change too. The further the location away from fiber's center, the messier the excited model groups shows [9]. It cannot be guaranteed that the laser enters into fiber's center coupled through lens, and at the same time, it's susceptible to external influences. All of these lead to a result that the seed cannot be good and stable. So as to get a stable seed with low-order mode, we make a core-aligned welding between single-mode fiber and multimode fiber in view that single-mode fiber has slim core diameter and allows only one mode spreading. Welding ensures the stable excitation further core-aligned welding guarantees that the Stokes beam reflected from MMF1 could excite a low-order mode in MMF2 through coupling into SMF. As a result, we get a good seed shown in Figure 2(a).

It's not conducive to observe beam cleanup if the original laser is used as pump involved in Stimulated Brillouin amplification directly because of its Gaussian distribution. Thus we let the original laser pass through MMF3 to damage its beam quality then get a bad pump composed by messy high-order modes shown in Figure 2(b) due to the mode excitation effect. If a good Stokes beam can be obtained after Stimulated Brillouin amplification, it can be confirmed that beam cleanup effect occurs.

In this experimental setup of Figure 1, MMF1 and MMF2 are GI multimode fiber with identical type, 105μm core diameter, 125μm cladding, 0.24 NA. MMF1 is longer than 1km while MMF2 is just 2m. SMF is single-mode fiber of step-index (SI) with 6μm core diameter, 125μm cladding, 0.14 NA, 0.5m length. MMF3 is SI multimode fiber with 200μm core diameter, 220μm cladding, 0.22 NA, 2m length.

III. EXPERIMENTAL RESULT

In this experiment, original Gaussian beam passes through Faraday isolators, beamsplitters and attenuators then couples into MMF1 with 180uJ energy. Self SBS occurs and a Stokes beam with 160uJ energy is reflected. This Stokes beam passes through beamsplitter and lens then couples into SMF, exciting the seed. As the efficiency coupling into single-mode fiber is very low, the seed is only 4uJ, though it's enough for Stimulated Brillouin amplification. The intensity distribution of seed with low-order mode is shown in Figure 2(a). Another part of original laser is firstly damaged by MMF3 then under the action of beamsplitters, lens and attenuators, about 180uJ energy is coupled into MMF2 used to be the pump for Stimulated Brillouin amplification. The pump pulse width is about 8ns, calculating a 20kW peak power. The intensity distribution of pump composed by messy high-order modes is shown in Figure 2(b). The pump and the seed meet in MMF2 at the location 0.5m away from the end of multimode fiber, then a 42uJ Stokes beam shown in Figure 2(c) is obtained after Stimulated Brillouin amplification. It's apparent that the Stokes beam is composed by low-order mode, and its M^2 factor is about 1.5 measured through knife-edge method. As a result, we obtain beam cleanup successfully, transforming the beam of messy high-order modes to a good one of low-order mode, although the efficiency is not very high, about only 21%.

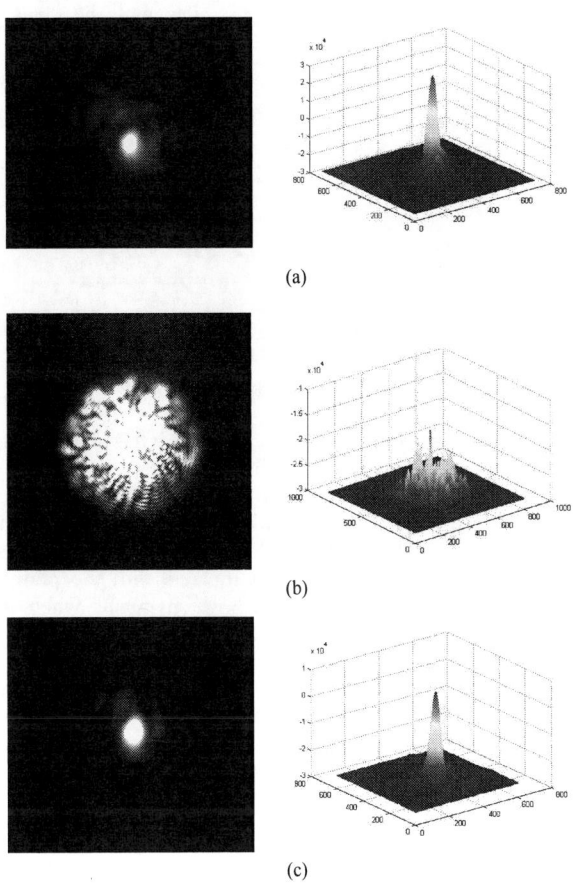

Figure 2. Beam patterns and intensity distribution: (a)seed; (b)pump; (c)Stokes.

As we have seen in Figure 2, beam cleanup has been achieved in GI fiber of 105μm core diameter through Stimulated Brillouin amplification, whose pump has 20kW peak power. A good beam with $M^2 \approx 1.5$ has been transformed from messy pump. If we observe the result carefully, we will find that the intensity distribution of Stokes is very similar to the seed even in details. Additionally, the waveform of Stokes and seed, which are shown in Figure 3, are similar to each other too. Actually, another comparative experiment has been completed, that no seed is provided, and it gets a result that beam cleanup effect doesn't appear. After analysis we judge that in current experimental conditions, appearance of beam cleanup effect is essentially based on the holding of good seed's nature when Stimulated Brillouin amplification occurs. Different modes in GI fiber have different Brillouin gain, while Brillouin gain of the modes which the seed composed by will increase rapidly if adding a seed. Then energy of partial pump's modes which are same with seed will transform to seed. As the result of mode coupling effect in GI fiber, energy of these modes in pump will be supplemented quickly by other modes, keeping on transformation into seed. At last, an effect called beam cleanup realizes. Essentially, it's a transformation in energy from pump to seed, through mode coupling and Stimulated Brillouin amplification in GI fiber, when the pump is composed by messy high-order modes and the seed is composed by good low-order mode.

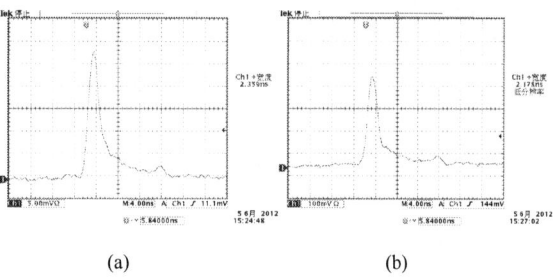

Figure 3. Waveforms: (a)seed; (b)Stokes.

Subsequent analysis is carried out to solve the problem that why efficiency of only 21% has been obtained under current experimental conditions. The light source we used is a pulsed laser with only 8ns pulse width, which is even compressed to less than 3ns by self SBS in MMF1 using as seed. The pulse width is so short that it can be compared with the phonon lifetime when SBS occurs in fiber, which means when the pulsed pump and pulsed seed meet in MMF2, there is extremely limited length (L) for their interaction. It leads to reduction of Brillouin gain ($G=gIL$) in process of amplification, ultimately makes the conversion efficiency low. In case of continuous or quasi-continuous laser, there will be enough interaction between pump and seed when similar experiment is conducted, and it could be expected that the efficiency must be significantly improved.

IV. CONCLUSION

In conclusion, we achieve beam cleanup in 105μm core diameter fiber by using Stimulated Brillouin amplification when the 20kW peak power pulsed pump is provided, and obtain a $M^2 \approx 1.5$ output beam after cleanup with 21% efficiency, then give a positive evaluation of using nonlinear effects to improve the beam quality in fiber laser with larger then 100μm core diameter. In the future, further experiments will be carried out in two aspects, one of which is research on holding of seed's nature when Stimulated Brillouin amplification occurs, another of which is studying under the conditions of continuous or quasi-continuous laser. Higher efficiency can be expected.

ACKNOWLEDGMENT

This work is supported by the National Natural Science Foundation of China (Grant No. 61008004), China Postdoctoral Science Foundation funded project (Grant No. 201104397), the Research Fund for the Doctoral Program of Higher Education 20102302120034, the Fundamental Research Funds for the Central Universities (Grant No. HIT. KLOF. 2010035).

REFERENCES

[1] A. Galvanauskas, M.–Y. Cheng, K.-C. Hou, and K-H. Liao, "High peak power pulse amplification in large-core Yb-doped fiber amplifiers," IEEE journal of selected topics in quantum electronics, vol. 13, pp. 559–566, 2007.

[2] H. Bruesselbach, in conference on lasers and electro-optics (Optical Society of America), vol. 11, pp. 424–426, 1993.

[3] Blake C. Rodgers, Timothy H.Russell and Won B. Roh, "Coherent and incoherent beam combining and cleanup via stimulated Brillouin scattering in multi-mode optical fiber," Cat. No. 99. TH8464, vol. 3, pp. 1026–1027, 1999.

[4] Blake C. Rodgers, Timothy H. Russell, and Won B. Roh, "Laser beam combining and cleanup by stimulated Brillouin scattering in a multimode optical fiber," Optics Letters, vol. 24, pp. 1124–1126, 1999.

[5] L. Lombard, A. Brignon, J.-P. Huignard, and E. Lallier, "Beam cleanup in a self-aligned gradient-index Brillouin cavity for high-power multimode fiber amplifiers," Optics Letters, vol. 31, pp.158–160, 2006.

[6] B. Steinhausser, A. Brignon, E. Lallier, J. P. Huignard, P. Georges, "High energy, single-mode, narrow-linewidth fiber laser source using stimulated Brillouin scattering beam cleanup," Opt. Express, vol. 15, pp. 6464–6469, 2007.

[7] Kirk C. Brown, Timothy H. Russell, Thomas G. Alley, and Won B. Roh, "Passive combination of multiple beams in an optical fiber via stimulated Brillouin scattering," Optics Letters, vol. 32, pp. 1047–1049, 2007.

[8] L. Lombard, C. Delezoide, G. Canat, V. Jolivet, P. Bourdon, "Laser beam combining by beam cleanup in a gradient index Brillouin ring cavity," 2008 conference on lasers and electro-optics and quantum electronics and laser science conference, vol. 19, pp. 1292-1293, 2008.

[9] C. P. Tsekrekos, R. W. Smink, B. P. de Hon, A. G. Tijhuis and A. M. J. Koonen, "Near-field intensity pattern at the output of silica-based graded-index multimode fibers under selective excitation with a single-mode fiber," Optics Express. vol. 15, pp. 3656–3664, 2007.

A practical way of selective mode group excitation in large core graded-index multimode fibers

C.Y. Zhu*, Y.B. Yuan, J. Yan, J.H. Zhang, D.X. Ba and Z.W. Lu

Harbin Institute of Technology,
Harbin, China
zhuchy@hit.edu.cn

Abstract—**In this paper, we proposed a easy and stable way to selectively excite mode group in graded-index multimode fibers(GI-MMF).It was experimentally achieved by the welding of a single-mode fiber(SMF) and a GI-MMF with core/cladding diameter of 105/125 μm and adjusting the radial offset of the two fiber axis. The experiment result was discussed by observing the near-field pattern at the output of the GI-MMF. The effect of fiber length and bending curvature was also discussed.**

Keywords-graded-index multimode fibers, selective mode excitation, radial offset, mode mixing.

I. INTRODUCTION

Multi-mode fibers were widely used in data transmission and high power fiber laser systems. The ability of selectively excite and transmit various mode group in MMF means the potential of loading different information on different mode group, which would have lots of important application prospect. In the year of 1978, L Jeunhomme and J P Pocholle proposed a way of side launching to excite different mode groups in GI-MMF. The input beam was focus through a microscope objective onto the input plane of the fiber. By varying entering positions, the experiment observed far field radiation patterns of different mode groups. Though the way of side launching to achieve selective excitation was proved practicable, but it was severely restricted to the stability and beam quality of light source. In addition, directly coupling by optical lens needs accurate control and alignment, which means it was susceptible by external factor. Thus side launching technique was seldom used in practical use. We proposed a different way in this paper by the welding of fibers and got a new technique of selective mode group excitation that is practical and stable to implement.

II. THEORY ANALYSIS

When multi-mode fibers was injected with light energy by external light source, some of the mode field would be excited in MMFs. Since the total optical fields in MMF were the linear combination of each guided mode field, the issue of mode excitation in MMF is equaled to the distribution of input optical field energy to each guided mode field.

Considering at the input plane of MMF, the optical field of input light is E_{in}, since the guided mode field in MMF compose a complete orthonormal system, E_{in} could be expanded as guided mode field:

$$E_{in} = \sum_i c_i E_i \qquad (1)$$

Where E_i is the i-th guided mode field and c_i is the excitation coefficient of the i-th guided mode field.

Multiplying the conjugate of E_i at both side of equation (1) ,then integrating at the whole fiber core end face and using the orthonormal condition of guided mode field, equation(1) becomes:

$$c_i = \iint_\Omega E_{in} E_i d\Omega \qquad (2)$$

Normalize the equation above, we get:

$$|c_i|^2 = \frac{(\iint E_{in} \bullet E_i^* \mathrm{d}x\mathrm{d}y)^2}{(\iint E_{in} \bullet E_{in}^* \mathrm{d}x\mathrm{d}y)(\iint E_i \bullet E_i^* \mathrm{d}x\mathrm{d}y)} \qquad (3)$$

Apparently, $0 \leq |c_i|^2 \leq 1$. Equation (3) is the general situation of mode excitation in MMF. where $|c_i|^2$ means that $|c_i|^2$ times of input light power is injected into the i-th guided mode field. So, given that the input light field and the distribution of every transverse mode in MMF is known, the power distribution of each order modes could compute by this equation.

In a SMF, the light energy is transmitted as fundamental mode, which could approximated with a Gaussian field:

$$\mathrm{E}_t = A \exp[-(\frac{r}{w_0})^2] \qquad (4)$$

Where w_0 is the radius of fundamental mode in SMF, A is the normalized constant. In practical application, there is a equation to approximate w_0 . As to a step-index SMF, it is:

$$\frac{\mathrm{w}_0}{a} = 0.65 + 2.319 V^{-\frac{3}{2}} + 3.879 V^{-6} \quad (\ 2.2 < V < 3) \qquad (5)$$

Where a is the core diameter of SMF, V is normalized frequency parameter, which is defined as:

978-1-4799-1215-5/13 $31.00 © 2013 IEEE

$$V = k_0 a \sqrt{n_1^2 - n_2^2} = k_0 a * NA \qquad (6)$$

With the equation above and the parameter of SMF, we compute that V=2.48and w_0 =3.8μm.

In the GI-MMF, the distribution of transverse mode field is:

$$E_t(x, y) = A_{m,l} (\sqrt{2} \frac{r}{w_0})^l L_m^l (\frac{2r^2}{w_0^{'2}}) \exp(-\frac{r^2}{w_0^{'2}}) \cos l\phi \qquad (7)$$

Where $A_{m,l}$ is the normalized parameter of transverse mode field, $w_0^{'2}$ is the diameter of fundamental mode in MMF.

$$w_0^{'} = \sqrt{\frac{2a^{'}}{n_1 k_0 \sqrt{2\Delta}}} = \sqrt{\frac{2a^{'}}{NA^{'} * k_0}} \qquad (8)$$

Substituting the mode field expression of SMF and GI-MMF in equation (3), then we can get the power distribution situation of selective mode excitation in MMF.

Figure 1 is the welding sketch map of a SMF to a MMF that we proposed in this paper. On the process of operating welding splicer, excitation of different mode group could be obtained by setting radial offset of the core axis between two fibers.

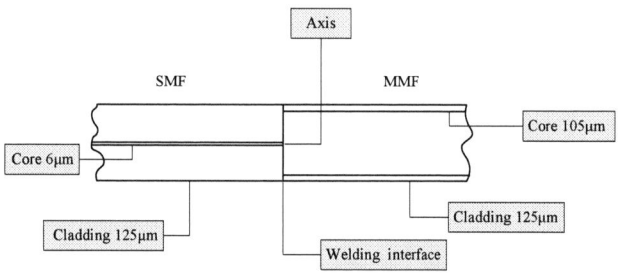

Figure 1. The welding of a SMF to a GI-MMF

III. EXPERIMENT SCHEME

In this experiment, we fuse a SMF with a large core GI-MMF to explore a practical way to selectively excite mode group in GI-MMFs. The effect of mode mixing in this type of fiber was also examined. The result could provide a reference for latter use or research. The optical path was showed by figure 2.

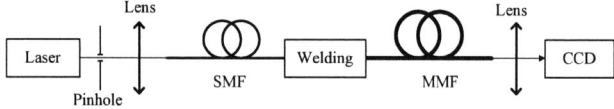

Figure 2. Experiment setup

The YAG laser device export laser pulse with the wavelength of 1064nm and the pulse width of 8ns. The SMF used in the experiment is a step-index fiber, with core/cladding diameter of 6/125 μm and central numerical aperture (NA)

0.14. The GI-MMF has core/cladding diameter of 105/125 μm and NA= 0.24. Three aspect were conducted in the experiment: (1) selective excitation in GI-MMF by changing the radial offset of core axis between two fibers; (2) Changing the length of MMF to explore the effect of mode mixing; (3) Exploring the effect of bending curvature on mode transmitting by circle the GI-MMF.

Comparing to the conventional way of using lens to focus input light into the MMF, the way of fusing a SMF with a MMF and couple the input light into the SMF to achieve selective mode group excitation has lots of advantage. First, since the SMF can only contain one guided mode, it isolated the influence of light source and the spatial optical path; At the same time, the inject plane between SMF and MMF is fixed by fusing process, which guarantees the stability of excitation condition; Moreover, due to the beam radius and beam divergence of the transmitted light field in SMF are both very small, the number of mode group excited in MMF are also relatively small. Thus it can be seen that by fusing a SMF to a MMF, we could get a much more stable light field at the output plane of GI-MMF. However, because of the core diameter of SMF is pretty small, it is quite difficult to couple the light form light source into the SMF core and the couple efficiency is very low. So that this way is suit for the situations that strictly require with the optical field distribution but do not need high transmitted power in MMF.

IV. EXPREMENT RESULT

A. *influnce of radial offset on excitaed mode group*

The theory of selective mode excitation indicates that radial offset and tilt are equivalent to mode excitation. In this experiment, the input beam angle was restricted with a very small range and it could be approximately considered as vertical injecting. Operating the welding splicer and gradually increasing the radial offset of the two fiber axis, the changing relationship between output optical field and radial offset distance was observed by a CCD camera.

The length of the SMF and the MMF are both 1m. since the length of GI-MMF is very short, the effect of mode mixing that may produce could be ignored.

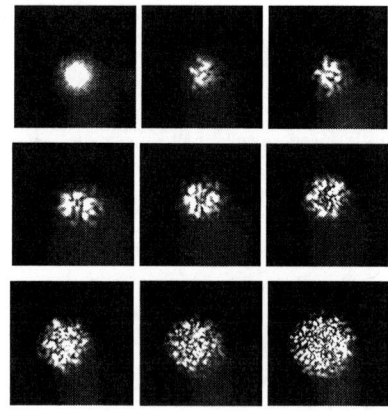

978-1-4799-1215-5/13 $31.00 © 2013 IEEE   252

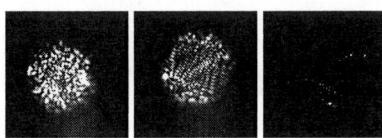

Figure 3. The near-field pattern at the output of GI-MMF while incresing the radial offset

Figure 3 was the near-field pattern at the output of GI-MMF. From left to right, up to below, the radial offset were gradually increase. From the tendency of light speckles, it could tell that the distribution of output fields are gradually disperse and complicated., the radius of light speckle gradually becomes lager. Which correspond to the variation feature of changing from low order to high order in a GI-MMF. When the radial offset exceed to a certain value, the input light beam located at somewhere nearby the cladding layer and the light field at the output side rapidly vanished.

B. Influence of fiber length on mode group transmission

The experiment above only uses 1 meter length of GI-MMF to avoid the possible influence of mode mixing. Now add the fiber length to 500 meters to explore the effect of mode mixing in GI-MMF. Observe the near-field pattern of the output light at some certain experiment condition. Meanwhile, maintain all the experiment conditions except cutting out the fiber length to 1 meter and take the picture of exporting light field again.

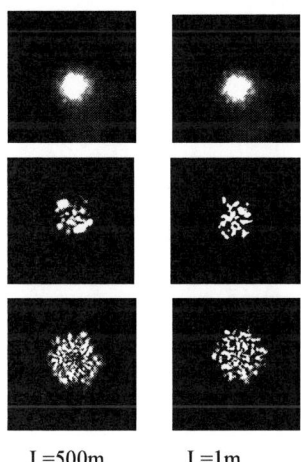

L=500m   L=1m

Figure 4. The near-field pattern at the output of GI-MMF with the fiber length of 500m and 1m

Figure 4 shows three group of experiment results. Each group of picture was measured under different radial offset. The left picture was taken with fiber length of 500m and the right was of 1m. The image of figure 4 indicate that the length of fiber does not affect the overall near-field pattern, which remains confined within a disk. The radial offset of the SMF determines the radius of the disk. This indicates that mode mixing is weak in this type of GI-MMF. Because in the presence of strong mode mixing, light would coupled to almost each mode group and expanded to most of the area of the fiber core in GI-MMF after a long distance of propagation.

C. Influence of bending curvature on mode group transmission

On this step of experiment, it was designed to circle the GI-MMF at different curvature to explore the effect of bending curvature on mode group transmission.

The length of GI-MMF is 1m. The radius of fiber circle is respectively $r_1 = 10mm$, $r_2 = 5mm$, $r_3 = 3mm$. We circle the GI-MMF around 3 times and conduct the selective mode group excitation experiment. The experiment result is showed below.

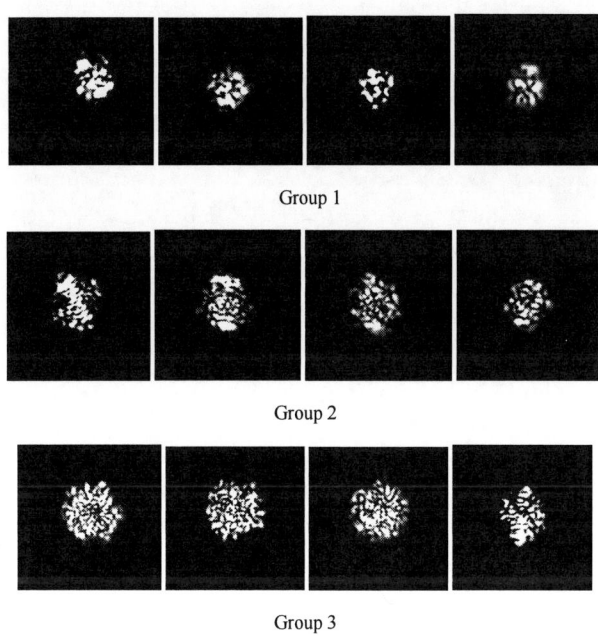

Group 1

Group 2

Group 3

Figure 5. Influence of bending curvature on mode group transmission

The radial offset were different among 3 group figure. In each group figure, from left to right, the first picture is the near-field pattern at the output of GI-MMF with no circle, the other 3 pictures are obtained respectively with the circle radius of r_1, r_2, r_3.

Form the comparison of each group pictures in figure 5, it can be seen that the impact turn out to be small at any radial offset when the circle radius is r_1 or r_2. But the influence becomes obvious when the circle radius is r_3. Comparing to other pictures in the same group, the radius of the disk becomes smaller . one reasonable interpretation is that when the bending curvature of GI-MMF is too big, the guided modes of higher order would couple into radiation modes and get vanished quickly. According to the experiment result, the radius of bending of GI-MMF should less than at least 5mm when it is normally used.

V. CONCLUSION

In this paper, experimental results of the near-field pattern at the output of GI-MMFs have been presented and compared.

By fusing a SMF to a GI-MMF and changing radial offset of the core axis of two fibers, we achieved a practical way of selective mode group excitation. The experiment also explored the influence of fiber length and bending curvature on mode group transmitting. The result indicates that the varying of fiber length doesn't change the overall field pattern at the output of GI-MMF. The field pattern also will not get influenced too much when the radius of bending is no less than a few millimeters.

ACKNOWLEDGMENT

This work is supported by the National Natural Science Foundation of China (Grant No. 61008004), China Postdoctoral Science Foundation funded project (Grant No. 201104397), the Research Fund for the Doctoral Program of Higher Education 20102302120034, the Fundamental Research Funds for the Central Universities (Grant No. HIT. KLOF. 2010035).

REFERENCES

[1] A. Galvanauskas, M.-Y. Cheng, K.-C. Hou, and K.-H. Liao. High Peak Power Pulse Amplification in Large-Core Yb-Doped Fiber Amplifiers[J]. IEEE J. Sel. Top. Quantum Electron., 2007, 13(3):559-566.

[2] H. R. Stuart. Dispersive Multiplexing in Multimode Optical Fiber[J]. Science, 2000, 289(5477):281-283.

[3] Dubois F, Emplit P H, Hugon. Selective mode excitation in graded-index multimode fiber by a computer-generated optical mask[J]. Optical Letters, 1994, 19(7):433-435.

[4] Stuart H R. Dispersive multiplexing in multimode fiber[J]. Science, 2000 , 289 (5477) , 281-283.

[5] Jeunhomme L, Pocholie J P. Selective mode excitation of graded-index optical fibers[J]. Applied Optics,1978, 17(3):463-468.

[6] Saijonmaa J, Sharma A B, Halme S. Selective excitation of parabolic-index optical fibers by Gaussian beams[J]. Applied Optics, 1980, 19(14):2442-2452.

An improved location algorithm in Wireless Sensor Network

Jinlong Liu
School of Electronics and Information Engineering
Harbin Institute of Technology
Harbin, China
E-mail: yq20@hit.edu.cn

Zhilu Wu,Zhendong Yin
School of Electronics and Information Engineering
Harbin Institute of Technology
Harbin, China
E-mail: wuzhilu@hit.edu.cn, yinzhendong@hit.edu.cn

Abstract—**Localization, for a sensor to determine its location information has become an attractive research issue in Wireless Sensor Network(WSN). At present, Time Difference of Arrival Algorithm (TDOA) has the higher significance of the research. According to improve the precision of the algorithm, in this paper, By the study of the chan algorithm, least squares algorithm, the optimization algorithm of least square, according to improve the precision of the algorithm, based on the three algorithm ,the author puts forward a kind of improved TDOA algorithm. By using the simulation software MATLAB, to set up the network, to make the simulation experiments, the author get a large number of experimental data. Most of experimental data prove this improved location algorithm can improve the position precision.**

Keywords-Wireless Sensor Networr;TDOA;Centroid weighted; Anchor node

I. INTRODUCTION

Wireless sensor network (WSN) is composed of a large number of sensor nodes with wireless communications, computations, and sensing capabilities, which are densely deployed either inside the monitoring phenomenon or very close to it[1]. These tiny WSN nodes which are capable of communicating with each other, acquiring various physical values, performing computations can cooperatively achieve a desired task through specific protocol. This makes it possible for the WSN to a variety of monitoring and tracking applications in military, agriculture, and industry, etc[2].

An important aspect in most of the WSN applications is the accurate localization of the individual node[3]. WSN nodes are usually placed in different environment to accomplish different tasks, in which the locations of these nodes are random and unknown in advance. However, data acquired by these nodes are only available when connected with location information. Therefore the node localization technology has aroused widespread attention and become one of the hotspots of WSN study. Some scholars put forward the classic TDOA algorithm, but the timeliness and accuracy of algorithm needs to be improved, this paper will put forward a kind of improved TDOA algorithm.

II. ESTABLISHING MODEL

A. Mathematical model for Wireless Sensor Network

Suppose the coordinates of unknown and known nodes are (Z,Y), According to the principle of TDOA algorithm, the mathematical expressions of model is

$$R_{i,1} = \left[(Z_i - z)^2 + (Y_i - y)^2 \right]^{1/2} - \left[(Z_i - z)^2 + (Y_i - y)^2 \right]^{1/2} \quad (1)$$

The anchor node to the unknown node time delay in $f_{i,1}$,for example, $f_{i,1}$ says the delay time of the anchor node i to the unknown node 1. $R_{i,1}$ says the distance between the anchor node and the node location, for example, $R_{i,1}$ says the delay time of the anchor node i to the unknown node 1.

By formula (1) can be derived linear expressions:

$$\begin{aligned} R_i &= \sqrt{(Z_i - z)^2 + (Y_i - y)^2} \\ R_i^2 &= (Z_i - z)^2 + (Y_i - y)^2 = K_i - 2Z_i z - 2Y_i y + z^2 + y^2 \\ K_i &= Z_i^2 + Y_i^2 \end{aligned} \quad (2)$$

According to the principle of description:

$$R_i^2 = (R_{i,1} + R_1)^2 = R_{i,1}^2 + 2R_{i,1}R_1 + R_1^2 \quad (3)$$

Merge (2) and (3):

$$\begin{aligned} R_{i,1}^2 + 2R_{i,1}R_1 + R_1^2 &= K_i - 2Z_i z - 2Y_i y + z^2 + y^2 \\ R_1^2 &= K_1 - 2Z_1 z - 2Y_1 y + z^2 + y^2 \end{aligned} \quad (4)$$

The two expressions of (4) make subtraction:

$$\begin{aligned} R_{i,1}^2 + 2R_{i,1}R_1 &= K_i - 2Z_{i,1}z - 2Y_{i,1}y - K_1 \\ -2Z_{i,1}z - 2Y_{i,1}y &= R_{i,1}^2 + 2R_{i,1}R_1 - K_i + K_1 \end{aligned} \quad (5)$$

$$\begin{cases} Z_{i,1} = Z_i - Z_1 \\ Y_{i,1} = Y_i - Y_1 \quad i = 2 \cdots\cdots l \\ K_i = Z_i^2 + Y_i^2 \end{cases} \quad (6)$$

According to the derivation of a few steps above, reference to analytical theory of knowledge, the conclusion can be drawn that if there are 10 anchor node exists in the system, then there are 10 hyperbolic model. If the time delay estimation and the anchor node coordinates both are accurate, then the intersection

978-1-4799-1215-5/13 $31.00 © 2013 IEEE

of analytic graphics hyperbolic equations is the unknown node that we need to locate[4].

B. The experimental model for WSN

Suppose the plane for the square, the length is 50metres, the coordinates of 4 anchor nodes are $K(0,0), L(0,30), H(30,0)$, $I(30,30)$, the working anchor nodes are independent of each other, and electromagnetic from the anchor nodes constantly waves to $A(16,6), B(10,20)$, electromagnetic wave propagation velocity is the speed of light C, signal waveform is sine wave. The propagation time between K and A subtracts the propagation time between L and A, the results is τ, through the simulation results, calculating the amplitude of cross-correlation function. Using this experimental model, we can get 100 sets of simulation data.

III. PRINCIPLE OF THE ALGORITHM AND SIMULATION EXPERIMENTS

A. Chan algorithm and simulation experiments

Expression of Chan algorithm(Chan) is:

$$\begin{bmatrix} z \\ y \end{bmatrix} = -\begin{bmatrix} Z_{2,1} & Y_{2,1} \\ Z_{3,1} & Y_{3,1} \end{bmatrix} \times \left\{ \begin{bmatrix} R_{2,1} \\ R_{3,1} \end{bmatrix} R_1 + \frac{1}{2} \begin{bmatrix} R_{2,1}^2 - K_2 + K_1 \\ R_{3,1}^2 - K_3 + K_1 \end{bmatrix} \right\} \tag{7}$$

(7) analyses the two anchor nodes, when selecting a plurality of anchor nodes，(8) analyses a plurality of anchor nodes, finds multiple approximation, narrows the distance with the correct value. Fig.1 shows the simulation results of the Chan algorithm.

$$\begin{bmatrix} z \\ y \end{bmatrix} = \begin{bmatrix} Z_{2,1} & Y_{2,1} \\ Z_{3,1} & Y_{3,1} \\ \vdots & \vdots \\ Z_{l,1} & Y_{l,1} \end{bmatrix} \times \left\{ \begin{bmatrix} R_{2,1} \\ R_{3,1} \\ \vdots \\ R_{l,1} \end{bmatrix} R_1 + \frac{1}{2} \begin{bmatrix} R_{2,1}^2 - K_2 + K_1 \\ R_{3,1}^2 - K_3 + K_1 \\ \vdots \\ R_{l,1}^2 - K_l + K_1 \end{bmatrix} \right\} \tag{8}$$

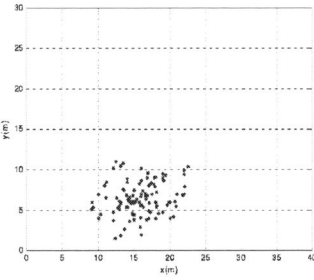

Figure 1.   Simulation results of the Chan (A)

B. Least square algorithm and simulation experiments

The least square algorithm (LSA) is a classical algorithm, it contains a variety of uses, in wireless sensor network, it can also be applied, to complete the approximate estimation using the least square algorithm, the expression is:

$$BZ_o = a \tag{9}$$

in which

$$B = -2\begin{bmatrix} Z_{2,1} & Y_{2,1} \\ Z_{3,1} & Y_{3,1} \end{bmatrix}, Z_o = \begin{bmatrix} z \\ y \end{bmatrix} \quad a = \begin{bmatrix} R_{2,1}^2 - Z_2^2 - Y_2^2 + Z_1^2 + Y_1^2 \\ R_{3,1}^2 - Z_3^2 - Y_3^2 + Z_1^2 + Y_1^2 \end{bmatrix} \tag{10}$$

It be required to find out the estimated value of Z_o, According to the principle of least squares, the estimated value of B is subtracted from the measured values, and to do the square operation at least three anchor nodes[5]. Because the algorithm has calculation error, so introduce error vector M in (9):

$$BZ_o + M = a \tag{11}$$

$$M = a - BZ_o \tag{12}$$

calculate the minimum M vector ,modulus (11):

$$\|M\|^2 = \|a - BZ_o\|^2 \tag{13}$$

According to the principle of least square algorithm, the differential treatment (13), calculate the estimate Z_o, get the following expression:

$$\hat{Z}_{o/s} = (B^T B)^{-1} B^T a \tag{14}$$

Fig.2 shows the simulation results of the least square algorithm:

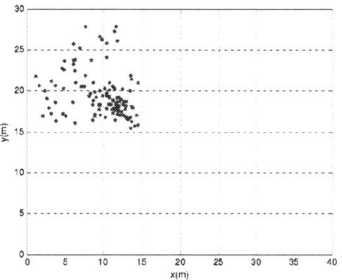

Figure 2.   Simulation results of LSA (B)

C. Optimization of the least squares algorithm （OLSA） and simulation experiments

In the least squares algorithm (LSA) of TDOA ranging technology of wireless sensor network (WSN), weighted LSA, improve algorithm and precision, introduce a weight matrix X to (14):

$$\hat{Z}_{ox/s} = (B^T X B)^{-1} B^T X a \tag{15}$$

(15) is the least squares expression through weighted processing, if need to compute minimum variance unbiased estimation of $Z_{ox/s}$, The fact proved that, when $X = N-1$, OLSA contains the highest precision, the expression $N = E\{MM^T\}$ represents a modified covariance matrix.

Fig.3 shows the simulation results of Optimization of the least squares algorithm:

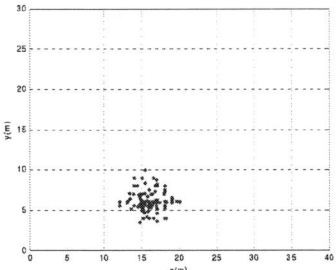

Figure 3.   Simulation results of OLSA (A)

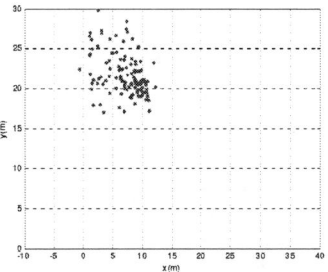

Figure 4.   Simulation results of the improved Chan (B)

D.   An improved algorithm on location

Chan, LSA, OLSA, the three algorithms are occasionally occur some positioning data error[6]. Give an example, Chan calculate 100 data, the node positions accounted for 22% of all nodes out 10 m, in wireless sensor network, this algorithm can not be applied. In mathematical analysis, the deviation of Chan approximately meets Cramer-Rao bound requirement. LSA and OLSA also appeared in varying degrees of the position deviation, causing a lot of nodes is invalidly, energy consumption is also relatively high, so scientists discard these inaccurate data when the algorithms are simulated. In order to obtain the best precision of location, the paper will introduce one algorithm to eliminate the deviation of the three algorithms. The simulation data obtained by these three algorithms are did centroid weighted processing. These data will be constrained in closer range accuracy values. Specific processing algorithm is the centroid calculation of experimental data, reduce the error of the multiple passes through the centroid calculation. Refer to (16):

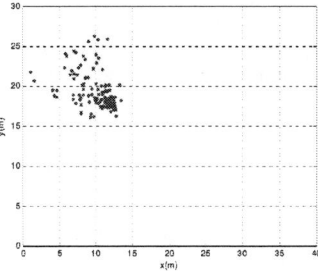

Figure 5.   Simulation results of the improved LSA (B)

$$z = \frac{\dfrac{z_1}{l_1+l_2} + \dfrac{z_2}{l_2+l_3} + \dfrac{z_3}{l_1+l_3}}{\dfrac{1}{l_1+l_2} + \dfrac{1}{l_2+l_3} + \dfrac{1}{l_1+l_3}}$$

$$y = \frac{\dfrac{y_1}{l_1+l_2} + \dfrac{y_2}{l_2+l_3} + \dfrac{y_3}{l_1+l_3}}{\dfrac{1}{l_1+l_2} + \dfrac{1}{l_2+l_3} + \dfrac{1}{l_1+l_3}}$$

(16)

According to the mathematical formula (16), accurately calculate the coordinates *(z,y)* of unknown nodes[7]. Using the empirical formula and approximate estimation, l_1, l_2, l_3, separately denote from the three known anchor node to the distance of the unknown node. If calculated positioning node coordinates obtained by each algorithm is *(z₁,y₁)*, *(z₂,y₂)* *(z₃,y₃)* separately introduce three influencing factors, $1/(l_1+l_2)$, $1/(l_2+l_3)$, $1/(l_1+l_3)$, the three factor is used to describe the anchor nodes and the unknown nodes accuracy of matching correction.

The simulation results of Chan, LSA, OLSA is Fig.4, Fig.5, Fig.6:

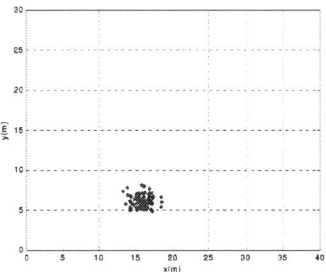

Figure 6.   Simulation results of the improved OLSA (A)

IV.   CONCLUSION

According to the analysis of the experimental data of the two tables, in Chan algorithm experimental simulation, 52% of nodes fell within 4 meters radius of circular area, data distribution is discrete, in LSA experimental simulation, the ratio is increased to 61%, in OLSA experimental simulation, the ratio is increased to 72%.

TABLE I.   THE THREE ALGORITHM SIMULATION DATA ANALYSIS

Algorithm	Average coordinate	Nodes in 4m of the circle
Chan	(4.6,24.73)	52%
LSA	(9.15,16.43)	61%
OLSA	(16.86,6.15)	72%

TABLE II.   IMPROVED ALGORITHM SIMULATION DATA ANALYSIS

Algorithm	Average coordinate	Nodes in 4m of the circle

Improved Chan	(7.01, 22.45)	57%
Improved LSA	(9.84, 18.01)	79%
Improved OLSA	(15.81, 6.10)	96%

Obtaining 100 data of the position coordinates of unknown nodes of the three algorithms, the paper proposed an improved algorithm, put the data convergence in the controllable range, avoid excessive dispersion of node. After the improved algorithm optimize Chan, 57% of nodes fell within 4 meters radius of circular area; after the improved algorithm optimize LSA, 79% of nodes fell within 4 meters radius of circular area; after the improved algorithm optimize OLSA, 96% of nodes fell within 4 meters radius of circular area. So the improved algorithm that the paper proposes can be implemented, is one kind of wireless sensor network localization algorithm with high accuracy.

ACKNOWLEDGMENT

This work is supported by the Professor Zhilu Wu of School of Electronics and Information Engineering in Harbin Institute of Technology.

REFERENCES

[1] Y. F. Wang, "Computational Algorithms for Inverse Problems and Their Applications," Beijing, Higher Education Press, 2007, pp. 76-77.

[2] Gigl, T.; Janssen, G.J.M.; Dizdarevic, V.; Witrisal,K.; Irahhauten, Z.; , "Analysis of a UWB Indoor Positioning System Based on Received Signal Strength," Positioning, Navigation and Communication, 2007. WPNC '07. 4th Workshop on, pp.97-101, 22-22 March 2007

[3] Jackson, B.R.; Wang, S.; Inkol, R.; "Received signal strength difference emitter geolocation least squares algorithm comparison," Electrical and Computer Engineering (CCECE), 2011 24th Canadian Conference on, pp.001113-001118, 8-11 May 2011.

[4] Dabin, J.A.; Nan Ni; Haimovich, A.M.; Niver, E.; Grebel, H.; , "The effects of antenna directivity on path loss and multipath propagation in UWB indoor wireless channels," Ultra Wideband Systems and Technologies, 2003 IEEE Conference on, pp. 305-309, 16-19 Nov. 2003.

[5] Y. Q. Ding, Y. G. Sun, and T. X. Li, "Sensor localization algorithm for wireless sensor network based on multidimensional scaling and multidimensional calibration," Chinese Journal of Scientific Instrument, vol.30(5), pp. 1002-1008, 2009.

[6] L. Wang, H. M. Li, and X. T. Du, "WSN Multilateral Localization Algorithm Based on Tikhonov Regularization Algorithm," Proceedings of SPIE 2008, Shen Yang, China, vol.1, pp. 404-413, Sep 2008.

[7] L. Wang, Y. T. Liu, and X. H. Xu, "WSN Multilateration Algorithm Based on Landweber Iteration,"Proceedings of ICEMI2009, vol.1, pp.250-254, Aug.2009.

Author Index

B

Bai Yuanyuan 227
bei Cao 94
Bei Cao 138
Beng Kang Tay 91
Biao Yang 160
Biao Yang 177
Bin Zhou..138
Bing Zhu..62
Bingfang Zhang....................................53
Bingxiu Zhang......................................53
Bo Sun..10

C

C.Y. Zhu 248
C.Y. Zhu 251
Cao Zhigang 146
Chen En...83
CHEN Ling 67
Chengguo Tong 29
Chunlian Lu..189
Chunqiu You.......................................163
Cuiling Wang......................................189

D

D.X. Ba...248
D.X. Ba...251
Dan Bu 138
De Ming Ren 155
Dejian Meng 112
Dongdong Zi...29
DunLiang Ren.....................................215

F

Fanming Liu 5
Fengjun Tian......................................234
Fuzhen Zhu..62

G

GAO Huaihui 67
Guangchun Li 14
Gui Yonglei 146
Guoli Song...125

H

Haijiao Yu..183
Haijiao Yu..186
Haijiao Zhou 101
Haitao Chen 117
Haitao Chen 128
Haiyong Zheng 211
Hao Tian 197
He Lv 189
He Zhang...79
Hong Bo Bai.......................................142
Hong Jia 197
Hong Li 203
Hong Qi...50
Hong Wang 91
Hongquan Zhang.................................180
Hongsong Mei 200
HongXin Shi.......................................215
Hui Liu 215
Hui Wang...98
Huiyu Liu...24

J

J. Yan 248
J. Yan 251
J.F.Sun 151
J.H. Zhang 248
J.H. Zhang 251
Jian Gao 112
Jian Xing 231
Jian.Gao 151
Jiandong Jin 50
Jianfeng Sun 112
Jian-feng Sun 242
Jianhui Zeng 24
Jiannan Yu 117
Jianqi Yao 5
Jie Tang 134
Jiejiang Xing 177
Jing Dai 203
Jing Dai 207
Jinghua Yin 33
Jinhui Shi 245
Jinlong Liu 255
Jinping Li 121
Jinyu Dong 33
Jin-Zhong Yu 160
Jitao Han 18
Jun Du 155
Jun Hu 134
Jun Li 197
Jun Yang 219
Junhai Zhang 200
JunQing Li 215

K

Kunpeng He 18
Kunpeng He 24

L

Li Jia Ma 142
Li Jie Geng 155
Li Li 203
Li Li 207
Li Tian 79
Li Yuxiang 109
LIAN Tongli 67
Liang Sun 197
Libo Yuan 53
Libo Yuan 219
Libo Yuan 234
Li-Bo Yuan 238
Lijie Chen 117
Lijie Chen 121
Li-Jie Chen 167
Likai Sun 117
Lin Jiping 146
Lin Zhang 215
Lingli Zhang 71
Liqiu Wei 62
Liu Yunhui 227
Lufei Hong 125

M

MaoJun Fan 134
Mei-Yu Zhang 167
Meng Hong 121
Meng Shanshan 170
Mingwei Wang 50
Mingxin Song 46

N

Na Zhang 211

P

Peihua Li 62

Peijing SUN 75
Peng Zhang 98
Peng Zhang 128
Ping Yuan 87
Ping Yuan 105

Shiyin Guan 33
Shouchen Chai 98
Shuangqiang Liu 193
Shuting Gao 33
Souaci Farida 238
Sun Weimin 109

Q

Q.L. Gao 248
Qi Li 131
Qi Wang 131
Qi Wang 242
Qi Yan 183
Qi Yan 186
Qi.Wang 151
Qiang Dai 142
Qiang Huang 200
Qiang Xu 131

T

Tang Miao 1
Tao Geng 29
Tao Zhang 142

W

Wang Jingran 83
Wang Shuai 109
Wei GAO 75
Wei Jiang Zhao 155
Wei Xin-lao 1
Weimin Sun 174
Weimin Sun 180
Weimin Sun 183
Weimin Sun 186
Weimin Sun 189
Weimin Sun 193
Weimin Sun 200
Weimin Sun 231
Wenjiang Zou 56
Wenjuan Wang 94
Wenjun Sun 101
Wenlei Yang 29

R

Rongbin Hu 38
Rongbin Hu 42
Rongke Ye 38
Rongke Ye 42
Ruichen WANG 174
Ruichen Wang 180

S

Shaozhi PU 75
She Li 215
Shen Tao 1
Shenbo Zhu 14
Shengnan LIU 75
Shi Xin 121
Shijie Wang 189
Shimin Fu 117
Shi-Ning Wang 167

X

Xi Xiao 160
Xiang Yang Cheng 155
Xiangbao Yin 71
Xiao Liang Zhu 142

Xiaobo HU	75
Xiaoliang Zhu	183
Xiaoliang Zhu	186
Xiaoqi Liu	87
Xiaoqi Liu	105
Xiaowei Han	79
Xiaowei Liu	79
XIE Minxiang	67
Xie Xin	227
Xin Lv	91
Xin Shi	128
Xingchen Liu	245
Xingyue Xu	56
Xingzhi Zhang	24
Xinlu Zhang	203
Xinlu Zhang	207
Xinping Dong	98
Xiong Yan-ling	1
XU Jiren	67
Xu Yunfei	227
Xuelian YU	75
Xuenan Zhang	87
Xuenan Zhang	105
Xue-song Tian	242
Xuyou Li	10

Yingying Yu	10
Yiqun Xu	245
Yong-Hai Cao	167
Yonghe Qin	56
Yongjun Liu	71
Yongjun Liu	174
Yongjun Liu	180
Yongpeng Zhao	131
Yongsheng Wang	94
Yongsheng Wang	138
Yu Zhang	219
Yuandong Sui	231
Yuanyuan Wang	125
Yude Yu	177
Yu-De Yu	160
Yue Li	33
Yue Li	46
Yuelan Lu	224
Yuling Li	50
Yu-Ming Wu	91
Yundong Zhang	87
Yundong Zhang	105
Yunfei Du	94
Yunfeng He	14
Yunxiang Yan	174
Yunxiang Yan	180
Yuping Shao	18

Y

Y.B. Yuan	248
Y.B. Yuan	251
Yan Chen Qu	155
Yang Li	197
Yang Qian-ru	1
Yang Wen-long	1
Yang Yang	46
Yanhui Wei	14
Yanwei Dou	138
Yanzhe Cao	14
Yao Xie	131
Ya-Xian Fan	238
Yingjie Qiao	56

Z

Z.W. Lu	248
Z.W. Lu	251
Zhang Jin	83
Zhanjiang Gong	121
Zhen Lei Chen	155
Zhendong Yin	255
Zheng Zhu	245
Zhengping Wang	245
Zhenguo Zhang	203
Zhenqi Zhao	98
Zhifang Wang	62

Zhi-gang Fan 242

Zhihai Liu 219

Zhilu Wu 255

Zhiyong Li 177

Zhi-Yong Li 160

Zhong Meng 101

Zhongyang Liu 101

Zong Fabao 83

Zongjun Huang 200

Zou Pengyi 83